Statistics for Science and Engineering

Statistics for Science and Engineering

John J. Kinney

Professor Emeritus of Mathematics
Rose-Hulman Institute of Technology

Addison
Wesley

Boston San Francisco New York
London Toronto Sydney Tokyo Singapore Madrid
Mexico City Munich Paris Cape Town Hong Kong Montreal

Sponsoring Editor: Deirdre Lynch
Project Editor: Ellen Keohane
Associate Production Supervisor: Julie LaChance
Marketing Manager: Brenda Bravener
Manufacturing Buyer: Evelyn Beaton
Prepress Services Buyer: Caroline Fell
Senior Designer: Barbara Atkinson
Cover Design: Night & Day Design
Cover Art: Annie S. Carter
Interior Design: Sandra Rigney
Production Services: TechBooks
Composition: TechBooks
Art Studio: Scientific Illustrators

Library of Congress Cataloging-in-Publication Data
Kinney, John J.
        Statistics for science and engineering / John J. Kinney.
                p.   cm.
        Includes bibliographical references and index.
        ISBN 0-201-43720-1
        1. Mathematical statistics.   I. Title

        QA276.12.K56   2002.   2001046066
        519.5—dc21

MINITAB is a registered trademark of Minitab.

Many of the designations used by manufacturers and sellers to distinguish their products are claimed as trademarks. Where those designations appear in this book, and Addison-Wesley was aware of a trademark claim, the designations have been printed in initial caps or all caps.

To Cherry, Kaylyn and James

Preface

Why Study Statistics?

S TATISTICS IS VERY LIKELY TO BE THE CENTER of applied mathematics in the twenty-first century. In part this is because occupational arenas as diverse as medical research, earthquake prediction, development and testing of building materials, design of underground and undersea structures, weapons development, public opinion analysis, design and production of electronic devices, actuarial science, and urban planning regularly use statistical reasoning and statistical procedures. Statistics has long been a part of undergraduate education in mechanical engineering.

Scientists and engineers of all specializations conduct experiments in order to obtain knowledge about the natural, physical, and social worlds. These experiments are usually only done on a sample, often a small sample, of all the observations that might possibly be obtained in a given situation. Yet the goal is to draw conclusions concerning the characteristics of the entire population from which the sample is selected. How is it that we can draw conclusions about the population of all possible observations from only a small sample of them? This question, a central one in experimentation, is answered by statistics. This is why statistics is an essential course in the education of scientists and engineers.

Statistics is a relatively new field of mathematical study. While Gauss (1777–1855) and others knew some notions of probability and elementary data analysis, modern statistics really began with the discovery of the t distribution by W. G. Gossett in 1908. So what we study here is essentially brand new in relation to the 5000-year-old subject of mathematics.

Statistics is now an active intellectual field of study, so much so that fundamental discoveries in statistics may alter statistical practice during the lifetime of the readers of this book. This means, among other things, that while there are numerous questions we cannot answer at the moment, some questions may be answered soon, and the solutions to others will probably continue to elude us. I urge students who wish to become statisticians or statistically literate engineers and scientists to continue to study this rapidly expanding and fascinating field.

About This Book

Readers should have some knowledge of both derivatives and integrals and should appreciate their meaning. Some elementary matrix algebra is included in the chapter on multiple regression, but it is not necessary to have studied this previously.

This book is an introduction to the statistical analysis of data that arise from experiments, sample surveys, or other observational studies. The text is not meant to be encyclopedic; it is a relatively brief introduction to a large and rapidly expanding subject. It focuses on topics that are frequently used by scientists and engineers, particularly regression, the design of experiments, and statistical process control. This text explains how we know what we know about these methodologies involving the analysis of sample data.

Thus the content of the book is very practical. Many topics are introduced with an example drawn from an actual scientific or engineering situation. These situations serve to demonstrate the reasons for studying the statistical subject matter presented. The problems left to the reader to solve are an integral part of the book's design and further exemplify the practical situations introduced in each section.

The purpose of this book is to explain as clearly as possible some of the most widely used statistical techniques and to show how they are used, not their mathematical origins. The examples and the problems contain a wide variety of actual data drawn from many different fields. While the basis of statistics lies in mathematics, very few proofs or mathematical explanations are included because they rarely explain the result, the importance of the result, or how the result can be applied. The interested reader can easily find books on the mathematics of statistics; several are noted in the Bibliography.

There are, however, numerous computer experiments included in the book; each is designed to illustrate a statistical result and to make it tenable. It is important that these computer experiments be carried out because they establish the intellectual plausibility of results and significantly aid in understanding the conclusions drawn. Without examining these experiments, the reader may feel compelled to accept conclusions that have no basis whatsoever. From a teaching perspective, these computer experiments often take the place of proofs in that they demonstrate the basis of the experimental results.

Computers play another extremely important role in learning and doing statistics. There are, in fact, many statistical routines that are impractical and incredibly lengthy if carried out by hand. Several of these routines are used in this book. It is important that the reader have access to computing facilities and a standard statistical package such as Minitab®, SPSS®, or JMP®, to mention only three of many. Some statistical routines can now be carried out with graphing calculators such as the TI-83 Plus®. We will often use *Mathematica*®, a computer algebra system, to illustrate results graphically and to do complex calculations. Other computer algebra systems such as Macsyma® or Gauss® could be used in place of *Mathematica*. Appendix A gives some instruction in the use of *Mathematica* and illustrates how some of the calculations were done and how graphs were drawn in this text. Appendix B provides guidance on using Minitab. These appendices are not intended to be exhaustive, but are brief introductions to what are extensive, and widely applicable, computer programs.

For many students, it can be hard to accept the fact that samples can be the basis for conclusions about the population from which they are drawn. They may also find it difficult to appreciate the fact that randomness can be well-behaved and is not at all chaotic. It may surprise some to find the normal curve in so many apparently different situations. Demonstrations, explanations, and applications of these and other interesting facts are part of the marvelous intellectual adventure awaiting the reader.

Note to the Instructor

This book intentionally differs from many other texts written for the post-calculus introductory course in probability and statistics. There are very few proofs here; rather, the results are made plausible by computer experiments designed to explicate the theory. Each chapter begins with one or more introductory examples drawn from an applied situation. These examples set the stage for the material of the chapter and provide a rationale for studying the material that chapter contains. The exercises also frequently contain published data sets with references. Many of the papers from which the data sets are drawn are readable by students; assigning them will further underscore the practicality of the material studied here.

Other distinctive features of the book include the following.

- A large number of homework problems is provided.
- Exercises in each section range in difficulty and are presented in order of increasing difficulty.
- There is systematic use of *Mathematica* and its graphic capability to illustrate statistical concepts.
- Minitab is featured, but other statistical packages can be used with the book.
- An appendix on *Mathematica* and its uses in this text is included.
- An appendix on Minitab and its uses in this text is included.

Statistics is not an easy subject to teach. We grow up as deterministic thinkers and are not asked to consider random factors in outcomes, even though we experience randomness in many situations every day and we frequently depend upon its predictability. Students rarely come to a statistics course with an understanding of randomness. Nor do students have a structured understanding of sampling despite informal use of the technique in everyday situations. Throughout the course, students will encounter the frequency with which randomness occurs in daily situations and, more importantly, the occurrence of randomness in experimentation. The computer experiments are designed to be helpful in making plausible some probabilistic facts; I highly recommend their use.

This book contains more material than can possibly be covered in a single course. Because different courses have different needs enough material is included to permit the instructor to alter emphases in the course in order to serve students with particular interests or requirements and to cover topics of choice in considerable depth. It is a challenge to cover regression, statistical process control, and experimental design in sufficient depth in a single course, but it is important that we present the basics of these topics because our audience will use this material with great frequency.

In most courses, some hard choices must be made in selecting the material one has time to address. As a professor of this course, I have had 2000 minutes of my students' time in which to discuss topics of great importance to scientists and engineers. This volume is the result of my own systematic thought about how to use this very limited time to the best advantage of our future scientists and engineers. I encourage others to consider this challenge as well.

I sincerely hope that teachers of statistics will find the text a useful tool in their endeavors. I encourage you to contact me at the Web site listed below with any comments, suggestions, or errors found.

Supplements

Instructor's Solutions Manual
0-201-74201-2
Contains worked-out solutions to all of the exercises in the text.
Student's Solutions Manual
0-201-74200-4
Provides detailed, worked-out solutions to the odd-numbered text exercises.
Web site
This Web site may be accessed at www.aw.com/kinney. It provides dynamic resources for students. Some of the resources include data downloads, instructions on using Minitab and *Mathematica,* and more.

Acknowledgments

It is a pleasure to acknowledge the considerable assistance I have received from the people of Addison-Wesley. I am particularly indebted to Deirdre Lynch, Ellen Keohane, Julie LaChance, and Joe Vetere who have made the editorial and production process as easy as it can be for an author. In addition, I would like to thank the project manager, Denise Keller, for ensuring the composition of this book ran smoothly and efficiently.

I owe a particular debt of gratitude to my friend and former colleague, Dr. Ralph Grimaldi of Rose-Hulman Institute of Technology, who read the manuscript carefully and whose suggestions added considerably to the readability of the book. I count myself fortunate to know this outstanding teacher and scholar.

I would like to express my appreciation to the following reviewers whose comments and suggestions were invaluable in preparing this text:

Rick Cleary, *Cornell University*
Gary Herrin, *University of Michigan, Ann Arbor*

Wei-Min Huang, *Lehigh University*
Ralph Grimaldi, *Rose-Hulman Institute of Technology*
Marvin H. J. Gruber, *Rochester Institute of Technology*
Herman Senter, *Clemson University*

I would also like to thank Marvin H. J. Gruber, Paul Lorczak, Holly Zullo and Gigi Williams for their help in accuracy checking the manuscript. Any remaining errors are my responsibility. I am very pleased to thank Annie Carter for the artwork on the cover.

My greatest debt, however, is to my wife, Cherry, who endured my long days in my study with particular grace and understanding. She has been, simply, indispensable.

John Kinney
Colorado Springs

Contents

5 Multiple Linear Regression   223

6 Design of Science and Engineering Experiments   277

7 Statistical Process Control   351

1 Graphs and Statistics

1.1 Introduction

ANYONE WHO HAS TAKEN a sip of a cup of hot liquid and decided to let it cool for a time knows something about sampling. We presume that the liquid in the cup is fairly uniform in temperature and that our sample is typical. This conclusion may be correct, but it may also be incorrect.

Sampling occurs in many other situations we encounter on a daily basis. For example, when we see a preview of a film, it is a sample of the film. On the basis of this sample, which may or may not be typical of the film, we may decide whether or not to attend the film.

Consider as well a situation in which a student needs to decide when to leave home in order to arrive at a class on time. Her travel time is based on a sample of travel times of previous trips; these times may be typical or they may be atypical. An estimate of our travel time to a new destination may be based upon a sample of travel times to similar destinations; these may be accurate or inaccurate.

In the presidential election of 2000, predictions made on the night of the election were partially based upon samples of voters as they left voting places. Some of these samples were faulty, and subsequent erroneous predictions were made from them. As we will see, samples are not always accurate, but the rate with which erroneous samples occur can be controlled. We just never know which samples are accurate and which are inaccurate.

Everyone, whether aware of it or not, uses the results of sampling with some frequency in daily life. Sampling plays a very important, indeed a central, role in science and engineering where it is done very systematically and purposefully.

An important use of sampling in science and engineering occurs in the process called the **scientific method.** When we wish to discover new knowledge, we commonly find theoretical results and then seek to verify them through observations taken in the laboratory. These laboratory observations comprise a sample of all the possible observations

1

that could be made. Anyone who has done laboratory experiments knows that these observations can be variable, so a fundamental task for researchers is to decide when the observations support the theory and when they are so variable or so far from expectation that they do not support the theory.

When we perform a laboratory experiment in order to discover a model for a situation in which we have no relevant theory, we again are dependent upon a sample of all the possible observations that could be made.

So sampling occurs both when we know the theory and wish to verify it experimentally, and when the theory is not known and we wish to gather experimental evidence in search of a theory. When we use the scientific method, we generally propose a theory and then test it by using a sample of all of the possible observations. This sample usually leads to a refinement of the theory, which requires more sampling, which in turn may further refine the theory. And so it goes, endlessly, in search of accurate explanations of the behavior of our surroundings.

Engineers who design or test products use a similar repetitive process. A product is first designed and then it is tested through data gathered as a sample. The results of the sample may lead to design improvements, which may lead to more samples. The process is repeated until a final design is accepted.

Statistics and the statistical analysis of data are concerned with sampling, both the manner in which the sampling is done and the inferences, or conclusions, that can be drawn from the sampling. We intend to show how we can draw conclusions from the results of sampling and further, to explain why it is not necessary to take all the possible observations, even if we could do so. We will discuss how it is that this small view (the **sample**) can tell us something about the much larger **population** (the set of all possible observations from the experiment).

As we often do in this text, we begin with a real example from industrial experimentation. Our example involves not a single sample, but two samples, which we wish to compare. We want, in particular, to decide whether or not these two samples could have been selected from the same population, a basic scientific problem.

Example 1.1.1   The production of integrated circuits (IC) must take place under conditions where the air is free of particulates because the presence of such particulates in the circuits is a major cause of their failure. Hewlett-Packard constructed a clean room using ultra-low-particulate filters to maintain uniform air flow over areas in which production occurred.

Hall and Carpenter [7] report on a study comparing the variability in air velocity through clean-room air filters to the variability stated in the purchaser's specifications. The data report the average of three flow rates by two technicians at eight sites on 11 filters. We begin by comparing only two of the sites. We will ignore the averaging that has been done, the rounding of the numbers in the data, and the fact that the flow rates were compared by two technicians. The data are shown in Table 1.1.

It is difficult to tell much about the data by looking at the table. It is much more informative to draw some pictures of the data. Figure 1.1 shows comparative **dotplots** of the data. We have used a horizontal scale to show the flow rates in the data and then plotted each observation with a dot at the appropriate place on the horizontal scale. The sites are separated so that comparisons can be made. (The graphs were produced using the statistical data analysis program **Minitab.**)

Filter	Site 3	Site 8
1	44	23
1	36	19
2	40	25
2	34	12
3	49	40
3	47	42
4	49	45
4	44	43
5	52	40
5	51	42
6	51	29
6	48	22
7	51	32
7	46	31
8	35	27
8	31	15
9	54	50
9	55	54
10	49	39
10	44	34
11	43	30
11	46	27

Table 1.1

Two characteristics of the data are now apparent: The flow rates at the two sites appear to be *centered* at different places, and the data for site 8 appear to be much more *variable* about their center than are the data for site 3 about their center. One might presume that samples would show considerable individual characteristics, so it is difficult for us to predict how other samples from these sites would behave if we had them. Nevertheless, we must deal with the samples we have. That is always the case in reality.

The basic question is: "Could the samples have been selected from the same population?"; that is, are we dealing with expected sampling variation from the same

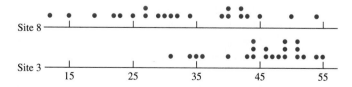

Figure 1.1 Comparative Dotplots of the Data from Example 1.1.1

population, or were the samples in fact chosen from different populations? Although it is impossible to answer this question with absolute certainty, we can say here that the samples were almost certainly *not* from the same population. Just how we reached that answer requires a fairly long explanation, which we will return to in Chapter 3.

Other examples of the decision process illustrated in this example include (1) comparing an old production process to a new one, (2) determining the effectiveness of a drug when we compare the results of people tested with the drug to the results of people given a placebo, and (3) comparing the strengths of two steels of different chemical composition. Examples of other types of problems considered by statistics are given in Section 1.2.

For now we need to explore sampling a bit more and draw more graphs of samples. We shall characterize samples by numbers in Section 1.3.   ■

EXERCISES 1.1

1. An engineer is testing two different types of gear designs, A and B. A sample of 10 gears from each design is tested. The data shown below give the life of the gear in units of 10^6 cycles.

A	120 150 210 250 260 270 510 870 980 1140
B	110 160 350 400 410 430 500 520 530 570

a. Draw comparative dotplots of the two samples.
b. What conclusions concerning centering and variability of the samples can be drawn from the dotplots?
c. If you were to choose the better gear, which type would you choose?

2. An oil company wants to advertise its gasoline as being particularly effective in the Super Fire Cat automobile. An experiment is conducted on three different blends of gasoline, A, B, and C. The following data represent mileages (in miles per gallon).

A	B	C
24.0	25.3	26.3
25.0	26.5	24.0
24.3	26.4	24.7
25.5	27.0	21.2
	27.6	23.5
	28.1	
	19.4	
	20.5	

a. Draw comparative dotplots of the data.
b. Compare the three blends of gasoline with respect to centering and variability.
c. What implications can be drawn concerning the three blends of gasoline?

1.2   What Is Statistics About?

Example 1.1.1 illustrates a very important and commonly occurring statistical application; comparing samples is a very frequent use of the statistical analysis of data. Before proceeding with the development of some results useful in the analysis of the data in Example 1.1.1, let us show three more examples of the use of statistical analysis. This will allow us to anticipate some of the problems frequently encountered by scientists and engineers that we will solve in this book.

Example 1.2.1   A manufacturer of coils for portable heaters is testing the quality of the coil. If the coils are tested under normal operating temperatures to determine how long they last, normally several years, the coils might well become obsolete while they are still being tested because of the speed of technical advances. Moreover, during this lengthy testing period, the manufacturer would not know how long the coils might be expected to last.

Therefore manufacturers often use a process called **accelerated life testing** in which the product is purposely tested under severe conditions. The data in Table 1.2 are from a test on heater coils showing the length of their useful life in hours at various very high temperatures (in degrees Fahrenheit).

Temperature (X)	500	600	700	800	900
Length of Life (Y)	804	791	658	599	562

Table 1.2

It appears that increasing the testing temperature (X) decreases the length of life of the coil (Y), but exactly *how* the temperature influences the length of life of the coil is far from clear.

We might presume that X and Y are linearly related, say by the equation $Y = \alpha + \beta X$. The points given in Table 1.2 do not all lie on a straight line, however, so the linear equation represents a reasonable, not an exact, relationship between the variables. If this is so, what are α and β? What is an estimate for the length of life had the temperature 625° F been observed? Can we give a reasonable estimate in the form of an interval for this length of life? Other questions undoubtedly will occur to the reader. This is an example of **simple linear regression,** where the length of life (Y) is the dependent variable and temperature (X) is the single independent, or predictor, variable. We will later consider **multiple linear regression,** where there are two or more independent variables, these problems being the subjects of Chapters 4 and 5. We next consider an example of some data arising from an experiment designed to discover whether or not some factors are important in configuring a computer. ∎

Example 1.2.2 The time in which a computer performs a complex mathematical process is of concern to a group of investigators. It is presumed that two factors, processor speed (Sp) and random access memory size (RAM), are important. Two processor speeds, $Sp_1 = 133$ MHz and $Sp_2 = 400$ MHz, are studied together with two sizes of RAM, namely, $RAM_1 = 128$ MB and $RAM_2 = 256$ MB. The mathematical program is run with each combination of processor speed and RAM size and the time taken to perform the program is measured in thousandths of a second. Three observations are taken with each of the four possible combinations of processor speed and RAM size, so the observations are **replicated.** Each of the observations is, of course, a sample and subject to all the variation that is encountered when sampling is done. The data are given in Table 1.3.

The values in parentheses in Table 1.3 are the means of the three observations (the sum of the observations divided by 3) in each cell (the combinations of the factors Sp and RAM), and the mean of all the data is 15.

It is clear from an examination of the data that each of the combinations of processor speed (Sp) and size of the random access memory (RAM) produces a somewhat different computation time. The values observed when the values of the factors Sp and RAM are held fixed—that is, those values within each cell—vary as well.

	Sp₁		Sp₂	
	30		16	
RAM₁	26		9	
	16	(24)	11	(12)
	22		6	
RAM₂	12		10	
	14	(16)	8	(8)
				[15]

Table 1.3

In this example, we see from the data that increasing the processor speed (Sp), decreases the time taken to run the mathematical program. We see that increasing the size of the random access memory (RAM) decreases the computing time as well. But, since we have not observed these factors individually, it is difficult to tell exactly, or even to estimate, the amount of the gain in computing time each of these factors contributes. ∎

We note here that rather than losing information when the two factors are considered together, information in fact is *gained* under these conditions, and it is possible not only to estimate the amount of the decreases in computing time in Example 1.2.2, but to estimate the interaction as well. Further, we will see that more than two factors can be considered together and, perhaps surprisingly, not all the factor combinations need to be observed. Showing how to design efficient experiments, that is to determine the observations to be taken, has become a major contribution of statistics to science and engineering. (We will consider the design of experiments further in Chapter 6.)

Our final example in this section comes from manufacturing. It raises some interesting questions concerning **statistical process control.**

Example 1.2.3   A quality control engineer is monitoring a production plant that produces washers whose interior diameters are supposed to average 50 mm. The engineer takes samples of four washers every half-day. Following are the means (the sum of the diameters divided by four) of the samples chosen over a period of five days.

54.2584   33.5034

49.5806   66.1872

51.8335   46.2587

53.6065   53.6817

48.1792   49.2067

These samples show variability, as do the samples in our previous examples. The engineer, however, recognizing that samples will vary even if the process is producing washers with mean 50, is concerned that the process mean may have changed from the target value of 50. Has it changed? This is not an easy question! We consider it, and many related questions, in Chapter 7 (on statistical process control) when we show how statistics improves the quality of products manufactured and sold.   ■

Each of the examples here contains **samples.** In order to answer any of the questions we have raised, we must learn how samples chosen from some universe, or population, of possible observations can vary. We must also learn how to describe samples, both visually and numerically, and we must consider the mathematical basis for the answers to the questions we have raised. It may strike the reader as peculiar that randomness is in reality quite well behaved and that samples chosen randomly actually follow some mathematical laws!

We hope that these examples show some of the rich variety and practical applications of statistics in science and engineering. Now we begin creating a basis for considering these problems and answering our questions.

1.3   What Happens When We Take Samples?

In order to answer this question, we will take an entire population, select samples from it, and see what happens. This will be a useful exercise despite the fact that we usually select samples in order to *discover* some characteristics of a population that are unknown. If an entire population is known, so usually are its characteristics, so sampling becomes an idle and uninformative exercise. This procedure is useful because it will inform us about those concepts of centering and variability that we mentioned in Example 1.1.1.

First, we define a measure of the *center* of a sample, which we call the **mean.**

DEFINITION

Suppose a sample consists of the set of observations $\{x_1, x_2, x_3, \ldots, x_n\}$. The **mean** of the sample is denoted by $\overline{x}$ where

$$\overline{x} = \frac{x_1 + x_2 + x_3 + \cdots + x_n}{n} = \frac{\sum_{i=1}^{n} x_i}{n} = \frac{1}{n} \sum_{i=1}^{n} x_i.$$

As shown in the definition box, we find the mean of a sample by adding the observations in the sample and dividing by the sample size, which is n in this case. The mean treats every observation in the sample equally. There are instances in which we might use a **weighted mean,** where the sample observations are not weighted equally, but we don't need to do that now. There are also other measures of the center of a sample, and we will define some of them later. The mean is sometimes described as the **average** of the sample observations, but many averages are used in statistics, so we try to avoid using this word because it does not have a unique meaning.

For a measure of the *variability* of the sample, we define a quantity we call the **range.**

DEFINITION

The **range** of a sample of the set of observations $\{x_1, x_2, x_3, \ldots, x_n\}$ is

$$\text{range} = \text{Max}\{x_1, x_2, x_3, \ldots, x_n\} - \text{Min}\{x_1, x_2, x_3, \ldots, x_n\}.$$

So the range is simply the difference between the largest value in the sample and the smallest value in the sample. There are other measures of variability and we will discuss them later in this chapter.

Now we have a measure of center and a measure of variability. These are examples of quantities calculated from samples; any such quantity is called a **statistic.** We were about to choose some random samples from a population and see what happens when we do that.

Example 1.3.1   A scanner at a supermarket checkout does not always scan items properly. Sometimes it undercharges by 1 or 2 cents per item, sometimes it charges the correct amount per item, and sometimes it overcharges by 1 or 2 cents per item. One hundred items are run through the scanner with the following results: 30 items were undercharged by 2 cents, 20 items were undercharged by 1 cent, 10 items were charged correctly, 20 items were overcharged by 1 cent, and 20 items were overcharged by 2 cents. These 100 values of $-2, -1, 0, 1,$ and 2 comprise our population, where the minus signs indicate undercharging and the positive signs indicate overcharging.

To show the population we construct a **histogram,** or **bar chart,** which is shown in Figure 1.2. This histogram is easy to construct. We simply show a bar indicating the frequency with which each population value occurs. So we have a bar with height 30 at -2, a bar with height 20 at -1, and so on.

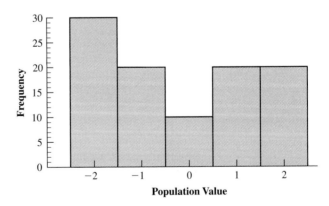

Figure 1.2 Histogram of the Population

Now for the sampling. From the population we select 100 samples of size 5 each. The sampling is done with replacement; that is, an item drawn is replaced in the population

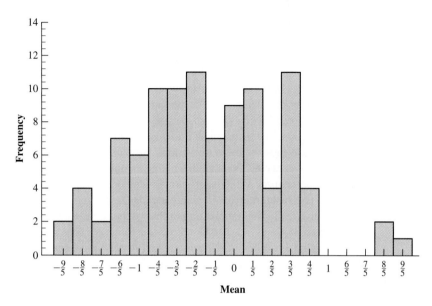

Figure 1.3 Histogram of Sample Means

before the next item is drawn. (We used **Mathematica** to select the samples, but many other computer programs could be used, including Minitab.) Then, for each sample, we compute the sample mean and the sample range. Figures 1.3 and 1.4 show histograms indicating the frequencies with which the various values of the sample means and sample ranges occurred.

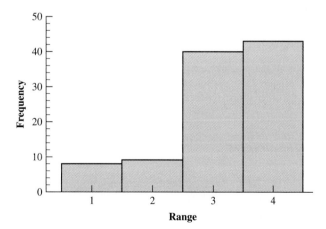

Figure 1.4 Histogram of Sample Ranges

Some observations can be made now. The shape of the histogram of the population can be seen in Figure 1.2. The histogram for the sample means, shown in Figure 1.3, does not resemble the shape of the histogram for the population. The values for the sample means vary from $-9/5$ to $9/5$ in steps of $1/5$. These values tend to bunch up in the middle of the histogram. The reason for this is not difficult to detect. Samples will usually exhibit a range of values, some large and some small. Finding the mean allows the large values to be offset by the small values, giving a value near the center of the range of possible values for the sample mean. (We will explore this behavior more fully in Chapter 2 when we consider the *Central Limit Theorem.*)

We also see that sample means greater than $4/5$ or less than $-6/5$ are relatively rare, because most of the observations must be very large or very small in order for these means to occur. The samples giving sample means in these ranges are then atypical of the population. While such samples can occur, they seldom do.

The histogram of the sample ranges, shown in Figure 1.4, resembles neither the histogram of the population nor that of the sample means. This histogram indicates that about 82 times out of 100, the sample range will be either 3 or 4. Why does this occur? Again the explanation lies in the fact that a typical sample will likely include either -2 or -1 and either 1 or 2, giving ranges of 3 or 4.   ■

So the histograms for the population, the sample means, and the sample ranges are quite distinct and exhibit individual characteristics. However, this behavior is not at all peculiar to this example and we will see it again.

In the next section we will consider other ways of exhibiting the data in a sample graphically and show some other summary measures of the data as well.

EXERCISES 1.3

1. Samples of size 2 are selected without replacement (so the first number selected is not replaced before the second number is selected) from the set $\{1, 2, 3, 4, 5, 6, 7, 8\}$.
   a. Write a list of all the possible samples. The order in which the sample items are selected may be ignored so that the ordered samples (8, 2) and (2, 8) are identical.
   b. Show a histogram of the population.
   c. Calculate the mean for each sample and draw a histogram of these sample means.
   d. Calculate the range for each sample and draw a histogram of these sample ranges. Explain why this histogram does not resemble the histogram in part (c).

COMPUTER EXPERIMENT

1. Take all the possible samples of size 5, selected without replacement, from the set $\{1, 2, 3, 4, 5, 6, 7, 8, 9, 10\}$. The order in which the sample items appear may be ignored.
   a. Show a histogram of the population.
   b. Calculate the mean for each sample and draw a histogram of these sample means.
   c. Calculate the range for each sample and draw a histogram of these sample ranges.

1.4 Some Graphic Displays of Samples

We have used dotplots and histograms of sample means and sample ranges to exhibit their characteristics and general shape. These graphs were easy to draw; most statistical programs for computers produce them as well. Other graphs that are commonly used are *stem-and-leaf plots* and *boxplots*. We now consider drawing histograms, stem-and-leaf plots, and boxplots.

Histograms

The histograms we drew in Figures 1.2, 1.3, and 1.4 are fairly simple in that we could construct a bar for each and every value in the population or every value of the sample mean or every value of the sample range.

Samples, however, are often not so simple to deal with since a bar for each value might not show any variation at all and hence would not reveal the shape of the distribution. To solve this problem, the sample data are first divided into classes and then counts are made of the frequency with which the data occur within those classes. As an example, consider the data from Site 8 in Table 1.1. In Table 1.4 we divide this data into classes and show the frequency with which an observation falls within each class. The histogram is then drawn and appears as in Figure 1.5.

Class	Frequency
7.5–12.5	1
12.5–17.5	1
17.5–22.5	2
22.5–27.5	4
27.5–32.5	4
32.5–37.5	1
37.5–42.5	5
42.5–47.5	2
47.5–52.5	1
52.5–57.5	1

Table 1.4

Histograms are easy to construct provided the sample is not too large. One should be careful, of course. In order to be able to place an observation into one and only one class, we make the boundaries of the classes more precise than the data itself. The data in Table 1.1 were measured to the nearest integer, so we have taken the class boundaries half-way between integers. Had we used classes of 10–15, 15–20, etc., for example, we would not have known into which class to place an observation of 15.

We also need to be careful when selecting the number of classes; selecting too many classes will hide the shape of the data, while selecting too few classes may make the data appear to be less variable than they actually are. Both errors lead to inaccurate displays

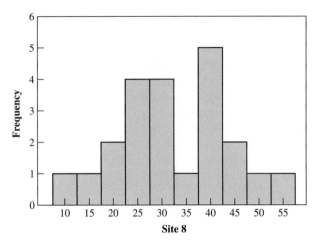

Figure 1.5 Histogram for Site 8

of the data. Selecting between 10 and 15 classes at most is a fairly good general rule to follow.

Stem-and-Leaf Displays

A stem-and-leaf display for the data in Site 8 is shown in Table 1.5.

Stem	Leaf	
1	925	
2	359277	
3	21940	
4	025302	2|9 means 29
5	04	

Table 1.5

The two-digit numbers in the data for Site 8 have tens and units digits that produce, respectively, the stems and leaves of the diagram. Here is how this display is constructed.

■ The tens digit for each of the observations must be 1, 2, 3, 4, or 5. We display these digits in the labeled Stem above.

■ The units digits comprise the column labeled Leaf. The first observations are 19, 12, and 15. The corresponding leaves are 9, 2, and 5. The remainder of the display is constructed in a similar way. The stem-and-leaf display provides a quick way in which to classify the sample observations in this example.

Notice too that the display contains every one of the data points; nothing is lost, except the order in which the data points were observed. Frequently, the time order for the data is of little interest (unless one wants to produce a quality control chart, a topic we will consider later). In contrast, the histogram generally destroys the data and only exhibits the shape of the data set.

However, if the data were more complex, say observations to four decimal places such as 13.4368, a computer program might then use a stem of 1 and a leaf of 3, truncating the data and making its recovery from the stem-and-leaf display impossible.

Note that a legend (2|9 means 29) indicating how the data are to be recovered from the display is given to the right of the column of leaves. Every stem-and-leaf display should have a legend.

It is customary to arrange the leaves in numerical order, from smallest to largest. This results in the display shown in Table 1.6.

Stem	Leaf	
1	259	
2	235779	
3	01249	2|9 means 29
4	002235	
5	04	

Table 1.6

If the stem-and-leaf display is rotated 90°, a crude histogram results (with only five bars in this case).

Some computer programs, such as the display from Minitab shown in Table 1.7, make further alterations in the stem-and-leaf display.

	Stem	Leaf	
1	1	2	
3	1	59	
5	2	23	
9	2	5779	2|9 means 29
(4)	3	0124	
9	3	9	
8	4	00223	
3	4	5	
2	5	04	

Table 1.7

In Table 1.7, several things have happened to our display. First, Minitab repeated the stems for each value, separating the leaves into parts that are 4 or below and those that are 5 or greater. This results in a somewhat more detailed view of the data than we had before.

The left-most column exhibits cumulative frequencies, accumulated from both the top of the display and from the bottom of the display. Thus the entry 5 opposite the first stem of 2 indicates that there are 5 observations less than or equal to the maximum value in this category, which is 23. The 8 opposite the stem value 4 indicates that there are 8 observations greater than or equal to 40. The entry (4) will be explained in the next section.

For more information on constructing stem-and-leaf diagrams (as well as many other types of graphs), see Tukey[12].

Now we consider other measures of centrality for data sets.

Quartiles

We have seen that the mean is a single-number measure of the centrality of a sample. There are other measures of centrality that are commonly used. We consider the *median* first.

DEFINITION

The **median** of a sample is a value in the middle of the sample when the sample observations are arranged in numerical order.

The median is one of the observations if the sample size, n, is odd; otherwise, the median is the mean of the two observations in the middle when the data are arranged in order. The median is denoted by M_d.

If we arrange the data for Site 8 in order we have:

$$12, 15, 19, 22, 23, 25, 27, 27, 29, 30, 31,$$

$$32, 34, 39, 40, 40, 42, 42, 43, 45, 50, 54.$$

Since $n = 22$ observations here, the median is the mean of the two middlemost observations, so

$$M_d = \frac{31 + 32}{2} = 31.5.$$

One advantage of using the median as a measure of the centrality of the data is that it is insensitive to **outliers,** or extreme values. For example, if the largest value in the data set for Site 8 were 74 rather than 54, the mean would be 33.68, a fair increase over its current value of 32.77, but the median remains at 31.5. The largest value is simply the largest value; its size does not affect the value of the median at all.

Now we can explain the entry (4) in Table 1.7. The parentheses indicate that the median is in this group and that the group contains 4 observations. Since the mean of the 11th and the 12th observations is the median, we see again that $M_d = (31 + 32)/2 = 31.5$.

The median divides the sample observations into two equal parts. It is customary to divide each of these parts in half again, to obtain quartiles. We make the following definition.

DEFINITION

The **quartiles** are the medians of the two groups separated by the median.
It is common to denote the quartiles by Q_1, Q_2, and Q_3. The median is then Q_2.

In our example, $Q_1 = 25$ and $Q_3 = 42$. It is also customary to divide really large data sets, such as scores on the Scholastic Aptitude Test, into 100 equal parts. The values that make these divisions are called **percentiles.**

We have defined quartiles as unique values, but there is, perhaps, an added difficulty. If the sample size, n, is even, then there are two middlemost observations. We have defined the median to be the mean of these two observations. If, however, we seek only to divide the data set into two equal parts, *any* value between the two middlemost values will accomplish that. We point this out since different computer programs, while almost always agreeing on the median of a data set, may well differ on the values of the quartiles. Minitab, in fact, produces $Q_1 = 24.5$ and $Q_3 = 42.25$ in the example above. This need not cause too much concern, and it explains why different computer programs may produce somewhat different results for the same data set.

Boxplots

The final graphic display we will discuss is the boxplot. A **boxplot** is a very nice way in which to exhibit the quartiles of the data set, the maximum, the minimum, and (under a convention to be discussed below) *outliers* in the data set.

To draw a boxplot, we draw a vertical scale of the values in the sample data set. We indicate Q_1, Q_2, and Q_3 by horizontal lines and then join the lines made next to Q_1 and Q_3 forming a box with an interior line denoting Q_2.

Next we draw lines, sometimes called **whiskers,** to the minimum and maximum of the sample values, forming the complete boxplot. Boxplots for the sample values from Sites 3 and 8 are shown in Figure 1.6.

Boxplots are often used to compare two samples, which is helpful in answering the question "Do the samples come from the same population?"

In Figure 1.6, note that the boxplots show that the medians have different values. The positions of the medians within the boxes with respect to Q_1 and Q_3, the boundaries of the boxes, also differ, indicating some difference in the amount of data above and below the median. The median occurs in the middle of the box only if the data are symmetric about the median.

The heights of the two boxes also differ. Note that the box, *regardless of its height,* contains exactly 50% of the data. Small heights, then, indicate data that are not very disperse while large heights indicate data that are more widely dispersed. Our comparative

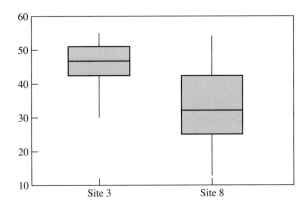

Figure 1.6 Comparative Boxplots for Sites 3 and 8

boxplots indicate that the data for the sample from Site 8 are considerably more disperse than the data from Site 3.

These observations might well lead us to conclude that the two samples do not, in fact, arise from the same population. This will be formally substantiated in Chapter 3 when we establish some statistical theory. For now, however, our graphic display will have to suffice.

Before we conclude our discussion of boxplots, we discuss extreme observations, or *outliers*. Minitab's convention on outliers is this:

Rule

Any observation greater than $Q_3 + 1.5 \cdot (Q_3 - Q_1)$ or any observation smaller than $Q_1 - 1.5 \cdot (Q_3 - Q_1)$ is an **outlier.**

This rule is fairly easy to remember because $Q_3 - Q_1$ is the height of the box, so we look for observations that are beyond one and one-half box heights from the first or third quartiles. (The quantity $Q_3 - Q_1$ is called the **interquartile range.**) Minitab detects outliers and indicates them on the boxplot by the symbol*. Then, rather than connecting these extreme values to the box, the maximum and minimum values, after outliers are deleted, are joined to the box by whiskers.

Figure 1.7 shows a boxplot of a data set with outliers. The data, which can be found in Minitab as the file Cholest.mtw, are the blood cholesterol levels of 28 heart-attack victims, four days after the attack. The outliers are observations of 116 and 352.

There are many conventions regarding what constitutes an outlier and many points of view concerning whether or not such values should be excluded from a statistical analysis of the data set in which they are found. We will only suggest here, that if an outlier is found, the conditions under which it was recorded should be thoroughly investigated since outliers often arise from experimental anomalies.

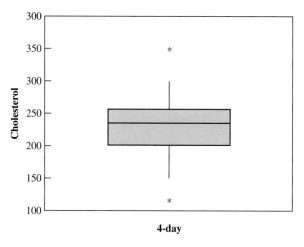

4-day

Figure 1.7 Data with Outliers

EXERCISES 1.4

1. A midwestern state reported the following motor vehicle fatalities by month for two consecutive years.

Month	Jan.	Feb.	Mar.	Apr.	May	June
Year 1	29	29	33	45	72	77
Year 2	36	42	37	45	53	66

Month	July	Aug.	Sept.	Oct.	Nov.	Dec.
Year 1	61	73	60	55	75	41
Year 2	54	61	60	59	56	53

a. Make comparative boxplots of the data and interpret the results.

b. Show stem-and-leaf displays of the data for each year.

2. A process producing transistors is sampled every four hours. At each time, 50 transistors are randomly sampled and the number of flawed transistors is observed. These numbers are as follows.

3, 1, 4, 2, 0, 2, 3, 3, 5, 4, 1, 1, 1, 2, 0

a. Draw a stem-and-leaf display of the data.

b. Draw a boxplot of the data.

c. What conclusions can be drawn from the graphs in parts (a) and (b)?

3. The following data represent the length of time 30 artificial hip joints lasted beyond ten years

2.0	3.0	0.3	3.3	1.3	0.4
0.2	6.0	5.5	6.5	0.2	2.3
1.5	4.0	5.9	1.8	4.7	0.7
4.5	0.3	1.5	0.5	2.5	5.0
1.0	6.0	5.6	6.0	1.2	0.2

a. Draw a histogram of the data.

b. Draw a boxplot of the data.

c. What conclusions can be drawn from the graphs in parts (a) and (b)?

4. The manager of a plant cafeteria is studying the volume of coffee sold in vending machines and gathers the following data.

508.1	831.8	498.4	841.1	568.2	792.1	577.3
787.6	651.7	758.9	657.0	755.3	713.4	697.5

a. Draw a stem-and-leaf display of the data.

b. Draw a boxplot of the data.

5. A study of machine performance was conducted on 20 machines with similar characteristics. The time between machine failures was recorded with the following results.

21.6	21.7	21.9	23.3	23.1
22.7	21.6	23.6	24.2	24.7
21.2	24.8	23.0	25.5	26.2
21.9	22.5	22.3	22.5	24.7

Make a histogram of the data and interpret the result.

6. The following data represent the lifetime, in hours, of a certain type of incandescent lamp.

61, 62, 66, 74, 88, 96, 97, 98, 100, 101,
102, 103, 106, 109, 109, 112, 114, 120, 120

a. Make a stem-and-leaf display of the data.

b. Make a boxplot of the data.

c. What conclusions can be drawn from the graphs in parts (a) and (b)?

7. A sample of LC50 measurements for DDT is given as follows.

5, 10, 8, 7, 4, 9, 0, 13, 8, 23, 4, 5, 9, 6, 10

a. Make a boxplot of the data.

b. Are there any outliers?

8. Three workers with different experience manufacture brakewheels for a magnet brake. Worker A has four years of experience, worker B has seven years of experience, and worker C has one year of experience. Product quality is measured by the difference between the specified diameter and the actual diameter of the brakewheel. On a given day, a supervisor selects nine brakewheels from the output of each worker. The following table shows the absolute values of the differences between the specified and actual diameters of the brakewheels.

A	B	C
2.0	1.5	2.5
3.0	3.0	3.0
2.3	4.5	2.0
3.5	3.0	2.5
3.0	3.0	1.5
2.0	2.0	2.5
4.0	2.5	2.5
4.5	1.0	3.5
3.0	2.0	3.5

a. Which of the workers is the most consistent?

b. Which of the workers produces the least acceptable product?

9. A study of 250 students measured the intelligence quotient (IQ) for each student. The following stem-and-leaf display shows the data gathered.

```
Stem-and-Leaf of IQ, N = 250

   1      6   1
   5      6   6667
  11      7   122344
  19      7   55568899
  39      8   00111111222333344444
  60      8   555667778889999999999
  83      9   000001122222223333444444
 116      9   5555556666677777777777888888999
 (33)    10   000000000011111122233333333334444
 101     10   555555555555556777888888889999
  72     11   000000000011111122222222233444
  43     11   5555666666678889999
  24     12   0111111123334
  11     12   567889
   5     13   04
   3     13   557
```

9|5 means 95

a. Find the quartiles for the data.

b. Show that there are no outliers for this data set by determining the IQ scores that would be outliers.

c. Make a boxplot of the data.

10. The following stem-and-leaf display shows mileages on 200 automobiles recently registered in a state motor vehicle bureau.

```
Stem-and-Leaf of Miles, N = 200
(Leaf Unit = 1.0)

  19      1   0012333334477788889
  39      2   02334555566777888999
  55      3   1111134556677999
  68      4   2222455567789
  83      5   122333345588999
 (18)     6   011223344455566889
  99      7   001223345566788
  84      8   112222244678999
  69      9   11122444679
  58     10   00111245555666788999
  39     11   00011124556666678889999
  15     12   001122333344455
```

6|4 means 64

a. Find the quartiles for the data.

b. Determine which mileages would be outliers.

c. Make a boxplot of the data.

11. The accompanying boxplot arises from a set of 100 observations, each of which is a positive integer.

   a. Estimate the quartiles.

   b. What percentage of the data is below the median? What percentage of the data is above the median?

   c. Use the boxplot to draw a graph approximating the complete set of observations.

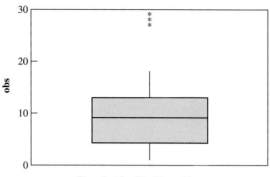

Boxplot for Problem 11

1.5   The Variance

We have shown some measures of the centrality of a data set, namely, the mean and the median. We have used the range as a measure of the variability, or dispersion, of the data set. We have also defined the interquartile range as a measure of the dispersion of the data.

The dotplots in Figure 1.3 of the sample data from Sites 3 and 8 indicate that the data from Site 8 are much more disperse than the data from Site 3. We will now consider another way of measuring this dispersion.

We observe that the dispersion might be measured by calculating the deviations of the sample values from the mean, that is, by looking at all the values of $x_i - \bar{x}$, and it might make sense to add these up. The intent here is to make use of the fact that large deviations from the mean produce dispersion in the data. This is true enough, but we cannot use this fact to measure the dispersion of the data set. The reason for this is that

$$\sum_{i=1}^{n}(x_i - \bar{x}) = \sum_{i=1}^{n} x_i - \sum_{i=1}^{n} \bar{x} = n\bar{x} - n\bar{x} = 0,$$

regardless of the data set. For any data set, the sum of the positive deviations then always exactly offsets the sum of the negative deviations.

One way to prevent this from happening is to square the deviations and add them up. This prevents the sum of the negative deviations from offsetting the sum of the positive deviations, but, as we will see later in this book, the resulting quantity has many other remarkable properties as well. It is also customary to take a variant of the mean of these and divide the result, not by n, but by $n - 1$. The result is called the **sample variance** and is denoted by s^2.

DEFINITION

The **sample variance** for a data set $\{x_1, x_2, x_3, \ldots, x_n\}$ is given by

$$s^2 = \frac{1}{n-1} \sum_{i=1}^{n}(x_i - \bar{x})^2$$

and the non-negative square root of s^2, s, is called the **standard deviation**.

We will explain later why the divisor in this definition is $n - 1$ rather than n. For the moment, be assured that there are good and sufficient reasons for this. (We note here, however, that if an entire population, say of size N, is known, then the divisor used is N, but this is a rare circumstance.) We can obtain another formula for the variance, s^2, by simplifying the sum of the expanded square in the definition. The result is:

$$s^2 = \frac{n \sum_{i=1}^{n} x_i^2 - \left(\sum_{i=1}^{n} x_i \right)^2}{n(n-1)}.$$

This is a good formula to use if the sample variance is calculated by hand and if the data set is not too large. All computer statistical packages will calculate s^2 as well as many other descriptive statistics for a data set. Table 1.8 shows the result when Minitab is asked to provide descriptive statistics for the data in Sites 3 and 8.

Variable	N	Mean	Median	TrMean	StDev	SE Mean
Site 3	22	45.41	46.50	45.65	6.65	1.42
Site 8	22	32.77	31.50	32.75	11.17	2.38

Variable	Minimum	Maximum	Q1	Q3
Site 3	31.00	55.00	42.25	51.00
Site 8	12.00	54.00	24.50	42.00

Table 1.8

In the Minitab display in Table 1.8, TrMean refers to a **trimmed mean,** where the largest 5% and the smallest 5% of the data have been excluded; the trimmed mean is the mean of the remaining observations. We will explain SE Mean later. For now we turn to an interpretation of the standard deviation.

Chebyshev's Inequality

It is perhaps intuitively clear that the standard deviation is a measure of the dispersion of the data. Values of x_i far away from $\bar{x}$ lead to large values of $x_i - \bar{x}$, which in turn contribute more heavily to the size of s than do values of x_i, which are close to $\bar{x}$. It is also clear, if we begin at $\bar{x}$ and move to the right or left of this value, that we take in more and more of the data set. Since data sets vary so widely in their characteristics, it is quite difficult to be very specific about exactly how much of the data set is encompassed in an interval about the sample mean. If we add or subtract multiples of s to $\bar{x}$, the following fact sheds some light on how much of the data set is encompassed.

Chebyshev's Inequality

Suppose that $\bar{x}$ is the mean of a data set with standard deviation s. Choose a positive quantity, $k > 1$. Then at least $1 - 1/k^2$ of the data lie in the interval $(\bar{x} - k \cdot s, \bar{x} + k \cdot s)$.

The inequality is meaningful only for values of $k > 1$ since values of $k \leq 1$ produce negative or 0 values of $1 - 1/k^2$, the proportion of the data in the interval $(\overline{x} - k \cdot s, \overline{x} + k \cdot s)$.

For example, if $k = 2$, then at least $1 - 1/2^2 = 3/4$ of the data lie in the interval $(\overline{x} - 2s, \overline{x} + 2s)$ for any data set. If $k = 3$, then at least $1 - 1/3^2 = 8/9$ of the data is in the interval $(\overline{x} - 3s, \overline{x} + 3s)$ for any data set.

It is also clear that as k increases $1 - 1/k^2$ increases, so wider intervals contain more of the data than do narrower intervals. Data sets vary widely in their shape when histograms are drawn. Chebyshev's Inequality is one of few statements we will make about *any* data set. The inequality involves the standard deviation, indicating that s is indeed a measure of dispersion. The price to be paid, however, for a statement involving any data set is a considerable lack of precision. A proof of the inequality is given so that the approximations that must be made, which produce this lack of precision, can be seen clearly.

Proof.

We begin with the definition of the sample variance.

$$s^2 = \frac{1}{n-1} \sum_{i=1}^{n} (x_i - \overline{x})^2$$

Now consider a positive quantity, k, and two mutually exclusive sets of points, A and B:

$$A = \{x_i |\ |x_i - \overline{x}| < k \cdot s\} \quad \text{and} \quad B = \{x_i |\ |x_i - \overline{x}| \geq k \cdot s\}.$$

Note that the points in set A are those in the interval $(\overline{x} - k \cdot s, \overline{x} + k \cdot s)$ while the points in set B are those outside this interval. The inequality we wish to establish concerns the points in set A.

Now we can write the variance as

$$s^2 = \frac{1}{n-1} \sum_{i=1}^{n} (x_i - \overline{x})^2 = \frac{1}{n-1} \sum_{A} (x_i - \overline{x})^2 + \frac{1}{n-1} \sum_{B} (x_i - \overline{x})^2$$

since A and B are mutually exclusive sets. We now replace each x_i in A by $\overline{x}$ so that

$$\frac{1}{n-1} \sum_{A} (x_i - \overline{x})^2$$

becomes 0.

In set B, $|x_i - \overline{x}| \geq k \cdot s$, so if we replace each $|x_i - \overline{x}|$ in set B by $k \cdot s$, then in the second sum on the right, we can write

$$\frac{1}{n-1} \sum_{B} (x_i - \overline{x})^2 \geq \frac{1}{n-1} \sum_{B} (k \cdot s)^2$$

$$\geq k^2 s^2 \frac{1}{n-1} \sum_{B} 1$$

But $\sum\limits_{B} 1$ represents the number of data points in set B, so

$$\frac{1}{n-1}\sum_{B}(x_i - \bar{x})^2 \geq k^2 s^2 \cdot \frac{\text{Number of points in } B}{n-1}$$

$$\geq k^2 s^2 \cdot (\text{Percent of the data in set } B),$$

where we have used

$$\frac{\text{Number of points in } B}{n-1}$$

as an approximation of the percentage of the data points in set B.

Now since

$$s^2 = \frac{1}{n-1}\sum_{i=1}^{n}(x_i - \bar{x})^2 = \frac{1}{n-1}\sum_{A}(x_i - \bar{x})^2 + \frac{1}{n-1}\sum_{B}(x_i - \bar{x})^2,$$

it follows that

$$s^2 \geq \frac{1}{n-1}\sum_{B}(x_i - \bar{x})^2 \geq k^2 s^2 \cdot (\text{Percent of the data in set } B).$$

Dividing both sides by s^2, we see that $1/k^2 \geq (\text{Percent of the data in set } B)$, so

$$1 - \frac{1}{k^2} \leq 1 - (\text{Percent of the data in set } B) = \text{Percent of the data in set } A.$$

So (Percent of the data in set A) $\geq 1 - 1/k^2$, which is the inequality we set out to prove. ∎

When we replace each $|x_i - \bar{x}|$ in set A by 0 and each $|x_i - \bar{x}|$ in set B by $k \cdot s$, we are clearly making very crude approximations. The lack of precision in Chebyshev's Inequality then becomes quite evident, but it must be stressed that the result holds for any data set, regardless of its form. If more is known about the form of the data set, then much more accurate statements can be made as we will see subsequently.

EXERCISES 1.5

1. The following data are yields of 20 runs in a chemical plant.

71.5	69.5	74.5	77.1
72.7	65.9	72.0	77.3
69.8	71.6	69.3	69.7
71.3	63.9	72.5	72.0
77.1	78.5	77.1	73.0

   a. Find $\bar{x}$ and s^2 for the data.
   b. What percentage of the data lies in the interval $(\bar{x} - s, \bar{x} + s)$? Compare this with the percentage indicated by Chebyshev's Inequality.

2. Samples of 150 are periodically selected from a production line and tested for conformity to customer specifications. Twenty-five samples showed the following number of nonconforming items:

6, 4, 9, 3, 4, 6, 4, 8, 8, 5, 3, 8, 6,

12, 3, 9, 1, 10, 8, 4, 5, 5, 4, 5, 4

   a. Find $\bar{x}$ and s^2 for the data.
   b. Find the mean and the variances for the percentage of nonconforming items in the samples.

3. A sample of manufactured products is tested for its length of life. The lengths, in hours, are as follows:

$$47, 92, 87, 59, 63, 72, 81, 52, 63, 54, 92, 91, 68.$$

  a. Find $\overline{x}$ and s^2 for the data.
  b. What percentage of the data lies in the interval $(\overline{x} - 2s, \overline{x} + 2s)$? Compare this with the percentage indicated by Chebyshev's Inequality.

4. A machine is producing metal pieces that are cylindrical in shape. A sample of 10 pieces is taken. The diameters are as follows:

$$2.07, 2.20, 2.05, 2.04, 1.80, 1.98, 1.90, 1.98, 2.02, 2.02.$$

  a. Find $\overline{x}$ and s^2 for the data.
  b. What percentage of the data lies within 1.5 standard deviations of the mean? Compare this with the percentage indicated by Chebyshev's Inequality.

5. The starting salaries, in thousands of dollars, for high school graduates were as follows:

$$22, 17, 21, 20, 14, 20, 20, 16, 14, 13, 9, 14$$

  a. Find $\overline{x}$ and s^2 for the data.
  b. Find k so that about $2/3$ of the salaries are in the range $(\overline{x} - ks, \overline{x} + ks)$.

1.6 A Look Ahead

In Section 1.2 we presented several examples of the important statistical problems that we will discuss later in the text. They included simple regression, the design of an experiment, and an example from statistical process control. We shall now begin to establish some tools with which to solve these complex problems.

In the course of looking at two samples in this chapter, we discussed several statistics. Their values, however, are totally dependent upon the sample that is chosen, and in all probability, another sample would yield somewhat different values for the sample mean, the range, the sample variance, and the sum of the absolute deviations. When we drew samples from the grocery-store-scanner data, we actually computed the sample means and sample ranges and drew dotplot graphs of these sample statistics, which we showed in Figure 1.4. This gave us some idea of how these sample statistics varied from sample to sample.

It is far less clear how the data in the examples from regression, the design of an experiment, or statistical process control might vary, or what this variation might mean, when we select additional samples.

As we will explain later, we know a great deal about the variability of the sample mean when samples are drawn from any population. The result, called the **Central Limit Theorem,** says, in brief, that the sample mean has a normal distribution, which we will soon encounter. We know, at the present time, much less about the distribution of $\sum_{i=1}^{n} |x_i - \overline{x}|$. This may serve to explain, for the moment at least, why we do not make much use of $\sum_{i=1}^{n} |x_i - \overline{x}|$. However we can determine the distribution of s^2 under certain circumstances, and we will also explain this later.

Since the sample statistics vary from sample to sample, they are called **random variables.**

To continue our study of statistics, we move on to a discussion of the relative chances with which values of statistics might occur. In the next chapter, we study probability and random variables which will lead to a discussion of the solutions of the examples selected from regression, design of experiments, and statistical process control.

Chapter Review

This chapter introduced the subject of statistics. Statistics is concerned with the analysis of data arising from experiments or sample surveys. Usually these experiments or sample surveys are dependent upon a *sample*, or a small portion, of all the possible observations. In this chapter, we simply *described* the data arising from a sample.

Key Examples

- Two samples arising from an industrial experiment
- Life length of heater coils using accelerated life testing
- The influence of RAM size and processor speed on the time a computer takes to do a calculation
- Monitoring the quality of a product produced by an industrial process

Key Concepts

- Drawing histograms, dotplots, stem-and-leaf displays, and boxplots
- Comparing two samples arising from an industrial experiment using comparative dotplots
- Investigating the behavior of the sample mean and range
- Exhibiting the shape of a data set using graphic displays

Key Terms

Sample Mean   The *sample mean* of a data set $\{x_1, x_2, x_3, \ldots, x_n\}$ is

$$\bar{x} = \frac{x_1 + x_2 + x_3 + \cdots + x_n}{n} = \frac{\sum\limits_{i=1}^{n} x_i}{n} = \frac{1}{n}\sum_{i=1}^{n} x_i$$

Sample Range   The *sample range* of a data set $\{x_1, x_2, x_3, \ldots, x_n\}$ is

$$\text{sample range} = \text{Max}\{x_1, x_2, x_3, \ldots, x_n\} - \text{Min}\{x_1, x_2, x_3, \ldots, x_n\}$$

Quartiles   *Quartiles* divide the data into four equal parts and are denoted by Q_1, Q_2 and Q_3

Median Q_2 divides the data into two equal parts and is called the *median*

Interquartile Range   The quantity $Q_3 - Q_1$ is called the *interquartile range*

Outlier   An *outlier* of a data set is defined as any observation greater than $Q_3 + 1.5 \cdot (Q_3 - Q_1)$ or any observation smaller than $Q_1 - 1.5 \cdot (Q_3 - Q_1)$

| **Sample Variance** | The *sample variance* is |

$$s^2 = \frac{1}{n-1}\sum_{i=1}^{n}(x_i - \overline{x})^2 = \frac{n\sum_{i=1}^{n}x_i^2 - \left(\sum_{i=1}^{n}x_i\right)^2}{n(n-1)}.$$

Key Theorem

Chebyshev's Inequality — Suppose that $\overline{x}$ is the mean of a data set with standard deviation s. Choose a positive quantity, $k > 1$. Then at least $1 - \frac{1}{k^2}$ of the data lies in the interval $(\overline{x} - k \cdot s, \overline{x} + k \cdot s)$.

2 Random Variables and Probability Distributions

2.1 Introduction

I N CHAPTER 1 WE PRESENTED several problems that can be solved using statistical data analysis. They included the comparison of two samples, linear regression, the design of an experiment, and statistical process control. We saw that some descriptive *statistics,* such as the *mean, range, variance,* and *standard deviation,* were useful. These statistics are highly dependent upon the random sample selected; that is, another random sample will most probably yield different values for these statistics. Since these statistics vary from sample to sample, they are examples of what we will call **random variables.** While these statistics arose in an example regarding the comparison of two samples, other statistics and random variables will arise when we consider linear regression, the design of an experiment, and statistical process control. As we will see, it is the variability in these statistics that is important to us.

In order to investigate these random variables, we will first consider all the possible samples we might select from a given population. Since the number of these samples is likely to be very large in any practical situation, we will first examine some simpler situations, in order to establish some ideas. This will lead us to the subject of *probability.* So in this chapter, we will examine random variables and probability and then, in later chapters, we will return to the statistical analysis of some real data sets.

2.2 An Example of Random Sampling: Permutations and Combinations

We consider first a situation in which randomness is evident so that we can discuss the ideas of randomness and variability.

Example 2.2.1   Samples of size 3 are selected from the set {1, 2, 3, 4, 5, 6, 7, 8, 9, 10}. We choose these samples without replacement; that is, once a number is chosen, it is removed from the set and cannot be selected again. Industrial sampling is frequently done without replacement because it makes little sense to select an item, inspect it, and then place it back into the population from which it was drawn.

Examples of some of the possible samples are {1, 4, 7}, {2, 4, 9}, and {6, 7, 10}. The order in which the sample items are drawn is irrelevant; we are interested only in the items that occur in the sample. There are industrial situations in which the order in which the items are drawn is important; we will see some examples of this when we study quality control charts in Chapter 7, but for the moment, order is ignored. ■

It makes sense in any situation involving randomness to consider all the possibilities. The set of all the possible samples comprises what we call the **sample space** for an experiment. A more general definition of a sample space will be given later. The sample space does not show us what *will* happen; it is a list of what *can* happen.

To count the number of samples in the sample space, we must consider *permutations* and *combinations*. We begin with sets in which order is important.

DEFINITION

A **permutation** is a linear ordering of distinct objects.

In a permutation, the order in which the items occur is important. For example, if we have the set {a, b, c}, then the complete set of the possible permutations of all these letters is:

$$a, b, c \quad a, c, b$$
$$b, c, a \quad b, a, c$$
$$c, a, b \quad c, b, a$$

We see that there are six permutations of these three letters, but before we show how to count permutations in general, we state an important counting principle.

Counting Principle

If events A and B can occur in n and m ways, respectively, then both events can occur in $n \cdot m$ ways.

To see that this is so, examine the tree diagram (Figure 2.1). For an event A, we show a branch for each of the n ways in which A can occur. Then, after a choice has been made for A, we show m branches after each possible choice for A. These branches indicate the choice that can be made for B. There are then $n \cdot m$ branches in this diagram, each one corresponding to a unique way in which events A and B can occur together. It follows that A and B can occur together in $n \cdot m$ ways.

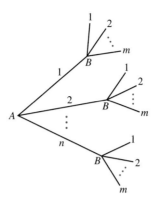

Figure 2.1 Tree Diagram Showing the Counting Principle

The counting principle can easily be extended to more than two events.

Now we can count the number of permutations of three distinct objects. Think of a particular permutation as either the branches of a tree diagram with three levels of branches or as the contents of three boxes, which have been numbered as follows.

$$\text{Box 1} \quad \text{Box 2} \quad \text{Box 3}$$

There are three possible choices for the content of the first box. After the first choice is made, there are two possible choices for the content of the second box, leaving only one possibility for the content of the third box. There are then $3 \cdot 2 \cdot 1 = 6$ possible permutations of these three objects.

It is easy to generalize this situation for n objects, assuming that the objects are different. We now have n boxes to fill and this can be done in

$$n \cdot (n-1) \cdot (n-2) \cdots 3 \cdot 2 \cdot 1$$

different ways. We denote this expression by $n!$ so that

$$n! = n \cdot (n-1) \cdot (n-2) \cdots 3 \cdot 2 \cdot 1.$$

Values for $n!$ increase very rapidly, as Table 2.1 shows.

Now suppose that we have n distinct objects and we wish to permute r of them where $r \le n$. We now have r boxes to fill. This can be done in

$$n \cdot (n-1) \cdot (n-2) \cdots [n-(r-1)] = n \cdot (n-1) \cdot (n-2) \cdots (n-r+1)$$

ways. If $r < n$, this expression is not a factorial, but can be expressed in terms of factorials by multiplying and dividing by $(n-r)!$ We see that

$$n \cdot (n-1) \cdot (n-2) \cdots (n-r+1)$$
$$= \frac{n \cdot (n-1) \cdot (n-2) \cdots (n-r+1) \cdot (n-r)!}{(n-r)!}$$

$$= \frac{n!}{(n-r)!}.$$

n	$n!$
1	1
2	2
3	6
4	24
5	120
6	720
7	5040
8	40,320
9	362,880
10	3,628,800

Table 2.1

with this formula we can now count the number of samples that can be chosen from a population: If we have n distinct objects and we choose only r of them, we denote the number of possible samples, *where the order in which the sample items are selected is of no importance,* by $\binom{n}{r}$, which we read as "n choose r." We want to find a formula for this quantity, and first we consider a special case. Return to the problem of counting the number of samples of size 3 that can be chosen from the set $\{1, 2, 3, 4, 5, 6, 7, 8, 9, 10\}$. We denote this number by $\binom{10}{3}$. Let us suppose that we have a list of all these $\binom{10}{3}$ samples. Each sample contains three distinct numbers, and each sample could be permuted in 3! different ways. Were we to do this, the result might look like Table 2.2 in which only some of the possible samples are listed; then each sample is permuted in all possible ways, so each sample gives 3! permutations.

Sample	Permutations					
{1,4,7}	1,4,7	1,7,4	4,1,7	4,7,1	7,1,4	7,4,1
{2,4,9}	2,4,9	2,9,4	4,2,9	4,9,2	9,2,4	9,4,2
{6,7,10}	6,7,10	6,10,7	7,6,10	7,10,6	10,6,7	10,7,6
⋮	⋮	⋮	⋮	⋮	⋮	⋮

Table 2.2

There are two ways in which to view the contents of the table, which, if shown in its entirety, would contain all the permutations of ten objects taken three at a time. First, using our formula for the number of permutations of ten objects taken three at a time, the table must contain $10!/(10 - 3)!$ permutations. However, since each of the $\binom{10}{3}$ combinations can be permuted in 3! ways, the total number of permutations must also be $3! \cdot \binom{10}{3}$. It follows then that

$$3! \cdot \binom{10}{3} = \frac{10!}{(10 - 3)!}.$$

From this we see that

$$\binom{10}{3} = \frac{10!}{7! \cdot 3!} = \frac{10 \cdot 9 \cdot 8 \cdot 7!}{7! \cdot 3 \cdot 2 \cdot 1} = 120.$$

This process is easily generalized. If we have $\binom{n}{r}$ distinct samples, each of these can be permuted in $r!$ ways, yielding all the permutations of n objects taken r at a time, so

$$r! \cdot \binom{n}{r} = \frac{n!}{(n-r)!}$$

or

$$\binom{n}{r} = \frac{n!}{r! \cdot (n-r)!}.$$

We have presumed that $r < n$, but if $r = n$ or $r = 0$, the formula works if we make the presumption that $0! = 1$.

The number of samples that can be drawn from a population increases very rapidly, as Table 2.3 shows.

n	$\binom{n}{3}$
10	120
100	161,700
200	1,313,400
300	4,455,100
400	10,586,800
500	20,708,500
600	35,820,200
700	56,921,900
800	85,013,600
900	121,095,300
1000	166,167,000

Table 2.3

Clearly, it is impossible to enumerate the sample space when a small sample—much less a sizeable sample—is selected from a large population, but we can find the total number of such samples. In any event, the sample space provides a frame in which we can think about the situation.

The quantities $\binom{n}{r}$ count the number of *combinations* of n objects chosen r at a time. In combinations, unlike permutations, order is not considered. The quantities $\binom{n}{r}$ are also called **binomial coefficients** since they occur in the binomial theorem, a fact that we will consider later when we study the *binomial probability distribution*. It is possible to extend the definition of $\binom{n}{r}$; see Kinney [18].

Now we know how to count the number of samples chosen without replacement from a population of n distinct items. These samples comprise the sample space for our experiment. In the next section we will return to our study of random variables. In particular, we will consider the mean of the sample as a random variable.

EXERCISES 2.2

1. a. Show the sample space when all possible samples of size 2 are selected without replacement from the set $\{1, 2, 3, 4, 5\}$. Ignore the order in which the sample items are selected.

   b. Show the sample space when all possible samples of size 2 are selected with replacement from the set $\{1, 2, 3, 4, 5\}$.

2. Samples of size r are selected with replacement from the population $\{1, 2, 3, \ldots, n\}$. How many samples are there?

3. How many committees of size 3 can be chosen from a class of 24 students?

4. In how many ways can six people line up at a theater box office?

5. How many different orders of a deck of 52 cards are there?

6. An experimental scientist conducts an experiment in which the outcome is a function of three factors: A, B, and C. Factor A can be set at three levels, factor B at six levels, and factor C at four levels. If all the possible levels of each factor are to be observed with all the possible levels of each of the other factors, how many experimental runs will have to be made?

7. A sample space consists of all the possible samples of size 6 selected without replacement from the population $\{1, 2, 3, \ldots, 10\}$.

   a. How many points are there in the sample space?

   b. What proportion of the total number of samples contains the number 7?

   c. What proportion of the total number of samples does not contain the number 7?

8. Explain why $n! = n \cdot (n-1)!$

9. The outcome of an experiment is dependent upon n factors, each of which can be set at one of two levels. If the experiment is to be run with every level of every factor observed with every level of every other factor, how many experimental runs must be made?

10. A sample space consists of all the permutations of the integers $1, 2, 3, \ldots, 15$.

   a. How many points are there in the sample space?

   b. What proportion of the sample points places 7 in the seventh place?

   c. What proportion of the sample points places every integer in its own place?

11. A sample of size 6 is selected from the set $\{1, 2, 3, \ldots, 15\}$ and then a sample of size 3 is chosen from the set $\{a, b, c, d\}$. If the sample space consists of ordered pairs (x, y) where x is the result of the first sample and y is the result of the second sample, how many points are there in the sample space?

12. A scientist is investigating a response, y, from an experiment. It is thought that y is dependent upon one or more variables, $x_1, x_2, x_3, \ldots, x_n$. A number of models can be considered, such as $y = f(x_1, x_2)$ or $y = g(x_4, x_7, x_9)$, and so on. Supposing that each of the functions is linear—that is, each x_i appears to the 0th or first power only and there are no terms involving products of the x's—how many possible models are there?

COMPUTER EXPERIMENT

Use your computer to calculate the following numbers and explain what each represents in a sampling situation.

a. $\dbinom{100}{15}$

b. $\dbinom{1000}{150}$

c. $\dbinom{120}{7} \cdot \dbinom{250}{6}$

d. $\dfrac{42!}{16!}$

e. 2^{15}

f. $3^5 \cdot 6^{10} \cdot 5^7 \cdot 2^6$

2.3 Distribution of Sample Means

In Example 2.2.1, we counted the total number of samples of size 3 that could be chosen from the set $\{1, 2, 3, \ldots, 10\}$. We saw that there are $\binom{10}{3} = 120$ of these. Now suppose that we are interested in choosing a sample of size 3 and in computing the mean of the sample. Some samples and their means are exhibited in Table 2.4.

Sample	Mean
{1,2,3}	2
{1,2,4}	$\frac{7}{3}$
{3,5,7}	5
{4,6,10}	$\frac{20}{3}$
{3,6,8}	$\frac{17}{3}$
$\vdots$	$\vdots$

Table 2.4

This table could be completed showing all $\binom{10}{3} = 120$ samples and their means. The result is shown in Table 2.5, where the frequency with which each mean occurs is given as well as the *relative frequency* (that is, the frequency divided by 120) of each mean. A graph representing these means and their relative frequencies is shown in Figure 2.2. An explanation of how the samples were found and Figure 2.2 was produced can be found in Appendix A.

We can draw some interesting conclusions from an examination of Table 2.5 and Figure 2.2.

■ From an examination of the extreme values, or tails, of the figure, we can conclude that sample means of 2, $\frac{7}{3}$, $\frac{8}{3}$, $\frac{25}{3}$, $\frac{26}{3}$, or 9 occur with total frequency $\frac{1}{120} + \frac{1}{120} + \frac{2}{120} + \frac{2}{120} + \frac{1}{120} + \frac{1}{120} = \frac{8}{120}$, or only about 7% of the time. One of these means would be quite unusual if the samples were drawn repeatedly. Many other observations can be made.

■ The various values the sample mean can assume comprise a set $\{2, \frac{7}{3}, \frac{8}{3}, \ldots, 9\}$. It is clear that these values are not equally likely to occur if we select a sample randomly. For example, a sample mean of $\frac{16}{3}$ occurs with relative frequency $\frac{10}{120}$ while a sample mean of $\frac{11}{3}$ occurs with relative frequency $\frac{5}{120}$, only half the relative frequency of a sample mean of $\frac{16}{3}$.

■ The population from which we are sampling is the set $\{1, 2, 3, 4, 5, 6, 7, 8, 9, 10\}$, so we know that any value in the set is as likely to occur as any other value in the set.

Mean	Frequency	Relative Frequency
2	1	$\frac{1}{120}$
$\frac{7}{3}$	1	$\frac{1}{120}$
$\frac{8}{3}$	2	$\frac{2}{120}$
3	3	$\frac{3}{120}$
$\frac{10}{3}$	4	$\frac{4}{120}$
$\frac{11}{3}$	5	$\frac{5}{120}$
4	7	$\frac{7}{120}$
$\frac{13}{3}$	8	$\frac{8}{120}$
$\frac{14}{3}$	9	$\frac{9}{120}$
5	10	$\frac{10}{120}$
$\frac{16}{3}$	10	$\frac{10}{120}$
$\frac{17}{3}$	10	$\frac{10}{120}$
6	10	$\frac{10}{120}$
$\frac{19}{3}$	9	$\frac{9}{120}$
$\frac{20}{3}$	8	$\frac{8}{120}$
7	7	$\frac{7}{120}$
$\frac{22}{3}$	5	$\frac{5}{120}$
$\frac{23}{3}$	4	$\frac{4}{120}$
8	3	$\frac{3}{120}$
$\frac{25}{3}$	2	$\frac{2}{120}$
$\frac{26}{3}$	1	$\frac{1}{120}$
9	1	$\frac{1}{120}$

Table 2.5

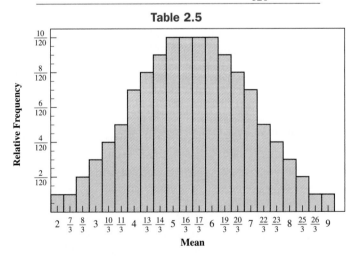

Figure 2.2 Distribution of Sample Means

A graph of the relative frequencies of these population values is therefore uniform, as is shown in Figure 2.3.

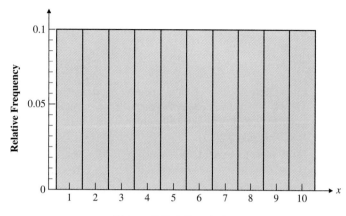

Figure 2.3 A Population

- The shape of the graph in Figure 2.2 is far from flat. We will soon come to recognize that this shape resembles what we will call a "normal" curve. It rises from small values until it reaches a maximum and is symmetric about that maximum value. The normal curve occurs repeatedly in statistical inference and, rather than regard its occurrence as unusual, we will soon come to expect it in a variety of situations.

- The mean of the population is $\frac{1+2+\cdots+10}{10} = 5.5$. The mean value of the sample means is also 5.5, so we say that the mean or expected value of the sample mean is the population mean. This is not a unique feature of this example, but is true in general.

- The set of sample means has a variance, which can be calculated as $\frac{77}{36} = 2.13889$. This is much less than the variance of the population, which is 8.25. This is also to be expected; averaging reduces the variability and is a common consequence of sampling. We will explore this thoroughly, as well as some other consequences of sampling, in this book.

The relative frequencies here reflect the relative frequencies with which the sample means would occur were the sampling to be done repeatedly. This is a luxury that is rarely afforded; rather, we have only one sample to deal with. If repeated, the outcomes would usually differ from the theoretical distribution.

Some further conclusions can be drawn from the theoretical table, however. For example, to find the relative frequency of a mean between 3 and 8 inclusive, we add the relative frequencies in that range to obtain $\frac{112}{120} \doteq 93\%$ (or we could calculate $1 - \frac{8}{120}$, as found above). If a sample gave a mean outside this range, in the tails of the graph in Figure 2.2, we might decide that we had received a very unusual sample. There is, however, another possibility: If our sample mean is outside the range from 3 to 8 inclusive, we might conclude we were sampling from a set for which this mean is more likely rather than from the set $\{1, 2, 3, 4, 5, 6, 7, 8, 9, 10\}$. We are considering, then, a fundamental problem of statistical inference to which we will return in Chapter 3.

The theoretical relative frequencies we have been considering here are called **probabilities.** The sampling distribution of the sample mean is called the **probability distribution of the sample mean.** In the next section we consider some properties of probabilities and probability distributions in general and explain these ideas in more depth.

EXERCISES 2.3

1. a. Show the sample space consisting of all the possible samples of size 3, selected without replacement, from the population {1, 2, 3, 4, 5, 6}. Ignore the order in which the items were selected.

  b. Calculate the mean value for each of the samples in part (a).

  c. Calculate the mean of the sample means and show that this is the mean of the population.

d. Calculate the variance of the sample means. Show that this is

$$\frac{N-n}{N-1} \cdot \frac{\sigma^2}{n}$$

where N is the size of the population, n is the size of the sample, and σ^2 is the variance of the population.

2. Draw graphs of the population and the sample means. What shape do the sample means exhibit?

COMPUTER EXPERIMENT

  a. Make a list of all the possible samples of size 3, selected without replacement, from the population {1, 2, 3, ..., 20}.

  b. Calculate the mean value for each of the samples in part (a) and draw a graph of these mean values.

  c. Calculate the mean and the variance of the population of mean values in part (a).

2.4   Probability

We referred to the set of all possible samples in Example 2.2.1 as the *sample space* for the experiment of selecting a sample of size 3 from the population of integers from 1 to 10. We can now make a more general definition.

DEFINITION

The **sample space, S,** is a set of all the possible outcomes from an experiment.

We also make another definition.

DEFINITION

An **event** is any subset of the sample space.

In our example, the set of means that are even integers is an event, since these means comprise the subset {2, 4, 6, 8} of the population.

In order to reflect the long-range relative frequencies of the sample points and events, we first consider the sample space of all the possible means. It is common to assign *probabilities,* or the expected relative frequencies of each mean if the experiment were to be repeated indefinitely, to the sample points. In our example, we would expect each mean to occur with relative frequency $\frac{1}{120}$, so we say each mean occurs with probability $\frac{1}{120}$. It will be easiest to use some notation, so we let A denote an event and $P(A)$ denote the probability of the event. If $A = \{2, 4, 6, 8\}$, and since no two of the sample points in A can occur together, it makes sense to add the relative frequencies of the sample points to find that

$$P(A) = \frac{1}{120} + \frac{7}{120} + \frac{10}{120} + \frac{3}{120} = \frac{21}{120} = \frac{7}{40}.$$

In general, we would like the probabilities in the sample space to have the following properties:

1. $P(A) \geq 0$ where A is an event.
2. $P(A \text{ or } B) = P(A) + P(B)$ where A and B are mutually exclusive events; that is, they cannot occur together.
3. $P(S) = 1$ where S is the sample space.

Property 1 indicates that since the relative frequency of an event must be non-negative, an event cannot have a negative probability. If events A and B cannot occur together, then it is sensible to add their relative frequencies to find the relative frequency with which one or the other could occur. Property 3 says that the relative frequency of the total sample space must be 1; something in the sample space must occur.

Example 2.4.1 What is the probability that the sample mean will be less than 6 ?

To find this probability, we use property 2 and add the probabilities of the sample points where the sample mean is less than 6. Since these points are mutually exclusive, we find that

$$P(\text{sample mean} < 6) = \frac{1}{120} + \frac{1}{120} + \frac{2}{120} + \cdots + \frac{10}{120} = \frac{70}{120}.$$

It would have been a little easier in this example to find the probability that the sample mean was 6 or more and subtract this probability from 1. Then

$$P(\text{sample mean} < 6) = 1 - P(\text{sample mean} \geq 6)$$

$$= 1 - \left(\frac{10}{120} + \frac{9}{120} + \frac{8}{120} + \cdots + \frac{1}{120} + \frac{1}{120} \right) = \frac{70}{120}. \qquad \blacksquare$$

This example leads us to the following fact about probabilities:

$$P(\overline{A}) = 1 - P(A)$$

where $\overline{A}$ denotes the event "A does not occur."

Example 2.4.2   What is the probability that the sample contains either 1 or 2 or both? (It is common in the study of probability to use simply the word "or" and to omit the "or both," so that the word "or" is used in an inclusive sense.) The events "the sample contains 1" and "the sample contains 2" are not mutually exclusive, since the sample may contain both 1 and 2.

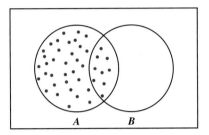

Figure 2.4 Two Events

Figure 2.4 shows a sample space and two symbolic events, A and B, which are not mutually exclusive. We can think of each of the events A and B as being comprised of mutually exclusive points (shown by the dots in set A in Figure 2.4). Suppose, for example, that event A is composed of individual points, so that a typical point has probability p_i. Since these points are mutually exclusive, property 2 is used. We find the probability of event A by adding up the probabilities of the individual points in A, so:

$$\sum_{i \in A} p_i = P(A).$$

If we find $P(A) + P(B)$, we have included the probabilities of the points in the intersection, $A \cap B$, twice, so we must subtract these probabilities once in order to have counted the probabilities in event A or B exactly once. We denote the set of points in A or in B by $A \cup B$.   ■

We conclude this in the addition theorem for two events.

Addition Theorem for Two Events

$$P(A \cup B) = P(A) + P(B) - P(A \cap B)$$

This formula also works if A and B are mutually exclusive since in that case $A \cap B$ is the null set and has probability 0 since it contains no sample points. So the answer to our problem (that is, what is the probability the sample contains items 1 or 2?) is $P(1) + P(2) - P(1$ and $2)$. At this point, we could write out the entire sample space of 120 points and evaluate the needed probabilities directly. We can avoid it and in the process make even more difficult problems possible—by the following process.

There are $\binom{10}{3} = 120$ possible samples. How many of them contain 1? These points can be counted by creating the sample in the following way: Place 1 in the sample and then choose two items from the remaining nine items. This can be done in $\binom{9}{2} = 36$ ways. So $P(1) = \frac{36}{120}$. By symmetry, this is also $P(2)$. To count the points where the sample contains both 1 and 2, note that if these items are in the sample, we have only eight choices for the remaining single item, so $P(1 \text{ and } 2) = \frac{8}{120}$. So the probability that the sample contains items 1 or 2 or both is

$$P(1 \text{ or } 2) = \frac{36}{120} + \frac{36}{120} - \frac{8}{120} = \frac{64}{120} = \frac{8}{15} = 0.53333.$$

We would therefore expect about 53% of our samples to contain one or both of these items. ∎

EXERCISES 2.4

Exercises 1 through 5 refer to the following situation: An inspector for a pharmaceutical firm is inspecting a box containing five pills, denoted by $a, b, c, d,$ and e. Underfilled pills, that is pills with a lesser amount of medication than they should contain, pose an unwanted risk to the patient consuming them. Unknown to the inspector, pills $a, b,$ and c contain the proper amount of medication while pills d and e are underfilled.

1. Show the sample space if the inspector selects two pills, the first pill not being replaced before the second pill is selected. Ignore the order in which the sample items occur.

2. If each of the samples in Exercise 1 has the same probability, what is the probability that the inspector selects at least one underfilled pill?

3. Find the probability that the sample contains only underfilled pills.

4. What is the probability that at least one pill containing the proper level of medication is selected?

5. If we know that at least one of the selected pills contains the proper level of medication, what is the probability that the remaining pill in the sample is underfilled?

Exercises 6 through 8 refer to the following situation: A laboratory will test six items (denoted by $A, B, C, D, E,$ and F). The order in which the items are tested is important.

6. In how many orders may the items be tested?

7. Assuming that the orders in Exercise 6 are equally likely, what is the probability that item E will be tested among the first three tests?

8. What is the probability that items $A, B, C,$ and D are tested before items E and F?

9. In a sample space, suppose that $P(A) = 1/2$, $P(B) = 1/3$, and $P(A \cup B) = 2/3$. What is $P(A \cap B)$?

10. In a certain neighborhood of a city, 15% of the households have above-average incomes, 32% have college educations, and 60% have neither above-average incomes nor college educations. What percent of the households have above-average incomes and college educations?

11. A producer of automobile batteries knows that 3% of the batteries have defective terminals, 4% have defective plates, and 5% have both defective terminals and defective plates. What percent of the batteries have neither of these defects?

12. Explain why $P(A \cup B) \geq P(A) + P(B)$.

2.5   Conditional Probability and Independence

Again we use the sample space consisting of all the possible samples of size 3 that can be selected, without replacement, from the set $\{1, 2, 3, \ldots, 10\}$.

Example 2.5.1   Suppose we know that a sample contains the number 5. What then is the probability that the sample also contains the number 4?

Since we know the sample contains the number 5, the original sample space of $\binom{10}{3}$ sample points is no longer appropriate since many of these points, in fact most of them, do not involve the number 5. We must restrict ourselves to those points containing the number 5. To count these points, note that, after choosing 5, we must choose two of the remaining nine integers, and since there are no other restrictions, there are $\binom{9}{2} = 36$ of these points. It would be sensible to suppose that each of these is equally likely, so we assign a probability of $\frac{1}{36}$ to each of these points. Now we must count the number of these points that also contain the number 4. If 4 is contained in the sample (and 5 has also been included) then there are eight remaining sample values possible. It appears then that the answer to our question is $\frac{8}{36} = \frac{2}{9}$. ■

The stated condition, in this case that the sample contains the number 5, reduces the size of the original sample space from 120 points to 36 points. Problems such as this can always be solved by considering a reduced sample space. However, the problem can also be done using the original sample space. We now show how to do this.

Figure 2.5 shows again a general sample space with two events, A and B. Suppose we want the probability that event A occurs, knowing, or given, that event B has already occurred. We denote this probability by $P(A|B)$ and read this as "the probability that A occurs, given that B has occurred."

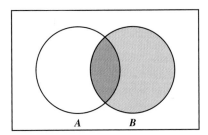

Figure 2.5 Conditional Probability

It might at first glance appear that we simply want $P(A \cap B)$, but this cannot be correct since we don't use the fact that event B has occurred. It is clear, however, that the sample space has become the set of points for which B has occurred. In the figure, we want the doubly shaded points (those in $A \cap B$) as a proportion of the shaded points (those in B). The problem here is that B is not a proper sample space, since the probabilities in the event B sum to $P(B)$, not to 1 as they should for a proper sample space.

This can be resolved in the following way. Suppose that event B is composed of individual mutually exclusive points whose probabilities are, say, p_i for some range of values of i. We know that $\sum_{i \in B} p_i = P(B)$. Now suppose we multiply each of the probabilities p_i by some positive constant, k. This has the following advantage: If an event (in B) had twice the probability of, say, another event (in B), then the multiplication by k will not alter the *ratio* of the probabilities of these two events. But now we want k

so that

$$\sum_{i \in B} k \cdot p_i = 1.$$

This can be written as

$$k \cdot \sum_{i \in B} p_i = 1$$

or

$$k \cdot P(B) = 1 \quad \text{so} \quad k = \frac{1}{P(B)} \quad \text{if } P(B) \neq 0$$

It is easy, then, to make B a sample space; divide the probabilities of each of the points in B by $P(B)$. It is then clear that

$$P(A \mid B) = \frac{\sum\limits_{i \in A \cap B} p_i}{P(B)} = \frac{P(A \cap B)}{P(B)} \quad \text{if } P(B) \neq 0$$

so

$$P(A \mid B) = \frac{P(A \cap B)}{P(B)} \quad \text{if } P(B) \neq 0.$$

Note here that the probabilities in the preceding fractions are calculated using the original sample space, not the reduced one.

We have previously calculated each of these probabilities, so now we find that

$$P(4 \mid 5) = \frac{P(4 \text{ and } 5)}{P(5)} = \frac{8/120}{36/120} = \frac{8}{36} = \frac{2}{9},$$

just as before.

Let us now consider a somewhat more complicated example of conditional probability.

Example 2.5.2   What is the probability that the sample contains a 6 given that it contains a 6 or a 9?
We need, then, $P(6 \text{ or } 9)$ and $P(6)$.

$$P(6) = \frac{\binom{9}{2}}{\binom{10}{3}} = \frac{3}{10}$$

and

$$P(6 \text{ or } 9) = P(6) + P(9) - P(6 \text{ and } 9) = \frac{3}{10} + \frac{3}{10} - \frac{8}{120} = \frac{8}{15} = 0.53333.$$

So

$$P(6 \mid 6 \text{ or } 9) = \frac{P[6 \text{ and } (6 \text{ or } 9)]}{P(6 \text{ or } 9)} = \frac{P(6)}{P(6 \text{ or } 9)} = \frac{3/10}{8/15} = \frac{9}{16}. \qquad \blacksquare$$

Independence

Since $P(A \mid B) = P(A \cap B)/P(B)$, provided $P(B) \neq 0$, it follows that

$$P(A \cap B) = P(B) \cdot P(A|B).$$

Since $P(A \cap B) = P(B \cap A)$, we can also write this as

$$P(B \cap A) = P(A \cap B) = P(A) \cdot P(B \mid A).$$

If $P(B \mid A) = P(B)$, then events A and B are called **independent.** In that case, the occurrence of A does not alter the probability that B will subsequently occur and we see that

$$P(A \cap B) = P(A) \cdot P(B) \quad \text{if } A \text{ and } B \text{ are independent.}$$

We note here that the ideas of mutually exclusive events and independent events are quite different. Mutually exclusive events cannot occur together while independent events must be able to occur together. (We are excluding events of probability zero from our discussion for the moment. Perhaps surprisingly, such events will soon be of great interest to us.) It is a fact that if such events are mutually exclusive, then they cannot be independent, and, equivalently, if events are independent, then they cannot be mutually exclusive.

Example 2.5.3 Referring once more to our example of selecting a sample of size 3, are the events "the sample contains 6" and "the sample contains 9" independent?

The events are independent if $P(6 \mid 9) = P(6)$. But

$$P(6 \mid 9) = \frac{\binom{8}{1} / \binom{10}{3}}{\binom{3}{2} / \binom{10}{3}} = \frac{2}{9}$$

and

$$P(6) = \frac{\binom{9}{2}}{\binom{10}{3}} = \frac{3}{10},$$

so the events are not independent. ∎

EXERCISES 2.5

1. Events A and B in a sample space have $P(A) = 1/4$, $P(B) = 1/3$ and $P(A \cap B) = 1/9$.
  a. Find $P(A \cup B)$.
  b. Find $P(A \mid B)$.
  c. Find $P(B \mid A)$.
  d. Are A and B independent? Why or why not?

2. A quality control inspector knows that the items she inspects have at most two types of defects: Type A and type B. Past experience has shown that $P(A) = 0.02$, $P(B) = 0.03$, and $P(A \cup B) = 0.06$.
  a. Find $P(A \mid B)$.
  b. Are A and B independent? Why or why not?

3. A plant manufacturing transistors has two shifts: shift I and shift II. It is known that shift I produces 60% of the transistors made and that 3% of its production will not meet customer specifications. Five percent of shift II's production will not meet customer specifications.

  a. What percent of the plant's production will not meet customer specifications?
  b. A transistor is chosen at random and found not to meet customer specifications. What is the probability that it was produced by shift I?

4. A quality control inspector selects two motors from motors A, B, C, D, and E and subjects them to destructive testing. Find
  a. the probability that motor C is chosen.
  b. the probability that motor B is chosen given that motor C is chosen.

5. Box I contains 6 good and 2 defective light bulbs. Box II contains 3 good and 3 defective light bulbs. A box is selected at random and a light bulb is chosen from it. Find
  a. the probability that the light bulb is defective.
  b. the probability that the light bulb came from box II given that it is defective.

6. Suppose, in Exercise 5, that it is known that box II contains 6 light bulbs, but the composition of the box is unknown. If the probability of choosing a box at random and selecting a defective light bulb is 5/24, what is the composition of box II?

7. A test for the presence of the HIV virus is such that it detects the virus in patients actually having the virus with certainty. Among patients not actually having the virus, the test (falsely) indicates the presence of the virus with probability 1/20,000. If 1 person in 10,000 in a population actually has the virus, what is the prob-ability that a patient actually has the virus if the test indicates the presence of the virus?

8. Use the conditional probability formulas to solve Exercise 5 in Section 2.4.

9. Find $P(A \cup B)$ if
  a. A and B are mutually exclusive.
  b. A and B are independent.

10. Find $P(A \mid B)$ if
  a. A and B are mutually exclusive.
  b. A and B are independent.

2.6 Random Variables and Probability Distributions

In Section 2.3 we established a histogram showing the means of all the samples of size 3 that could be chosen from the set $\{1, 2, 3, 4, 5, 6, 7, 8, 9, 10\}$ and the relative frequencies with which these means occurred. We now recognize the relative frequencies as probabilities. The histogram, Figure 2.2, shows what we will call the **probability distribution of the sample mean.** We want to make the idea of probability distribution more specific and to extend our understanding of random variables as well. We first show another example from industrial sampling.

Example 2.6.1 Production from an industrial process is subject to quality control inspection. Suppose a factory has produced 100 items in a day. The production actually contains 10 items that do not meet customer specifications, but the factory management does not know there are 10 such items, or which items they are. An *acceptance sampling* scheme is adopted according to the following plan: A random sample of 15 items will be selected and each item inspected for compliance with customer specifications. If the sample contains 3 or fewer items not meeting customer specifications, the entire production lot (with any defective items found in the sample replaced by acceptable items) will be shipped to the customer.

We know there are $\binom{100}{15} = 253{,}338{,}471{,}349{,}988{,}640$ possible samples; these samples comprise our sample space, but clearly we cannot write them all down. Since we are interested in the number of items not meeting customer specifications, let us denote this quantity by X. The quantity X is a function of the particular sample, and since X varies from sample to sample, X is called a *random variable*. We make the following general definition.

DEFINITION

A **random variable** is a real-valued function that takes on values on the points of a sample space.

It is clear in this case that X can assume the values $0, 1, 2, \ldots, 10$, but the relative frequencies with which they occur is not so clear. We are now interested in the probabilities with which X assumes each of these values. Let us begin with $P(X = 3)$. For X to be 3, we must choose 3 of the 10 unacceptable items, and the remaining 12 items must be selected from the 90 acceptable items. These two events can occur in $\binom{10}{3} \cdot \binom{90}{12}$ ways by using the counting principle. We assume that each of the $\binom{100}{15}$ samples is equally likely, so

$$P(X = 3) = \frac{\binom{10}{3} \cdot \binom{90}{12}}{\binom{100}{15}} = \frac{616,980,915}{4,755,579,521} = 0.129738.$$

(The expression for $P(X = 3)$ reduces quickly to a quantity that can be calculated with a pocket calculator.) We can similarly find the probabilities for all the remaining values of X. We show the result of these calculations in Table 2.6.

X	Probability
0	0.18077
1	0.35678
2	0.29191
3	0.12974
4	0.03449
5	0.00569
6	0.00059
7	0.00004
8	0.00000
9	0.00000
10	0.00000

Table 2.6

Table 2.6 lists the values of $P(X = x)$ for all the possible values of x. To put it another way, the table shows the *probability distribution function* of the random variable X. Note that a random variable could be defined as π, regardless of the point involved in some sample space, making the random variable neither random nor variable. We realize that we are in fact referring to a *function* taking values on the points of a sample space, but the term "random variable" is very common in the literature on probability and statistics and so we continue to use it.

DEFINITION

The function $f(x) = P(X = x)$, defined on the points of a discrete sample space, is called the **probability distribution function** for the random variable X.

Probability distribution functions can be defined by either a table or a mathematical expression. In our example, we can give a formula for $f(x)$. Since we must choose x

from 10 items and then $15 - x$ from 90 items, the formula is

$$f(x) = P(X = x) = \frac{\binom{10}{x} \cdot \binom{90}{15-x}}{\binom{100}{15}}, \qquad x = 0, 1, 2, \ldots, 10.$$

We can state some properties of a general probability distribution function, $f(x)$:

1. $f(x) \geq 0$.
2. $0 \leq f(x) \leq 1$.
3. $\sum_x f(x) = 1$ where the sum extends over all the values x in the sample space, S.

Property 1 says that probabilities must be non-negative, while property 2 says that probabilities must be between 0 and 1. Property 3 assures us that the probabilities over the entire sample space add up to 1.

A graph of this probability distribution function is shown in Figure 2.6. See Appendix A for an explanation of how the probabilities and Figure 2.6 are produced using Mathematica.

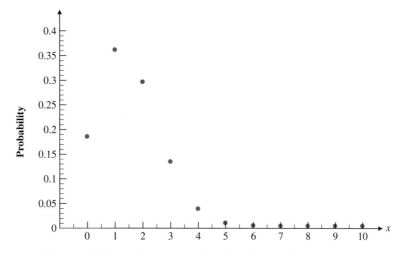

Figure 2.6 Probability Distribution Function for Example 2.6.1

The random variable X in this example is called a **hypergeometric random variable.**

It is obvious that the sampling scheme adopted in this example serves to increase the quality of the product sold to the customer since it is probable that some of the unacceptable items, which will be replaced by acceptable items before delivering the product to the consumer, will be discovered in the sampling process. We will return to this random variable in Chapter 7 and demonstrate specifically how the quality of the product sold is increased. ∎

We have described data sets by measures we call *means* and *variances.* Now we turn to similar descriptions of random variables.

Means and Variances of Discrete Random Variables

Consider a set of values, say $\{x_i \mid i = 1, 2, \ldots, n\}$, each of which occurs with frequency f_i for $i = 1, 2, \ldots, k$. The mean value is the sum of all the x's, here $\sum_{i=1}^{k} x_i \cdot f_i$, divided by the total number of x's, $\sum_{i=1}^{k} f_i = n$, or

$$\frac{\sum_{i=1}^{k} x_i \cdot f_i}{\sum_{i=1}^{k} f_i} = \frac{\sum_{i=1}^{k} x_i \cdot f_i}{n}.$$

But notice that this could be written as

$$\frac{\sum_{i=1}^{k} x_i \cdot f_i}{n} = \sum_{i=1}^{k} x_i \cdot \frac{f_i}{n}.$$

The relative frequencies f_i/n could be regarded as probabilities or values of $f(x)$. This explains why we make the following definition.

DEFINITION

If random variable X has the discrete probability distribution function $f(x)$, we define its **mean** or **expected value** or **expectation** as

$$E(X) = \mu = \sum x \cdot f(x)$$

where the summation extends over all the values x.

In our sampling inspection example, we find that

$$\begin{aligned} E(X) = {} & 0 \cdot (0.18077) + 1 \cdot (0.35678) + 2 \cdot (0.29191) + 3 \cdot (0.12974) + 4 \cdot (0.03449) \\ & + 5 \cdot (0.00569) + 6 \cdot (0.00059) + 7 \cdot (0.00004) + 8 \cdot (0.00000) + 9 \cdot (0.00000) \\ & + 10 \cdot (0.00000) = 1.50. \end{aligned}$$

So, on average, we expect our sampling scheme to detect 1.5 unacceptable items. It can be shown that the mean value for the hypergeometric random variable is $n \cdot p$ where n is the sample size and p is the proportion of unacceptable items in the population. In this case, $\mu = 15 \cdot 10/100 = 1.50$, verifying the result obtained above.

Random variables also have *variances,* which are defined as follows.

DEFINITION

The **variance** of a random variable X with probability distribution $f(x)$ is given by

$$\sigma^2 = \sum (x - \mu)^2 \cdot f(x)$$

where the summation extends over all the values of x.

The positive square root of σ^2, σ, is called the **standard deviation.** We often denote the variance of X by $\text{Var}(X)$.

It is often easier to use the following formula for σ^2 :

$$\sigma^2 = \sum_x x^2 \cdot f(x) - \mu^2.$$

In our example we find, using the second formula given for σ^2, that

$$\begin{aligned}\sigma^2 = &\ 0^2(0.18077) + 1^2(0.35678) + 2^2(0.29191) + 3^2(0.12974) + 4^2(0.03449) \\ &+ 5^2(0.00569) + 6^2(0.00059) + 7^2(0.00004) + 8^2(0.00000) + 9^2(0.00000) \\ &+ 10^2(0.00000) - 1.50^2 = 1.1574.\end{aligned}$$

It can be shown for the hypergeometric random variable that

$$\sigma^2 = n \cdot \frac{\overline{M}}{N} \cdot \frac{N - \overline{M}}{N} \cdot \frac{N - n}{N - 1}$$

where n is the sample size, N is the population size, and $\overline{M}$ is the number of items in the sample not meeting customer specifications. In this case, we have

$$\sigma^2 = 15 \cdot \frac{10}{100} \cdot \frac{100 - 10}{100} \cdot \frac{100 - 15}{100 - 1} = 1.1591.$$

(*There is some roundoff error in our first calculation*)

Our examples to this point have all involved **discrete random variables,** that is, variables that assume values in a finite or countably infinite set. There are situations, however, in which the variables can assume values in an interval or intervals producing random variables we call **continuous.**

Continuous Random Variables

Suppose an experiment consists of throwing a dart at an interval 10 units long, the random variable X denoting where the dart lands in the interval. It is then natural to use a sample space consisting of the interval $[0, 10]$. Any number in that interval now can be the result of the experiment. What probability should be assigned to each of the points?

Suppose we choose to assign our favorite small probability, say $0.0000000000000001 = 10^{-16}$ to each point. It is, however, easy to show that the interval $[0, 10]$ contains more than 10^{16} points, so we have used up more than the allowable total probability of 1 by this assignment. The only reasonable conclusion is that each point should be assigned a probability of 0 and that the probability distribution function, $f(x)$, should be $f(x) = 0$.

Now suppose that the dart thrower is three times as likely to land the dart in the interval $[5, 10]$ as she is in the interval $[0, 5]$. We are still faced with assigning a probability of zero to each individual point in the interval, giving $f(x)$ in this case as $f(x) = 0$ again.

Our assignment fails to distinguish a fair dart thrower from an unfair one!

The fault here lies not in the answer, but in the question. It is clear that *any* random variable assuming values in an interval or interval of real numbers has a probability distribution function of 0. We need another function to describe behavior in the continuous case.

It would, however, in the fair dart thrower's case, be natural to assign a probability of 1/2 to the interval [5, 10] since we would argue that, perfectly random tosses of the darts being made, the dart should land in that interval, or indeed in any interval of length 1/2, about half the time. It then becomes sensible to assign probabilities to intervals rather than to specific values.

DEFINITION

We define a **probability density function, $f(x)$,** where X is a continuous random variable, to have the following properties:

1. $f(x) \geq 0$.
2. $\int_a^b f(x)\, dx = P(a \leq X \leq b)$.
3. $\int_{-\infty}^{\infty} f(x)\, dx = 1$.

Property 2 reflects the fact that probabilities can be assigned to intervals rather than to points as is the case with discrete random variables. Property 1 says that the function must be non-negative (since negative values might lead to negative probabilities), while property 3 reflects the fact that the entire sample space must have probability 1.

The word *density* is an indicator that we are dealing with a continuous random variable, which has a probability density function. Discrete random variables have probability *distribution* functions. The words density and distribution serve to distinguish the two cases, which we deal with mathematically in quite different ways.

Example 2.6.2   What probability density function should be used in the case of the fair dart thrower?

Since *any* interval of length say 1/4 should have probability 1/4, and keeping in mind the other properties of a probability density function, we conclude that

$$ f(x) = \begin{cases} \dfrac{1}{10} & \text{if } 0 \leq x \leq 10, \\[2mm] 0 & \text{otherwise.} \end{cases} $$ ∎

A graph of $f(x)$ for the general situation where $a \leq x \leq b$, is shown in Figure 2.7 X here is called a **uniform random variable** and its probability distribution is called the **uniform probability density function.** Note that $f(x)$ is discontinuous and is 0 except for the interval $0 \leq x \leq 1$. In our particular example, $a = 0$ and $b = 10$.

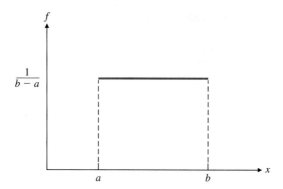

Figure 2.7 Uniform Probability Density Function

Example 2.6.3   What probability density function should be used in the case of the unfair dart thrower whose probability of landing in the interval [5, 10] is three times that of landing in the interval [0, 5]?

One—but not the only—probability density function satisfying the conditions is given by

$$f(x) = \begin{cases} \dfrac{x}{50} & \text{if } 0 \le x \le 10, \\[2mm] 0 & \text{otherwise.} \end{cases}$$

A graph of $f(x)$ is given in Figure 2.8.

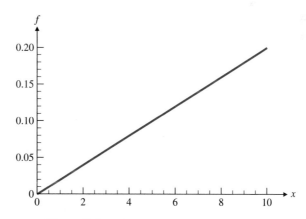

Figure 2.8 $f(x)$ for the Unfair Dart Thrower

Let us check that $f(x)$ satisfies the conditions for the unfair dart thrower. We find that $f(x) \ge 0$ and that

$$P(0 \le X \le 10) = \int_0^{10} \frac{x}{50}\, dx = 1;$$

further,

$$P(5 \leq X \leq 10) = \int_5^{10} \frac{x}{50} \, dx = \frac{3}{4}$$

so that

$$P(0 \leq X \leq 5) = \frac{1}{4}.$$

So

$$P(5 \leq X \leq 10) = 3 \cdot P(0 \leq X \leq 5).$$

Example 2.6.4   Fluorescent light bulbs have life length given by the probability density function

$$f(x) = \begin{cases} \dfrac{1}{500} e^{-x/500}, & x \geq 0, \\ \\ 0 \text{ otherwise.} \end{cases}$$

A graph of $f(x)$ is given in Figure 2.9.

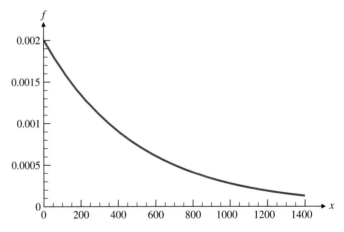

Figure 2.9 Exponential Probability Density Function

It is easy to check that $f(x) > 0$ and that $\int_0^\infty (1/500) \, e^{-x/500} \, dx = 1$ so that $f(x)$ satisfies the conditions for a probability density function. It is called an **exponential probability density function.**   ∎

Here is a surprise. Suppose that our fluorescent fixture has lasted h hours. What is the probability that it will last an additional j hours?

We must calculate $P(X \geq h + j \mid X \geq h)$. It will be useful to calculate $P(X \geq a)$ since this quantity will occur repeatedly in our calculation.

We see that

$$P(X \geq a) = \int_a^\infty \frac{1}{500} e^{-x/500}\, dx = -e^{-x/500} \big|_a^\infty = e^{-a/500}.$$

Now

$$P(X \geq h + j \mid X \geq h) = \frac{P(X \geq h + j)}{P(X \geq h)} = \frac{e^{-(h+j)/500}}{e^{-h/500}} = e^{-j/500}$$

$$= P(X \geq j).$$

So we find that the probability that the fixture lasts an additional j hours after having lasted h hours is the same as the probability that the fixture lasts j hours initially. This unexpected result is known as the *memoryless* property of the exponential random variable. It is the only continuous random variable with this property.

We now turn to finding means and variances for continuous random variables.

Means and Variances for Continuous Random Variables

The following definition will come as no surprise.

DEFINITION

If the continuous random variable X has probability density function $f(x)$, then

$$\mu = \int_{-\infty}^\infty x \cdot f(x)\, dx \quad \text{and} \quad \sigma^2 = \int_{-\infty}^\infty (x - \mu)^2 \cdot f(x)\, dx.$$

It is not difficult to show that

$$\sigma^2 = \int_{-\infty}^\infty (x - \mu)^2 \cdot f(x)\, dx = \int_{-\infty}^\infty x^2 \cdot f(x)\, dx - \mu^2,$$

a result similar to that holding for discrete random variables.

Example 2.6.5   If

$$f(x) = \begin{cases} \dfrac{1}{b - a} & \text{if } a \leq x \leq b, \\[2mm] 0 \text{ otherwise}, \end{cases}$$

then X is called a **uniform random variable** on the interval $[a, b]$. So

$$\mu = \int_a^b x \cdot \frac{1}{b - a}\, dx = \frac{a + b}{2}$$

and

$$\sigma^2 = \int_{-\infty}^{\infty} x^2 \cdot f(x)\, dx - \mu^2 = \int_{a}^{b} x^2 \cdot \frac{1}{b-a}\, dx - \left(\frac{a+b}{2}\right)^2 = \frac{(b-a)^2}{12}.$$

So in the case of the fair dart thrower where $a = 0$ and $b = 10$,

$$\mu = \frac{a+b}{2} = \frac{0+10}{2} = 5$$

and

$$\sigma^2 = \frac{(b-a)^2}{12} = \frac{(10-0)^2}{12} = \frac{25}{3}.$$ ∎

Example 2.6.6   For the general exponential random variable defined as

$$f(x) = \begin{cases} \dfrac{1}{\lambda} e^{-x/\lambda}, & x \geq 0, \quad \lambda > 0, \\[2mm] 0 \text{ otherwise}, \end{cases}$$

$$\mu = \int_{0}^{\infty} x \cdot \frac{1}{\lambda} e^{-x/\lambda}\, dx = \lambda \quad \text{and} \quad \sigma^2 = \int_{0}^{\infty} x^2 \cdot \frac{1}{\lambda} e^{-x/\lambda}\, dx - \lambda^2 = \lambda^2$$

so we see that the mean and the standard deviation for the exponential random variable are equal. ∎

Many different random variables can be created from the samples of size 3 drawn from the set $\{1, 2, 3, 4, 5, 6, 7, 8, 9, 10\}$. We began to investigate the sample mean in the previous sections of this chapter, and we will learn more about this important statistic as we proceed. We saw that the sample mean has a "normal" sampling distribution. The sampling distributions of statistics such as the variance, range, and standard deviation—as well as many other statistics—have, in general, much more complex probability distributions. However, statistical inference and data analysis are based upon these distributions, so they are very important to our understanding.

Chebyshev's Inequality

Chebyshev's Inequality, first stated in Section 1.5, also holds for both discrete and continuous random variables as follows:

Suppose a random variable, X, has mean μ and standard deviation σ. Choose $k > 1$. Then

$$P(\mu - k\sigma \leq X \leq \mu + k\sigma) \geq 1 - \frac{1}{k^2}.$$

The proof is very similar to that given in Section 1.5. ∎

We now turn to important specific discrete and continuous probability functions and show the applications of each.

EXERCISES 2.6

1. Samples of size 5 are selected, without replacement, from a population of 75 items, 25 of which do not meet customer specifications. Let X denote the number of items in the sample that do not meet customer specifications.
   a. Find $P(X = 3)$.
   b. Find $P(X \geq 3)$.
   c. Find the probability distribution function for X.

2. A random variable X has probability distribution function $f(x) = 1/5$ for $x = 1, 2, 3, 4, 5$. (X here is called a *uniform discrete* random variable.)
   a. Find $P(X \geq 2)$
   b. Find the mean, μ, and the variance, σ^2, for X.
   c. Find $P(X = 3 \mid X \geq 2)$.

3. Random variable X has probability distribution function

$$f(x) = k \cdot (7 - x) \quad \text{for } x = 1, 2, 3, 4, 5, 6.$$

   a. Write out the probability distribution function and show that k must be $1/21$.
   b. Find the mean, μ, and the variance, σ^2, for X.

4. Random variable X has probability distribution function

$$f(x) = k \cdot (x^2 - x) \quad \text{for } x = 1, 2, 3, \ldots, 10.$$

   a. Show that $k = 1/330$.
   b. Find the mean, μ, and the variance, σ^2, for X.

5. If random variable X has probability distribution function

$$f(x) = \frac{e^{-\mu} \cdot \mu^x}{x!}$$

where $\mu > 0$ and $x = 0, 1, 2, \ldots$, then X is said to have a *Poisson* distribution. It can be shown that both the mean and the variance of X are μ.

   a. If $\mu = 3$, show that the mean of X is 3.
   b. If $\mu = 3$, show that the variance of X is 3. (*Hint:* Find $E[X(X - 1)]$.)

6. Suppose X is a (continuous) uniform random variable on the interval $[3, 7]$.
   a. Show that the mean of X is 5.
   b. Calculate $P(X \leq 6)$.
   c. Show that the variance of X is $4/3$.

7. Random variable X has an exponential probability density with $\lambda = 4$ so that

$$f(x) = \frac{1}{4} e^{-x/4}, \quad x \geq 0.$$

   a. Show that the mean of X is 4.
   b. Show that the standard deviation of X is 4.
   c. Calculate $P(X \geq 4)$.
   d. Calculate $P(X \geq 4 \mid X \geq 1)$.

8. A (continuous) uniform random variable on the interval $[a, b]$ has mean 7 and variance 3. Find a and b.

9. Suppose that $f(x) = (1/2)^x$ for $x = 1, 2, 3, \ldots$.
   a. Show that $f(x)$ is a discrete probability distribution function.
   b. Find $P(X > 3 \mid X > 2)$.

10. Suppose that $f(x) = (3/8)x^2$ for $0 < x < 2$.
   a. Show that $f(x)$ is a continuous probability density function.
   b. Find $P(\mu - 2\sigma < X < \mu + 2\sigma)$ and compare with the result given by Chebyshev's Inequality.

11. Suppose that

$$f(x) = \frac{1}{3} \left(\frac{2}{3} \right)^{x-1} \quad \text{for } x = 1, 2, 3, \ldots.$$

   a. Show that $f(x)$ is a discrete probability distribution function.
   b. Find $P(X > 4)$.

COMPUTER EXPERIMENTS

1. a. Select a sample of 1000 integers from the uniform discrete distribution on the set of integers {1, 2, 3, 4, 5, 6, 7, 8}.
   b. Draw a histogram showing the frequency with which each integer occurred and compare these frequencies with the expected frequencies for each integer.
   c. Find the mean value of the sample and compare this to the mean of the population sampled.

2. a. Select a sample of 1000 observations from the uniform continuous distribution on the interval [1, 8].
  b. Draw a histogram showing the frequency with which each integer occurred and compare these frequencies with the expected frequencies for each integer.
  c. Find the mean value of the sample and compare this to the mean of the population sampled.

2.7 Binomial Random Variable

Example 2.7.1   Consider an industrial production line that is producing a very large number of transistors. The transistors are subject to a quality inspection and are classified as either meeting customer specifications or not meeting customer specifications. For the sake of brevity, we will refer to these two classes as "M" and "NM". From long past experience with the production process it has been learned that about 80% of the production is classified as M while the remaining 20% is classified as NM. Items in the two classifications appear in random order as the production line is run. Assuming that the production process produces items independently, what is the probability that of the next five transistors, at least three will be in the M class?

To solve this problem, first consider the sample space and an appropriate random variable. We are producing five items, each of which can be M or NM, in any order. A possible sample space consists of all the possible arrangements of these objects. Some of the sample points are shown below:

$$\{M\ M\ M\ M\ M\}, \{M\ M\ NM\ M\ M\}, \{NM\ NM\ NM\ M\ M\}, \dots$$

Since each of the five positions can be occupied by one of two letters, it follows by the counting principle that there are $2^5 = 32$ distinct points in the sample space. They are not equally likely, however. For example, since we presume the items are produced in an independent manner,

$$P\{M\ M\ M\ M\ M\} = 0.8^5 = 0.32768$$

while

$$P\{NM\ NM\ NM\ M\ M\} = (0.2)^3 \cdot (0.8)^2 = 0.00512.$$

This inequity in the probabilities assigned to the sample points is due to the different probabilities assigned to the two classes of production items. The problem requires us to find the probability that the sample contains at least three M items, that is, the probability that it contains exactly three or exactly four or exactly five M items. These are mutually exclusive events, so we can find the probabilities of each one and add them.

There is only one way in which all five items can be in the M class; we calculated this probability above and found it to be

$$P\{M\ M\ M\ M\ M\} = 0.8^5 = 0.32768.$$

Here are the sample points in which exactly four of the five items are in the M class:

$$\{M\,M\,M\,M\,NM\}, \{M\,M\,M\,NM\,M\}, \{M\,M\,NM\,M\,M\},$$
$$\{M\,NM\,M\,M\,M\}, \{NM\,M\,M\,M\,M\}.$$

These sample points are easy to enumerate. There are exactly five of them since there are five positions in which the NM item can occur. Moreover, each point has the same probability, namely,

$$(0.8)^4 \cdot (0.2) = 0.08192.$$

It follows that

$$P(\text{exactly four of the five items are M}) = 5 \cdot (0.8)^4 \cdot (0.2)$$
$$= 5 \cdot 0.08192 = 0.4096.$$

Finally, we need the sample points where exactly three of the five items are M; these are more numerous. Here are some of them:

$$\{M\,NM\,M\,NM\,M\}, \{NM\,NM\,M\,M\,M\}, \{M\,M\,NM\,M\,\text{'}NM\}, \ldots.$$

How many of these points are there? We see that to construct such a sample point we must choose two positions among the five possible positions for the NM items, so there are $\binom{5}{2} = 10$ possible sample points. Each of these has probability $(0.8)^3 \cdot (0.2)^2 = 0.02048$, so

$$P(\text{exactly three of the five items are M}) = \binom{5}{2} \cdot (0.8)^3 \cdot (0.2)^2 = 0.2048.$$

We put all these results together to find that

$$P(\text{at least three of the five items are M}) = P(\text{exactly three of the five items are M})$$
$$+ P(\text{exactly four of the five items are M})$$
$$+ P(\text{exactly five of the five items are M})$$

$$= 0.32768 + 0.4096 + 0.2048$$

$$= 0.94208.$$

The process here is easily generalized to a larger sample. A natural random variable is the number of M items produced, so we will denote this by X. The problem above asks us to find $P(X \geq 3)$. Let us consider now a sample of size n and suppose that we want the probability that exactly r of these items are in the M class; that is, we want $P(X = r)$.

There are $\binom{n}{r}$ positions for which exactly r are M's, giving us $\binom{n}{r}$ sample points with exactly r M's. Each of these has probability $(0.8)^r \cdot (0.2)^{n-r}$, so

$$P(X = r) = \binom{n}{r} \cdot (0.8)^r \cdot (0.2)^{n-r}, r = 0, 1, 2, \ldots, n.$$

Note that the probability that at least three of the items are M is

$$\sum_{r=3}^{5} \binom{5}{r} (0.8)^r (0.2)^{5-r} = 0.94208. \qquad \blacksquare$$

One more generalization is in order. Suppose that, for an individual item, $P(M) = p$ and $P(NM) = 1 - p = q$. Then we obtain the following:

$$P(X = r) = \binom{n}{r} \cdot p^r \cdot q^{n-r}, r = 0, 1, 2, \ldots, n, \quad q = 1 - p.$$

X is called a **binomial random variable,** and the resulting probability distribution is called a **binomial probability distribution.** The conditions under which our formula for $P(X = r)$ is valid are:

1. Exactly one of two outcomes can occur at each trial of the experiment. Call these outcomes "success" (S) and "failure" (F).
2. $P(S) = p$ and $P(F) = 1 - p = q$ at every individual trial of the experiment.
3. The trials are independent.
4. The random variable X is the number of successes in n repetitions of the experiment.

So:

$$P(X = r) = \binom{n}{r} \cdot p^r \cdot q^{n-r}, \quad r = 0, 1, 2, \ldots, n, \quad q = 1 - p.$$

This is the probability distribution function for a binomial random variable.

Do the probabilities over the entire sample space sum to 1? From the binomial theorem, $\sum_{r=0}^{n} \binom{n}{r} \cdot p^r \cdot q^{n-r} = (q + p)^n = 1^n = 1$, showing that the sum of the probabilities over the entire sample space is 1, as it should be.

We will not offer proofs of the following facts, but they can be found in various texts listed in the Bibliography. It can be shown that, if X is a binomial random variable,

$$E(X) = n \cdot p$$

and

$$\text{Var}(X) = n \cdot p \cdot q.$$

The formula for the expected value can be explained intuitively. If we have n trials and we expect $p\%$ of them to be successes, then it makes sense to think that on the average $n \cdot p$ of these trials will be successes. The formula for the variance has no corresponding intuitive explanation.

Example 2.7.2 Practical industrial examples use samples much larger than the 10 we used before. The theory of the solution is exactly as we have shown, but the arithmetic may become formidable. Suppose we have the production process described above, but now we allow 1000 items to be produced. What is the probability that at least 780 of these items meet customer specifications (that is, are classified as M)?

The answer here, letting X denote the number of M items in the sample, is

$$P(X \geq 780) = \sum_{r=780}^{1000} \binom{1000}{r} (0.8)^r \cdot (0.2)^{1000-r}.$$

This is challenging to compute, to say the least, but computer programs can do the calculation easily. We used Mathematica here to evaluate this probability as 0.9461428.

This is often approximated, and we will show a fairly easy and accurate approximation to this result later in the chapter. ∎

Example 2.7.3 In the situation above, X can assume 1001 values. Since these sum to 1, each is quite small. The calculations are difficult, but we will show an approximation to these probabilities later in this chapter. For the moment, suppose we want

$$P(X = 822) = \binom{1000}{822}(0.8)^{822} \cdot (0.2)^{178}.$$

We used Mathematica to evaluate $P(X = 822)$ as 0.00694972. Details can be found in Appendix A. ∎

Proportions

Often, in a binomial situation, a scientist or engineer is interested in the *proportion* of successes obtained in n trials. If X is the random variable giving the total number of successes in n trials, then X/n is the proportion of successes in the sample. We denote this by p_s, so $p_s = X/n$.

Since X is a random variable, so is p_s. Problems involving p_s are often translated into problems involving X. We consider some examples.

Example 2.7.4 A sample survey of 100 voters is taken from a population where actually 45% of the voters favor a certain candidate. What is the probability that the sample will contain between 40% and 60% of voters who favor the candidate?

Here we seek

$$P(0.40 \le p_s \le 0.60) = P\left(0.40 \le \frac{X}{100} \le 0.60\right) = P(40 \le X \le 60)$$

$$= \sum_{x=40}^{60} \binom{100}{x}(0.45)^x \cdot (0.55)^{100-x},$$

which can be calculated to be

$$P(0.40 \le p_s \le 0.60) = 0.864808.$$

If we assume now that the candidate does not in fact know the true proportion of voters who favor his candidacy, this survey of 100 voters may not be particularly enlightening since it indicates that he might lose badly or win by a large margin. Increasing the sample size will increase the probability that the sample proportion is in the range from 40% to 60%. Surveys are often conducted in order to determine the proportion of voters favoring a certain candidate. We will discuss this more in Example 2.7.6 and in Chapter 3 as well. ∎

Example 2.7.5 In Example 2.7.4, we can compute the probability that the sample proportion is at least 60% as follows:

$$P(p_s \ge 0.60) = P\left(\frac{X}{100} \ge 0.60\right) = P(X \ge 60)$$

$$= \sum_{x=60}^{100} \binom{100}{x}(0.45)^x \cdot (0.55)^{100-x} = 0.00182018.$$

So the probability that the sample proportion is at least 60% when the true proportion is 45% is not very great.   ■

Example 2.7.6   Now suppose that the true proportion of voters in the sample favoring a certain candidate is 60%. The sample proportion favoring the candidate will of course vary from sample to sample. How large a sample should be taken so that the probability the sample proportion is within 2% of the true proportion is about 0.90?

Here we want to find the sample size, n, so that

$$P(0.58 \le p_s \le 0.62) \doteq 0.90,$$

which is equivalent to

$$P(0.58 \le \frac{X}{n} \le 0.62) \doteq 0.90$$

or

$$P(0.58n \le X \le 0.62n) \doteq 0.90.$$

Consequently we seek n so that

$$\sum_{x=0.58n}^{0.62n} \binom{n}{x}(0.60)^x \cdot (0.40)^{n-x} \doteq 0.90.$$

Clearly, this is not a trivial equation to solve. We can, however, with the aid of a computer, try various values of n. We find that if $n = 1600$, this probability is 0.9028199. (We will show an easier procedure, although an approximation to the solution, in Section 2.9.)

The accuracy of the sample proportion, here to within 2% of the true proportion, is often called the **sampling error.** In this example, we permitted the sample size to vary, but if it is fixed, then the sampling error is a complex function of the sample size. These topics will be discussed more thoroughly in Chapter 3.   ■

Before returning to the binomial random variable, we pause here to calculate the mean and variance of the random variable p_s.

Means and Variances of Multiples of Random Variables

We begin with a general discrete random variable, X, whose possible values are $\{x_1, x_2, x_3, \ldots, x_n\}$. If each population value is multiplied by a constant, k, we then have the set $\{kx_1, kx_2, kx_3, \ldots, kx_n\}$.

It is sensible to guess that the mean value is also multiplied by k. To see that this is true, we find that

$$E(kX) = \frac{\sum_{i=1}^{n} kx_i}{n} = \frac{k \cdot \sum_{i=1}^{n} x_i}{n} = k \cdot E(X) = k \cdot \mu.$$

The effect of the multiplier on the variance is not so clear. We find that

$$\operatorname{Var}(kX) = \frac{1}{n-1}\sum_{i=1}^{n}(kx_i - k\overline{x})^2 = \frac{1}{n-1}\sum_{i=1}^{n}k^2(x_i - \overline{x})^2$$

$$= \frac{k^2}{n-1}\sum_{i=1}^{n}(x_i - \overline{x})^2 = k^2 \cdot \operatorname{Var}(X).$$

So the variance is multiplied by k^2.

We can now use these facts to find the mean and variance of p_s.

Since $p_s = X/n = (1/n)\,X$, we have here that $k = 1/n$ and it follows that

$$E(p_s) = E\left(\frac{X}{n}\right) = \left(\frac{1}{n}\right)E(X) = \left(\frac{1}{n}\right)\cdot n\cdot p = p,$$

which tells us that the expected value of the sample proportion is the true population proportion.

In addition,

$$\operatorname{Var}(p_s) = \operatorname{Var}\left(\frac{X}{n}\right) = \frac{1}{n^2}\operatorname{Var}(X) = \frac{1}{n^2}\cdot n\cdot p\cdot q = \frac{pq}{n}.$$

The derivation of these facts for continuous random variables is very similar, using integrals instead of sums. We will use these formulas in Section 2.9 where some approximations will be considered. As a final example of a binomial random variable in this section, we consider another situation in which we are called upon to interpret the results of a sample.

Example 2.7.7   Refer to Example 2.7.2. Suppose now that a sample of 100 items gave 71 classified as M. Since experience has shown that 80% of the production can be classified as M, this sample has caused some concern that the quality of the production has decreased.

First we evaluate the probability that the sample contains 71 or fewer M items, presuming that the true value of p is 0.80. The result (which was found using Mathematica) is

$$P(X \le 71) = \sum_{x=0}^{71}\binom{100}{x}(0.8)^x(0.2)^{100-x} = 0.0200202.$$

We conclude that if the process is in reality producing 80% M items, we would see this sample or one that contained even fewer M items, only about 2 times in 100. Our choices are then one of two: Either the production process is unchanged and we have observed a very unusual sample, or the production process has changed and should be investigated. While we can never be certain of such choices, we would probably conclude, since the sample would occur with great rarity if the production process was stable, that the production process had indeed changed.

Another way of looking at the sample result is to calculate, for example,

$$\sum_{r=73}^{87}\binom{100}{r}(0.8)^r \cdot (0.2)^{100-r} = 0.94052.$$

This means that if the production process is in reality producing 80% M items, then about 94% of the time the number of M items in a sample of 100 would be between 73 and 87, inclusive. Our sample value, 71, is outside this interval indicating an unusual sample if the production process continues to produce 80% M items.   ■

In this example, we selected an interval that is symmetric about the mean value, 80. This interval is called a **confidence interval.** Reaching a conclusion about the percentage of M items being produced is called **testing a hypothesis.** These are both matters of drawing conclusions from samples, known as **statistical inference.** We will return to these matters in Chapter 3. For now, we continue with our study of probability and examine some graphic displays.

Graphs of Binomial Distributions

It is interesting to examine graphs of various binomial probability distributions; we will be able to draw some conclusions from them.

Figures 2.10–2.13 show graphs of various binomial distributions, as indicated. The details for producing these graphs are given in Appendix A. It is evident that the histograms become increasingly symmetric as the probability approaches $1/2$. However, the symmetry always occurs as the sample size increases. In Figure 2.13, we have increased the sample size to 100 and retained the probability of 0.8. The symmetry appears again in this graph, but around 80, as one might expect. Figure 2.13 has a somewhat different X scale than the other graphs; note that it uses X values from 68 to 92.

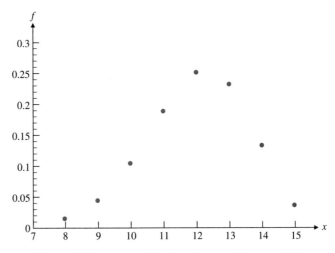

Figure 2.10 Binomial Distribution, $n = 15$, $p = 0.80$

All these figures indicate the "normal" appearance we show in Figure 2.13 with a continuous curve superimposed upon the discrete histogram. For this reason, we turn now to a consideration of the normal probability density function.

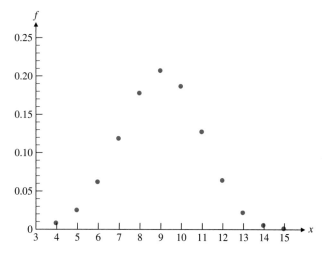

Figure 2.11 Binomial Distribution, $n = 15$, $p = 0.60$

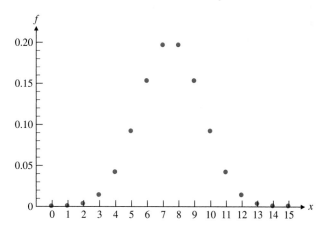

Figure 2.12 Binomial Distribution, $n = 15$, $p = 0.50$

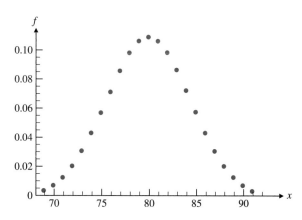

Figure 2.13 Binomial Distribution, $n = 100$, $p = 0.80$

EXERCISES 2.7

Assume a binomial model is appropriate for each of the following exercises.

1. Items from a production line can be classified as either meeting customer specifications or not meeting customer specifications. The probability that an item meets customer specifications is 0.80.
   a. Find the probability that exactly 10 of the next 15 items produced by the line meet customer specifications.
   b. Find the probability that at least 10 of the next 15 items produced by the line meet customer specifications.
   c. Find the probability that exactly 80% of the next 15 items produced by the line meet customer specifications.

2. Light bulbs produced by a manufacturer are known to last 600 hours or more with probability 0.92. A carton of 24 bulbs is purchased.
   a. What is the probability that exactly 20 of the bulbs last at least 600 hours?
   b. What is the probability that at most 20 of the bulbs last at least 600 hours?
   c. What is the probability that all of the bulbs last at least 600 hours?

3. A computer network is fairly unreliable. The probability that a log on to the network is successful is 0.87. Ten users attempt to log on, independently.
   a. Find the probability that all log on successfully.
   b. Find the probability that at least 8 log on successfully.
   c. Find the probability that between 6 and 8 log ons are successful.

4. A biologist studying the germination rate of a seed finds that 90% of the seeds of a given type germinate. She has a box of 45 seeds.
   a. What is the probability that all of the seeds germinate?
   b. What is the probability that at least 40 of the seeds germinate?
   c. What is the probability that at least one of the seeds germinates?

5. A commuter finds that he can go through a traffic light without stopping on his route with probability 0.50. He will make his commuter trip 30 times in the next month.
   a. What is the probability that he stops at the light exactly 20 times?
   b. What is the probability that he stops at the light at least 20 times?
   c. What is the probability that he stops at the light at least once?

6. A production line is producing items that meet customer specifications with probability 0.95. How large a sample of these items must be selected in order to make the probability that at least one of the items does not meet customer specifications 0.99?

7. a. Find the probability distribution function for a binomial random variable with $n = 20$ and $p = 0.60$.
   b. Draw a graph of the probability distribution function in part (a).

8. Production items are often inspected for quality. A production line produces items that meet customer specifications with probability 0.80. The inspection process, however, is not perfectly accurate; an item actually meeting customer specifications is classified in that way with probability 0.95. If an item does not meet customer specifications, it is classified that way with probability 0.85. Of the next 10 items produced and inspected, what is the probability that exactly 8 of them are classified correctly?

9. A binomial random variable has mean 16.2 and variance 6.48. Find n and p.

10. A binomial random variable, X, has $n = 50$ and $p = 0.75$. Find a and b such that $P(a \le X \le b) \doteq 0.95$.

11. A sample of 50 production items is selected from a production line, where 80% of the production items are acceptable to a customer. What is the probability that between 70% and 90% of the sample will be acceptable to the customer?

12. In Problem 11, find the probability that at least 70% of the sample items will be acceptable to the customer.

COMPUTER EXPERIMENT

Draw 200 samples, each of size 5, from a binomial distribution with $n = 20$ and $p = 0.60$.

a. Find the mean of each sample and draw a graph of the probability distribution of these sample means.

b. Find the mean of the distribution of sample means. What value should this approximate?

c. Find the standard deviation of each sample and draw a graph of the probability distribution of these standard deviations.

2.8 The Normal Probability Density Function

We have seen the normal curve previously, first when we examined the distribution of sample means in Section 2.3, and later in the graphs of the binomial distribution. All these histograms exhibit a shape that is similar to that of a normal curve, as we are about to show. The fact that these histograms are, in reality, well approximated by a normal curve is a concept that we will address later. This frequent occurrence of the normal curve is not due to coincidence, as we will show in Section 2.10 when we discuss the central limit theorem.

DEFINITION

A random variable, X, is said to have a *normal probability distribution* if its **probability density function** is

$$f(x) = \frac{1}{\sigma\sqrt{2\pi}} e^{-\frac{1}{2}\left(\frac{x-\mu}{\sigma}\right)^2}, \qquad -\infty < x < \infty, \qquad -\infty < \mu < \infty, \qquad \sigma > 0.$$

It can be shown that the parameters μ and σ are the mean and standard deviation of X, respectively.

If X has probability density function $f(x)$, we will write $X \sim N(\mu, \sigma)$ in order to have a shorthand notation for $f(x)$.

Figure 2.14 shows a normal probability density function with mean 80 and standard deviation 4. Although it can be shown, with some difficulty, that the total area under any normal curve is 1, it is impossible to calculate exactly the area under a normal curve between two arbitrary points. That is, it is impossible to evaluate

$$P(a \leq X \leq b) = \int_a^b f(x)\, dx$$

exactly. Approximations for these areas have been found and are given in Table 1 of Appendix C. Despite these difficulties, there are strong reasons for calling this probability density function "normal." A very useful fact is the following:

FACT

If $X \sim N(\mu, \sigma)$ and if $Z = (X - \mu)/\sigma$, then $Z \sim N(0, 1)$.

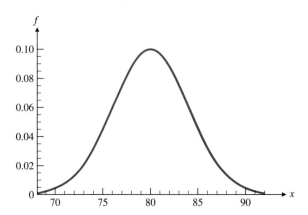

Figure 2.14 $N(80, 4)$ Probability Density Function

This fact gives us a transformation, or coding, of the X values so that a single normal distribution, the $N(0, 1)$ distribution, can be used for all calculations. Approximations then need only be found for this distribution. (They are given in Table 1 in Appendix C.) This distribution is often called the **standard normal distribution.** We show some examples of its use now. (All calculations using the normal curve in the following examples can be done by Mathematica and other computer algebra systems as well without using the coding indicated above; details are given in Appendix A.)

Example 2.8.1   Mathematics Scholastic Aptitude Test (SAT) scores are approximately normally distributed with mean 500 and standard deviation 100, so if X represents a mathematics SAT score, $X \sim N(500, 100)$. What is the probability that an individual mathematics SAT score is 550 or more?

We use the Z transformation to find that

$$P(X \geq 550) = P\left(\frac{X - 500}{100} \geq \frac{550 - 500}{100}\right),$$

which is equivalent to

$$P(Z \geq 0.5).$$

A word is in order concerning the use of Table 1 in Appendix C. This is a table of the *cumulative distribution function* for the standard normal random variable Z. This distribution, as its name implies, accumulates the area under the normal curve from $-\infty$ to the value of z. That is, the values in the table are approximations for $\int_{-\infty}^{z} f(z)\, dz$. For example, the table indicates that $P(Z \leq 0.5) = 0.6915$. It follows that $P(Z \geq 0.5) = 1 - 0.6915 = 0.3085$. So we would expect about 31% of Mathematics SAT scores to exceed 550. ∎

Example 2.8.2   Customers buying resistors supplied by a manufacturer require that the resistance be between 9.9 ohms and 10.5 ohms, inclusive. The manufacturing process is such that the actual resistances are well approximated by a normal distribution with mean 10.1 ohms and standard deviation 0.20 ohms. What percentage of the manufacturer's production is acceptable to the customer?

We need $P(9.9 \leq X \leq 10.5)$ where $X \sim N(10.1, 0.20)$. The Z transformation is again useful. We see that

$$P(9.9 \leq X \leq 10.5) = P\left(\frac{9.9 - 10.1}{0.20} \leq \frac{X - 10.1}{0.20} \leq \frac{10.5 - 10.1}{0.20}\right)$$

$$= P(-1 \leq Z \leq 2) = 0.9772 - 0.1587 = 0.8415.$$

(This calculation can be done directly without using the Z transformation by Mathematica. Details are given in Appendix A.) With this relatively small percentage of acceptable product, the manufacturer will surely try to improve the production process. ∎

Example 2.8.3 In Example 2.8.2, what resistance is exceeded by 95% of the manufacturer's product?

Here we seek a value of X, say x, so that $P(X \geq x) = 0.95$. Using the Z transformation again, we see that this is equivalent to

$$P\left(Z \geq \frac{x - 10.1}{0.20}\right) = 0.95.$$

But Table 1 in Appendix C indicates that $P(Z \geq -1.645) = 0.95$, so it follows that

$$\frac{x - 10.1}{0.20} = -1.645,$$

so $x = 9.771$. ∎

Example 2.8.4 The Z transformation, $Z = (X - \mu)/\sigma$, gives, for a particular value of X, the number of standard deviations X is from the mean, μ. A Z value of 2.5, for example, indicates that the value of X is 2.5 σ units from the mean. Working in Z values then allows us to think about the number of σ units we are away from the mean.

It is easy to use the standard table to establish that

$$P(-1 \leq Z \leq 1) = 0.6826,$$
$$P(-2 \leq Z \leq 2) = 0.9544,$$

and

$$P(-3 \leq Z \leq 3) = 0.9974.$$

This is often referred to as the "2/3, 95, 99" rule. It indicates that the standard deviation is, in fact, a measure of the dispersion of the data since more standard deviations include more of the area beneath the curve. ∎

Notice that Chebyshev's Inequality (from Section 2.6) gives, respectively, 0, 3/4, and 8/9 for these probabilities. We can always do better than Chebyshev's Inequality if we know the probability distribution or probability density function.

The rule also makes it possible to estimate normal probabilities without the use of tables. For instance, in Example 2.8.1, where we consider the $N(500, 100)$ normal distribution, we can estimate $P(250 \le X \le 650) = P(-2.5 \le Z \le 1.5)$ as

$$\left(\frac{1}{2}\right) P(-3 \le Z \le -2) + P(-2 \le Z \le 1) + \left(\frac{1}{2}\right) P(1 \le Z \le 2)$$

$$= \left(\frac{1}{2}\right) P(2 \le Z \le 3) + P(-2 \le Z \le 1) + \left(\frac{1}{2}\right) P(1 \le Z \le 2)$$

$$= P(-2 \le Z \le 1) + \left(\frac{1}{2}\right) P(1 \le Z \le 2) + \left(\frac{1}{2}\right) P(2 \le Z \le 3)$$

$$= P(-2 \le Z \le 1) + \left(\frac{1}{2}\right) P(1 \le Z \le 3).$$

Now, using the facts that $P(Z \le 3) = 0.9987$, $P(Z \le 1) = 0.8413$, and $P(Z \le -2) = 0.0228$, we approximate

$$P(-2.5 \le Z \le 1.5) = (0.8413 - 0.0228) + \left(\frac{1}{2}\right)(0.9987 - 0.8413) = 0.8972.$$

The exact answer is $0.4332 + 0.4938 = 0.9270$, so the approximation is crude, but then the linear interpolation used on the curve is very crude as well. Still, we do obtain an approximation, which may be useful for rough work.

EXERCISES 2.8

In the following exercises, Z denotes a standard normal variable, so Z ~ N(0, 1). Table 1 in Appendix C gives probabilities for the cumulative distribution function of Z.

1. Find:
  a. $P(0 \le Z \le 2)$.
  b. $P(-2 \le Z \le 0)$.
  c. $P(1 \le Z \le 2)$.
  d. $P(-2 \le Z \le 1)$.
  e. $P(-2 \le Z \le 2)$.
  f. $P(-2 < Z < 2)$.
  g. $P(-2 \le Z < 2)$.

2. Find a such that:
  a. $P(-a \le Z \le a) = 0.90$.
  b. $P(-a \le Z \le a) = 0.95$.
  c. $P(-a \le Z \le a) = 0.99$.
  d. $P(Z \le a) = 0.90$.
  e. $P(Z \le a) = 0.95$.
  f. $P(Z \le a) = 0.99$.
  g. $P(Z \ge a) = 0.90$.
  h. $P(Z \ge a) = 0.95$.
  i. $P(Z \ge a) = 0.99$.
  j. $P(Z \le a) = 0.05$.
  k. $P(Z \le a) = 0.01$.

3. Mathematics aptitude test scores, X, are approximately normally distributed with mean 500 and standard deviation 100.
  a. Find the probability that an individual score is at most 600.
  b. Find the probability that an individual score is at least 600.
  c. Find the probability that an individual score is between 450 and 650.
  d. Find an interval, symmetric around 500, in which 90% of the scores lie.

4. A machine that fills soft-drink cans advertised to contain 12 ounces of soft drink actually dispenses a volume per can, V, that is normally distributed with mean 11.8 oz and standard deviation 0.20 oz.
  a. Find the probability that a can contains at least 12 oz of soft drink.
  b. Government regulations require that at least 99% of the cans advertised as containing 12 oz of soft drink actually contain 11.6 oz or more. Does this machine meet government regulations?

5. Copper rods used in an electrical industry have diameters, D, that are approximately normally distributed with mean 0.75 in. and standard deviation 0.03 in.

A device using the rods requires that the rods have diameter no more than 0.80 in. What proportion of the rods can be used in the device?

6. In Exercise 5, suppose, for another application, that rods with diameters less than 0.70 in. must be discarded, that rods with diameters between 0.70 in. and 0.80 in. can be used, while rods with diameters more than 0.80 in. can be reworked to meet specifications. What proportion of the rods fall into each of the categories?

7. A customer buying reams of paper will purchase them only if the weight of a ream is between 2.6 lb and 2.8 lb. The weight of a ream is actually a normal random variable with mean 2.77 lb and standard deviation 0.82 lb.
   a. What is the probability that a ream of paper will be accepted by the customer?
   b. Find the weight, w, of a ream of paper, such that 90% of the reams produced weigh less than w.

8. Adult females have forearm lengths that are normally distributed with mean 17.75 in. and standard deviation 0.65 in.
   a. Find the probability that a female's forearm length is between 16.6 in. and 18.25 in.
   b. Find the *upper quartile* length of a female forearm— that is, a length that is exceeded by 25% of females.

9. Wood beams are submitted to a deflection test where the variable of interest is D, the amount of deflection a beam undergoes when a given weight is put on it. The deflection is a normal random variable with mean 2 in. and standard deviation 0.4 in. for a specific type of wood and a given weight.
   a. Find a deflection, d, so that $P(D \geq d) = 0.95$.
   b. Find a deflection, d, so that $P(D \leq d) = 0.99$.

10. Heights of adult males are normally distributed with mean 65 in. and standard deviation 2 in. Three adult males are chosen at random.
   a. What is the probability that all of them have height greater than 67 in.?
   b. What is the probability that at least two of them have heights greater than 67 in.?

11. Resistors purchased by a consumer have resistance that is normally distributed with mean 4.0 ohms and standard deviation 0.6 ohms. A circuit uses five of the resistors.
   a. What is the probability that at least one of the resistors has resistance less than 4 ohms?
   b. What is the probability that exactly three of the resistors have resistance of at least 5.2 ohms?

12. A chemical process requires a pH between 5.85 and 7.40. The pH of the process is a normal random variable with mean 6.0 and standard deviation 0.9. Assume that the process must be shut down if the pH falls outside the acceptable range of 5.85 to 7.40.
   a. What is the probability that the process will be shut down?
   b. Now suppose that the mean pH of the process can be adjusted, but that the standard deviation does not change. To what value should it be adjusted in order to maximize the probability that the process will not be shut down? (*Hint*: Make a table of the probabilities that the process will not be shut down for various values of the mean pH and draw a graph of these values.)

13. Lengths of "2 by 4" pieces of wood are normally distributed with mean 6.0 ft and standard deviation 0.10 ft. A customer requires pieces that are between 5.9 ft and 6.2 ft in length. (These limits are often called *lower and upper specification limits,* respectively.) To what value should the mean of the pieces be changed, assuming that the standard deviation does not change, in order to maximize the percentage of the wood pieces that are acceptable to the customer? (*Hint*: Make a table of the probabilities that the wood pieces are acceptable for various values of the mean and draw a graph of these values.)

COMPUTER EXPERIMENTS

1. Choose a sample of 100 observations from an $N(10, 1)$ distribution. Count the number of observations that are within 1 standard deviation of the mean and compare this to the expected number in this range.
2. In Exercise 13 above, choose 200 samples from the $N(6, 0.1)$ distribution.
   a. Count the number of observations in the interval [5.9, 6.2] and compare this number to the expected number in this interval.

b. Now change the distribution to $N(6, 0.075)$ and count the number of observations in the interval $[5.9, 6.2]$. How does this number compare to the expected number of observations in this interval?

2.9   Normal Approximation to the Binomial

The graphs in Figures 2.10–2.13 indicate that the normal curve could be used to approximate binomial probabilities. We now show how this is done. The approximation given here is generally used for very large values of n or when a computer system giving exact binomial probabilities is not available.

We found, in Example 2.7.7, that

$$\sum_{r=73}^{87} \binom{100}{r}(0.8)^r \cdot (0.2)^{100-r} = 0.94052,$$

a result obtained using Mathematica. The binomial distribution here has $n = 100$ and $p = 0.80$. We now show how this probability can be approximated using the normal curve.

Example 2.9.1   Figure 2.15 shows a histogram of the binomial distribution with $n = 100$ and $p = 0.80$ and a normal curve with mean 80 and standard deviation 4 superimposed upon it, for values of X between 73 and 87. The normal curve chosen has the same mean and standard deviation as the binomial distribution it approximates.

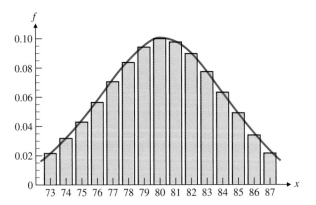

Figure 2.15 Binomial Distribution, $n = 100$, $p = 0.80$

The heights of the bars of the histogram represent individual probabilities; for example, the height of the bar at $X = 75$ is $P(X = 75)$. Since the width of each bar of the histogram is 1, the area of each bar represents an individual probability. Then $P(73 \le X \le 87)$ is the sum of the areas of the rectangles in the range from 73 to 87, inclusive. We want, then, to approximate the sum of these areas.

It follows that we should find the area between 72.5 and 87.5 on the normal curve. If we do this, we find that $P(72.5 \leq X \leq 87.5) = 0.9392$; a good approximation to the actual value, 0.94052.   ■

Note that the addition and subtraction of 0.5 to the upper and lower limits is due to the width of the bars of the histogram. The width is always 1 since the binomial values are consecutive positive integers. These 0.5 corrections make only a slight difference in the approximation if the value of n is fairly large.

A typical calculus problem requires that the area under a curve be approximated by the area of a histogram approximating the curve because the curve may be difficult, while the histogram is easy. Note that the problem here is exactly the inverse of this; we use the curve, which is easy, to approximate the histogram, which is hard.

Example 2.9.2   Using the binomial distribution with $n = 100$ and $p = 0.80$, we saw in Example 2.7.4 that $P(X \leq 71)$ was 0.0200202. To approximate this, we use the area between -0.5 and 71.5 under the normal curve with mean 80 and standard deviation 4. The situation is shown in Figure 2.16.

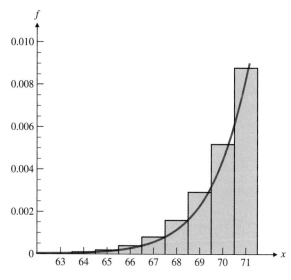

Figure 2.16 $P(X \leq 71)$

The lower limit here, -0.5, has a Z value of less than -20, so $P(Z < -20)$ can safely be ignored. We find that $P(X \leq 71.5) \doteq P(Z \leq -2.125) = 0.0167933$, not a bad approximation to the true value, 0.0200202.   ■

Example 2.9.3   Now we approximate $P(X = 822)$ for a binomial distribution with $n = 1000$ and $p = 0.80$. The situation is shown in Figure 2.17.

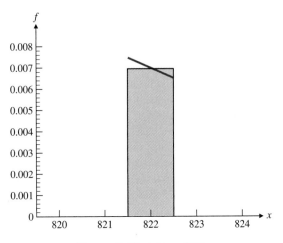

Figure 2.17 $P(X = 822)$

Here we see we need the area under the approximating normal curve between 821.5 and 822.5. (Without the 0.5 corrections, the approximation would be 0, obviously not a very good approximation.)

We calculate that $P(821.5 \le X \le 822.5) = 0.006954$, a good approximation to the actual value, 0.0069472. ∎

Example 2.9.4   In Example 2.7.7, we sampled from a population, 60% of which favored a certain political candidate. We found that a sample size of approximately 1600 made the probability about 0.90 that the proportion in the sample favoring the candidate was between 58% and 62%. Our solution there involved considerable trial and error using a computer. We can now use the material presented in this section to create another approximation.

We know that $(X - np)/\sqrt{npq}$ is approximately normal. But this can be written as

$$\frac{X - np}{\sqrt{npq}} = \frac{\dfrac{X}{n} - p}{\sqrt{\dfrac{pq}{n}}} = \frac{p_s - p}{\sqrt{\dfrac{pq}{n}}}.$$

So $(p_s - p)/\sqrt{p \cdot q/n}$ is approximately normal. In this case we know that $p = 0.60$. So in order to have

$$P(0.58 \le p_s \le 0.62) \doteq 0.90,$$

it follows that

$$P\left[\frac{0.58 - 0.60}{\sqrt{\dfrac{(0.6) \cdot (0.4)}{n}}} \le \frac{p_s - 0.60}{\sqrt{\dfrac{(0.6) \cdot (0.4)}{n}}} \le \frac{0.62 - 0.60}{\sqrt{\dfrac{(0.6) \cdot (0.4)}{n}}}\right] \doteq 0.90.$$

But

$$\frac{p_s - 0.60}{\sqrt{\frac{(0.6) \cdot (0.4)}{n}}} \doteq Z$$

and we know that $P(-1.645 \le Z \le 1.645) \doteq 0.90$, so we conclude that

$$\frac{0.62 - 0.60}{\sqrt{\frac{(0.6) \cdot (0.4)}{n}}} = 1.645.$$

Solving this for n gives $n = 1624$, so these two approximate answers are quite close.

∎

Note here that the population size never enters the calculation, so the election could be regional with only thousands of voters or national with millions of voters. Our sample size of 1624 then will estimate the proportion of voters favoring the candidate to within 2% if the true proportion favoring the candidate is 60%. We will continue to investigate this problem in Chapter 3 where it will not be necessary to assume that the true proportion is 60%.

EXERCISES 2.9

In each of the following problems, assume that the assumptions of the binomial random variable hold.

1. A manufacturer of cellular telephones inspects 1000 of these phones for a certain characteristic that should occur in the telephone with probability 0.92.
  a. Approximate the probability that exactly 920 of these telephones have the characteristic.
  b. Approximate the probability that between 900 and 940, inclusive, of these telephones have the characteristic.

2. Drivers at a rural intersection have been observed to drive through a stop sign 45% of the time. A study is made and 550 drivers traveling through the stop sign are observed.
  a. Approximate the probability that half or more of the drivers actually stop at the stop sign.
  b. Approximate the probability that exactly 240 of the drivers do not stop at the stop sign.

3. Approximate $P(X = 284)$ for a binomial random variable with $n = 550$ and $p = 0.47$.

4. Approximate $P(235 \le X \le 385)$ for a binomial random variable with $n = 450$ and $p = 0.55$.

5. Automobile tires are subject to inspection for flaws, which occur with probability 0.10. A lot of 200 tires is inspected.
  a. Approximate the probability that exactly 20 of the tires have a flaw.
  b. Approximate the probability that at least 20 of the tires have a flaw.

6. A random sample of 300 items is chosen from a production line where 80% of the items are acceptable to a consumer. Approximate the probability that between 77% and 83% of the items in the sample are acceptable to the consumer.

7. Approximate the probability that $p_s = 0.80$ in the situation described in Exercise 7.

8. In a binomial model where $p = 0.70$, a sample of $n = 250$ observations is made. Approximate $P(\mu - 2\sigma \le p_s \le \mu + 2\sigma)$.

COMPUTER EXPERIMENTS

Use a computer algebra system such as Mathematica, Derive, or Maple to find the exact answer in each of the above problems and compare these exact answers to their approximations using the normal curve.

2.10   The Central Limit Theorem

In the Computer Experiment following Section 1.2, we drew samples of size 5 from the set {1, 2, 3, 4, 5, 6, 7, 8, 9, 10} and examined a histogram of the means of these samples. This histogram, which we now recognize as the probability distribution of the sample mean, exhibits a normal shape. This is to be expected because of a theorem known as the **Central Limit Theorem.** Before stating this theorem, we will do another sampling experiment and examine the results.

Example 2.10.1   In the Computer Experiment following Section 1.3, samples were drawn from a discrete distribution, the set {1, 2, 3, 4, 5, 6, 7, 8, 9, 10}. The values in this set are *uniformly distributed* and are far from normal themselves. We now draw samples from a continuous distribution and again select a population which does not at all resemble the normal distribution, the continuous exponential distribution with mean 4. A graph is shown in Figure 2.18.

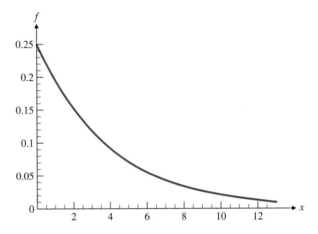

Figure 2.18 Exponential Probability Density Function

We used the computer program Mathematica to select 200 samples, each of size 5, from this distribution. The mean of each sample was computed and a histogram of the means constructed. This histogram is shown in Figure 2.19.

We see first that the graph of the means is much closer to a normal curve than the graph of the exponential population. The mean of the means in this case is 4.13506, fairly close to 4, the mean of the exponential population. The variance of the exponential population is 16; the variance of the means is 3.3288. We conclude that the means are approximately normal and that the mean of the means is approximately equal to the mean of the population, 4 in this case. The variance of the means is very close to $16/5 = 3.20$, the variance of the population divided by the sample size. These conjectures are in fact correct and would be borne out by other simulations which may be performed. ∎

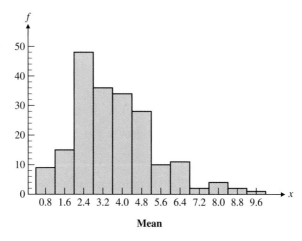

Figure 2.19 Distribution of Means Selected from an Exponential Density

We now formally state these results as the *Central Limit Theorem.*

Central Limit Theorem

If samples of size n are selected from a continuous population with mean μ and variance σ^2, the distribution of the sample means is approximately $N(\mu, \sigma/\sqrt{n})$.

This result holds for continuous populations. For discrete populations, the means are approximately normally distributed with mean μ, but the standard deviation of the means is

$$\sqrt{\frac{N-n}{N-1}}\,\frac{\sigma}{\sqrt{n}}$$

where N is the finite population sample size. For large populations, the *sampling fraction*, n/N, is usually not very important in the formula for the standard deviation, so the standard deviation is frequently approximated by $\sigma/\sqrt{n}$, the standard deviation in the continuous case.

Example 2.10.2 The central limit theorem states that sample means are approximately normally distributed when the sampling is done from an arbitrary population. If this population is itself normal, then the sample means are exactly normally distributed.

Suppose we are sampling from the $N(80, 4)$ distribution as shown in Figure 2.14 in Section 2.8. The probability that a single observation is within one standard deviation of the population mean is $P(-1 \leq Z \leq 1) = 0.6826$, so about 2/3 of the sample values are contained in the interval $[76, 84]$.

Now suppose samples of size 25 are chosen and their sample means calculated. The means, by the central limit theorem, are distributed (exactly) by $N(80, 4/5 = 0.8)$. The probability that a sample mean is in the interval [76, 84] is now

$$P(76 \le \overline{X} \le 84) = P\left(\frac{76-80}{0.80} \le \frac{\overline{X}-80}{0.80} \le \frac{84-80}{0.80}\right)$$

$$= P(-5 \le Z \le 5),$$

a virtual certainty.

The sample size obviously has a great influence on the distribution of sample means. ∎

Example 2.10.3 Sampling often is done from populations whose characteristics are unknown. Indeed, one of the primary purposes of sampling is to discover these characteristics. Suppose now that we are sampling from a population whose mean, μ, is unknown, but we can presume that the standard deviation, σ, is about 5. The central limit theorem tells us that the sample means, the $\overline{X}$'s, are distributed by the $N(\mu, \sigma/\sqrt{n})$ distribution, which in this case is the $N(\mu, 5/\sqrt{n})$ distribution.

We would like our sample mean, $\overline{X}$, to differ from the true population mean, μ, by no more than 2 units. We cannot, of course, be sure that this will be the case, but we could ask that it be the case with probability 0.95. So we want

$$P(-2 \le \overline{X} - \mu \le 2) = 0.95.$$

This means that

$$P\left(\frac{-2}{5/\sqrt{n}} \le \frac{\overline{X}-\mu}{5/\sqrt{n}} \le \frac{2}{5/\sqrt{n}}\right) = 0.95,$$

which is equivalent to $P(-0.4\sqrt{n} \le Z \le 0.4\sqrt{n}) = 0.95$. However, for a standard normal distribution, $P(-1.96 \le Z \le 1.96) = 0.95$. We conclude that $0.4\sqrt{n} = 1.96$ from which we find that $n = 24.01$. So a sample of size 25 is sufficient. ∎

We will explore the ideas in this example more fully in the next chapter. We conclude this chapter with a study of the distribution of sample variances.

EXERCISES 2.10

1. Mathematics Scholastic Aptitude Test (SAT) scores are normally distributed with mean 500 and standard deviation 100.

   a. Find the probability that an individual SAT score is between 480 and 540.

   b. Find the probability that the mean of a sample of 25 SAT scores is between 480 and 540.

   c. Find the probability that an individual score is at least 675.

   d. Find the probability that the mean of a sample of 25 SAT scores is at least 675.

2. Samples of size 36 are selected from the exponential distribution with probability density function $f(x) = \frac{1}{6}e^{-x/6}$ for $x \ge 0$; $\mu = 6$ and $\sigma = 6$ for this distribution.

   a. Find the probability that an individual observation is at least 6.2.

   b. Find the probability that the mean of the sample of 36 observations is at least 6.2.

3. Samples are selected from a uniform distribution on the interval [0, 12]; $\mu = 6$ and $\sigma = \sqrt{12}$ for this distribution.

a. Find the probability that the mean of a sample of 25 observations is at least 6.75.

b. Find the probability that the mean of a sample of 49 observations is at least 6.75.

4. Samples are chosen from a normal distribution whose standard deviation is 15, but whose mean, μ, is unknown. How large a sample is necessary so that the probability that the sample mean, $\overline{X}$, is within 3 units of μ is 0.95?

5. In Exercise 4, find the sample size required for the probability that the sample mean, $\overline{X}$, is within 1.5 units of μ with probability 0.95. Show that this is four times the sample size required in Exercise 4.

6. Actual volumes of "12 ounce" soft-drink cans are normally distributed with mean 11.8 oz and standard deviation 0.20 oz.

a. Approximate the probability that at least 80% of a sample of 50 cans has volume of at least 12 oz.

b. Find the probability that the mean volume of the sample of 6 cans is at least 12 oz.

7. A bridge crossing a river can support at most 85,000 pounds. Suppose that the weights of automobiles using the bridge have mean weight 3200 lb and standard deviation 400 lb. How many automobiles can use the bridge simultaneously so that the probability that the bridge is not damaged is at least 0.99?

8. Rivets are packaged in boxes advertised to contain 200 rivets. The boxes are actually composed by weight. The weights of individual rivets have mean 3 oz with standard deviation 0.5 oz. How much should the boxes weigh, on average, so the probability that the boxes contain at least 200 rivets is at least 0.99?

COMPUTER EXPERIMENTS

1. Select 200 samples, each of size 5, from a normal population with mean 50 and standard deviation 2.

a. Calculate the mean of each sample and draw a histogram of the means. Calculate the mean value of the distribution of sample means. What value should this approximate?

b. Calculate the standard deviation of the distribution of sample means. What value should this approximate?

2. Repeat Problem 1, sampling from the exponential distribution, $f(x) = \frac{1}{10}e^{-x/10}$ for $x \geq 0$.

2.11 Chi-Squared Distribution

In addition to the mean, many random variables based on samples are of importance in statistical inference, a topic we will turn to in the next chapter. Generally probability distributions, other than the normal distribution, arise when random variables other than the mean are considered. We will consider these probability distributions when they are needed; none are quite as simple as the normal distribution, which, as we will see, arises with great frequency.

The central limit theorem affords us the luxury of sampling from any probability distribution, no matter how far it might depart from a normal distribution; usually the statistical procedures discussed in this book require that the sampling be done from normal distributions. We close this chapter with a discussion of the sampling distribution of the sample variance when samples are selected from a normal distribution.

Five hundred samples, each of size 5, were chosen from a standard $N(0, 1)$ distribution; we calculated the sample variance of each sample. Details of these calculations can be found in Appendix A. The probability distribution of these variances is shown in Figure 2.20.

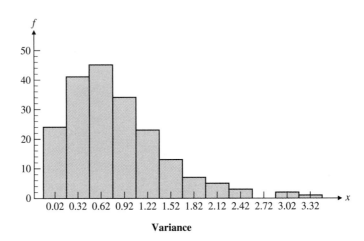

Figure 2.20 Distribution of Sample Variances

This distribution shows some characteristics that appear normal, but is decidedly not normal. Note that the sample variance cannot be negative, tying the distribution to the origin. In addition, Figure 2.20 exhibits a rather long right tail.

The probability distribution is actually a sample from a **chi-squared distribution** with $n - 1$ degrees of freedom, which we denote by χ^2_{n-1}. A graph of χ^2_4 is shown in Figure 2.21. The situation is explained with the following theorem.

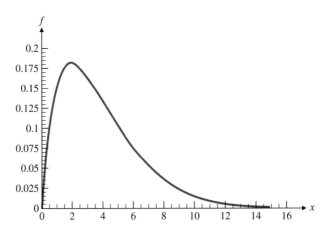

Figure 2.21 Chi-Squared Density with 4 Degrees of Freedom

> **Theorem**
>
> Let $x_1, x_2, x_3, \ldots, x_n$ denote a random sample from an $N(\mu, \sigma)$ distribution and let
>
> $$s^2 = \frac{1}{n-1} \sum_{i=1}^{n} (x_i - \bar{x})^2$$
>
> denote the variance of the sample. Then the random variable $(n-1)S^2/\sigma^2$ is distributed according to a chi-squared distribution with $n-1$ degrees of freedom.

It can also be shown that $E(S^2) = \sigma^2$, so the mean value of the sample variance is the population variance. This explains the divisor of $n-1$ rather than n in the formula for s^2; otherwise, the expected value would not be the population variance. It can also be shown that the variance of the sample variances is given by the formula

$$\text{Var}(S^2) = \frac{2\sigma^4}{n-1}.$$

These facts are true for samples from any probability density function, not just the normal density. The last formula indicates that the sample variance is very variable, its variance being a multiple of the fourth power of the population variance.

We show the use of these facts in some examples. A table of areas under various chi-squared distributions is given in Table 3 of Appendix C. The chi-squared distribution is difficult to use mathematically and so we strongly suggest the use of a computer program such as Minitab or Mathematica.

Example 2.11.1   Consider the 500 samples of size 5 chosen from a standard normal distribution in the beginning of this section. We would expect the distribution of sample variances to have variance

$$\frac{2\sigma^4}{n-1} = \frac{2 \cdot 1}{5-1} = 0.50.$$

The actual variance of our distribution of sample variances was 0.507478.   ∎

Example 2.11.2   A sample of size 9 is selected from an $N(14, 4)$ distribution. Find the probability that the sample variance exceeds 6.

$$P(S^2 > 6) = P\left(\frac{8S^2}{4} > \frac{8 \cdot 6}{4}\right) = P\left(\chi_8^2 > 12\right) = 1 - 0.848796 = 0.151204.$$

The calculation here was done by Mathematica. This would indicate that a sample variance of 6 or more is not unlikely if the normal population has a variance of 4.   ∎

Example 2.11.3   If samples of size 9 are selected from a $N(14, 4)$ distribution, find a likely interval in which the sample variances should lie.

Here we will interpret "likely interval" to be an interval of values to which we would assign a probability of 0.95. Consequently, we seek values a and b so that

$$P(a < S^2 < b) = 0.95$$

or that

$$P\left[\frac{(n-1)\cdot a}{\sigma^2} < \frac{(n-1)\cdot S^2}{\sigma^2} < \frac{(n-1)\cdot b}{\sigma^2}\right]$$

or that

$$P\left[\frac{8\cdot a}{4} < \chi_4^2 < \frac{8\cdot b}{4}\right].$$

Mathematica indicates that

$$\frac{8\cdot a}{4} = 2.17973 \quad \text{and} \quad \frac{8\cdot b}{4} = 17.5345$$

from which we find that $a = 1.089865$ and $b = 8.76725$. The resulting interval, $(1.089865, 8.76725)$, is called a **95% confidence interval.** We will study these intervals and their interpretation in the next chapter. ∎

EXERCISES 2.11

1. A sample of size 5 is selected from an $N(50, 5)$ distribution. Find the probability the sample variance is at most 10.

2. What is the probability that the variance of a sample of 20 observations from an $N(100, 5)$ distribution exceeds 15?

3. Find a number a so that $P(S^2 < a) = 0.95$ where S^2 is the variance of a sample of size 20 selected from an $N(100, 12)$ distribution.

4. Find numbers a and b so that $P(a < S^2 < b) = 0.99$ where S^2 is the variance of a sample of size 30 selected from an $N(300, 56)$ distribution.

5. A sample of size 10 is selected from an $N(45, 7)$ distribution. What is the probability that the variance of the sample will be at least twice the variance of the population?

COMPUTER EXPERIMENT

Select 200 samples, each of size 10, from an $N(100, 5)$ distribution and calculate the sample variance for each of the samples. Then draw a graph of the probability distribution of the sample variances. Find the mean and variance of the distribution of sample variances. What values should these approximate?

2.12 Bivariate Random Variables

Random samples often consist of two or more observations from the same probability distribution. It is useful to consider how these random variables vary together, so we consider the behavior of two such random variables, often called **bivariate random**

variables. (The ideas presented in this section can be extended to three or more random variables, most easily through the use of matrix algebra. We will not do this here.)

DEFINITION

If X and Y are both discrete random variables, the **joint probability distribution function** of X and Y, denoted $f(x, y)$, has the following properties:

1. $f(x, y) \geq 0$ for all x and y.
2. $f(x, y) = P(X = x \text{ and } Y = y)$, which is written as $P(X = x, Y = y)$.
3. $\sum_x \sum_y f(x, y) = 1$.

If X and Y are both continuous random variables, the **joint probability density function** of X and Y, denoted $f(x, y)$, has the following properties:

1. $f(x, y) \geq 0$ for all x and y.
2. $\int_a^b \int_c^d f(x, y) \, dy \, dx = P(a \leq X \leq b, c \leq Y \leq d)$.
3. $\int_{-\infty}^{\infty} \int_{-\infty}^{\infty} f(x, y) \, dy \, dx = 1$.

So in the discrete case, the values of $f(x, y)$ are probabilities while in the continuous case, probabilities are volumes under the joint density function. (We will not consider the case where one variable is discrete and the other is continuous.) These properties are in direct analogy to single, or univariate, random variables. We can now consider expectations of functions of the random variables X and Y; we look at a simple sum here.

Theorem

For jointly distributed random variables X and Y,

$$E(X \pm Y) = E(X) \pm E(Y).$$

Proof.

Since the integral of a sum or difference is the sum or difference of the integrals,

$$E(X \pm Y) = \int_{-\infty}^{\infty} \int_{-\infty}^{\infty} (x \pm y) f(x, y) \, dy \, dx$$

$$= \int_{-\infty}^{\infty} x \cdot f(x, y) \, dx \pm \int_{-\infty}^{\infty} y \cdot f(x, y) \, dy = E(X) \pm E(Y).$$

Example 2.12.1 X and Y are discrete random variables with probability distribution function

$$f(x, y) = \begin{cases} \dfrac{x+y}{21}, & x = 1, 2; \quad y = 1, 2, 3, \\ 0 \text{ otherwise.} \end{cases}$$

The sample space here consists of the points $\{(1, 1), (1, 2), (1, 3), (2, 1), (2, 2), (2, 3)\}$. The sample space, together with the probabilities assigned to the individual points, is shown in Figure 2.22.

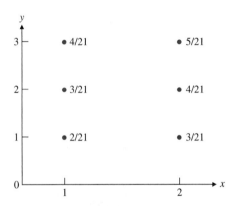

Figure 2.22 Sample Space for Example 2.12.1

So if we want $P(X = 1$ and $Y = 2)$, we find that this is $(1+2)/21 = 1/7$. We find that

$$P(Y \geq 2) = P\{(1, 2), (1, 3), (2, 2), (2, 3)\} = \frac{3+4+4+5}{21} = \frac{16}{21}.$$ ∎

Example 2.12.2   Consider the joint continuous random variables X and Y whose joint probability density function is

$$f(x, y) = \begin{cases} \dfrac{3}{8}(x + y^2), & 0 < x < 2, \quad 0 < y < 1, \\ 0 \text{ otherwise.} \end{cases}$$

We want to calculate $P(X > 1, Y > 1/2)$. The surface and sample space are shown in Figure 2.23.

We see that $f(x, y) \geq 0$, so if $f(x, y)$ is to satisfy the properties of a joint probability density function, we must check that $\int_{-\infty}^{\infty} \int_{-\infty}^{\infty} f(x, y)\, dy\, dx = 1$. In this case we must calculate

$$\int_0^2 \int_0^1 f(x, y)\, dy\, dx = \int_0^2 \int_0^1 \frac{3}{8}(x + y^2)\, dy\, dx.$$

This double integral is 1, as it should be.

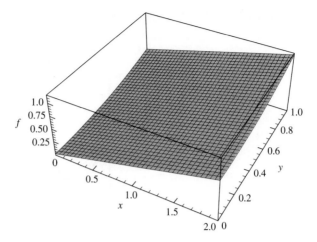

Figure 2.23 Probability Density Function for Example 2.12.2

Finally, we calculate

$$P\left(X > 1, Y > \frac{1}{2}\right) = \int_1^2 \int_{1/2}^1 \frac{3}{8}(x + y^2)\, dy\, dx = \frac{25}{64}.$$

∎

Marginal Distributions

It is often necessary to consider the random variables in a bivariate distribution as individual random variables. The resulting probability distributions or probability densities are called *marginal distributions*.

DEFINITION

If X and Y have joint probability distribution function $f(x, y)$, then the **marginal distributions** of X and Y are

$$h(x) = \sum_y f(x, y) \quad \text{and} \quad g(y) = \sum_x f(x, y)$$

where the summations are over all the possible values of X and Y, respectively.

If X and Y have joint probability density function $f(x, y)$, then the marginal distributions of X and Y are

$$h(x) = \int_{-\infty}^{\infty} f(x, y)\, dy \quad \text{and} \quad g(y) = \int_{-\infty}^{\infty} f(x, y)\, dx$$

These distributions are called marginal because they exist along the margins of the table in Figure 2.22. The probability, through the summations, has been collapsed along the margins of the table. In the continuous case, the joint density has been collapsed along the margins of the sample space.

Example 2.12.3   In Example 2.12.1, we find that

$$h(x) = \begin{cases} \displaystyle\sum_{y=1}^{3} \frac{1+y}{21} = \frac{9}{21} & \text{if } x = 1, \\[3ex] \displaystyle\sum_{y=1}^{3} \frac{2+y}{21} = \frac{12}{21} & \text{if } x = 2 \end{cases}$$

and

$$g(y) = \begin{cases} \displaystyle\sum_{x=1}^{2} \frac{x+1}{21} = \frac{5}{21} & \text{if } y = 1, \\[3ex] \displaystyle\sum_{x=1}^{2} \frac{x+2}{21} = \frac{7}{21} & \text{if } y = 2, \\[3ex] \displaystyle\sum_{x=1}^{2} \frac{x+3}{21} = \frac{9}{21} & \text{if } y = 3. \end{cases}$$

Note that both $h(x)$ and $g(y)$ have all the properties of univariate probability distribution functions.   ■

Example 2.12.4   For the probability density function used in Example 2.12.2, we find the marginal distributions to be

$$h(x) = \int_0^1 \frac{3}{8}(x + y^2)\,dy = \frac{3}{8}x + \frac{1}{8}, \qquad 0 < x < 2$$

and

$$g(y) = \int_0^2 \frac{3}{8}(x + y^2)\,dx = \frac{3}{4} + \frac{3}{4}y^2, \qquad 0 < y < 1.$$

It is easy to check that these functions are again individual probability density functions.   ■

Independence

The random variables X and Y are called *independent* if and only if the joint probability distribution or density function can be expressed as the product of the marginal distributions.

DEFINITION

Random variables X and Y are **independent** if and only if
$$f(x, y) = h(x) \cdot g(y) \quad \text{for all } x \text{ and } y$$
where $h(x)$ and $g(y)$ are the marginal distributions for X and Y, respectively.

It is easy to see that the random variables in either Example 2.12.3 or Example 2.12.4 are not independent.

Example 2.12.5   Consider the joint probability density function

$$f(x, y) = e^{-x-y}, \qquad x \geq 0, \qquad y \geq 0.$$

We find that $h(x) = \int_0^\infty e^{-x-y} \, dy = e^{-x}, x \geq 0$ and, by symmetry, $g(y) = e^{-y}, y \geq 0$. Since $f(x, y) = h(x) \cdot g(y)$, X and Y are independent. This probability distribution is sometimes called the **double exponential distribution.** ∎

Independent random variables are of particular importance to us since the assumption of independent observations is a common assumption concerning random sampling. We show now some properties of independent random variables.

Theorem

If $f_1(X)$ and $f_2(Y)$ are functions of the independent random variables X and Y, respectively, then $E[f_1(X) \cdot f_2(Y)] = E[f_1(X)] \cdot E[f_2(Y)]$.

Proof.

$E[f_1(X) \cdot f_2(Y)] = \int_{-\infty}^\infty \int_{-\infty}^\infty f_1(x) \cdot f_2(y) \cdot f(x, y) \, dy \, dx$ where $f(x, y)$ is the joint probability density function for the random variables X and Y. But, using the fact that the variables are independent, we can write

$$E[f_1(X) \cdot f_2(Y)] = \int_{-\infty}^\infty \int_{-\infty}^\infty f_1(x) \cdot f_2(y) \cdot h(x) \cdot g(y) \, dy \, dx.$$

The integrals are then separable, so

$$E[f_1(X) \cdot f_2(Y)] = \int_{-\infty}^\infty f_1(x) \cdot h(x) \, dx \cdot \int_{-\infty}^\infty f_2(y) \cdot g(y) \, dy = E[f_1(X)] \cdot E[f_2(Y)].$$

∎

A particularly important special case of the theorem is when the functions f_1 and f_2 are X and Y, respectively, for then $E[X \cdot Y] = E[X] \cdot E[Y]$. We can now consider the variance of a sum of random variables. We see that

$$\text{Var}(X + Y) = E[(X + Y) - (\mu_x + \mu_y)]^2 = E[(X - \mu_x) + (Y - \mu_y)]^2$$

$$= E[(X - \mu_x)^2 + 2 \cdot (X - \mu_x) \cdot (Y - \mu_y) + (Y - \mu_y)^2]$$

$$= E(X - \mu_x)^2 + 2 \cdot E[(X - \mu_x) \cdot (Y - \mu_y)] + E(Y - \mu_y)^2.$$

Now consider the middle term. By independence we can write

$$E[(X - \mu_x) \cdot (Y - \mu_y)] = E(X - \mu_x) \cdot E(Y - \mu_y).$$

Each of these expectations is 0 since each is the expected value of a variable minus its mean value. The result is

$$\mathrm{Var}(X + Y) = \mathrm{Var}(X) + \mathrm{Var}(Y) \quad \text{if } X \text{ and } Y \text{ are independent.}$$

It is also easy to see that $\mathrm{Var}(X - Y) = \mathrm{Var}(X) + \mathrm{Var}(Y)$ if X and Y are independent since the negative sign affects only the middle term, which is zero.

The middle term above is defined as the *covariance* of X and Y.

DEFINITION

The **covariance** of the random variables X and Y is

$$\mathrm{Cov}(X, Y) = E(X - \mu_x) \cdot E(Y - \mu_y).$$

The **correlation coefficient** of the variables X and Y is denoted by $\rho_{X,Y}$ and is defined as

$$\rho_{X,Y} = \frac{\mathrm{Cov}(X, Y)}{\sigma_X \cdot \sigma_Y}$$

If X and Y are independent, then $\mathrm{Cov}(X, Y) = 0$, but the converse of this is not true. (For an example, see Kinney[18].) If $\mathrm{Cov}(X, Y) = 0$, then X and Y are called **uncorrelated.**

It can be shown that $-1 \leq \rho_{X,Y} \leq 1$.

We will make considerable use of the correlation coefficient in Chapter 4 when we consider regression.

Normal Bivariate Random Variables

We conclude this section, and this chapter, with a very important bivariate probability density function, the *bivariate normal density function.*

DEFINITION

We say that the random variables X and Y have a **bivariate normal density** if

$$f(x, y) = \frac{1}{2\pi\sigma_x\sigma_y\sqrt{1 - \rho^2}} e^{-\frac{1}{2(1 - \rho^2)}\left\{\left(\frac{x - \mu_x}{\sigma_x}\right)^2 - 2\rho\left(\frac{x - \mu_x}{\sigma_x}\right)\left(\frac{y - \mu_y}{\sigma_y}\right) + \left(\frac{y - \mu_y}{\sigma_y}\right)^2\right\}}$$

for $-\infty < x < \infty$, $-\infty < y < \infty$, $-\infty < \mu_x < \infty$, $-\infty < \mu_y < \infty$, $\sigma_x > 0$, and $\sigma_y > 0$.

It can be shown that μ_x, μ_y, σ_x, and σ_y are the means and standard deviations of X and Y, respectively; ρ is the correlation coefficient.

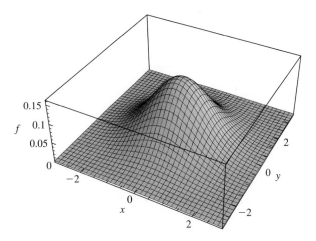

Figure 2.24 Bivariate Normal Distribution with $\rho = 0$

Figures 2.24 and 2.25 show two bivariate normal distributions. In Figure 2.24, the correlation coefficient is 0, and in Figure 2.25, the correlation coefficient is 0.6. It can be shown that the intersections of planes perpendicular to the X, Y plane of a bivariate normal surface are univariate normal distributions and that the marginal distributions are normal. It is very difficult to calculate volumes, representing probabilities, under a bivariate normal surface without the aid of a computer algebra system. We show one such calculation in an example.

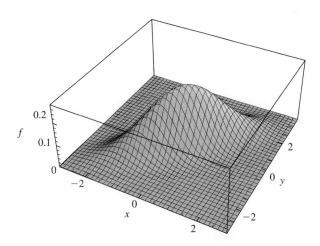

Figure 2.25 Bivariate Normal Distribution with $\rho = 0.6$

Example 2.12.6   A component of a shock absorber has two crucial measurements—diameter, X, and length, Y. Some components are subject to inspection, and these two measurements are made on samples. When these two measurements are made by a certain quality control inspector they are known to have a correlation coefficient 0.3. It is also known that $\mu_x = 1, \sigma_x = 0.5, \mu_y = 1$, and $\sigma_y = 0.5$. Assuming a bivariate normal distribution, what is $P(1/2 < X < 3/2, 1 < Y < 5/2)$?

Here we must integrate the bivariate normal surface

$$\frac{1}{2\pi \cdot 0.5 \cdot 0.5\sqrt{1 - 0.3^2}} e^{-\frac{1}{2(1 - 0.3^2)}\left\{\left(\frac{x-1}{0.5}\right)^2 - 0.6\left(\frac{x-1}{0.5}\right) \cdot \left(\frac{y-1}{0.5}\right) + \left(\frac{y-1}{0.5}\right)^2\right\}}$$

for X from $1/2$ to $3/2$ and Y from 1 to $5/2$. This is far from trivial to do! We used Mathematica to obtain the answer 0.562319.

EXERCISES 2.12

1. Suppose that X and Y are continuous random variables with joint probability density function

$$f(x, y) = \frac{1}{2}, \qquad 1 < x < 2, \qquad 2 < y < 4.$$

a. Show that $f(x, y)$ is a joint probability density function.
b. Find the marginal distributions, $f(x)$ and $g(y)$.
c. Find $P(X > 3/2, Y < 3)$.

2. A lot of video recorders contains three that have defective transistors, four that have defective sound systems, and five that have no defective parts. Three of the recorders are selected, the sampling being done without replacement. Let X denote the number in the sample with defective transistors and Y denote the number in the sample with defective sound systems.

a. Show that the joint probability distribution function for X and Y is

$$f(x, y) = \frac{\binom{3}{x}\binom{4}{y}\binom{5}{3 - x - y}}{\binom{12}{3}} \qquad x = 0, 1, 2, 3;$$

$$y = 0, 1, \ldots, 3 - x; \qquad x + y \le 3.$$

b. Find $P(X = 2, Y = 1)$.
c. Find the marginal distributions.

3. The random variables X and Y have joint probability density function

$$f(x, y) = \frac{1}{2\pi}e^{-\frac{1}{2}(x^2 + y^2)}, \qquad -\infty < x < \infty,$$

$$-\infty < y < \infty.$$

a. Show that the marginal densities are normal.
b. Find $P(-1 < X < 1, -2 < Y < 2)$.

4. Three quality control inspectors are selected from a group where three have one year of experience, two have two years of experience, and two have three years of experience. Let X denote the number of inspectors with one year of experience and Y denote the number of inspectors with two years of experience.

a. Find the joint probability distribution function for X and Y.
b. Find the marginal distributions from a table of values of $f(x, y)$.

5. The amount of particulate pollution in air samples of a given volume is measured over the smokestack of a coal-operated power plant. The random variable X denotes the amount of pollutant (by weight) per sample when a cleaning device on the stack is not in operation, and the random variable Y denotes the same amount when the cleaning device is in operation. It is known that the joint probability density function for X and Y is

$$f(x, y) = k, \qquad 0 \le x \le 2, \quad 0 \le y \le 1, \quad x > 2y.$$

a. Find k.
b. Find the marginal densities for X and Y.
c. Find the probability that the amount of pollutant with the cleaning device in operation is at most $1/3$ of the amount when the cleaning device is not in operation.

6. Let X and Y be random variables with joint probability density function

$$f(x, y) = \frac{1}{x}, \qquad 0 < y < x, \qquad 0 < x < 1.$$

a. Show that $f(x, y)$ is a probability density function.
b. Find $P(X > 1/2, Y < 1/4)$.

7. The random variables X and Y have joint probability distribution function

$$f(x, y) = \frac{1}{6}, \qquad x = 1, 2, 3 \quad \text{and}$$

$$y = 1, 2, \ldots, 4 - x.$$

a. Show that $f(x, y)$ is a probability distribution function.
b. Find the marginal distributions.
c. Are X and Y independent?

8. A box contains five good transistors and two defective transistors. Two transistors are selected, the sampling being done without replacement. Define random variables X and Y as follows: X is 1 if the first transistor is defective and is 0 otherwise. The random variable Y is 1 if the second transistor is defective and is 0 otherwise.

a. Find the joint distribution of X and Y.
b. Find the correlation coefficient.
c. Are X and Y independent? Explain.
d. Find $\text{Var}(X + Y)$.

COMPUTER EXPERIMENTS

1. Suppose that X and Y have a standard bivariate normal density (i.e., $\mu_x = 0$, $\mu_y = 0$ and $\sigma_x = \sigma_y = 1$) and that $\rho = 0.6$. Find $P(X > 1, Y < 1)$.
2. The guidance system for a missile is undergoing a test. The missile is aimed at the origin of a grid, but the actual landing point is a bivariate normal density with $\mu_x = 0$, $\sigma_x = 1$, $\mu_y = 0$, $\sigma_y = 6$, and $\rho = 0.42$. Find the probability the missile lands within two units of the origin.

Chapter Review

Our goal in this chapter was to establish some fundamental ideas about probability and random variables, since these ideas form the basis of statistical inference.

Key Terms	
Sample Space	A set of all the possible outcomes from an experiment is called a *sample space.*
Probability	The theoretical relative frequency of each of the points in the sample space is the *probability* of the sample point.
Axioms of Probability	If S denotes a sample space and A and B denote events in that sample space we assume 1) $P(A) \geq 0$ 2) $P(A \text{ or } B) = P(A) + P(B)$ where A and B are *mutually exclusive* events, that is they cannot occur together 3) $P(S) = 1$
Permutation	A linear arrangement of objects.
Combination	A selection from a set without regard for order.

Event	Any subset of a sample space is called an *event*.
Conditional Probability	The *conditional probability* of an event, A, given that another event, B, has occurred is denoted by $P(A\|B)$.
Random Variable	A *random variable* was defined as a variable taking on values on the points of a sample space. Random variables are denoted by capital letters such as $X, Y, Z, \ldots$ while the values they can assume are denoted by small letters such as $x, y, z, \ldots$.
Discrete Random Variable	A *discrete random variable* takes on values on the points of a finite or countably infinite sample space.
Probability Distribution Function	The function giving $P(X = x)$ is called the *probability distribution function,* and is denoted by $f(x)$, for a discrete random variable, X. Properties of $f(x)$ include 1) $f(x) \geq 0$ 2) $0 \leq f(x) \leq 1$ 3) $\sum f(x) = 1$ where the sum extends over all the values x
Continuous Random Variable	A *continuous random variable* is a random variable assuming values on an interval or interval.
Probability Density Function	The *probability density function*, $f(x)$, for a continuous random variable, X, has these properties: 1) $f(x) \geq 0$ 2) $\int_a^b f(x)dx = P(a \leq X \leq b)$ 3) $\int_{-\infty}^{\infty} f(x)dx = 1$.
Covariance	The *covariance* of two random variables, X and Y is $\mathrm{Cov}(X, Y) = E[(X - \mu_x)(Y - \mu_y)]$

Key Theorems and Facts

Counting Principle	If an event A can occur in n ways and an event B can occur in m ways then the events A and B can occur in $n \cdot m$ ways.
Addition Theorem for Two Events	$P(A \cup B) = P(A) + P(B) - P(A \cap B)$

Complementary Events	$P(\overline{A}) = 1 - P(A)$ where $\overline{A}$ denotes the set of sample points where the event A does not occur.
Conditional Probability	$P(A\|B) = \dfrac{P(A \cup B)}{P(B)}$
Independent Events	*Events* A and B are *independent* if $P(A \cap B) = P(A) \cdot P(B)$
Mean of a Random Variable	The *mean of a* discrete *random variable* X is $\mu = \sum_x x \cdot f(x)$ and the mean or a continuous random variable is $$\mu = \int_{-\infty}^{\infty} x \cdot f(x)dx.$$
Variance of a Random Variable	The *variance of a* discrete *random variable* X is $$\sigma^2 = \sum_x (x-\mu)^2 \cdot f(x) = \sum_x x^2 \cdot f(x) - \mu^2$$ and the variance of a continuous random variable X is $$\sigma^2 = \int_{-\infty}^{\infty} (x-\mu)^2 \cdot f(x)\,dx$$ $$= \int_{-\infty}^{\infty} x^2 \cdot f(x) - \mu^2.$$
Uniform Random Variable	The probability density function for a *uniform random variable* on the interval $[a, b]$ is $f(x) = 1/(b-a)$ for $a \le x \le b$.
Exponential Random Variable	The probability density function for an *exponential random variable* $f(x) = ae^{-ax}$ for $x \ge 0$ and $a > 0$.
Chebyshev's Inequality	Suppose that μ is the mean of a random variable X with standard deviation σ. Choose a positive quantity, $k > 1$. Then $$P(\mu - k \cdot \sigma \le X \le \mu + k \cdot \sigma) \ge 1 - \frac{1}{k^2}.$$
Combinations	The number of *combinations,* or samples, of size r that can be selected from a population of n items is denoted by the symbol $\binom{n}{r}$ and $\binom{n}{r} = \frac{n!}{r!(n-r)!}$

Binomial Random Variable

These conditions must hold:

1) Exactly one of two outcomes can occur at each trial of the experiment. Call these outcomes "success"(S) and "failure"(F).

2) $P(S) = p$ and $P(F) = 1 - p = q$ at each individual trial of the experiment.

3) The trials are independent

4) The random variable X is the number of successes in n repetitions of the experiment.

Under these conditions

$$P(X = r) = \binom{n}{r} \cdot p^r \cdot q^{n-r},$$

$$r = 0, 1, 2, \ldots, n$$

Binomial Mean and Variance

The *mean* of a *binomial* random variable is $\mu = n \cdot p$ and the *variance* of a *binomial* random variable is $\sigma^2 = n \cdot p \cdot q$.

Normal Random Variable

The probability density function for a *normal random variable* is

$$f(x) = \frac{1}{\sigma\sqrt{2\pi}} e^{-\frac{1}{2}\left(\frac{x-\mu}{\sigma}\right)^2}, \quad -\infty < x < \infty,$$

$$-\infty < \mu < \infty, \sigma > 0.$$

If X has probability density function $f(x)$ we say $X \sim N(\mu, \sigma)$.

Standard Normal Variable

If $X \sim N(\mu, \sigma)$ and if $Z = (X - \mu)/\sigma$ then $Z \sim N(0, 1)$. The random variable Z is called a *standard normal variable*.

Central Limit Theorem

If samples of size n are selected from a continuous population with mean μ and variance σ^2, the distribution of the sample means is approximately $N(\mu, \sigma/\sqrt{n})$.

Chi-Squared Distribution

Let $\{x_1, x_2, x_3, \ldots, x_n\}$ denote a random sample from a $N(\mu, \sigma)$ distribution and let $s^2 = \frac{1}{n-1} \sum_{i=1}^{n}(x_i - \overline{x})^2$ denote the variance of the sample. The random variable $\frac{(n-1)s^2}{\sigma^2}$ is distributed according to a *chi-squared distribution* with $n - 1$ degrees of freedom.

Bivariate Random Variables	If $f(x, y)$ is the joint probability distribution function for discrete random variables X and Y then $f(x, y)$ has the following properties 1) $f(x, y) \geq 0$ 2) $f(x, y) = P(X = x \text{ and } Y = y)$ 3) $\sum_x \sum_y f(x, y) = 1$. If $f(x, y)$ is the joint probability density function for continuous random variables X and Y then $f(x, y)$ has the following properties: 1) $f(x, y) \geq 0$ 2) $\int_{-\infty}^{\infty} \int_{-\infty}^{\infty} f(x, y) \, dx \, dy = 1$ 3) $\int_c^d \int_a^b f(x, y) \, dx \, dy = P(a \leq X \leq b, c \leq Y \leq d)$
Marginal Distributions	If X and Y are continuous random variables with joint probability density function $f(x, y)$ then the *marginal distributions* of X and Y are $$f(x) = \int_{-\infty}^{\infty} f(x, y) \, dy \quad \text{and}$$ $$g(y) = \int_{-\infty}^{\infty} f(x, y) \, dx.$$ If X and Y are discrete random variables with joint probability distribution function $f(x, y)$ then the *marginal distributions* of X and Y are $$f(x) = \sum_y f(x, y) \text{ and } g(y) = \sum_x f(x, y).$$
Independent Random Variables	If X and Y are continuous *random variables* with joint probability density function $f(x, y)$ then X and Y are *independent* if $f(x, y) = f(x) \cdot g(y)$.
Sums and Variances of Bivariate Random Variables	For random variables X and Y, $E(X \pm Y) = E(X) \pm E(Y)$. and $\text{Var}(X \pm Y) = \text{Var}(X) \pm 2 \, \text{Cov}(X, Y) + \text{Var}(Y)$ If X and Y are independent then $\text{Var}(X \pm Y) = \text{Var}(X) + \text{Var}(Y)$

Correlation Coefficient

The *correlation coefficient*, $\rho_{X,Y}$ between the random variables X and Y is

$$\rho_{X,Y} = \frac{\text{Cov}(X,Y)}{\sigma_X \cdot \sigma_Y}.$$

Bivariate Normal Density

We say that random variables X and Y have a *bivariate normal density* if

$$f(x,y) = \frac{1}{2\pi\sigma_x\sigma_y\sqrt{1-\rho^2}} e^{-\frac{1}{2(1-\rho^2)}\left\{\left(\frac{x-\mu_x}{\sigma_x}\right)^2 - 2\rho\left(\frac{x-\mu_x}{\sigma_x}\right)\left(\frac{y-\mu_y}{\sigma_y}\right) + \left(\frac{y-\mu_y}{\sigma_y}\right)^2\right\}}$$

3 Estimation and Hypothesis Testing

3.1 Introduction

I N CHAPTER 2 WE CONSIDERED probability and the consequences of selecting random samples from a population whose characteristics are known. Practical situations, however, almost always concern the discovery of unknown characteristics, such as the population mean or standard deviation, or even the form of the distribution itself. This, in fact, is the reason sampling is done. After all, if the population characteristics are completely known, sampling is an interesting exercise, but a pointless one. So we now turn to situations in which some, or all, of the characteristics of a population are unknown. We shall see that the basis of discovering unknown characteristics of populations depends entirely upon probability distributions whose properties were discussed in Chapter 2.

We have seen that statistics—random variables computed from samples—such as the mean or range or standard deviation, can vary when samples are selected from a population. This variation, however, is far from random. It may seem to be somewhat of a contradiction to discover that random samples are not at all chaotic, but are in fact rather well-behaved. To cite two examples, the central limit theorem shows that the sample mean follows, approximately, a normal distribution and a function of the sample variance follows a chi-squared distribution when the sampling is done from a normal distribution. It is knowledge of these sampling distributions, as well as others we are yet to discuss, that enables us to draw inferences or conclusions from individual samples.

Statistical inference—the subject of drawing conclusions from samples—comprises two primary parts: **estimation** and **hypothesis testing.** We sample, almost always, from a population whose characteristics are unknown. We usually do not know either the mean or the variance of the population from which we sample. Estimation allows us

93

to make estimates of these unknown parameters, although with some uncertainty, while hypothesis testing allows us to conclude, again with some uncertainty, whether or not a conjecture about a population parameter is true.

We will consider estimation and hypothesis testing in some depth in this chapter, these topics will pervade the remainder of the book as we address statistical inference in various experimental situations. We begin with estimation.

Estimation

We start with some typical examples of the kinds of problems we are going to consider in this chapter. Each of these problems will be solved and questions concerning several other types of problems will be raised as well.

Example 3.1.1   **a.** The following data were reported by Mitchell, Hegemann, and Liu [20] in a wire-pull experiment. Gold or aluminum wire is used to connect an integrated circuit to a lead frame. In order to prevent disconnection of the integrated circuit and the frame, the wire must have a sufficient pull strength. A sample of ten wires was used with pull strengths measured in grams. Since the testing process destroys the unit being tested, a common situation in industrial testing, the sample size is fairly small. The data are as follows.

11.6   11.3   10.1   10.9   9.7

10.7   10.8   9.9   9.7   9.8

We want to use the sample to estimate the unknown true population mean pull strength, μ. The mean of the sample is $\bar{x} = 10.45$ g. What does this tell us about μ? Can we estimate μ by a specific value? Can we in some sense find an interval of estimates for μ?

The next example comes from a production line.

b. A production process is routinely inspected for quality by selecting samples of size 15 every four hours from the process. Recent samples gave the following numbers of items not meeting customer specifications.

2   2   0   3   1

3   5   2   2   0

The samples exhibit some variability. Taken together, what do these samples tell us about p, the true proportion of items not meeting customer specifications? Can we estimate p by a specific value? Can we in some sense find an interval of estimates for p?

We show a final example.

c. Two production lines, X and Y, are making thermostats to be used in toaster ovens. Samples are selected from each production line, the thermostats are set at $350°$, and then the actual temperature in the toaster oven is measured. The results of the sampling are shown in the following table.

X	Y
347.2	326.4
351.6	338.9
352.4	355.4
346.1	351.6
348.9	350.2
356.6	356.9
354.9	352.4
350.2	349.6
	351.2
	344.2

Did the samples arise from the same population?

Examples similar to each of the examples in parts (a), (b), and (c) will be discussed in this chapter. ■

Point Estimation Suppose, in Example 3.1.1 (a), that we want to estimate the unknown mean, μ, by a single value. Such an estimate is called a **point estimate.** What statistic should we choose as our estimator? The answer to this question is not simple and depends upon the properties we want our estimator to possess.

Selecting the sample mean as an estimator of the population mean is an intuitive answer to the question, but we can go beyond intuition. We begin with a series of independent observations on the random variable X, say $X_1, X_2, \ldots, X_n$, which gives the set of sample values $\{x_1, x_2, \ldots, x_n\}$. Suppose X has a probability distribution with true mean μ and true variance σ^2 and that the sample mean is $\overline{X}$ and the sample variance is S^2. Now using formulas from Sections 2.7 and 2.12, we find that

$$E(\overline{X}) = E\left(\frac{X_1 + X_2 + \cdots + X_n}{n}\right)$$

$$= \frac{1}{n}[E(X_1) + E(X_2) + \cdots + E(X_n)]$$

$$= \frac{1}{n} \cdot n\mu$$

$$= \mu$$

so the expected value of the sample mean is the population mean.

We can also find the variance of $\overline{X}$:

$$\mathrm{Var}(\overline{X}) = \mathrm{Var}\left(\frac{X_1 + X_2 + \cdots + X_n}{n}\right)$$

$$= \frac{1}{n^2}[\mathrm{Var}(X_1) + \mathrm{Var}(X_2) + \cdots + \mathrm{Var}(X_n)]$$

$$= \frac{1}{n^2} \cdot n\sigma^2$$

$$= \frac{\sigma^2}{n}.$$

Note that the facts that $E(\overline{X}) = \mu$ and $\text{Var}(\overline{X}) = \frac{\sigma^2}{n}$ were found using no assumptions about the form of the population from which the samples were selected; we simply assumed that the population parameters were μ and σ. These results are consistent with the central limit theorem which tells us that

$$\overline{X} \sim N\left(\mu, \frac{\sigma}{\sqrt{n}}\right).$$

We have shown that while the expected value and variance of $\overline{X}$ can be found easily, the central limit theorem establishes the normality of the sample mean, so we know its probability density function as well.

Since we know that the distribution of values of $\overline{X}$ is centered about the unknown population mean, μ, we would use a specific value of $\overline{X}$, say $\overline{x}$, as a point estimate for μ. The fact that the values of $\overline{X}$ are centered about μ is a good property for an estimator to have. Loosely speaking, we see that $\overline{X}$ as an estimator is correct on the average for the unknown mean, μ. We will come to describe this estimator of μ, $\overline{X}$, as an *unbiased estimator* for μ. The fact that the standard deviation, $\sigma/\sqrt{n}$, approaches 0 as $n \to \infty$ is also a comforting property for an estimator to have. Such estimators are often called **consistent.**

Let us now consider Example 3.1.1 (b). Here we want to estimate an unknown population parameter, p, the unknown true proportion of items not meeting customer specifications, where these items are being produced in a production process. The ten samples given in that example have a total of 20 items not meeting customer specifications. We might then estimate the unknown p as $20/150 = 0.13333$. We have called this estimator p_s, the **sample proportion.** In Chapter 2 we established that $E(p_s) = p$ and that $\text{Var}(p_s) = p \cdot q/n$ so we see that p_s is both an unbiased and consistent estimator for p.

These examples lead us to make the following definition.

DEFINITION

Suppose that $\widehat{\theta}$ is an estimator for a parameter, θ, of a population. Then $\widehat{\theta}$ is an **unbiased estimator** for θ if

$$E(\widehat{\theta}) = \theta.$$

The central limit theorem tells us that $\overline{X}$ is an unbiased estimator for μ, and the binomial distribution indicates that p_s is an unbiased estimator for μ.

Example 3.1.2   The sample variance is defined as

$$s^2 = \frac{\sum\limits_{i=1}^{n}(x_i - \overline{x})^2}{n - 1}.$$

We also know that

$$\frac{(n - 1)s^2}{\sigma^2} = \chi_{n-1}^2$$

when the sampling is done from a normal population.

It is beyond the scope of this book to prove, but it is known that $E(\chi_{n-1}^2) = n - 1$. It follows then that

$$E\left[\frac{(n - 1)s^2}{\sigma^2}\right] = \frac{n - 1}{\sigma^2}E(s^2) = n - 1$$

from which we see that $E(s^2) = \sigma^2$. While we will not demonstrate it here, $E(s^2) = \sigma^2$ for samples from any population with variance σ^2.

This explains the denominator of $n - 1$ rather than n in the definition of the sample variance. Had a denominator of n been used, then the sample variance would have been a biased estimator of the population variance. Here is a detailed explanation:

$$E\left[\frac{\sum\limits_{i=1}^{n}(x_i - \overline{x})^2}{n}\right] = E\left[\frac{n-1}{n} \cdot \frac{\sum\limits_{i=1}^{n}(x_i - \overline{x})^2}{n-1}\right]$$

$$= \frac{n-1}{n} \cdot E\left[\frac{\sum\limits_{i=1}^{n}(x_i - \overline{x})^2}{n-1}\right] = \frac{n-1}{n}E(s^2) = \frac{n-1}{n} \cdot \sigma^2$$

so the sample variance, had we used a denominator of n, would be a biased estimator for the population variance. The factor $(n - 1)/n = 1 - (1/n)$ is small if n is large. However, we often have small samples and then the factor is of significance.   ■

Example 3.1.3   The principle of unbiasedness assures us that estimates are correct, but only on the average. An unbiased estimate may or may not be close to the parameter being estimated. Estimators often are sought that are in some sense "close" to the parameter being estimated. Let's suppose that we have a random variable X and that we seek a so that $E(X - a)^2$ is as small as possible. Values that achieve this minimum, if they can be found, are called **least squares estimators.**

Let

$$Q = E(X - a)^2.$$

To minimize Q, we differentiate and find that

$$\frac{dQ}{da} = -2E(X - a).$$

Note that the preceding equation is true whether X is a discrete or a continuous random variable. When we equate the derivative to 0 we have an equation in an estimator, which we call $\widehat{a}$. It follows then that

$$E(X - \widehat{a}) = 0,$$

so

$$E(X) - \widehat{a} = 0,$$

from which we find that

$$\widehat{a} = E(X).$$

We have shown that the mean minimizes $E(X - a)^2$. For this reason, and since we know that the mean is an unbiased estimator of the population mean, we call the mean a **minimum variance unbiased estimator** of the population mean, μ.   ∎

We will find when we study regression that the principle of least squares is a particularly important one.

COMPUTER EXPERIMENTS

1. Draw 200 samples, each of size 5, from the $N(10,5)$ distribution. Calculate the sample mean, $\overline{x}$, for each sample and then verify that the mean of the $\overline{x}$'s $\doteq 10$.

2. Select 200 samples, each of size 10, from a binomial distribution with $p = 0.75$. Calculate the sample proportion, p_s, of successes for each sample, p_s, and verify that the mean of the p_s's $\doteq 0.75$.

3. For the sample drawn in Computer Experiment 1, calculate the sample variance, s^2, for each sample and verify that the mean of the sample variances is approximately 5.

4. For the sample drawn in Computer Experiment 1, draw a graph of the function

$$f(\mu) = \sum_{i=1}^{1000}(x_i - \mu)^2,$$

where x_i denotes one of the sample values, and show that the minimum of the function occurs when $\mu = 10$ (approximately).

3.2 Maximum Likelihood

In addition to unbiasedness and minimum variance (when such estimators can be found), there are other properties that are desirable for an estimator to have. In this section we therefore consider **maximum likelihood estimators** and their properties. This will give us a mathematical principle commonly used to establish some estimators. We begin with an example.

Example 3.2.1 A laboratory technician is testing blood samples for the presence of a factor in a disease. Two groups have been tested. In one of the groups, 5 of 15 samples show the factor, while in another group, 10 of 15 samples show the factor. Regrettably, the technician has mixed up the two groups and does not know which group is which. Since the testing is expensive, a sample of size 3 (chosen without replacement) was selected from one of the groups. The sample contains one sample that contains the factor and two samples that do not contain the factor. From which group was the sampling done?

This is certainly a different question from the ones we have dealt with so far in this book. We have commonly known the population from which we were sampling and have raised questions about the samples that might be chosen. Now we have the sample and are inquiring about the population from which it was selected. This is an example of **statistical inference.**

We understand that we cannot be certain of the decision we will make here. How should we proceed? One way to consider the problem would be to presume that we were sampling from a particular one of the groups and to calculate the probability that the sample we actually saw occurred. We could then compare the probabilities, assuming the sampling was done from each of the groups, and choose the one that is larger. This is an example of using the principle of *maximum likelihood*. Let us do this.

If the sampling were from the group of 15 samples, 5 of which show the factor and 10 of which do not show the factor, then the probability that our sample of size 3 contained exactly one with the factor and two without the factor is

$$\frac{\binom{5}{1} \cdot \binom{10}{2}}{\binom{15}{3}} = \frac{45}{91}.$$

If the sampling were from the group of 15 samples, 10 of which show the factor and 5 of which do not show the factor, then in this case the probability that our sample of size 3 contained exactly one with the factor and two without the factor is

$$\frac{\binom{10}{1} \cdot \binom{5}{2}}{\binom{15}{3}} = \frac{20}{91}.$$

If we choose the group with the larger probability, we conclude that we were sampling from the group having 5 samples with the factor and 10 without the factor. We have chosen the group with the larger probability, or **likelihood.**   ∎

Example 3.2.2 Now suppose, in Example 3.2.1, rather than use the specific sample used there, that we wish to consider *all* the possible samples of size 3 that could be chosen from each of the two groups. Which samples lead us to conclude we were sampling from the group with five samples showing the factor rather than the group with ten samples showing the factor? We call the group having five samples showing the factor group I for convenience.

Let X denote the random variable giving the number of samples showing the factor. The possible values of X are clearly 0, 1, 2, or 3. We see that

$$P_I(X = x) = \frac{\binom{5}{x} \cdot \binom{10}{3-x}}{\binom{15}{3}}, \qquad x = 0, 1, 2, 3.$$

However, if we were sampling from the group with 10 samples showing the factor, which we call group II, then

$$P_{II}(X = x) = \frac{\binom{10}{x} \cdot \binom{5}{3-x}}{\binom{15}{3}}, \qquad x = 0, 1, 2, 3.$$

When does P_I exceed P_{II}? That is, which samples lead us to believe we were sampling from group I and which samples lead us to believe we were sampling from group II?

We could compare the probabilities above for all possible values of x, but it is simpler to look at the ratio of the probabilities:

$$\frac{P_I}{P_{II}} = \frac{\binom{5}{x} \cdot \binom{10}{3-x}}{\binom{10}{x} \cdot \binom{5}{3-x}}, \qquad x = 0, 1, 2, 3.$$

The ratio P_I/P_{II} can also be simplified to

$$\frac{(10-x)!(2+x)!}{(5-x)!(7+x)!}.$$

Following is a table of these ratios.

x	$\dfrac{P_I}{P_{II}}$
0	12
1	2.25
2	0.44
3	0.08

So our decision rule now becomes: Decide the sampling is from group I if the number of samples showing the factor is 0 or 1, since $P_I > P_{II}$ for such samples; otherwise, decide that the sampling is from group II, since if $x = 2$ or 3, $P_I < P_{II}$ for such samples. ∎

Example 3.2.3   Unlike the small numbers in Examples 3.2.1 and 3.2.2, real situations usually involve large numbers. We have therefore replaced the numbers 5 and 10 in those examples by 50 and 100, respectively, and changed the sample size to 10. We then used Mathematica to plot a graph of P_I/P_{II}, which is shown in Figure 3.1. (The value at $x = 0$, namely 1685.15, is omitted from the graph.)

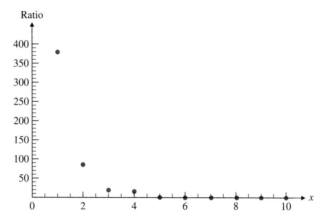

Figure 3.1 Maximum Likelihood Ratio for Example 3.2.3

Since $\binom{n}{r} = \binom{n}{n-r}$, the graph exhibits some symmetry due to the nature of P_I/P_{II}. In fact, denoting this ratio for a given value of x as $f(x)$, we see that $f(x) = 1/f(10-x)$. From this we conclude that $f(5) = 1$, and so in this case we cannot distinguish between the two groups at all. If x is 4 or less, $P_I > P_{II}$, and so in this case we are more likely to be sampling from the first group, while if x is 6 or greater, $P_I < P_{II}$ and we are most likely sampling from the second group. ■

Now we turn to an example from manufacturing.

Example 3.2.4 A sample of 100 items is chosen from a production line. Five of the items are found to have manufacturing defects. We want to estimate p, the probability that an item has a manufacturing defect. It would appear that 5/100 is a sensible choice, but it is not clear why this choice is optimal.

We assume a binomial process, so that items are produced independently, each with probability p of having a manufacturing defect. Call the probability that the sample occurred $L(p)$. Then

$$L(p) = \binom{100}{5} p^5 (1-p)^{95}.$$

We want to find the value of p that makes $L(p)$ most likely. Since the logarithm of $L(p)$ is an increasing function of $L(p)$, it is easiest to maximize the natural logarithm of $L(p)$:

$$\ln L(p) = \ln \binom{100}{5} + 5 \ln p + 95 \ln(1-p)$$

$$\frac{d \ln L(p)}{dp} = \frac{5}{p} - \frac{95}{1-p}.$$

Equating this to 0 gives a solution for $\widehat{p}$, our estimate of p, as $\widehat{p} = 5/100$. Since $\dfrac{d^2 \ln L(p)}{dp^2}$ is negative when $p = 5/100$, our estimate is a maximum. ∎

It is comforting to produce this result on the basis of a principle. Here we have chosen $\widehat{p}$ as our estimate since it maximizes the probability that the sample we saw occurred. It is called a **maximum likelihood estimate.** In previous examples, we also chose estimates that maximized the probability that the sample we saw occurred.

An alternative way to locate the maximum would be to plot the graph of $L(p)$ and note the maximum. The function $L(p)$ is called the *likelihood function.* Its graph is shown in Figure 3.2.

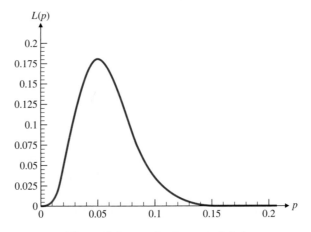

Figure 3.2 $L(p)$ for Example 3.2.4

We must now generalize these ideas so they can be applied to a large variety of situations, including those involving continuous random variables. We begin with a definition.

DEFINITION

Let $\{x_1, x_2, x_3, \ldots, x_n\}$ be a random sample from a probability distribution or a probability density function with unknown parameter θ. If $f(x;\theta)$ denotes the probability distribution, the **likelihood function $L(X;\theta)$** is defined as

$$L(X;\theta) = f(x_1;\theta) \cdot f(x_2;\theta) \cdot f(x_3;\theta) \cdots f(x_n;\theta).$$

If X is a discrete random variable, then, since $f(x;\theta) = P(X = x)$, $L(X;\theta)$ becomes

$$P(X = x_1) \cdot P(X = x_2) \cdot P(X = x_3) \cdots P(X = x_n),$$

and this is simply the probability of the sample that occurred. This reasoning was used in Example 3.2.4.

If X is a continuous random variable, then $L(X; \theta)$ is a point on the joint probability density function.

This leads directly to the *principle of maximum likelihood.*

Principle of Maximum Likelihood

The **principle of maximum likelihood** estimates an unknown population parameter θ by that value (or values) of θ, $\widehat{\theta}$, that maximize $L(X; \theta)$.

We give some additional examples of maximum likelihood estimation.

Example 3.2.5   A random sample, $x_1, x_2, x_3, \ldots, x_n$, is selected from an exponential distribution,

$$f(x) = \frac{1}{\theta} e^{-x/\theta}, \qquad x \geq 0.$$

We want the maximum likelihood estimator, $\widehat{\theta}$, of θ. Here

$$L(X; \theta) = \frac{1}{\theta} e^{-x_1/\theta} \cdot \frac{1}{\theta} e^{-x_2/\theta} \cdot \frac{1}{\theta} e^{-x_3/\theta} \cdots \frac{1}{\theta} e^{-x_n/\theta}$$

$$= \left(\frac{1}{\theta}\right)^n e^{-1/\theta \sum_{i=1}^{n} x_i}.$$

It will be simplest to differentiate the logarithm of the likelihood function so

$$\ln L(X; \theta) = -n \ln \theta - \frac{1}{\theta} \sum_{i=1}^{n} x_i.$$

Then

$$\frac{d \ln L(X; \theta)}{d\theta} = -\frac{n}{\theta} + \frac{1}{\theta^2} \sum_{i=1}^{n} x_i.$$

When this is equated to 0, we find that

$$\widehat{\theta} = \frac{\sum_{i=1}^{n} x_i}{n} = \overline{x}.$$

Also

$$\frac{d^2 \ln L(X; \theta)}{d\theta^2}$$

is negative when $\theta = \overline{x}$, so our result is a maximum. This is a pleasing result since the mean of the exponential distribution used here is θ. In terms of the random variable then $\widehat{\theta} = \overline{x}$.

If a computer algebra system is available, we can plot the likelihood function and note where its maximum occurs. We show one likelihood function where $n = 10$ and $\sum_{i=1}^{10} x_i = 30$ in Figure 3.3.   ■

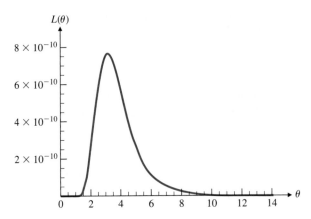

Figure 3.3 A Likelihood Function for Example 3.2.5

Example 3.2.6   We now consider a normal distribution whose standard deviation, σ, is known, but whose mean is unknown. Again we have a random sample $\{x_1, x_2, x_3, \ldots, x_n\}$ on which to base our estimate.

We know that

$$f(x; \mu) = \frac{1}{\sigma\sqrt{2\pi}} e^{-\frac{1}{2}\left(\frac{x-\mu}{\sigma}\right)^2}$$

and it follows that

$$L(X; \mu) = \frac{1}{\sigma\sqrt{2\pi}} e^{-\frac{1}{2}\left(\frac{x_1-\mu}{\sigma}\right)^2} \cdot \frac{1}{\sigma\sqrt{2\pi}} e^{-\frac{1}{2}\left(\frac{x_2-\mu}{\sigma}\right)^2} \cdots \frac{1}{\sigma\sqrt{2\pi}} e^{-\frac{1}{2}\left(\frac{x_n-\mu}{\sigma}\right)^2}$$

or

$$L(X; \mu) = \left(\frac{1}{\sigma\sqrt{2\pi}}\right)^n e^{-\frac{1}{2}\sum_{i=1}^{n}\left(\frac{x_i-\mu}{\sigma}\right)^2}.$$

Now the factor $(1/\sigma\sqrt{2\pi})^n$ will affect the size of the maximum (in which we are not interested), but not the value where the maximum occurs so, since it is positive, we denote it by k. Taking logarithms and differentiating, we have

$$\ln L(X; \mu) = \ln k - \frac{1}{2} \cdot \sum_{i=1}^{n} \left(\frac{x_i - \mu}{\sigma}\right)^2$$

and so

$$\frac{d \ln L(X; \mu)}{d\mu} = \frac{1}{\sigma^2} \sum_{i=1}^{n} (x_i - \mu).$$

When this is equated to 0, we find that $\widehat{\mu} = \bar{x}$. This should come as no surprise, since we found that $\widehat{\mu} = \bar{x}$ minimized $\sum_{i=1}^{n}(x_i - \mu)^2$ (and maximized the likelihood function) when we considered the principle of least squares in Example 3.1.3. ■

Example 3.2.7   A wheel has a pointer in its center. When the pointer is spun, it stops at one of the real numbers in the interval from 0 to θ; random variable X is the number on which the pointer stops. Assume that the wheel is fair so that X has a uniform distribution on the interval $[0, \theta)$; that is $f(x) = 1/\theta$ for $0 \leq x < \theta$. (Since the random variable X here is continuous, it doesn't matter if both 0 and θ are included in the domain of values of X or not.)

We want to find the maximum likelihood estimator for θ based on a random sample $\{x_1, x_2, x_3, \ldots, x_n\}$. This is an interesting situation, since the likelihood function is

$$L(X; \theta) = \begin{cases} \left(\dfrac{1}{\theta}\right)^n, & \theta \geq \text{Max}\{x_1, x_2, x_3, \ldots, x_n\}, \\ 0 \text{ otherwise.} \end{cases}$$

The function $(1/\theta)^n$ does not involve the random sample!

A graph of the situation may shed some light. A typical likelihood function (where $n = 10$ was selected) is shown in Figure 3.4. Since $\widehat{\theta}$ must necessarily be at least $\text{Max}\{x_1, x_2, x_3, \ldots, x_n\}$ (otherwise the largest value in the sample could not be observed), it is clear from the graph that $\widehat{\theta} = \text{Max}\{x_1, x_2, x_3, \ldots, x_n\}$ since the likelihood is largest there. ∎

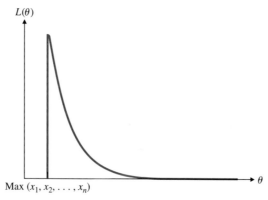

Figure 3.4 Likelihood Function for Example 3.2.7

Properties of Maximum Likelihood Estimators

Maximum likelihood estimators have large sample, or asymptotic, properties that make them very attractive. (It is beyond the scope of this book to show the following facts; proofs may be found in Hogg and Craig[15].) In each statement below, $\widehat{\theta}$ is the maximum likelihood estimator of a population parameter, θ, and the statement refers to asymptotic behavior, not necessarily to the behavior for small samples.

1. $E(\widehat{\theta}) = \theta$, so maximum likelihood estimators are unbiased.
2. $\widehat{\theta}$ is approximately normally distributed.
3. $\widehat{\theta}$ is a minimum variance estimator for θ.
4. If $f(\theta)$ is a function of the parameter θ, then $\widehat{f(\theta)} = f(\widehat{\theta})$.

More examples of the normal curve and maximum likelihood estimation are given in the exercises.

EXERCISES 3.2

1. Suppose the laboratory technician in Example 3.2.1 has mixed up two groups, one of which contains 7 of each type of sample and another with 8 samples having the factor and 12 that do not have the factor.
   a. If a sample contains three of each type of sample, which group was being sampled?
   b. Find a decision rule, as we did in Example 3.2.2, to distinguish the two groups.

2. It is known that a group of people is susceptible to a disease according to a binomial process. Ten people are observed; let X denote the number who are found to be susceptible to the disease. It is known that either $p = 1/10$ or $p = 1/5$.
   a. If four people in the sample are found to be susceptible to the disease, which value of p is more likely to be correct?
   b. Find a decision rule for distinguishing the two values of p based on a sample of 10 individuals in the group.

3. A random sample $\{x_1, x_2, x_3, \ldots, x_n\}$ is chosen from the discrete Poisson probability distribution function

$$f(x) = \frac{e^{-\mu} \cdot \mu^x}{x!}.$$

   a. Find the likelihood function.
   b. Show that the maximum likelihood estimator for μ is $\bar{x}$. (Note that the product of factorials that occurs in the likelihood function can be ignored.)

4. A normal random variable has mean 0 but its variance, σ^2, is unknown. We have a random sample $\{x_1, x_2, x_3, \ldots, x_n\}$ selected from this distribution.
   a. Find the likelihood function.
   b. Find the maximum likelihood estimator for σ^2.

5. A normal random variable has unknown mean, μ, and unknown variance, σ^2. We have a random sample $\{x_1, x_2, x_3, \ldots, x_n\}$ selected from this distribution.
   a. Find the likelihood function.
   b. To find the maximum likelihood estimators of the unknown parameters, take the partial derivatives first with respect to μ, and then with respect to σ^2, of the likelihood function. Equate each of these to 0 and solve them simultaneously.

6. A production process produces acceptable parts according to a binomial process with p being the probability that a part is acceptable. Parts are sampled from the line until an unacceptable part is found. If X de-

notes the number of good parts found before the first unacceptable part is found, the probability distribution function for X is

$$f(x) = P(X = x) = p^x \cdot (1 - p) \quad \text{for } x = 1, 2, 3, \ldots$$

Find the maximum likelihood estimator for p based on a sample $\{x_1, x_2, x_3, \ldots, x_n\}$.

7. A binomial process has n trials with probability of success p. The random variable X, the number of successes in n trials, is observed to be x. Find the maximum likelihood estimate for p.

8. A binomial process has probability of success p. It is observed that n_1 trials produced x_1 successes, n_2 trials produced x_2 successes, and n_3 trials produced x_3 successes.
   a. Find the maximum likelihood estimate for p using the three samples separately.
   b. Find the maximum likelihood estimator for p by combining the three samples so that $x_1 + x_2 + x_3$ successes were observed in $n_1 + n_2 + n_3$ trials.
   c. Compare the estimators in parts (a) and (b). Which has the smaller variance?

9. The lifetime of a component of a video camera in hours, X, follows the probability density function

$$f(x) = \frac{x}{\alpha^2} e^{-x/\alpha} \quad \text{for } x > 0.$$

Four cameras were observed to last 320, 450, 600, and 340 hours, respectively. Find the maximum likelihood estimator for α.

10. Suppose that X_1, X_2, and X_3 are independent random variables defined on a population with mean μ and variance σ^2.
   a. To find coefficients, a_1, a_2, and a_3 so that $E[a_1 X_1 + a_2 X_2 + a_3 X_3] = \mu$, show that $a_1 + a_2 + a_3 = 1$.
   b. Show that

$$\text{Var}\,[a_1 X_1 + a_2 X_2 + a_3 X_3] = a_1^2 \sigma^2 + a_2^2 \sigma^2 + a_3^2 \sigma^2.$$

   c. Write

$$\text{Var}[a_1 X_1 + a_2 X_2 + a_3 X_3]$$
$$= a_1^2 \sigma^2 + a_2^2 \sigma^2 + (1 - a_1 - a_2)^2 \sigma^2.$$

Find the partial derivatives with respect to a_1 and a_2 and conclude that the expression of the form $a_1 X_1 + a_2 X_2 + a_3 X_3$ that is an unbiased estimator of the

mean with minimum variance is the mean of the X_i's.
    b. $P_I > P_{II}$ if $x \leq 4$.

11. In Example 3.2.3, verify that
    c. $P_I = P_{II}$ if $x = 5$.

  a. $f(x) = \dfrac{1}{f(10-x)}$.
    d. $P_I < P_{II}$ if $x \geq 6$.

COMPUTER EXPERIMENTS

1. Select a sample of size 20 from the exponential distribution

$$f(x) = \left(\frac{1}{10}\right)e^{-x/10} \quad \text{for } x \geq 0.$$

Then plot the likelihood function

$$\left(\frac{1}{\theta}\right)^{20} e^{-\sum_{i=1}^{20} \frac{x_i}{\theta}}$$

and determine its maximum. At what value do you expect this to occur?

2. Select a sample of 200 observations from the $N(10,1)$ distribution. Then plot the function $f(\mu) = \text{Exp}[-(1/2)\sum_{i=1}^{200}(x_i - \mu)^2]$ as a function of μ, where the x_i are the sample values. (Note that $f(\mu)$ is a multiple of the likelihood function.) Where do you expect the maximum to occur?

3.3 Confidence Intervals

Having considered point estimates and some of their optional properties, in this section we now turn to the calculation of estimates by means of **intervals** for unknown population parameters.

Example 3.3.1 Consider a normal random variable X, whose variance, σ^2, is known, but whose mean, μ, is unknown. The Central Limit Theorem tells us that the random variable representing the mean of a sample of n random observations, $\overline{X}$, has a $N(\mu, \sigma/\sqrt{n})$ distribution. This means, to choose one example from an infinite number of possibilities, that

$$P\left(-1.96 \leq \frac{\overline{X} - \mu}{\sigma/\sqrt{n}} \leq 1.96\right) \doteq 0.95.$$

Other possibilities include altering the limits of -1.96 and 1.96, which could be done in an infinite number of ways. The probability, here 0.95, is often called the **confidence coefficient.** We could have chosen an infinity of other values. We also chose a symmetric interval, but we could have chosen a one-sided interval. Once we choose the confidence coefficient and a symmetric interval, the normal Z values (of ± 1.96 here) are obtained from the table of normal probability values (Table 1 in Appendix C).

We must also emphasize here that $(\overline{X} - \mu)/(\sigma/\sqrt{n})$ is a legitimate random variable; it will vary from sample to sample, and we know that the probability that it is between -1.96 and 1.96 is 0.95.

We next rearrange the inequality to read

$$P\left(-1.96\cdot\frac{\sigma}{\sqrt{n}}\leq\overline{X}-\mu\leq 1.96\cdot\frac{\sigma}{\sqrt{n}}\right)\doteq 0.95.$$

We see further that

$$P\left(-1.96\cdot\frac{\sigma}{\sqrt{n}}-\overline{X}\leq -\mu\leq 1.96\cdot\frac{\sigma}{\sqrt{n}}-\overline{X}\right)\doteq 0.95.$$

Now if we solve for μ and rearrange the inequalities, we find that

$$P\left(\overline{X}-1.96\cdot\frac{\sigma}{\sqrt{n}}\leq\mu\leq\overline{X}+1.96\cdot\frac{\sigma}{\sqrt{n}}\right)\doteq 0.95.$$

Note that the endpoints of the preceding inequality are both known, since $\overline{X}$ and n are known from the sample and we presume that σ is known.

The interval we just calculated is called a **95% confidence interval.**

It is important to proceed with some caution in interpreting the result. Despite the fact that we started with a probability statement, we *cannot* read the result as a probability statement! The reason is simple: While μ may be unknown to us, it is a *constant*, not a *random variable*. The fact that its constant value is unknown does not make it variable. The unknown mean, μ, is in the interval or it isn't in the interval; the probability that μ is in a given interval, such as the one calculated above, is then either 1 or 0.

What, then, does the 0.95 mean? The proper interpretation of our final result is this: 95% of all possible intervals calculated in this way will contain the unknown, constant value, μ, and 5% of all possible intervals calculated in this way will not contain the unknown, constant value, μ. ∎

Example 3.3.2   To illustrate the ideas presented in Example 3.3.1, we drew 20 samples from a normal distribution with mean 10 and standard deviation 5. A graph of the 20 confidence intervals generated is shown in Figure 3.5. As it happened, exactly 19 of the 20 confidence intervals contained the mean, indicated by the vertical line. This occurrence is not always to be expected. ∎

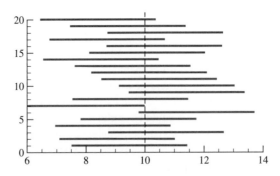

Figure 3.5 Confidence Intervals

Example 3.3.3   In Example 3.1.1(b), we presented the following situation: A production process is routinely inspected for quality by selecting samples of size 15 every four hours from the process. Ten recent samples produced the following numbers of items not meeting customer specifications.

$$\begin{array}{ccccc} 2 & 2 & 0 & 3 & 1 \\ 3 & 5 & 2 & 2 & 0 \end{array}$$

We then raised the following questions: What do these samples tell us about p, the true proportion of items not meeting customer specifications? Can we estimate p by a specific value? Can we find an interval of estimates for p?

We now see that we were asking first for a point estimate of p and then for a confidence interval for p. We can now answer these questions as follows: If we combine the samples, we find 20 items that do not meet customer specifications out of 150 items sampled. Our sample proportion is therefore

$$p_S = \frac{20}{150} = 0.13333.$$

If X is a binomial random variable representing the number of items in the sample not meeting customer specifications, and if there are n trials with probability of success p, then

$$p_S = \frac{X}{n}$$

and

$$E(p_S) = E\left(\frac{X}{n}\right) = \frac{E(X)}{n} = \frac{n \cdot p}{n} = p,$$

showing that the random variable p_S is an unbiased estimator of the true population proportion, p. Recall that we did this calculation in Section 2.7. We also found there that

$$\text{Var}\left(\frac{X}{n}\right) = \frac{\text{Var}(X)}{n^2} = \frac{n \cdot p \cdot q}{n^2} = \frac{p \cdot q}{n}.$$

It can also be shown that p_S is a maximum likelihood estimator for p.

Consider now finding a confidence interval for p. We know, from Section 2.9, that

$$\frac{p_S - p}{\sqrt{p \cdot q / n}}$$

is approximately normally distributed, the approximation becoming increasingly good as n increases. However, unlike in the previous example where we presumed that the population standard deviation, σ, was known, we notice now that the variance of p_S is $p \cdot q / n$, an expression that involves p! We show three ways of dealing with this difficulty. There is considerable empirical evidence to suggest that, for a wide range of values of n and p,

$$P\left(-2 \leq \frac{p_s - p}{\sqrt{p \cdot q / n}} \leq 2\right) \geq 0.95.$$

(See Kinney[18].)

Solution 1   We rearrange the inequality to find

$$P\left(-2\sqrt{\frac{p \cdot q}{n}} \leq p_s - p \leq 2\sqrt{\frac{p \cdot q}{n}}\right) \geq 0.95.$$

A further rearrangement gives

$$P\left(p_s - 2\sqrt{\frac{p \cdot q}{n}} \leq p \leq p_s + 2\sqrt{\frac{p \cdot q}{n}}\right) \geq 0.95.$$

Consider now the function $p \cdot q = p \cdot (1 - p)$. A graph of this function is shown in Figure 3.6. We see that, for a large range of values, namely, $0.3 \leq p \leq 0.7$, $p \cdot q = p \cdot (1 - p) \doteq 0.25$. Using this approximation gives

$$P\left(p_s - \frac{1}{\sqrt{n}} \leq p \leq p_s + \frac{1}{\sqrt{n}}\right) \geq 0.95.$$

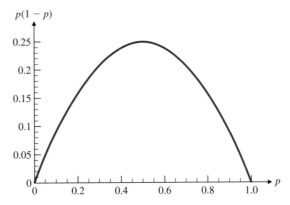

Figure 3.6 Graph of $p (1 - p)$

Our example has $n = 150$ and $p_s = 2/15$, so this confidence interval becomes

$$P\left(0.0517 \leq p \leq 0.2150\right) \geq 0.95.$$

Since our sample proportion, $2/15$, is not in the range from 0.3 to 0.7, the above approximation for $p \cdot q$ is probably not a very good one. We seek another solution that does not use this approximation.

Solution 2   We begin again with

$$P\left(-2 \leq \frac{p_s - p}{\sqrt{p \cdot q / n}} \leq 2\right) \geq 0.95.$$

Each of the inequalities can be solved for p. We show the result for the inequality

$$\frac{p_S - p}{\sqrt{p \cdot q/n}} \leq 2$$

or $n \cdot (p_S - p)^2 \leq 4 \cdot p \cdot (1 - p)$. The solution is

$$p \leq \frac{2 + n \cdot p_S + 2\sqrt{1 + n \cdot p_S - n \cdot p_S^2}}{n + 4}$$

with a similar result for the lower inequality. We have then that

$$P\left(\frac{2 + n \cdot p_S - 2\sqrt{1 + n \cdot p_S - n \cdot p_S^2}}{n + 4} \leq p\right.$$

$$\left. \leq \frac{2 + n \cdot p_S + 2\sqrt{1 + n \cdot p_S - n \cdot p_S^2}}{n + 4}\right) \geq 0.95.$$

Substituting $n = 150$ and $p_S = 2/15$ gives

$$P\,(0.08725 \leq p \leq 0.19846) \geq 0.95.$$

This result is a bit different from the approximation given in Solution 1, but this was to be expected.

Solution 3 Once more we begin with

$$P\left(-2 \leq \frac{p_S - p}{\sqrt{p \cdot q/n}} \leq 2\right) \geq 0.95.$$

If the sample size, n, is large, p_S is often substituted for the unknown p in the denominator. This produces the interval

$$P\left(p_S - 2\sqrt{\frac{p_S \cdot (1 - p_S)}{n}} \leq p \leq p_S + 2\sqrt{\frac{p_S \cdot (1 - p_S)}{n}}\right) \geq 0.95.$$

Using $n = 150$ and $p_S = 2/15$, the confidence interval becomes

$$P(0.077822 \leq p \leq 0.18884) \geq 0.95.$$

The different sets of assumptions made in these three solutions produce somewhat different confidence intervals. ∎

Example 3.3.4 A common problem is the determination of sample size in order to estimate a population parameter to a certain accuracy. Suppose a population has an unknown mean, μ, but the standard deviation, σ, is known to be 5. How large a sample is necessary so that $\overline{X}$ and μ differ by at most 1.5 units with probability 0.95?

Since the confidence interval is

$$P\left(\overline{X} - 1.96 \cdot \frac{\sigma}{\sqrt{n}} \leq \mu \leq \overline{X} + 1.96 \cdot \frac{\sigma}{\sqrt{n}}\right) \doteq 0.95,$$

we can write

$$P\left(-1.96 \cdot \frac{\sigma}{\sqrt{n}} \leq \overline{X} - \mu \leq 1.96 \cdot \frac{\sigma}{\sqrt{n}}\right) \doteq 0.95.$$

So here we want $1.96 \cdot \sigma/\sqrt{n} \leq 1.5$. Using $\sigma = 5$, we solve for n to find that $n \geq (1.96 \cdot 5/1.5)^2 = 42.684$, so a sample of 43 is necessary.   ∎

Example 3.3.3, where σ is not known, raises a question about finding a confidence interval under these conditions. This problem, fortunately, can be solved. In the next section we digress to consider another probability density function, the Student t distribution, that plays a key role in the solution.

EXERCISES 3.3

1. Scores on an examination are known to have variance 125. A sample of scores is: 72, 91, 38, 64, and 75. Find a 95% confidence interval for the true mean of the scores, μ.

2. A random sample of 50 measurements of the breaking strength of cotton threads gave $\overline{x} = 210$ grams. If $\sigma^2 = 300$ g^2 for this thread, find a 90% confidence interval for μ, the true mean breaking strength for the thread.

3. To determine the average speed of cars along a certain road, a random sample of 100 cars is taken and the mean speed in the sample is found to be 58.4 mph. If the true standard deviation of the speeds is known to be 15 mph, find a 95% confidence interval for μ, the true average speed.

4. A sample of 10 air pollution measurements for carbon monoxide was taken. Assume that the measurements have an unknown mean, μ, but that the population standard deviation, σ, is known to be 8. The sample mean was 42; find a 95% confidence interval for μ.

5. A telephone company is trying to decide whether some new lines in a large community should be installed underground. Because some of the installation cost must be passed on to customers, the company conducted a sample survey of 160 customers, 113 of whom favored the underground installation despite the added cost. Find a 95% confidence interval for p, the true proportion of customers favoring the underground installation.

6. A random sample of 36 production workers found that 12 were willing to join a union. Based on this sample, find a 95% confidence interval for p, the true proportion of these workers willing to join a union.

7. In an air pollution study, the following amounts of suspended benzene-soluble organic matter (in micrograms per cubic meter) were obtained at an experiment station for eight different samples of air: 2.3, 1.9, 3.2, 2.1, 2.4, 2.1, 2.2, and 1.3. Find a two-sided 90% confidence interval for the true mean, μ, assuming that $\sigma = 0.25$.

8. A sample of 22 showed that the average pitch diameter of the thread of a fitting was 0.3990 cm. Assume that the true standard deviation of the pitch diameters is 0.0004 cm.
   a. Find a 90% confidence interval for the true mean pitch diameter of the threads.
   b. How large a sample is necessary to estimate the true average pitch diameter of the threads by $\overline{X}$ to within 0.001 with probability 0.95?

9. A sample of 125 items from a production process showed that 60% were within the specification limits required by a customer.
   a. Find a 90% confidence interval for p, the true proportion of items within the specification limits.
   b. How large a sample is necessary to estimate p to within 0.06 units with probability 0.90?

10. The average zinc concentration recovered from a sample of zinc measurements in 22 different locations was found to be 2.6 grams per milliliter. Assume the population standard deviation is 0.3.
    a. Find a 95% confidence interval for μ, the true mean of the population.
    b. Suppose 100 samples are taken and confidence intervals are computed for each sample, as in part (a). What is the probability that at least 90 of these contain the true mean, μ?

11. Verify all the calculations in Example 3.3.3.

12. Consider a binomial situation with parameters n and p. Suppose a single trial has X successes. Show that the sample proportion, $p_s = X/n$ is a maximum likelihood estimator for p.

COMPUTER EXPERIMENTS

1. Select 300 samples, each of size 6, from a $N(50, 20)$ distribution. Calculate $\bar{x}$ for each and a 95% confidence interval for μ for each sample. Count the number of these confidence intervals that actually contain the true mean, 50.

2. Select 200 samples, each of size 10, from a binomial distribution with $p = 0.40$. For each sample, calculate the sample proportion of success, p_s. Then construct a histogram of the statistic

$$\frac{p_s - p}{\sqrt{p \cdot q/n}} = \frac{p_s - 0.40}{\sqrt{0.40 \cdot 0.60/10}} = \frac{p_s - 0.40}{0.15492}$$

and compare the histogram with a standard normal curve. (Approximating the denominator by $\sqrt{0.50 \cdot 0.50/10} = 0.15811$ will not make an appreciable difference in this case.)

3.4 Student *t* Distribution

The Central Limit Theorem assures us that the random variable $\overline{X}$ has approximately a normal distribution with mean μ and standard deviation $\sigma/\sqrt{n}$, where μ and σ are the mean and standard deviation of X. We used this fact to calculate confidence intervals for μ, but to do so requires knowledge of σ.

In sampling from populations, we generally do not know any of the population parameters, including σ, and, indeed, we may not even know the form of the distribution. Fortunately, a solution is known. We state the following theorem, whose proof can be found in Hogg and Craig [15].

Theorem

The ratio of a standard normal random variable and the square root of an independent chi-squared random variable divided by its degrees of freedom, n, follows a Student *t* distribution with n degrees of freedom.

We express this symbolically as

$$\frac{N(0, 1)}{\sqrt{\chi^2/n}} = t_n.$$

Note that we now require that the sampling be done from a normal distribution.

We know that $(\overline{X} - \mu)/(\sigma/\sqrt{n})$ is exactly $N(0, 1)$ if the sampling is from a normal distribution. We also know that

$$\frac{(n - 1) \cdot s^2}{\sigma^2}$$

follows a χ^2_{n-1} distribution. In addition, it can be shown that the random variables $\overline{X}$ and s^2 are independent. (The proof is beyond the scope of this book.) The theorem then says that

$$\frac{\dfrac{\overline{X} - \mu}{\sigma/\sqrt{n}}}{\sqrt{\dfrac{(n - 1) \cdot s^2}{\sigma^2 \cdot (n - 1)}}}$$

follows a Student t distribution with $n - 1$ degrees of freedom.

But the fraction simplifies, the unknown σ conveniently disappearing, to give

$$\frac{\overline{X} - \mu}{s/\sqrt{n}} = t_{n-1}.$$

It would appear that we have simply substituted S for σ in the variable $(\overline{X} - \mu)/(\sigma/\sqrt{n})$, but in fact a great deal more is going on. It is, of course, dangerous to substitute the value of a random variable from a sample for an unknown population parameter.

The t distribution was discovered by W. G. Gossett whose employer forbid him to use his real name, so he published papers under the pseudonym Student.

Values for various Student t distributions are given in Table 2 in Appendix C. Minitab and other statistical and computer algebra systems, and some pocket computers, will also calculate values for this distribution.

As the sample size, n, increases one would expect the t distribution to approach the normal distribution, and this indeed happens. Some t distributions, are shown in Figure 3.7.

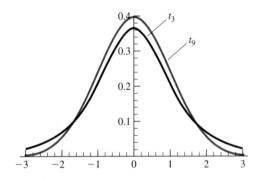

Figure 3.7 Some Student t Distributions

The Student t distribution is a bit more dispersed than the standard normal distribution. For example, we found the "2/3, 95, 99" rule to apply to areas between -1 and 1, -2

and 2, and -3 and 3 for the standard normal curve. If we find the same areas under t_8, we find that they are, respectively, 0.653406, 0.919484, and 0.982928, indicating that it takes values somewhat larger than 1, 2, and 3 to obtain the probabilities of "2/3, 95, 99." This is evidence of a somewhat larger dispersion of the t_8 distribution than that of the standard normal.

While the normal distribution is a good approximation to t_n when n is large, computer programs allow the exact calculation of t values for any number of degrees of freedom so it is never necessary to use the normal approximation.

Example 3.4.1 Suppose a random sample of ten observations is selected from a normal distribution whose mean and standard deviation are both unknown. We want a 95% confidence interval for μ, the unknown mean of the population.

The point on the t_9 distribution that has 97.5% of the distribution to its left is 2.26216. Therefore a two-sided confidence interval can be constructed by starting with the statement

$$P\left(-2.26216 \leq \frac{\overline{X} - \mu}{S/\sqrt{n}} \leq 2.26216\right) = 0.95.$$

This can be simplified to

$$P\left(\overline{X} - 2.26216 \cdot \frac{S}{\sqrt{n}} \leq \mu \leq \overline{X} + 2.26216 \cdot \frac{S}{\sqrt{n}}\right) = 0.95.$$

An actual sample of size 10 from the $N(10, 5)$ distribution gave $\overline{x} = 11.9307$ and $s = 4.1047$. This gives the 95% confidence interval (8.99437, 14.867). Repeated samples were taken, and the results are shown as 20 confidence intervals in Figure 3.8. Due to the variability in the standard deviation from sample to sample, the confidence intervals vary in length as well as in position.

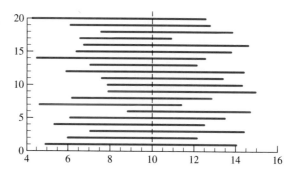

Figure 3.8 Confidence Intervals, σ Unknown

The proper interpretation of the confidence interval is again that 95% of these intervals will contain the unknown mean, μ. ∎

EXERCISES 3.4

1. A tensile strength test was performed in order to determine the strength of a particular adhesive for a glass-to-glass assembly. The data are: 16, 14, 19, 18, 19, 20, 15, 18, 17, 18.
   a. Find a 90% confidence interval for the true mean, μ, assuming that the observations are selected from a normal population.
   b. Compare the solution in part (a) to the solution where the true standard deviation, σ, is known to be 2.

2. It is thought that a vacuum cleaner expends an average of 46 kilowatt-hours per year. A random sample of 12 homes in a study shows an average of 42 kilowatt-hours per year with sample standard deviation 11.9 kilowatt-hours.
   a. Find a 95% confidence interval for μ, assuming that the observations are selected from a normal population.
   b. Find a 95% confidence interval for μ, assuming that $\sigma = 12$.

3. Eight samples of concrete were tested for compressive strength. The data are as follows: 1872, 1904, 2106, 1756, 1855, 1835, 1900, and 1872. Find a 90% confidence interval for the true compressive strength, μ.

4. "Twenty ohm" resistors are sampled and their actual resistance measured. A sample of 12 resistors gave $\bar{x} = 21.2$ ohms and $s = 0.36$ ohms. Find a constant a and a 90% confidence interval for the true mean resistance, μ, of the form $P(\mu \geq a) = 0.90$.

5. A well-known brand of peanut butter claims to contain 17% saturated fat. A sample of 20 jars of peanut butter gave $\bar{x} = 17.4\%$ and $s = 0.06\%$. Find a 90% confidence interval for the true percentage of saturated fat and decide whether or not the brand's claim is correct.

6. Computer printer cartridges from a manufacturer claim to have a useful life of 1,000,000 characters. To test this claim, a sample of 9 cartridges was tested and the number of useful characters printed showed $\bar{x} = 984{,}776.2$ and $s = 4{,}933.2$. Find a 95% confidence interval for the true mean number of useful characters typed and decide whether or not the manufacturer's claim is correct.

7. A sample of size 12 is selected from a normal population whose true mean is 84. The sample values gave $\bar{x} = 86.17$ and $s = 3.42$. What is the probability that a sample will give this mean or a mean greater than 86.17?

8. Soil pH is an important variable in the design of structures that will contact the soil. Nine test samples of soil from a construction site gave readings with $\bar{x} = 6.39$ and $s = 0.4428$.
   a. If it is important that the true mean pH level be no more than 8, is this a good construction site?
   b. Find a constant a so that a 90% confidence interval for the mean of the soil pH of the form $P(\mu < a) = 0.90$.

9. An article in *Chemical Engineering Progress* gave data similar to the following readings of the purity (in percent) of oxygen being delivered to two industrial air products suppliers.

$$n_1 = 5,\ \bar{x}_1 = 99.636,\ s_1 = 0.07765$$
$$n_2 = 5,\ \bar{x}_2 = 98.943,\ s_2 = 0.08473$$

   a. Find 95% confidence intervals for the true mean percent of oxygen being delivered by each of the suppliers.
   b. What meaning can be attached to the fact that the confidence intervals in part (a) overlap?

10. The drying times of epoxy substances are assumed to be normally distributed with mean 3 minutes. A sample of 9 drying times gave $\bar{x} = 3.7$ min and $s = 0.10$ min. If the assumptions about the drying times are correct, is this a typical sample or an unusual one?

COMPUTER EXPERIMENT

Select 200 samples, each of size 7, from a standard normal distribution. Calculate the mean and the standard deviation for each sample. Then for each sample, calculate the statistic $\bar{x}/(s/\sqrt{7})$, which should approximate a t_6 distribution. Construct a histogram of these values and compare the histogram to the t_6 distribution.

3.5 **Difference Between Proportions**

A common scientific problem is to distinguish between proportions arising from two samples. We have all seen advertising telling us that a group brushing with brand X had 24% fewer cavities than those brushing with brand Y. The question of course is whether this difference is significant or could be the simple result of chance. It is, in any event, highly dependent upon the sample sizes, which advertisers rarely reveal. We develop some of the theory and then show some applications.

Suppose we have two samples of sizes n_1 and n_2 from which we observe sample proportions of p_{s_1} and p_{s_2} for some attribute. Assuming these samples were selected from populations with true proportions p_1 and p_2, we want to use the samples to construct confidence intervals for the difference between the true proportions, $p_1 - p_2$.

We know that p_{s_1} and p_{s_2} are approximately normal with means p_1 and p_2 and variances $p_1 \cdot q_1/n_1$ and $p_2 \cdot q_2/n_2$. It is then true that

$$E(p_{s_1} - p_{s_2}) = E(p_{s_1}) - E(p_{s_2}) = p_1 - p_2$$

and, assuming the samples to be independent,

$$\text{Var}(p_{s_1} - p_{s_2}) = \text{Var}(p_{s_1}) + \text{Var}(p_{s_2}) = \frac{p_1 \cdot q_1}{n_1} + \frac{p_2 \cdot q_2}{n_2}.$$

We have then that

$$\frac{(p_{s_1} - p_{s_2}) - (p_1 - p_2)}{\sqrt{\dfrac{p_1 \cdot q_1}{n_1} + \dfrac{p_2 \cdot q_2}{n_2}}}$$

is approximately a standard normal random variable.

The construction of confidence intervals begins, as usual, with a statement about a standard normal random variable such as

$$P\left[-1.645 \leq \frac{(p_{s_1} - p_{s_2}) - (p_1 - p_2)}{\sqrt{\dfrac{p_1 \cdot q_1}{n_1} + \dfrac{p_2 \cdot q_2}{n_2}}} \leq 1.645\right] = 0.90.$$

Further progress, however, is impeded by the fact that the denominator depends upon the very parameters that are to be estimated. We will give two possible, approximate, solutions to this difficulty.

Solution 1 If we replace the true proportions in the denominator by the observed proportions, we have

$$P\left[-1.645 \leq \frac{(p_{s_1} - p_{s_2}) - (p_1 - p_2)}{\sqrt{\dfrac{p_{s_1} \cdot q_{s_1}}{n_1} + \dfrac{p_{s_2} \cdot q_{s_2}}{n_2}}} \leq 1.645\right] = 0.90$$

and it is easy to rearrange this to obtain

$$P\left[\begin{array}{l}(p_{s_1} - p_{s_2}) - 1.645 \cdot \sqrt{\dfrac{p_{s_1} \cdot q_{s_1}}{n_1} + \dfrac{p_{s_2} \cdot q_{s_2}}{n_2}} \le p_1 - p_2 \\[4mm] \le (p_{s_1} - p_{s_2}) + 1.645 \cdot \sqrt{\dfrac{p_{s_1} \cdot q_{s_1}}{n_1} + \dfrac{p_{s_2} \cdot q_{s_2}}{n_2}}\end{array}\right] = 0.90. \qquad \blacksquare$$

We show an example before giving another approximate solution.

Example 3.5.1   Two different types of iron ore are tested for the presence of a radioactive element. Ten samples of type 1 showed that 10% contained the element while 20 samples of type 2 showed that 15% contained the element. Find a 90% confidence interval for the difference between the true proportions, $p_1 - p_2$.

Applying the previous result, we have

$$P\left[\begin{array}{l}(0.10 - 0.15) - 1.645 \cdot \sqrt{\dfrac{0.10 \cdot 0.90}{10} + \dfrac{0.15 \cdot 0.85}{20}} \le p_1 - p_2 \\[4mm] \le (0.10 - 0.15) + 1.645 \cdot \sqrt{\dfrac{0.10 \cdot 0.90}{10} + \dfrac{0.15 \cdot 0.85}{20}}\end{array}\right] = 0.90$$

This simplifies to $P(-0.2540 \le p_1 - p_2 \le 0.1540) = 0.90$. Since 0 is included in the confidence interval, it appears very possible that the true proportions are equal. In any event, the confidence interval is quite wide due to the small sample sizes involved.

Solution 2   If it is thought that the true proportions, p_1 and p_2, may be equal, a common approximation is to replace the true proportions in the denominator,

$$\sqrt{\frac{p_1 \cdot q_1}{n_1} + \frac{p_2 \cdot q_2}{n_2}},$$

by an overall average proportion by combining the two samples.

Here we see that there was one sample of the first ten that showed the radioactive element while three samples of the 20 sampled of type 2 showed the element. This gives a combined percentage of $(1 + 3)/(10 + 20) = 0.13333$. Using this for both p_1 and p_2 in the denominator gives

$$P\left[\begin{array}{l}(0.10 - 0.15) - 1.645 \cdot \sqrt{\dfrac{0.13 \cdot 0.87}{10} + \dfrac{0.13 \cdot 0.87}{20}} \le p_1 - p_2 \\[4mm] \le (0.10 - 0.15) + 1.645 \cdot \sqrt{\dfrac{0.13 \cdot 0.87}{10} + \dfrac{0.13 \cdot 0.87}{20}}\end{array}\right] = 0.90,$$

giving the 90% confidence interval $(-0.26426, 0.16426)$, so the two solutions do not differ markedly. $\blacksquare$

EXERCISES 3.5

1. Two industrial processes are being compared with respect to the percentage of product each produces that is not within a customer's specifications. Samples are taken from each process. A sample of 30 items produced by the first process showed 2 items not meeting specifications, while a sample of 20 from the second process showed 4 not meeting specifications. Find a 95% confidence interval for the difference between the true proportions not meeting specifications.

2. Two processes for producing computer memory chips are being compared. The first process produced 22 non-functioning memory chips out of a sample of 150, while the second process produced 12 nonfunctioning memory chips out of a sample of 50.
   a. Find a 90% confidence interval for the difference between the true proportions of nonfunctioning memory chips.
   b. Are the processes equivalent?

3. A corporation conducts a survey concerning the percentage of its workers who favor joining a union. In one plant 45 of 200 workers favor joining the union, and in another plant 36 of 174 workers favor joining the union. Find a 90% confidence interval for the difference between the true proportions favoring joining the union. Is it likely that the proportions of workers favoring union membership in the two plants are equal?

4. A consumer is comparing companies writing computer software with respect to the number of errors made per line of computer code. Company A had 34 errors in 1000 lines of computer code, and company B had 68 errors in 1500 lines of computer code. Find a 95% confidence interval for the difference between the true proportions of errors in computer code produced by the two companies.

5. Sample surveys are done in two contiguous states regarding the imposition of a sales tax. In one state 34% of 100 residents favored the sales tax, and in the other state 45% of 150 voters favored the sales tax. Find a 90% confidence interval for the difference between the true proportions favoring the sales tax.

6. Evaluating the effectiveness of drugs in combating a disease is an important job for a drug company. Tests are commonly done on one group whose patients receive the drug and on a second group whose patients receive a placebo, thought to be ineffective in combating the disease. In a recent test, 76% of 100 patients in the group receiving the drug improved while 56% of 75 patients in the group receiving the placebo improved. Construct a 90% confidence interval and decide whether or not the drug is effective.

7. Brand A of toothpaste was tested on a group of 46 dental patients of whom 3 showed incidence of cavities. A group of 55 patients participating in a test of brand B showed 5 with incidence of cavities.
   a. Find a 99% confidence interval for the difference between the true proportions of patients showing incidence of cavities.
   b. Is brand A really better than brand B?

8. Recovery rates for patients in different age groups are thought to differ. To investigate this question, 40 patients aged 30 to 40 with the disease were studied while 20 patients aged 50 to 60 with the disease were studied. In the 30-to-40 age group 90% of the patients recovered, while in the 50-to-60 age group 80% of the patients recovered.
   a. Find a 90% confidence interval for the difference between the true recovery rates from the disease for the two groups.
   b. Does the study establish a difference between recovery rates for the two age groups?

9. It is possible to select equal size samples from two groups in order to compare some percentage in each group. Find the sample size so that a 95% confidence interval for the difference between the percentages is of length at most 0.16.

COMPUTER EXPERIMENT

Take 200 pairs of samples, each of size 40, from two binomial populations: one with $p = 0.75$ and the second with $p = 0.50$. For each sample, record the sample percentages of successes, p_{s_1} and p_{s_2}. Then calculate

$$\frac{p_{s_1} - p_{s_2} - (p_1 - p_2)}{\sqrt{\dfrac{p_1 \cdot q_1}{n_1} + \dfrac{p_2 \cdot q_2}{n_2}}} = \frac{p_{s_1} - p_{s_2} - 0.25}{\sqrt{\dfrac{0.75 \cdot 0.25}{40} + \dfrac{0.50 \cdot 0.50}{40}}} = \frac{p_{s_1} - p_{s_2} - 0.25}{0.10458}$$

for each pair of samples. Construct a histogram of the results and compare the histogram with a standard normal curve.

3.6   Confidence Intervals for the Variance

The chi-squared distribution was introduced in Section 2.11. We showed a simulation and did some calculations with various chi-squared curves. We also hinted at the construction of confidence intervals.

The fact that $(n-1) \cdot S^2/\sigma^2$ follows a χ^2_{n-1} distribution can be used to construct confidence intervals for the population variance based on a sample. It must be emphasized, however, that the sample must come from a normal distribution. Figure 3.9 shows some chi-squared distributions for various degrees of freedom.

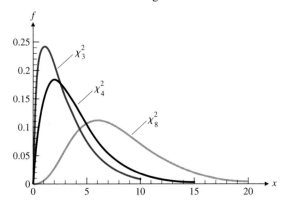

Figure 3.9 Some Chi-Squared Distributions

It is known that $E(\chi^2_{n-1}) = n-1$. This fact was used previously to show that $E(S^2) = \sigma^2$. It is also known that $\mathrm{Var}(\chi^2_{n-1}) = 2(n-1)$, so

$$\mathrm{Var}\left[\frac{(n-1)S^2}{\sigma^2}\right] = 2(n-1) \quad \text{or} \quad \frac{(n-1)^2}{\sigma^4}\mathrm{Var}(S^2) = 2(n-1)$$

so

$$\mathrm{Var}(S^2) = \frac{2\sigma^4}{n-1}.$$

This shows that the sample variance is highly variable if σ is large since it is in direct proportion to the fourth power of σ. As a consequence, confidence intervals are then often very wide, especially when the sample size is small.

Example 3.6.1   A sample of 10 observations from a normal distribution has sample variance 50. We want a 95% confidence interval for the true population variance, σ^2.

Table 3 in Appendix C gives some upper and lower points on chi-squared distributions with various numbers of degrees of freedom. Minitab, Mathematica, or Maple can give far more extensive results than any set of tables can provide, so we encourage their use.

To construct a one-sided confidence interval, we start with the probability statement

$$P\left(\chi^2_9 > 3.32511\right) = 0.95$$

so

$$P\left[\frac{(n-1)S^2}{\sigma^2} > 3.32511\right] = 0.95$$

4. $x^4 \geq 1$

5. $(x - 3)(x + 5)x > 0$

123

$\sigma^2 = 9$ from population

$df = 18$

c

$$\chi^2 = \frac{(n-1)s^2}{\sigma^2}$$

$$\chi^2_{Lower} = \frac{18.4}{9} = 8$$

$$\chi^2_{upper} = \frac{18.16}{9} = 3^2$$

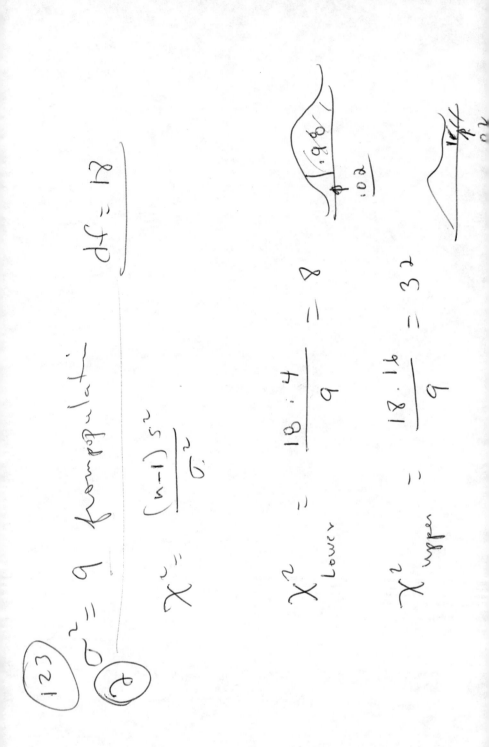

so

$$P\left(\frac{9\cdot 50}{\sigma^2} > 3.32511\right) = 0.95 \quad \text{or} \quad P\left(\sigma^2 < \frac{9\cdot 50}{3.32511}\right) = 0.95,$$

which becomes

$$P(\sigma^2 < 135.33) = 0.95.$$

The result is of course subject to our interpretation of a confidence interval: 95% of intervals constructed in this way will contain the unknown constant σ^2.

Upper or lower limits on the true variance are often of great interest since they give some bounds on the variation that might be expected to occur in a practical situation. Note that the initial inequality must be the reverse of the one that is desired in the end due to the occurrence of σ^2 in the denominator.

A two-sided 90% confidence interval can be found as follows:

$$P\left[3.32511 < \frac{(n-1)S^2}{\sigma^2} < 16.919\right] = 0.90,$$

and this can be written as

$$P\left[3.32511 < \frac{9\cdot 50}{\sigma^2} < 16.919\right] = 0.90$$

or

$$P(26.597 < \sigma^2 < 135.33) = 0.90. \qquad \blacksquare$$

Now suppose we want a confidence interval for the standard deviation, σ. Can we take square roots throughout the confidence interval above? Unfortunately, this gives only an approximate confidence interval. The reason is that the sample standard deviation, S, has quite a different probability distribution than that for the function of S^2, which has a chi-squared distribution. (More details can be found in Kinney[18].) Moreover, S is a biased estimator for σ, the size of the bias decreasing as the sample size increases.

When we compare two samples, it is often necessary to compare their variances. Fortunately, it is possible to do this, but we must first consider one more probability density function: the F distribution.

3.7 The *F* Distribution

The following theorem is known. Refer to Hogg and Craig [15] for a proof.

Theorem

If χ_n^2 and χ_m^2 are independently distributed chi-squared random variables, then the random variable

$$\frac{\chi_n^2/n}{\chi_m^2/m} = F(n, m),$$

where $F(n, m)$ is the F distribution with n and m degrees of freedom, respectively.

Table 4 in Appendix C gives some upper tail values on F distributions for some combinations of degrees of freedom. Only upper tail values are necessary since, if we wished to determine a so that $P[F(n, m) > a] = \alpha$, we could write

$$P\left[\frac{1}{F(n, m)} < \frac{1}{a}\right] = \alpha$$

but the reciprocal of $F(n, m)$ is $F(m, n)$ by the theorem above. So we equivalently determine

$$P\left[F(m, n) < \frac{1}{a}\right] = \alpha$$

showing that the upper point on $F(n, m)$ is the reciprocal of the lower point on $F(m, n)$.

Statistical programs or computer algebra programs can be used to find values not commonly found in tables. Figure 3.10 shows some F distributions.

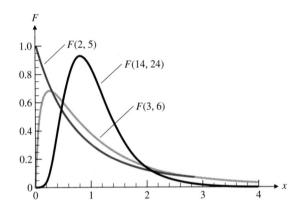

Figure 3.10 Some F Distributions

The ratio of the variances occurs in the F distribution specifically in this way: We know that $(\chi_n^2/n)/(\chi_m^2/m) = F(n, m)$ and that $(n-1) \cdot S^2/\sigma^2 = \chi_{n-1}^2$, so

$$F(n - 1, m - 1) = \frac{\dfrac{(n - 1) \cdot S_1^2}{\sigma_1^2 \cdot (n - 1)}}{\dfrac{(m - 1) \cdot S_2^2}{\sigma_2^2 \cdot (m - 1)}} = \frac{\dfrac{S_1^2}{\sigma_1^2}}{\dfrac{S_2^2}{\sigma_2^2}}.$$

We can now use these facts to determine confidence intervals for the ratio of two population variances when the samples are drawn independently from two normal populations.

Example 3.7.1 Two independent samples from normal distributions gave the following summary statistics: $n_1 = 15$, $s_1^2 = 25$, $n_2 = 25$, $s_2^2 = 50$. Use a confidence interval to decide whether or not the true population variances could be equal.

We find that a 95% confidence interval is

$$P\,[0.358576 < F(14, 24) < 2.46766] = 0.95,$$

but

$$F(14, 24) = \dfrac{\dfrac{s_1^2}{\sigma_1^2}}{\dfrac{s_2^2}{\sigma_2^2}} = \dfrac{\dfrac{25}{\sigma_1^2}}{\dfrac{50}{\sigma_2^2}} = \dfrac{\sigma_2^2}{2 \cdot \sigma_1^2},$$

so

$$P\left(0.358576 < \dfrac{\sigma_2^2}{2 \cdot \sigma_1^2} < 2.46766\right) = 0.95,$$

which reduces to

$$P\left(0.71715 < \dfrac{\sigma_2^2}{\sigma_1^2} < 4.9353\right) = 0.95.$$

Since the ratio 1 is contained in the confidence interval, we conclude that it is possible, despite the fact that one of the sample variances is twice the other, that the population variances could, in fact, be equal. Note that other ratios for the variances are also included in this confidence interval. ∎

EXERCISES 3.7

1. Use Table 4 in Appendix C to find lower 5% points on the following F distributions.
   a. $F(13, 15)$
   b. $F(20, 25)$
   c. $F(10, 14)$
   d. $F(14, 25)$

2. A random sample of 19 observations is selected from a normal distribution with $\sigma^2 = 9$. Find the probability that the variance of the sample, s^2, is between 4 and 16.

3. Resistances supplied by a manufacturer are claimed to be normal with $\mu = 200$ ohms and variance $\sigma^2 = 81$ ohms^2. Find the probability that s^2, the variance of a sample of size 12, is greater than 112.03.

4. A sample of 25 bulbs is taken from a large lot of 40-watt bulbs. The standard deviation of the sample is 1544.7 hours. Find a 90% confidence interval for the true variance, σ^2.

5. Records are made of the thickness of coatings on steel sheets. The data are in pounds per box. A sample of 5 from a plant's production gave $\bar{x} = 4.09$ lb and $s = 31.63$ lb. Find a 95% confidence interval for σ^2.

6. An experiment compared two different processes for producing steel. A sample of size 5 gave $s^2 = 102.4$. Find the probability that a sample of size 8 will have a sample variance 16 times that of the first sample. Assume that the true variances for each process are equal.

7. The variance of an industrial measurement is known to be 225. If a random sample of 14 measurements is taken, what is the probability that the sample variance is twice the true variance?

8. Thirteen batch yields of a plastic produced using a catalyst (Method 1) are to be compared with 13 batch yields of the plastic produced without the catalyst (Method 2). The data are as follows.

   Method 1: $n_1 = 13, \quad \bar{x}_1 = 6.40, \quad s_1^2 = 2.4264$
   Method 2: $n_2 = 13, \quad \bar{x}_2 = 6.02, \quad s_2^2 = 1.9560$

   a. Find a 90% confidence interval for σ_1^2/σ_2^2.
   b. Based on the result of part (a), would it be sensible to conclude that the population variances are in fact equal?

9. A sample of 10 brand A light bulbs showed a sample mean lifetime of 1400 hours and a sample standard deviation of 120 hours. A sample of 12 brand B light bulbs showed a sample mean of 1250 hours and a sample standard deviation of 80 hours.
  a. Find a 98% confidence interval for σ_A^2/σ_B^2.
  b. On the basis of the confidence interval in part (a), is it likely that $\sigma_A^2 = 2 \cdot \sigma_B^2$?

10. In testing the strength of chains under heat treatments A and B, the following two sets of strengths were found.

A: 3610, 3610, 3671

B: 3560, 3400, 3399

  a. Find a 90% confidence interval for σ_A^2/σ_B^2.
  b. On the basis of the confidence interval in part (a), do you think it likely that $\sigma_A^2 = \sigma_B^2$?

COMPUTER EXPERIMENT

Select two sets of 200 samples, each of size 25, from a normal distribution with mean 0 and variance 24. For each sample, calculate the sample variance, and for each pair of samples, calculate s_1^2/s_2^2.

a. Show a histogram of values of $(n-1) \cdot s^2/\sigma^2$ for all 400 samples and compare the result to a chi-squared distribution with 24 degrees of freedom.

b. For the 200 pairs of samples, show a histogram of the statistic s_1^2/s_2^2 and compare this with a graph of the $F(24,24)$ distribution.

3.8 Hypothesis Testing

We have considered many problems of point and interval estimation involving unknown population parameters such as means and variances, and we have used the principles of unbiasedness and maximum likelihood. There are other principles used in estimation, but we will not discuss them here. We now consider problems in which we must come to some conclusion about unknown population parameters; such decision-making processes in statistics are known as **hypothesis testing.**

We begin with two examples, the first for a discrete random variable and the second for a continuous random variable. Each example has some aspects which are artificial and are discussed here so that the structure of the decision-making process can be brought out clearly. Then, having established some ideas, we will turn to problems of a more practical nature.

Example 3.8.1 An issue about to be voted on in an election is known to be favored by either 50% of the voters or 80% of the voters. A sample survey of 20 voters is conducted before the election to determine which is the case.

The statements that 50% of the voters favor the issue or 80% of the voters favor the issue are called **hypotheses.** They are conjectures or suppositions concerning a parameter of a probability distribution. In this case, we assume a binomial model with $n = 20$ and $p = 0.50$ or $p = 0.80$. We must decide, then, on the basis of our sample which statement is correct. We will call the statement $p = 0.50$ the **null hypothesis** and the statement $p = 0.80$ the **alternative hypothesis.** If we assume a binomial model, the conjectures or hypotheses concern the value of p in that model. The hypotheses are then statements about parameters of a probability distribution or density. These hypotheses will be abbreviated this way:

$$H_0: p = 0.50$$
$$H_1: p = 0.80.$$

The process of deciding which hypothesis to believe is called **testing the hypothesis.**
In considering problems in probability, we used the sample space, which showed all
the possible outcomes when an experiment involving randomness was performed. In
this case we will make a similar consideration by showing all the possibilities and their
probabilities. If X is the random variable denoting the number of voters of the 20 voters
interviewed who favor the issue, then X is binomial and

$$P(X = x) = \binom{20}{x} p^x (1 - p)^{20-x} \quad \text{for } x = 0, 1, 2, \ldots, 20$$

where p is either 0.50 or 0.80. We show a table showing each possibility in Table 3.1.
(The probabilities were calculated using Minitab; the columns do not sum exactly to 1
due to round-off error.)

x	$p = 0.50$	$p = 0.80$
0	0.0000	0.0000
1	0.0000	0.0000
2	0.0002	0.0000
3	0.0011	0.0000
4	0.0046	0.0000
5	0.0148	0.0000
6	0.0370	0.0000
7	0.0739	0.0000
8	0.1201	0.0001
9	0.1602	0.0005
10	0.1762	0.0020
11	0.1602	0.0074
12	0.1201	0.0222
13	0.0739	0.0546
14	0.0370	0.1091
15	0.0148	0.1746
16	0.0046	0.2182
17	0.0011	0.2054
18	0.0002	0.1369
19	0.0000	0.0576
20	0.0000	0.0115

Table 3.1

Now consider the result of the decision process. We say, if we decide that H_0 is true, that we *accept* H_0 and, equivalently, we *reject* H_1. If we *reject* H_0, then we *accept* H_1.

The result of the sample survey will be some value of X. We must decide in advance which values of X cause us to accept the null hypothesis and which values of X cause us to reject the null hypothesis. The values of X that cause us to reject H_0 comprise what we call the **critical region** for the test. In this case, large values of X are more likely to come from a distribution with $p = 0.80$ than from a distribution with $p = 0.50$. An examination of the values in Table 3.1 shows that

$$P(X \geq 15 \text{ if } p = 0.80) = 0.8042$$

while

$$P(X \geq 15 \text{ if } p = 0.50) = 0.0207.$$

So it is reasonable to conclude that if $X \geq 15$, then $p = 0.80$.

Now the result of this action is not certain. We calculated that $P(X \geq 15 \text{ if } p = 0.50) = 0.0207$, so if we use the values of $X \geq 15$ as the critical region, we will reject the null hypothesis if it is true about 2% of the time. This is called a **Type I error,** and its size is denoted by α.

$$\alpha = P(H_0 \text{ is rejected if it is true})$$

α is often called the **size** or the **significance level** of the test.

There is one more possibility that will lead to an incorrect decision: Suppose the survey results in a value of $X < 15$ but the alternative hypothesis, $p = 0.80$, is true. Under these circumstances we will incorrectly accept the null hypothesis. The values in Table 3.1 show that $P(X < 15 \text{ if } p = 0.80) = 0.1958$. So almost 20% of the time, we will conclude that $p = 0.50$ when in fact $p = 0.80$. This is called a **Type II error,** and its size is denoted by β.

$$\beta = P(H_0 \text{ is accepted if it is false})$$

In this case, with the critical region $X \geq 15$, we find $\alpha = 0.0207$ and $\beta = 0.1958$.

Note that α and β are calculated under quite different assumptions: the calculation of α presumes the null hypothesis is true and the calculation of β presumes the null hypothesis is false, so they bear no particular relationship to one another.

It is of course possible to decrease α by reducing the critical region to, say, $X \geq 16$. This produces $\alpha = 0.0059$, but, unfortunately, the Type II error increases to 0.3704. The only way to decrease both α and β simultaneously is to increase the sample size.   ■

We now apply some ideas established in this example to a continuous case.

Example 3.8.2   It is important that cables used in an elevator have a mean breaking strength of 20,000 pounds. A sample of 30 cables from a manufacturer is tested and the mean breaking strength is recorded. Here we form the null and alternative hypotheses:

$$H_0: \mu = 20,000$$

$$H_1: \mu \neq 20,000.$$

The statistic calculated from the sample is $\overline{X}$. The central limit theorem assures us that

$$\overline{X} \sim N\left(\mu, \frac{\sigma}{\sqrt{n}}\right)$$

where n is the sample size, 30 in this case, and σ is the population standard deviation. We will assume for illustrative purposes in this example that σ is known. This may seem somewhat artificial, but it is, nonetheless, a useful starting point. Our purpose now is to explore some ideas, and we will deal with this nontrivial issue later in the chapter. The solution will involve the Student t distribution considered in Section 3.4. So, assuming that $\sigma = 2500$ lb,

$$\overline{X} \sim N\left(\mu, \frac{2500}{\sqrt{30}}\right) = N(\mu, 456.44).$$

Where should the critical region be placed? In this case, since if, H_1, is true the mean breaking strength could be either under or over 20,000 lb, it appears that the critical region should be in either extreme of the normal curve. These areas are shown shaded in Figure 3.11. We arbitrarily choose the regions $\overline{X} > 20,800$ and $\overline{X} < 19,200$ for the critical region for the test. The size of the Type I error is calculated to be

$$\alpha = P(\overline{X} > 20,800 \text{ or } \overline{X} < 19,200 \text{ if } \mu = 20,000)$$

$$= P\left(Z > \frac{20,800 - 20,000}{456.44} = 1.75\right)$$

$$+ P\left(Z < \frac{19,200 - 20,000}{456.44} = -1.75\right) = 0.0801183.$$

So α is about 8% for this critical region. ∎

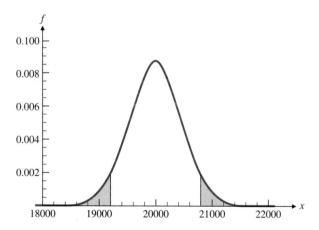

Figure 3.11 Normal Curve for Example 3.8.2

Finding a Critical Region for a Given α

In Example 3.8.2, we very arbitrarily selected the critical region as $\overline{X} > 20,800$ or $\overline{X} < 19,200$ and found that α is about 8%. Our purpose was to establish some ideas about critical regions and the relation of α to the critical region, but such arbitrary selections are seldom used in practice. It might be more natural to select a value of α and then determine the critical region from this value. This allows the experimenter to determine the size of the Type I error in advance and hence to control the size of this type of risk.

To illustrate the procedure, let us select α to be 5%. Since the alternative is two-sided ($\mu \neq 20,000$), we split α in half and put 2.5% in each of the extremities, or tails, of the distribution shown in Figure 3.11. Suppose now that the critical region consists of those points for which $\overline{X} > a$ or $\overline{X} < b$. We must determine a and b. From $P(\overline{X} > a) = 0.025$, we have

$$P\left(\frac{\overline{X} - 20,000}{456.44} > \frac{a - 20,000}{456.44}\right) = P\left(Z > \frac{a - 20,000}{456.44}\right) = 0.025.$$

Since $P(Z > 1.96) = 0.025$, it follows that $(a - 20,000)/456.44 = 1.96$, from which it is found that $a = 20,895$. Since b must be the same distance from the mean that a is, it follows that $b = 19,105$. This is the solution to $(b - 20,000)/456.44 = -1.96$ also. So a test with $\alpha = 5\%$ has the critical region $\overline{X} > 20,895$ or $\overline{X} < 19,105$.

β and the Power of a Test

What is β, the size of the Type II error, in this example? We recall that

$$\beta = P(H_0 \text{ is accepted if it is false})$$

or

$$\beta = P(H_0 \text{ is accepted if the alternative is true}).$$

We could calculate β easily in Example 3.8.1, since in that case we had a specific alternative to deal with (namely, that $p = 0.80$). In this case, however, we have an infinity of alternatives ($\mu \neq 20,000$) with which to deal. The size of β depends upon which of these specific alternatives is chosen. We show two examples. We use the notation $\beta_{\text{alt.}}$ to denote the value of β when a particular alternative is selected.

First, consider

$$\beta_{21,000} = P(19,105 < \overline{X} < 20,895 \text{ if } \mu = 21,000)$$

$$= P(-4.15 < Z < -0.23 \text{ if } \mu = 21,000)$$

$$= P(Z < -0.23 \text{ if } \mu = 21,000) = 0.409046,$$

since $P(Z < -4.15) \doteq 0$.

Now if we consider

$$\beta_{18,500} = P(19,105 < \overline{X} < 20,895 \text{ if } \mu = 18,500)$$

$$= P(1.32548 < Z < 5.24713) = P(1.326 < Z) = 0.0924198,$$

we see that β is highly dependent upon the alternative.

To illustrate this dependence, we show a graph of $\beta_{\text{alt.}}$ as a function of the alternative in Figure 3.12. This curve is often called the **operating characteristic curve** for the test. This curve is not a normal curve; in fact, it has no algebraic equation, each point on it being calculated as we have done in the previous two examples. ($\beta_{\text{alt.}}$ of course indicates the size of a Type II error as a function of the alternative.)

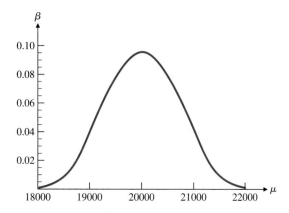

Figure 3.12 Operating Characteristic Curve for Example 3.8.2

We know that $\beta_{\text{alt.}} = P(\text{accept } H_0 \text{ if it is false})$, so $1 - \beta_{\text{alt.}} = P(\text{reject } H_0 \text{ if it is false})$. This is the probability that the null hypothesis is correctly rejected. $1 - \beta_{\text{alt.}}$ is called the **power of the test** for a specific alternative.

A graph of the power of the test is shown in Figure 3.13. Figure 3.13 also shows the power of the test if the sample size were increased from 30 to 100. The graph indicates that the sample of 100 is more likely to reject a false H_0 than is the sample of 30 and that the power of the test increases sharply as the distance between the alternative and the null hypothesis increases.

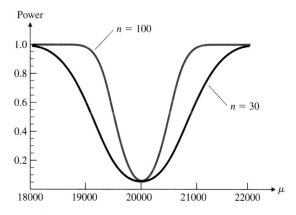

Figure 3.13 Power Curves for Example 3.8.2

We discuss one more idea before finishing this section.

p-Value for a Test

We have discussed selecting a critical region for a test in advance and the disadvantages of proceeding in that way. We abandoned that approach for what would appear to be a more reasonable one: that is, selecting α in advance and calculating the critical region that results. Selecting α in advance puts a great burden upon the experimenter: How is the experimenter to know what value of α to choose? Should 5% or 6% or 10% or 22% be selected? The choice often depends upon the sensitivity of the experiment itself. If the experiment involves a drug to combat a disease, then α should be very small, but, if the experiment involves a component of a nonessential mechanical device, then a somewhat larger value of α might suffice. Now we abandon that approach as well, only to confront a new problem: If we do not have a critical region, and if we do not have α either, then how can we proceed?

Suppose, to be specific, that we are testing

$$H_0: \mu = 22 \quad \text{against} \quad H_1: \mu \neq 22$$

with a sample of $n = 25$ and we know that $\sigma = 5$. The experimenter reports that the observed $\overline{X} = 23.72$. We could calculate the probability that a sample of 25 would give this result, or a result greater than 23.72, if the true mean were 22. This is found to be

$$P(\overline{X} \geq 23.72 \text{ if } \mu = 22)$$
$$= P(Z \geq 1.72)$$
$$= 0.0427162.$$

Since the test is two-sided, the phrase "a result greater than" is interpreted to mean $P(|Z| \geq 1.72) = 2 \cdot 0.0427162 = 0.085432$. This is called the **p-value for the test.**

The p-value allows the experimenter to make the final decision—to accept or to reject the null hypothesis, depending entirely upon the size of this probability. If the p-value is very large, one would normally accept the null hypothesis, while if it is very small, one would know that the result is in one of the extremities of the distribution and would therefore reject the null hypothesis. The decision, of course, is up to the experimenter.

Following is a set of rules governing the calculation of the p-value. We assume that z is the observed value of Z and that the null hypothesis is $H_0: \mu = \mu_0$.

Alternative	p-value
$\mu > \mu_0$	$P(Z > z)$
$\mu < \mu_0$	$P(Z < z)$
$\mu \neq \mu_0$	$P(|Z| > z)$

P-values have become popular because they can be easily computed. There are great limitations to the use of tables, because they generally allow only approximations to p-values. Statistical computer programs commonly calculate p-values.

EXERCISES 3.8

1. A quality control engineer is asked to inspect a large shipment of computer chips. A sample of 100 chips is selected. The chip manufacturer claims that 10% of its chips may not be functional.

  a. To test H_0: $p = 0.10$ against H_1: $p > 0.10$, the critical region $X > 12$ is chosen where X represents the number of chips in the sample that are not functional. Find α for this test.

  b. Now suppose that a critical region of size 0.04 is desired. Find an appropriate critical region in terms of X.

  c. Find the size of the Type II error, β, for the alternative $p = 0.13$, using the critical region $X > 12$.

2. A production process is regarded as satisfactory if a sample of 100 items yields 20 or fewer unusable items.

  a. What is the probability that the production line is judged to be satisfactory on the basis of the sample if in fact 15% of its items are unusable?

  b. What is the probability that the production line is judged to be unsatisfactory on the basis of the sample if in fact 10% of its items are unusable?

  c. Choose appropriate hypotheses and then identify each of the errors in parts (a) and (b) of this problem as α or β.

3. A drug is thought to be 60% effective for the cases in which it is used. A modified form of the drug is tested in the hope of relieving a larger percentage of the cases. The modified drug is tested on 15 patients.

  a. If it is decided to reject H_0: $p = 0.6$ in favor of H_1: $p > 0.6$ provided 13 or more patients improve, find α, the size of the Type I error.

  b. If the modified form of the drug is in fact 80% effective, what is the probability of committing a Type II error?

4. Specifications for a certain type of computer printer ribbon call for a mean breaking strength of 185 pounds. In order to monitor the process, a random sample of 30 pieces is taken each hour and the sample mean is used to decide if the mean breaking strength has shifted. The test is of the null hypothesis H_0: $\mu = 185$ against the alternative hypothesis H_1: $\mu < 185$. Here α is chosen as 0.05 and it can be assumed that $\sigma = 10$ lb.

  a. Find the critical region for the test in terms of $\overline{X}$.

  b. Find β for the alternative $\mu = 179.5$.

5. In a power-generating plant, pressure in a certain line is supposed to maintain an average of 100 psi over any 4-hour period. If the average pressure exceeds 103 psi in this period, some serious complications can evolve. During a 4-hour period, 30 random measurements are taken. For testing H_0: $\mu \leq 103$ against H_1: $\mu > 103$, α is to be 0.01.

  a. If $\sigma = 4$, find the critical region in terms of $\overline{X}$.

  b. Find β for the alternative $\mu = 104$.

6. Prestressing wire for wrapping concrete pipe is manufactured in large rolls. A quality control inspector required five specimens from a roll to be tested for ultimate tensile strength (UTS). The UTS measurements (in 1000 psi) were: 253, 261, 258, 255, and 256.

  a. Assuming that $\sigma = 2$ and $\alpha = 0.01$, decide whether to accept or reject H_0: $\mu \leq 255$ against H_1: $\mu > 255$.

  b. Calculate the size of β for the alternative $\mu = 258$.

7. A sample of 16 from a distribution whose variance is known to be 900 is used to test H_0: $\mu = 350$ against the alternative H_1: $\mu > 350$. The critical region is $\overline{X} > 365$.

  a. Calculate α for this test.

  b. Find β for the alternative $\mu = 372.5$.

8. The breaking strengths of steel wires used in elevator cables are assumed to come from a population with known $\sigma = 400$ lb. Before accepting a shipment of these steel wires, an engineer wants to be confident that $\mu > 10,000$ lb. A sample of 16 wires is selected and their mean breaking strength, $\overline{X}$, is measured.

  a. Find the critical region for testing H_0: $\mu = 10,000$ against H_1: $\mu < 10,000$ using $\alpha = 0.05$.

  b. Find the size of the Type II error for the alternative $\mu = 9800$.

  c. Find the p-value for the test if the sample mean is $\overline{x} = 9906$.

9. The manufacturer of a copying machine claims that the mean time to repair breakdowns is 93 minutes. To test this claim, 23 breakdowns of a model were observed, resulting in a mean repair time of 98.8 min. If it can be assumed that $\sigma = 26$ min,

  a. test H_0: $\mu = 93$ against H_1: $\mu > 93$ with $\alpha = 5\%$.

  b. find β for the alternative $\mu = 95$.

  c. find the p-value for the test.

10. A soft-drink machine is regulated so that the amount of drink dispensed should have $\mu = 200$ ml and $\sigma = 15$ ml. A sample of nine drinks is taken and gives $\bar{x} = 192.5$.

   a. Test $H_0: \mu = 200$ against $H_1: \mu \neq 200$ with $\alpha = 5\%$.

   b. Find the p-value for the test.

   c. Find β for the alternative $\mu = 215$.

11. Plot the operating characteristic curve for the situation in Exercise 10.

12. Plot the power curve for the situation in Exercise 9.

COMPUTER EXPERIMENT

In a binomial experiment, researchers want to test $H_0: p = 0.4$ against $H_1: p \neq 0.4$.

a. Using the probability distribution for $n = 100$ trials, find a critical region of size approximately 5%.

b. Plot the power curve for the test for a variety of alternatives.

c. Increase the sample size to $n = 200$ and plot the power curve for this sample size.

3.9 Confidence Intervals and Tests of Hypotheses

Confidence intervals are closely related to tests of hypotheses, as the following example will show.

Example 3.9.1 A sample of 25 observations from a normal distribution with $\sigma^2 = 100$ has a sample mean $\bar{x} = 342$. Use these data to test

$$H_0: \mu = 350 \quad \text{against} \quad H_1: \mu \neq 350.$$

Choosing $\alpha = 0.05$ in advance, the critical region becomes $|Z| > 1.96$ or

$$\left| \frac{\bar{X} - 350}{10/\sqrt{25}} \right| > 1.96,$$

giving $\bar{X} > 350 + 1.96 \cdot 2 = 353.92$ or $\bar{X} < 350 - 1.96 \cdot 2 = 346.08$. So, since $\bar{X} = 342 < 346.08$, the null hypothesis is rejected, although narrowly.

   We also notice that a 95% confidence interval for μ is the interval from $\bar{X} - 1.96 \cdot (\sigma/\sqrt{n}) = 342 - 1.96 \cdot 2 = 338.08$ to $\bar{X} + 1.96 \cdot (\sigma/\sqrt{n}) = 342 + 1.96 \cdot 2 = 345.92$.

   The 95% confidence interval $(338.08, 345.92)$ is the acceptance region for the hypothesis with $\alpha = 5\%$, and $\bar{x} = 342$ is not in this interval. ■

   This example is perfectly typical of the general situation, the confidence interval is the same as the acceptance region for the test of the hypothesis. Confidence intervals and tests of hypotheses are then equivalent in the sense that the same conclusions can be drawn from either.

3.10 Tests on μ, σ Unknown

We used Example 3.8.2 to introduce some ideas about hypothesis testing in general, but we said at the beginning of Section 3.8 that the examples given there have some artificial aspects. Our test of H_0: $p = 0.50$ against H_1: $p = 0.80$ involved only two values for p, but in most cases there is a range of values for p. In testing hypotheses on a single mean we assumed that σ was known. This is usually not the case. We will now show more examples, having established some ideas. First we remove the restriction that σ is known.

The Student t distribution, introduced in Section 3.4, can be used to test hypotheses about an unknown population mean if the population standard deviation is unknown with the restriction that the sampling now must be done from a normal distribution.

Example 3.10.1 A sample of 25 observations from a normal distribution has $\bar{x} = 126.7$ and $s = 24.2$. Use these data with $\alpha = 5\%$ to test

$$H_0: \mu = 115 \quad \text{against} \quad H_1: \mu \neq 115.$$

This is an example of using the Student t distribution to test a hypothesis. We cannot find a confidence interval here that is independent of the unknown population standard deviation, σ. We do know, however, that the null hypothesis is rejected if

$$|t_{24}| > 2.064.$$

In this case,

$$t_{24} = \frac{\bar{x} - \mu}{s/\sqrt{n}} = \frac{126.7 - 115}{24.2/\sqrt{25}} = 2.4174,$$

so the null hypothesis is rejected.

We could also calculate the p-value for the test. This is $P(|t_{24}| > 2.4174)$. One can only come to the conclusion that p is a little more than 2% from Table 2 in Appendix C. Mathematica calculates this exactly as $2 \cdot 0.0117974 = 0.0235948$. One would most probably reject the null hypothesis based on this calculation. ∎

Type II Errors and Sample Size

While finding β or determining the sample size given values of α and β are relatively straightforward tasks if the population variance is known, they become quite difficult matters when the population variance is unknown. For this reason, we delay considering the determination of β or the sample size for given values of α and β until Section 3.14.

Large Samples   Table 2 in Appendix C gives only a few values of the distribution function for the Student t distribution. If the table is to be relied upon, one has a very limited range of choices for α; in addition to this, the number of degrees of freedom is limited to 40. It is true as the sample size, n, increases that

$$t_{n-1} = \frac{\overline{X} - \mu}{S/\sqrt{n}} \longrightarrow \frac{\overline{X} - \mu}{\sigma/\sqrt{n}} = Z,$$

so we can make some approximate use of the normal tables when the test statistic is a Student t in reality. Computers, however, render this approximation unnecessary since any number of degrees of freedom can be used and p-values can be exactly calculated. To illustrate the difference between the actual values of t and the normal approximation, Table 3.2 shows the upper 0.025 points on t distributions with the indicated number of degrees of freedom. The approximating z value is 1.95996.

n	t_n
40	2.02108
50	2.00856
60	2.00030
70	1.99444
80	1.99006
90	1.98667
100	1.98397
⋮	⋮
2000	1.96115

Table 3.2

It is revealing to find that even for 2000 degrees of freedom, the approximating t value still does not exactly match the z value. The distinctions can be important when working with close tolerances.

Occasionally one is met with the argument that if we indeed had 2000 degrees of freedom, then the sample standard deviation would approximate the population standard deviation quite closely, so a normal approximation is highly justified. In that case one could, with similar logic, argue that the sample mean and the population mean were quite close, rendering the whole statistical exercise pointless.

EXERCISES 3.10

1. A researcher is investigating the weights of male college students. He wants to test H_0: $\mu = 68$ kg against H_1: $\mu \neq 68$ kg. A sample of 16 has $s = 4$ and $\bar{x} = 68.90$. Assume that the weight of male college students is normally distributed.
   a. Does the sample support the hypothesis?
   b. Find the p-value for the test.

2. Past experience indicates that the time for college sophomores to complete a standardized placement test is a normal random variable. A sample of 25 students gave a mean time to complete the test of 27.4 min with standard deviation 4 min.
   a. Using $\alpha = 5\%$, test H_0: $\mu = 25$ min against H_1: $\mu \neq 25$ min.

   b. Find the p-value for the test.
   c. What assumptions must be made regarding the data for the test in part (a) to be valid?

3. A manufacturer of sports equipment has developed a new synthetic fishing line that is claimed to have a mean breaking strength of 8 kg. To test H_0: $\mu = 8$ kg against H_1: $\mu \neq 8$ kg, a sample of 12 lines is tested. The sample had a mean breaking strength of 7.8 kg and a standard deviation of 0.5 kg.
   a. Is the manufacturer's claim supported by the sample if $\alpha = 5\%$?
   b. Find the p-value for the test.

4. A tensile test was performed to determine the strength of a particular adhesive for an assembly joining

pieces of glass. The data are: 16, 14, 19, 18, 19, 20, 15, 18, 17, 18.

a. Use the data to test H_0: $\mu = 19$ against H_1: $\mu \neq 19$ using $\alpha = 0.01$.

b. Find the p-value for the test.

5. It is crucial that fire prevention sprinkler systems become active quickly in case of fire. The times of activation for a series of tests were (in seconds): 27, 41, 22, 27, 23, 35, 30, 33, 24, 27, 28, 22, 24.

a. The system was designed so that the true mean activation time is at most 25 sec. Choosing α to be 5%, does the sample support this claim?

b. Find the p-value for the test.

COMPUTER EXPERIMENT

Select 200 samples, each of size 6, from a normal population with mean 20 and standard deviation 3. Calculate the sample mean and the sample standard deviation for each sample. Then calculate the statistic $(\overline{x} - \mu)/(s/\sqrt{n})$ for each sample. Construct a histogram of the results and compare the histogram to a Student t distribution with 5 degrees of freedom.

3.11   Tests on Proportions and Variances

Since we have established probability distributions for proportions, variances, and the ratio of variances, and since we have shown confidence intervals for these, we now show some examples of hypothesis tests regarding proportions and variances.

Example 3.11.1   A sample of 24 items selected from a lot of light bulbs was tested. Six of the bulbs did not last the "guarantee" time advertised. Test

$$H_0: p = 0.90 \quad \text{against} \quad H_1: p < 0.90$$

where p is the true proportion of bulbs lasting the guaranteed time.
    The statistic

$$\frac{p_s - p}{\sqrt{p \cdot q / n}} \doteq z.$$

Since $p_s = 18/24 = 3/4$ is a little above the upper limit of the range $(0.3, 0.7)$, it is sensible to replace p and q in the denominator with their sample approximations. (Recall from Section 3.3 that in the range from 0.3 to 0.7, $p \cdot q \doteq 1/2$.) This gives

$$z \doteq \frac{18/24 - 0.90}{\sqrt{\dfrac{18/24 \cdot 6/24}{24}}} = -1.69706,$$

and since $P(Z < -1.69706) = 0.0448427$, the null hypothesis most probably would be rejected.   ■

Example 3.11.2   Two drugs are being tested with respect to their effectiveness in combating a disease. Of 40 patients given drug A, 75% improved while 65% of 20 patients given drug B improved. Can the manufacturer of drug A claim that at least 20% more patients improve using drug A rather than drug B?

We test the null hypothesis

$$H_0: p_A - p_B \geq 0.20 \quad \text{against} \quad H_1: p_A - p_B < 0.20.$$

The test statistic here is

$$\frac{(p_{s_A} - p_{s_B}) - (p_A - p_B)}{\sqrt{\dfrac{p_A \cdot q_A}{n_A} + \dfrac{p_B \cdot q_B}{n_B}}} \doteq z.$$

It will not make an appreciable difference if the proportions in the denominator are replaced by $1/2$ or by their sample estimates. The latter are used to find that

$$\frac{(0.75 - 0.65) - 0.20}{\sqrt{\dfrac{0.75 \cdot 0.25}{40} + \dfrac{0.65 \cdot 0.35}{20}}} = -0.78903,$$

producing a p-value of 0.215047, so the null hypothesis is accepted.

A 95% confidence interval will give us a range of acceptable hypotheses as

$$(0.75 - 0.65) - 1.96 \cdot \sqrt{\frac{0.75 \cdot 0.25}{40} + \frac{0.65 \cdot 0.35}{20}} \quad \text{to}$$

$$(0.75 - 0.65) + 1.96 \cdot \sqrt{\frac{0.75 \cdot 0.25}{40} + \frac{0.65 \cdot 0.35}{20}}$$

or the interval $(-0.14841, 0.34841)$.

This range is so wide that the manufacturer of drug B could claim that it is more effective than drug A, a consequence of the small sample sizes for the importance of the tests conducted. ∎

Example 3.11.3   The length of a component in a manufactured product is of crucial importance if the assembly of the product is to proceed swiftly. The manufacturer would like to control the variance of the lengths if possible. A sample of 35 parts is selected and it is found that the sample variance is $s^2 = 16.09$. If the true variance is at least 15, the manufacturing process must undergo a costly change. We test

$$H_0: \sigma^2 < 15 \quad \text{against} \quad H_1: \sigma^2 \geq 15.$$

The test statistic in this case is

$$\frac{(n-1) \cdot S^2}{\sigma^2} = \chi^2_{n-1}.$$

We have then that $\chi^2_{34} = 34 \cdot 16.09/15 = 36.4707$. The rejection region is in the right tail of the chi-squared distribution; the critical 5% point is 48.6024, so the null hypothesis can be accepted here. The p-value for the test is $P(\chi^2_{34} > 36.4707) = 0.354525$, confirming our earlier conclusion. ∎

We conclude this section with an example regarding two variances. This will become a particularly important test for Section 3.13 when we consider comparing two samples.

Example 3.11.4 Two samples gave the following data: $n_1 = 22$, $s_1^2 = 104.78$, $n_2 = 13$, $s_2^2 = 219.44$. The fact that one sample variance is more than twice the other sample variance leads us to test the hypothesis:

$$H_0: \sigma_1^2 = \sigma_2^2 \quad \text{against} \quad H_1: \sigma_1^2 \neq \sigma_2^2.$$

Our test statistic is

$$F(n_1 - 1, n_2 - 1) = \frac{s_1^2/\sigma_1^2}{s_2^2/\sigma_2^2} = \frac{s_1^2}{s_2^2}.$$

In this case, we have $F(21, 12) = 104.78/219.44 = 0.47749$. The critical values, using $\alpha = 5\%$, calculated using Mathematica, are 0.379253 and 3.05746. The p-value for the test is 0.133165. Either consideration leads to the acceptance of the null hypothesis. ∎

Determining β, the size of the Type II error, in tests on one or two variances is not very easy since the distributions involved, known as noncentral distributions, are difficult to deal with. We will not show examples of these except for one noncentral distribution in Section 3.14.

Tests on the Difference Between Means Since we have compared proportions and variances, it is natural to inquire about tests of hypotheses concerning the difference between means arising from two samples. This important situation is more difficult than the ones we have considered so far; we will discuss it later, in Section 3.13.

EXERCISES 3.11

1. Times for high school seniors to complete a standardized test are known to be normally distributed. Using $\alpha = 5\%$, test $H_0: \sigma = 6$ against $H_1: \sigma < 6$ if a sample of 20 students has a sample standard deviation of 4.51.

2. A commonly prescribed drug to relieve nervous tension is believed to be only 60% effective. A sample of 100 patients showed that 70 received relief. Test $H_0: p = 0.6$ against $H_1: p > 0.6$ with $\alpha = 1\%$. What can be concluded?

3. An experiment to compare the viscosity of leaded and unleaded gasoline gave the following results.

    Leaded: $n_1 = 25$, $\bar{x}_1 = 35.84$, $s_1^2 = 116$

    Unleaded: $n_2 = 21$, $\bar{x}_2 = 38.42$, $s_2^2 = 57$

    Does the sampling support $H_0: \sigma_1^2 = \sigma_2^2$ against the alternative, $H_1: \sigma_1^2 \neq \sigma_2^2$ if $\alpha = 1\%$?

4. The following data represent running times of films produced by two motion picture companies.

    A: 102, 86, 98, 109, 92

    B: 81, 165, 97, 134, 92, 87, 114

    a. Test $H_0: \sigma_A^2 = \sigma_B^2$ against the alternative, $H_1: \sigma_A^2 \neq \sigma_B^2$ if $\alpha = 1\%$.

    b. What assumptions must be made for the test in part (a) to be valid?

5. Two different types of injection molding machines are used to form plastic bottles. Two random samples, each of size 500, are selected. The sample from machine I produces 21 defective bottles, and the sample from machine II produces 32 defective bottles. Using $\alpha = 5\%$, test $H_0: p_I = p_{II}$ against the alternative, $H_1: p_I \neq p_{II}$ where p_I and p_{II} are the true proportions of defectives produced by the two machines, respectively.

6. The standard deviation of the modulus of elasticity of a type of rubber is thought to be 18. The standard deviation of a sample of size 20 is found to be 23.3. Test H_0: $\sigma = 18$ against H_1: $\sigma > 18$ using $\alpha = 5\%$.

7. Nicotine determinations were made on each of four standard units of tobacco at each of two laboratories, A and B. The data are:

$$A: 26, 24, 28, 27$$

$$B: 28, 31, 23, 29.$$

Test H_0: $\sigma_A^2 = \sigma_B^2$ against the alternative, H_1: $\sigma_A^2 \neq \sigma_B^2$ if $\alpha = 2\%$.

8. Records are kept on the weight, in pounds per box, of tin coating on steel sheets. A sample of five units from plant 1's production gave $s_1 = 4.09$, while a sample of nine units from plant 2's production gave $s_2 = 2.99$. The manufacturer of the coating process is concerned that σ_1^2 could be as great as $2 \cdot \sigma_2^2$. Is this a reasonable concern?

9. In the manufacture of synthetic fiber, the material, still in the form of a continuous flow, is subjected to high temperatures in order to improve its shrinkage properties. The data for tests run at two temperatures are summarized as follows.

$$120°: n_1 = 16, \bar{x}_1 = 13.4, s_1^2 = 49.54$$

$$140°: n_2 = 9, \quad \bar{x}_2 = 36.0, s_2^2 = 127.32$$

Test H_0: $\sigma_1^2 = \sigma_2^2$ against the alternative, H_1: $\sigma_1^2 \neq \sigma_2^2$ if $\alpha = 2\%$.

10. A magazine advertisement claims that people who brush with Super Caries Fighter toothpaste have 18% fewer cavities than those who don't. Persistent research finds that the study is based on a sample of 9 people using Super Caries Fighter of whom only 1 developed a cavity and 12 people who did not use that toothpaste, of whom 3 developed cavities. Is the advertisement correct?

COMPUTER EXPERIMENTS

1. Select 200 samples of size 10 each from a binomial population with $p = 0.20$ and then select 200 samples of size 10 each from a binomial population with $p = 0.30$. In each case, calculate the difference between the sample proportions. Draw a histogram of the results and compare this to a normal curve.

2. Draw 200 samples of size 8 from a normal population with variance 16 and then draw 200 samples of size 8 from a normal population with variance 25. For each pair of samples, calculate the ratio of the sample variances. Draw a histogram of the result and compare the histogram to the appropriate F curve.

3.12 Constructing Best Tests: Neyman-Pearson Lemma

We want to consider more statistical tests in this chapter, but we need to pause and consider the construction of tests in general that have some optimal properties and hence are the "best" tests, in some sense, that can be found. We have given examples of tests on μ and we have used the statistic $\overline{X}$ to do this. While it may seem natural to use $\overline{X}$ in order to perform a test on μ, it is not obvious that this is an optimal thing to do. Perhaps the median, or the sample maximum, or some other statistic should be used for this purpose. We have also shown tests on σ^2, but we have not justified the use of a function of the sample variance in this test. In fact, we have not justified the critical regions used in any of these cases either, but have used only simple intuition. We now wish to show that the critical regions we have used are optimal in a sense, and we wish to show why the test statistics chosen were used as well. We begin with a specific example.

Example 3.12.1 A wheel has a pointer in its center. When the pointer is spun, it stops at one of the real numbers in the interval from 0 to 1; the pointer is spun and the random variable X is the number at which the pointer stops. How the wheel is loaded is not known, however, although it is known that either $f(x) = 2x$, $0 \le x \le 1$, or $g(x) = 2 - 2x$, $0 \le x \le 1$ is the appropriate probability density function. Given a single observation obtained in one spin of the wheel, how can one come to some conclusion about the probability density function?

The density functions are shown in Figure 3.14.

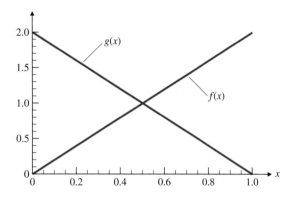

Figure 3.14 Competing Density Functions

As hypotheses, choose, for example,

$$H_0: f(x) = 2x, \qquad 0 \le x \le 1$$
$$H_1: g(x) = 2 - 2x, \qquad 0 \le x \le 1.$$

Since small values of X have more area under $g(x)$ than under $f(x)$, it might appear sensible to reject the null hypothesis if X is small, but it is not obvious that this is an optimal procedure. If α is chosen as 0.05, there are in fact any number of critical regions available. Some of these are considered here.

1. Suppose the critical region chosen is $X \le a$, for some number a. If α is to be 0.05, then $\int_0^a 2x \, dx = 0.05$, which gives $a = 0.2236$. Then the size of the Type II error, β, is $\beta = \int_a^1 (2 - 2x) \, dx = 0.6028$.

2. Now consider the critical region $X \ge b$, again where b is chosen so as to make the critical region of size 0.05. Then $\int_b^1 2x \, dx = 0.05$, giving $b = 0.9747$. Then the size of the Type II error, β, is $\beta = \int_0^b 2 - 2x \, dx = 0.9994$.

3. Finally, consider a critical region from 0.4 to c, where again c is chosen so as to make the critical region of size 0.05. Then $\int_{0.4}^c 2x \, dx = 0.05$, giving $c = 0.4583$. The size of the Type II error in this case is then

$$\beta = \int_0^{0.4} 2 - 2x \, dx + \int_c^1 2 - 2x \, dx = 0.9334.$$

We could continue with these calculations, but if we call the *best* test of size α that test that minimizes the size of the Type II error, β, then we would choose the critical region $X \le 0.2236$ as the best one among the three alternatives investigated since β is smallest there.

It is, in fact, the best critical region that can be chosen among *all* other critical regions of size 0.05. This fact will be shown later. ∎

It is not always possible to determine best tests for hypotheses, but some information is known. The following lemma, presented without proof, allows us to determine best tests when testing a simple hypothesis (which completely specifies a probability distribution or density function) against a simple alternative.

Neyman-Pearson Lemma

In testing the simple hypothesis H_0: $\theta = \theta_0$ against the simple alternative hypothesis H_1: $\theta = \theta_1$, if there exists a region R and a constant k such that

$$\frac{L(X;\theta_0)}{L(X;\theta_1)} \le k \quad \text{inside } R$$

and

$$\frac{L(X;\theta_0)}{L(X;\theta_1)} \ge k \quad \text{outside } R$$

where $L(X;\theta)$ is the likelihood function, then R is the best critical region of size α. (The likelihood function is the function used in Section 3.2.)

For a proof, see Hogg and Craig [15]. We will see that the lemma often allows us to construct best critical regions; the tests constructed form part of a class called **likelihood ratio tests.** We turn now to several examples.

Example 3.12.2   We return to Example 3.12.1.

The Neyman-Pearson Lemma says that the best critical region is that for which

$$\frac{2x}{2 - 2x} \le k.$$

Solving this gives $x \le k/(1 + k)$, indicating that the best critical region is, in fact, small values of X. The constant k is chosen so that the size of the test is α; that is,

$$\int_0^{\frac{k}{1+k}} 2x \, dx = \alpha$$

so that

$$k = \frac{\sqrt{\alpha}}{1 - \sqrt{\alpha}}.$$

If, for example, $\alpha = 0.05$, then $k = 0.288$.

So the critical region is $X \leq k/(1+k) = 0.2236$. This is exactly the result found previously in this section.

We also calculated the size of the Type II error, β, as 0.6028.

The results we found in Example 3.12.1 then are supported by the results given by the Neyman-Pearson Lemma.

The test is obviously not a very sensitive one, since it is based on only one value of X. The Neyman-Pearson Lemma does, however, tell us the test statistic to use as well as the form of the best critical region. ∎

Example 3.12.3 As an example of wider significance than the preceding one, consider a sample $\{x_1, x_2, \ldots, x_n\}$ selected from a normal distribution whose variance, σ^2, is known. The competing hypotheses are:

$$H_0: \mu = \mu_0 \quad \text{and} \quad H_1: \mu = \mu_1$$

where $\mu_1 > \mu_0$.

The Neyman-Pearson Lemma gives the best critical region as that for which

$$\frac{\prod_{i=1}^{n} \left(\frac{1}{\sigma\sqrt{2\pi}}\right) e^{-\frac{1}{2}\left(\frac{x_i-\mu_0}{\sigma}\right)^2}}{\prod_{i=1}^{n} \left(\frac{1}{\sigma\sqrt{2\pi}}\right) e^{-\frac{1}{2}\left(\frac{x_i-\mu_1}{\sigma}\right)^2}} \leq k.$$

This becomes

$$\frac{\left(\frac{1}{\sigma\sqrt{2\pi}}\right)^n e^{-\frac{1}{2}\sum_{i=1}^{n}\left(\frac{x_i-\mu_0}{\sigma}\right)^2}}{\left(\frac{1}{\sigma\sqrt{2\pi}}\right)^n e^{-\frac{1}{2}\sum_{i=1}^{n}\left(\frac{x_i-\mu_1}{\sigma}\right)^2}} \leq k,$$

which simplifies to

$$e^{-\frac{1}{2}\sum_{i=1}^{n}\left(\frac{x_i-\mu_0}{\sigma}\right)^2 + \frac{1}{2}\sum_{i=1}^{n}\left(\frac{x_i-\mu_1}{\sigma}\right)^2} \leq k.$$

Taking logs and simplifying, we see that this is equivalent to

$$\sum_{i=1}^{n}[(x_i - \mu_0)^2 - (x_i - \mu_1)^2] \geq k.$$

(The letter k simply denotes some constant and is not the same k as in the previous formula.) This again simplifies to

$$2\left(\sum_{i=1}^{n} x_i\right)(\mu_1 - \mu_0) \geq k$$

or, since $\mu_1 > \mu_0$,

$$\sum_{i=1}^{n} x_i \geq k$$

or

$$\overline{X} \geq k.$$

This tells us two facts: The best test is indeed based upon the sample mean (whose distribution is known by the central limit theorem), and the best critical region is in the right-hand tail of the normal curve, selected so as to have size α. This is the test we used previously.

Note that the test created here is independent of the particular alternative, μ_1, so the test is exactly the same regardless of the specific alternative selected. When β is minimized, the power of the test is maximized. Since the test minimizes β for any of the alternatives, it is called **uniformly most powerful.** Uniformly most powerful tests are not commonly available for many hypotheses; the example above is an exception to the usual state of affairs. ∎

Example 3.12.4 Suppose X has a $N(0, \sigma)$ distribution where σ is unknown. The Neyman-Pearson Lemma gives the critical region for testing

$$H_0: \sigma^2 = \sigma_0^2 \quad \text{against} \quad H_1: \sigma^2 = \sigma_1^2$$

where $\sigma_1^2 > \sigma_0^2$ as the set of points where

$$\frac{\left(\frac{1}{\sigma_0\sqrt{2\pi}}\right)^n e^{-\frac{1}{2}\sum_{i=1}^n \left(\frac{x_i}{\sigma_0}\right)^2}}{\left(\frac{1}{\sigma_1\sqrt{2\pi}}\right)^n e^{-\frac{1}{2}\sum_{i=1}^n \left(\frac{x_i}{\sigma_1}\right)^2}} \leq k.$$

This can be simplified to

$$\left(\frac{\sigma_1}{\sigma_0}\right)^n e^{-\frac{1}{2}\sum_{i=1}^n \left(\frac{x_i}{\sigma_0}\right)^2 + \frac{1}{2}\sum_{i=1}^n \left(\frac{x_i}{\sigma_1}\right)^2} \leq k$$

for some k, or

$$n \log\left(\frac{\sigma_1}{\sigma_0}\right) - \frac{1}{2}\sum_{i=1}^n x_i^2 \left(\frac{1}{\sigma_0^2} - \frac{1}{\sigma_1^2}\right) \leq k$$

for some k, or, since $\sigma_1^2 > \sigma_0^2$,

$$\sum_{i=1}^n x_i^2 \geq k.$$

To construct a test, we must be able to find k so that the test has a given size, say α. Now x_i/σ_0 has a $N(0, 1)$ distribution so $\sum_{i=1}^n (x_i/\sigma_0)^2$ has a χ_n^2 distribution, and this may be used to construct a test. To be specific, suppose that the hypotheses are:

$$H_0: \sigma^2 = 200 \quad \text{and} \quad H_1: \sigma^2 = 450$$

and that a random sample of size 10 is available. Then the rejection region is

$$\chi_{10}^2 \geq 18.307 \quad (\text{for } \alpha = 5\%)$$

or

$$\sum_{i=1}^{n} \left(\frac{x_i}{\sqrt{200}} \right)^2 \geq 18.307$$

or

$$\sum_{i=1}^{n} x_i^2 \geq 3661.4.$$

The null hypothesis, H_0, is rejected for any sample for which $\sum_{i=1}^{n} x_i^2 \geq 3661.4$. The size of the Type II error, β, is

$$\beta = P \left[\sum_{i=1}^{10} x_i^2 \leq 3661.4 \text{ if } \sigma^2 = 450 \right]$$

$$= P \left[\sum_{i=1}^{10} \left(\frac{x_i}{\sqrt{450}} \right)^2 \leq \frac{3661.4}{450} = 8.1364 \right]$$

$$= 0.384489. \qquad \blacksquare$$

A Word About Hypotheses

We have often used hypotheses and alternatives of the form $H: \theta = \theta_0$. Since such hypotheses completely specify the probability distribution, they are called **simple hypotheses.** Hypotheses of the form $H: \theta \neq \theta_0$ or $H: \theta > \theta_0$ or $H: \theta < \theta_0$ do not completely specify the probability distribution and are called **composite hypotheses.** Uniformly most powerful tests do not as a rule exist for composite hypotheses.

When testing $H: \theta \leq \theta_0$ against $H: \theta > \theta_0$ it is best to presume that $\theta = \theta_0$ in calculating α or a critical region or a p-value, since this hypothesis is closest to the alternative and hence is the most difficult situation to detect.

EXERCISES 3.12

1. a. Find the test statistic and the best critical region for testing $H_0: \mu = 340$ against the alternative $H_1: \mu < 340$ assuming that $X \sim N(\mu, 60)$ and $\alpha = 0.10$.

   b. Is H_0 accepted or rejected by the sample 304, 231, 208, 316, and 312?

2. a. Find the test statistic and the best critical region for testing $H_0: \sigma^2 = 3600$ against the alternative $H_1: \sigma^2 < 3600$ in Exercise 1.

   b. Is H_0 accepted or rejected by the data in Exercise 1?

3. If $f(x) = (1 + \lambda)x^{\lambda}$, $0 < x < 1$, find the test statistic and the best critical region for testing $H_0: \lambda = 5$ against the alternative $H_1: \lambda < 5$ based on a sample $\{x_1, x_2, \ldots, x_n\}$.

4. a. Find the test statistic and the best critical region for testing $H_0: \mu = \mu_0$ against the alternative $H_1: \mu = \mu_1 > \mu_0$ where $f(x) = e^{-\mu} \cdot \mu^x / x!$, $x = 0, 1, 2, 3, \ldots$. (X is called a *Poisson random variable* with parameter μ.)

   b. If $\alpha = 0.064$, is the null hypothesis $H_0: \mu = 3$ accepted or rejected by the sample 3, 8, 2, and 1? (Use the fact that the sum of n Poisson variables with parameter μ is Poisson with parameter $n\mu$.)

5. If $f(x) = (1/\lambda) e^{-x/\lambda}$, $x > 0$, $\lambda > 0$, find the test statistic and the best critical region for testing $H_0: \lambda = \lambda_0$ against the alternative $H_1: \lambda = \lambda_1 < \lambda_0$ based on a sample $\{x_1, x_2, \ldots, x_n\}$.

6. a. Find the test statistic and the best critical region for testing $H_0: p = p_0$ against the alternative $H_1: p = p_1 > p_0$ based on one observation of X, if $f(x) = p \cdot (1-p)^{x-1}$, $x = 1, 2, 3, \ldots, X$ is the waiting time for the first success to occur in a binomial situation with probability p. The random variable X is called a *geometric random variable*.

b. Answer part (a) of this question if two observations of X are available, X_1 and X_2.

7. a. One observation of a binomial random variable with probability p of success is available upon which the hypothesis $H_0: p = p_0$ is to be tested against the alternative $H_1: p = p_1 > p_0$. Construct the best test.

b. Repeat part (a) of this question if a random sample, $\{x_1, x_2, \ldots, x_n\}$, is available.

8. Suppose $\{x_1, x_2, \ldots, x_n\}$ is a random sample from a normal distribution for which μ is known and σ^2 is unknown, and suppose it is desired to test the hypothesis $H_0: \sigma^2 = 2$ against the alternative $H_1: \sigma^2 = 3$.

a. Show that, among all test procedures of size α, β is minimized by that test that rejects H_0 when $\sum_{i=1}^{n}(x_i - \mu)^2 > k$.

b. If $n = 8$ and $\alpha = 0.05$, find the value of k in part (a).

9. A single observation, X, is chosen from $f(x) = \theta x^{\theta - 1}$, $0 < x < 1$.

a. Construct the best test of $H_0: \theta = 1$ against the alternative $H_1: \theta = 2$ if $\alpha \doteq 0.05$.

b. Find β.

10. Let

$$f(x) = 1 + \theta^2 \cdot \left(x - \frac{1}{2}\right), \quad 0 < x < 1, \; 0 < \theta < \sqrt{2}.$$

Find the test statistic and the best critical region for testing $H_0: \theta = 0$ against the alternative $H_1: \theta = 1$ based on a single observation.

11. Two observations, X_1 and X_2, are chosen from $f(x) = ae^{-ax}$, $x > 0$. Find the test statistic and the best critical region for testing $H_0: a = 1$ against the alternative $H_1: a > 1$.

COMPUTER EXPERIMENT

Consider testing $H_0: \mu = 20$ against the alternative $H_1: \mu = 25$ for a normal random variable with variance 5 and critical regions of the form $a \leq \overline{X} \leq b$ where $P(a \leq \overline{X} \leq b) = 0.05$. Calculate β for a wide variety of choices of a and b. Graph β as a function of a and show that the minimum value of β occurs when $b = \infty$.

3.13 Tests on Two Samples

It is often necessary to compare two samples to each other. Recall that in Section 3.5 we compared proportions arising from two samples. We may wish, for example, to compare methods of formulating a product, or to compare different versions of standardized tests to decide if they in fact test equally able candidates equally, or to compare production methods with respect to the quality of the products each produces. Two samples may be compared by comparing a variety of sample statistics; we consider comparing only means and variances here.

Tests on Two Means

Example 3.13.1   We repeat part (c) of Example 3.1.1. Two production lines are making thermostats to be used in toaster ovens. Samples are selected from each production line, the thermostats are set at $350°$, and then the actual temperature in the toaster oven is measured. The results of the sampling are given in the following table.

X	Y
347.2	326.4
351.6	338.9
352.4	355.4
346.1	351.6
348.9	350.2
356.6	356.9
354.9	352.4
350.2	349.6
	351.2
	344.2

Graphs should be drawn of the data so that we may make an initial visual inspection. Comparative dotplots of the samples are shown in Figure 3.15. Looking at the figure, we can see that the data from production line Y appear to be much more variable than the data from production line X. We also see that the samples appear to be centered about different points. We calculate some sample statistics and find that

$$n_X = 8, \quad \overline{X} = 350.99, \quad s_X = 3.63$$
$$n_Y = 10, \quad \overline{Y} = 347.68, \quad s_Y = 9.11.$$

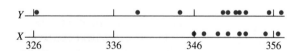

Figure 3.15 Comparative Dotplots for Example 3.13.1

The real question here is how these statistics would vary if we were to select large numbers of pairs of samples. The Central Limit Theorem can be used if assumptions can be made concerning the population variances.

We formulate the problem as follows. We wish to test the hypothesis

$$H_0: \mu_X = \mu_Y$$

against

$$H_1: \mu_X \neq \mu_Y.$$

We know that

$$E[\overline{X} - \overline{Y}] = \mu_X - \mu_Y$$

and that

$$\mathrm{Var}[\overline{X} - \overline{Y}] = \frac{\sigma_X^2}{n_X} + \frac{\sigma_Y^2}{n_Y}.$$

We also know that each of the variables $\overline{X}$ and $\overline{Y}$ is individually approximately normally distributed by the Central Limit Theorem. It can be shown that the difference between normal variables is also normal, so

$$Z = \frac{(\overline{X} - \overline{Y}) - (\mu_X - \mu_Y)}{\sqrt{\dfrac{\sigma_X^2}{n_X} + \dfrac{\sigma_Y^2}{n_Y}}}$$

is a $N(0, 1)$ variable.

The statistic Z can be used to test hypotheses or to construct confidence intervals if the variances are known. Consider for the moment that we know that the populations have equal variances, say $\sigma_X^2 = \sigma_Y^2 = 52.0$. Then

$$z = \frac{(350.99 - 347.68) - 0}{\sqrt{\dfrac{52}{8} + \dfrac{52}{10}}} = 0.96769.$$

The p-value for this test is approximately 0.33, so the null hypothesis would most probably be accepted.

We could also use Z to construct a confidence interval. Here

$$P\left[(\overline{X} - \overline{Y}) - 1.96\sqrt{\frac{\sigma_X^2}{n_X} + \frac{\sigma_Y^2}{n_Y}} \le \mu_X - \mu_Y \le (\overline{X} - \overline{Y}) + 1.96\sqrt{\frac{\sigma_X^2}{n_X} + \frac{\sigma_Y^2}{n_Y}}\right] = 0.95,$$

which becomes in this case the interval from -3.39423 to 10.0142. Since 0 is in this interval, the hypothesis of equal means is accepted. ∎

Larger samples reduce the width of the confidence interval. We will consider this topic in Section 3.14.

Knowledge of the population variances in the previous example may be regarded as artificial or unusual, although it is not infrequent, when data have been gathered over a long period of time, to have some idea of the size of the variance. In the next two examples we consider cases in which the population variances are unknown. In such cases, the unknown variances are equal or they are not equal. We give examples of the procedure in each case.

Example 3.13.2   If the population variances are known to be equal, with true value σ^2, then we form an estimate of this common value, which we denote by s_p^2, where

$$s_p^2 = \frac{(n_X - 1)s_X^2 + (n_Y - 1)s_Y^2}{n_X + n_Y - 2}.$$

Here s_p^2 is often called the **pooled variance**.

Now

$$E\left(s_p^2\right) = E\left[\frac{(n_X - 1)s_X^2 + (n_Y - 1)s_Y^2}{n_X + n_Y - 2}\right] = \frac{n_X - 1}{n_X + n_Y - 2}E\left(s_X^2\right) + \frac{n_Y - 1}{n_X + n_Y - 2}E\left(s_Y^2\right)$$

$$= \frac{n_X - 1}{n_X + n_Y - 2}\sigma^2 + \frac{n_Y - 1}{n_X + n_Y - 2}\sigma^2 = \sigma^2,$$

so s_p^2 is an unbiased estimator of σ^2.

Now

$$\frac{(n_X - 1)s_X^2}{\sigma^2} \quad \text{and} \quad \frac{(n_Y - 1)s_Y^2}{\sigma^2}$$

are each chi-squared random variables and their sum is a chi-squared random variable as well, which can be shown to have $n_X + n_Y - 2$ degrees of freedom.

First note that

$$Z = \frac{(\overline{X} - \overline{Y}) - (\mu_X - \mu_Y)}{\sqrt{\dfrac{\sigma_X^2}{n_X} + \dfrac{\sigma_Y^2}{n_Y}}}$$

is a $N(0, 1)$ variable. Then, since a Student t variable is the ratio of a $N(0, 1)$ variable to the square root of an independently distributed chi-squared variable divided by its degrees of freedom, it follows that

$$t_{n_X + n_Y - 2} = \frac{\dfrac{(\overline{X} - \overline{Y}) - (\mu_X - \mu_Y)}{\sqrt{\dfrac{\sigma^2}{n_X} + \dfrac{\sigma^2}{n_Y}}}}{\sqrt{\dfrac{\dfrac{(n_X - 1)s_X^2}{\sigma^2} + \dfrac{(n_Y - 1)s_Y^2}{\sigma^2}}{n_X + n_Y - 2}}}.$$

Note that the sampling must now be done from normal distributions. The fraction can be simplified to

$$t_{n_X + n_Y - 2} = \frac{(\overline{X} - \overline{Y}) - (\mu_X - \mu_Y)}{s_p\sqrt{\dfrac{1}{n_X} + \dfrac{1}{n_Y}}}$$

In this case, we find that

$$s_p^2 = \frac{7 \cdot 3.63^2 + 9 \cdot 9.11^2}{8 + 10 - 2} = 52.4480,$$

so

$$t_{16} = \frac{(350.99 - 347.68) - 0}{\sqrt{52.4480\left(\dfrac{1}{8} + \dfrac{1}{10}\right)}} = 0.96355.$$

The *p*-value for the test is then about 0.35 leading to the acceptance of the hypothesis that the population means are equal. ∎

Example 3.13.3   Now we must consider the case where the population variances are unknown and cannot be presumed to be equal. We do not know the exact probability distribution of any statistic involving the sample data in this case. This unsolved problem is known as the **Behrens-Fisher problem.** Several approximations are known; an approximation due to Welch [22] is given here.

The variable

$$T = \frac{(\overline{X} - \overline{Y}) - (\mu_X - \mu_Y)}{\sqrt{\dfrac{s_X^2}{n_X} + \dfrac{s_Y^2}{n_Y}}}$$

is approximately a *t* variable with ν degrees of freedom where

$$\nu = \frac{\left(\dfrac{s_X^2}{n_X} + \dfrac{s_Y^2}{n_Y}\right)^2}{\dfrac{\left(\dfrac{s_X^2}{n_X}\right)^2}{n_X - 1} + \dfrac{\left(\dfrac{s_Y^2}{n_Y}\right)^2}{n_Y - 1}}.$$

Using the data in Example 3.13.2, we find $\nu = 12.3037$ so we must use a *t* variable with 12 degrees of freedom. (We must always use the greatest integer less than or equal to ν; otherwise, the sample sizes are artificially increased.) This gives

$$T_{12} = 1.0495,$$

a result quite comparable to previous results. The Welch approximation will make a very significant difference if the population variances are quite disparate.

It is not difficult to see that, as $n_X \to \infty$ and $n_Y \to \infty$,

$$T = \frac{(\overline{X} - \overline{Y}) - (\mu_X - \mu_Y)}{\sqrt{\dfrac{s_X^2}{n_X} + \dfrac{s_Y^2}{n_Y}}} \longrightarrow \frac{(\overline{X} - \overline{Y}) - (\mu_X - \mu_Y)}{\sqrt{\dfrac{\sigma_X^2}{n_X} + \dfrac{\sigma_Y^2}{n_Y}}} = Z$$

as the sample sizes increase. Certainly, if each of the sample sizes exceeds 30, the normal approximation will be a very good one. However, it is very dangerous to assume normality for small samples if their population variances are quite different. In that case, the normal approximation should be avoided. Computer programs such as Minitab make it easy to do the exact calculations involved, regardless of sample size. This is a safe and prudent route to follow in this circumstance. ∎

The tests given in this section are heavily dependent upon the relationship between the variances. Confidence intervals for the ratio of two variances were considered in Section 3.7. The theory developed there can easily be used to test hypotheses on population variances.

Tests on Two Variances

In Section 3.7 we used the fact that if χ_a^2 and χ_b^2 are independent chi-squared variables, then the random variable

$$\frac{\chi_a^2/a}{\chi_b^2/b} = F(a, b),$$

where $F(a, b)$ denotes the F random variable with a and b degrees of freedom, respectively. Recall that we showed that

$$\frac{1}{F(a, b)} = F(b, a),$$

indicating that the reciprocal of an F variable is an F variable with the degrees of freedom interchanged. Some typical F curves were shown in Figure 3.10.

The F distribution can be used in testing the equality of variances in the following way. If the sampling is from normal populations, then

$$\frac{(n_X - 1)s_X^2}{\sigma_X^2} \quad \text{and} \quad \frac{(n_Y - 1)s_Y^2}{\sigma_Y^2}$$

are independent chi-squared variables. It follows then that

$$\frac{\dfrac{(n_X - 1)s_X^2}{\sigma_X^2(n_X - 1)}}{\dfrac{(n_Y - 1)s_Y^2}{\sigma_Y^2(n_Y - 1)}}$$

is the ratio of two independent chi-squared random variables each divided by its degrees of freedom. So this variable simplifies as

$$\frac{s_X^2/\sigma_X^2}{s_Y^2/\sigma_Y^2} = F(n_X - 1, n_Y - 1).$$

Now consider the hypotheses

$$H_0: \sigma_X^2 = \sigma_Y^2$$
$$H_1: \sigma_X^2 \neq \sigma_Y^2.$$

If the null hypothesis is true, then the F variable becomes

$$F(n_X - 1, n_Y - 1) = \frac{s_X^2}{s_Y^2}.$$

This is used as the test statistic with a two-tailed critical region.

Example 3.13.4   For an example, we use the data in the previous three examples where

$$n_X = 8, \qquad s_X^2 = 3.63^2 = 13.177$$

and

$$n_Y = 10, \qquad s_Y^2 = 9.11^2 = 82.992.$$

Here $F(7, 9) = \frac{13.177}{82.992} = 0.15877$. Minitab computes the p-value of this test as 0.0240, so the hypothesis that the variances are equal is supported in this case.

We have used a two-sided test here, but frequently experimenters worry that the variance has become too large, leading to a one-sided test. Small variances are rarely the subject of concern.   ■

We note here that the comparison of three or more means is a considerably harder task than the comparison of two means. Suppose we have m means to compare. It is tempting to consider all possible pairs of samples, but this would lead to $\binom{m}{2}$ tests. If each test had a significance level of 0.05, say, then the probability that at least one of the tests would incorrectly be rejected is $1 - P(\text{none of the tests are significant}) = 1 - (1 - 0.05)^{\binom{m}{2}}$. This number grows very rapidly with increasing m. If $m = 10$, for example, it is 0.9006, indicating that the overall significance level is very difficult to control unless each test is run at a very high level. Fortunately, there is a way to compare the means while controlling the significance level; we will return to this problem in Chapter 5.

We conclude this section with a special **experimental design** where the observations are paired with one another as part of the plan or *design* of the experiment. Such a design, planned in advance of taking the data, influences the conclusions that can be drawn from the data gathered in the experiment. Experimental designs considerably more complex than pairing will be considered in Chapter 6.

Paired Observations

Occasionally the experiment can be designed so that the observations can be *paired* with one another on some logical basis. Measurements on the quality of manufactured steels, for example, may be difficult to interpret because various steels were used in the measurements and they differ in chemical composition. Students taught to read by two different methods differ with respect to other characteristics, such as chronological age, intelligence, social background, and many other factors thought to influence their ability to learn to read. One way to remove the influence of these factors and to isolate the experimental one, is to pair the experimental subjects or materials in advance of the experiment.

We emphasize that this technique must be preplanned; it is not valid to analyze the data as if they had arisen from a paired experiment when in fact they did not. The occurrence of equal sample sizes, without the observations being paired, is not a sufficient reason to justify the analysis that follows.

Since the sample sizes are different, the experiment considered in the previous examples in this section obviously did not arise from a paired experiment.

Example 3.13.5   Two methods of teaching reading, method A and method B, are to be compared. Pairs of students are identified who are as alike as possible with respect to intelligence, chronological age, health, and social class. The pairs are formed so that within each pair these characteristics are as similar as possible, while the characteristics of the separate pairs may differ widely. Then one student from each pair is instructed by method A while the other is instructed by method B; the results of the instruction are measured by administering a standard reading achievement test. Subtracting one score from the other for each pair will remove all the influences from the factors identified on the score except for the method of instruction. The data are as follows, where $D = A - B$.

Pair	1	2	3	4	5	6	7
A	73	56	95	64	68	94	55
B	64	73	89	73	52	87	75
D	9	−17	6	−9	16	7	−20

Pair	8	9	10	11	12	13	14
A	84	73	92	99	68	44	53
B	88	85	96	91	86	59	67
D	−4	−12	−4	8	−18	−15	−14

The hypothesis $H_0: \mu_A = \mu_B$ is equivalent to $H_0: \mu_A - \mu_B = 0$, and, since the difference between means is the mean of the difference, this hypothesis is equivalent to $H_0: \mu_D = 0$. Various alternative hypotheses can be considered. Consider $H_1: \mu_D \neq 0$.

The t test on one sample (the D's) is appropriate here. The data give

$$\overline{D} = -\frac{67}{14} \quad \text{and} \quad s_D^2 = \frac{25{,}989}{182},$$

which leads to

$$t_{13} = \frac{-\dfrac{67}{14} - 0}{\sqrt{\dfrac{25{,}989}{182}}{14}} = -1.49848.$$

The p-value for this test is 0.1579, so the null hypothesis is accepted for any α greater than that value. ∎

The results of this paired comparison test can differ from those drawn from a two-sample test. Note that the two-sample test is usually inappropriate here since the samples are not independent. If the data above are used in a two-sample test with unknown variances that cannot be presumed to be equal, the result is quite different; now the p-value for the test is 0.426408. The degrees of freedom for the two tests also differ, the paired comparison test having 13 degrees of freedom while the two-sample test, assuming equal variances, has 26 degrees of freedom. We emphasize again, however, that only one of these tests is appropriate, the paired comparison test being valid only if the pairing is part of the statistical design of the experiment.

EXERCISES 3.13

1. An industrial engineer is studying the length of time it takes assembly line workers to perform a task. Independent samples are taken with these results.

   Men: $n_1 = 50$, $\bar{x}_1 = 42$ sec, $s_1^2 = 18$ sec

   Women: $n_2 = 59$, $\bar{x}_2 = 38$ sec, $s_2^2 = 14$ sec

   a. Test $H_0: \sigma_1^2 = \sigma_2^2$ against the alternative $H_1: \sigma_1^2 \neq \sigma_2^2$ using $\alpha = 0.02$.
   b. Assuming that $\sigma_1^2 = \sigma_2^2$, test $H_0: \mu_1 = \mu_2$ against the alternative $H_0: \mu_1 > \mu_2$ with $\alpha = 0.05$.

2. An experiment is conducted to compare the crash resistance of two types, A and B, of automobile bumpers. The data refer to repair costs after the cars are crashed into a wall at low speed.

   $$n_A = 11, \quad \bar{x}_A = 235, \quad s_A^2 = 421$$
   $$n_B = 9, \quad \bar{x}_B = 286, \quad s_B^2 = 511$$

   a. Test $H_0: \sigma_A^2 = \sigma_B^2$ against the alternative $H_1: \sigma_A^2 \neq \sigma_B^2$, using $\alpha = 0.02$.
   b. Assuming that $\sigma_A^2 = \sigma_B^2$, test $H_0: \mu_A = \mu_B$ against the alternative $H_0: \mu_A < \mu_B$ with $\alpha = 0.05$.

3. Golf scores of two competitors, A and B, are recorded over a period of ten days. This is a paired comparison experiment since weather conditions vary on different days. Golfer A claims that his true mean score is less than that for golfer B. The data follow.

Day	A	B
1	87	89
2	86	85
3	79	83
4	82	87
5	78	76
6	87	90
7	84	85
8	81	78
9	83	85
10	81	84

a. Do the data support the claim?
b. Discuss the assumptions that must be made concerning the golf scores for the test in part (a) to be valid.

4. In an experiment measuring the effects of liming and urea fertilizer applications on pesticide retention by foamy soil, the following numbers of parts per million of pesticide retention were observed.

   Soil treated with lime: 28.5, 24.7, 26.2, 23.9, 29.6

   Soil treated with urea: 38.7, 41.6, 35.9, 41.8, 43.2

   a. Test $H_0: \sigma_L^2 = \sigma_U^2$ against the alternative $H_1: \sigma_L^2 \neq \sigma_U^2$ using $\alpha = 0.02$.
   b. Assuming that $\sigma_L^2 = \sigma_U^2$, test $H_0: \mu_L = \mu_U$ against the alternative $H_0: \mu_L \neq \mu_U$ with $\alpha = 0.05$.

5. In a batch chemical process, two catalysts are being compared for their effect on the output of the process reaction. The sample results follow.

   Catalyst 1: $n_1 = 11$, $\bar{x}_1 = 85$, $s_1^2 = 16$
   Catalyst 2: $n_2 = 9$, $\bar{x}_2 = 81$, $s_2^2 = 25$

   Assuming that $\sigma_1^2 = \sigma_2^2$, test $H_0: \mu_1 = \mu_2$ against the alternative $H_0: \mu_1 > \mu_2$ with $\alpha = 0.05$.

6. A large transportation company is trying to decide whether using radial tires rather than regular tires improves fuel economy. Twelve cars, of a variety of brands and models, were equipped with radial tires and driven over a test course. Without changing drivers, and under the same weather conditions, the cars were equipped with regular tires and driven again over the test course. The gasoline consumption, in kilometers per liter, is as follows.

   Radial Tires: 4.2, 4.7, 6.6, 7.0, 6.7, 4.5, 5.7, 6.0, 7.4, 4.9, 6.1, 5.2

   Regular Tires: 4.1, 4.9, 6.2, 6.9, 6.8, 4.4, 5.7, 5.8, 6.9, 4.7, 6.0, 4.9

   Analyze the data and draw a conclusion for the company about the use of radial tires.

7. In comparing times to failure (in hours) of two different types of light bulbs, two samples gave the following data.

$$n_1 = 13, \quad \bar{x}_1 = 984, \quad s_1^2 = 8742$$
$$n_2 = 15, \quad \bar{x}_2 = 1121, \quad s_2^2 = 9411$$

Assuming that $\sigma_1^2 = \sigma_2^2$, test H_0: $\mu_1 = \mu_2$ against the alternative H_1: $\mu_1 < \mu_2$ with $\alpha = 0.05$.

8. A random sample of size $n_1 = 35$ has $s_1 = 5.2$ and $\bar{x}_1 = 81$. A second random sample has $n_2 = 44$ with $s_2 = 3.4$ and $\bar{x}_2 = 76$. Assuming that $\sigma_1^2 \neq \sigma_2^2$, test H_0: $\mu_1 = \mu_2$ against the alternative H_0: $\mu_1 > \mu_2$ with $\alpha = 0.05$.

9. Over a long period of time, ten patients are selected at random and each is given two different treatments for arthritis. The results of standard tests are as follows.

Patient	Treatment 1	Treatment 2
1	47	52
2	38	35
3	50	52
4	33	35
5	47	46
6	23	27
7	40	45
8	42	41
9	15	17
10	36	41

Test H_0: $\mu_1 = \mu_2$ against the alternative H_1: $\mu_1 < \mu_2$ with $\alpha = 0.05$.

10. The following data have been gathered by an industrial engineer who is comparing the time (in seconds) taken by an operator to complete a task using two different methods.

Operator	Method A	Method B
1	45	39
2	88	71
3	40	42
4	32	27
5	29	28
6	34	30
7	59	50
8	55	60
9	62	51
10	50	48

Test an appropriate hypothesis and decide which method is best.

11. Two manufacturers of chain links are being evaluated to determine if one manufacturer's chain links have significantly greater breaking strength than those of the other. The following data were gathered.

$$n_1 = 5, \quad \bar{x}_1 = 310, \quad s_1 = 25$$
$$n_2 = 9, \quad \bar{x}_2 = 235, \quad s_2 = 47$$

Assuming that the population variances are not equal, decide which manufacturer makes chain links with the higher breaking strength.

12. Two new methods for producing automobile tires have been proposed. The manufacturer believes that there will be no appreciable difference in the lifetimes of the tires produced by these methods. The data show the lifetimes of the tires in units of 1000 miles.

Method 1 : 66.4, 61.6, 60.5, 59.1, 63.6, 61.4, 62.5, 64.4, 60.7

Method 2 : 58.2, 60.4, 55.2, 62.0, 57.3, 58.7, 56.1.

Are the methods equivalent?

COMPUTER EXPERIMENT

Select 200 pairs of samples of size 10 each from normal distributions with means 100 and 105 and variances 20 and 30, respectively. Calculate the means and variances of each sample and calculate the statistic

$$T = \frac{(\bar{X} - \bar{Y}) - (\mu_X - \mu_Y)}{\sqrt{\dfrac{s_X^2}{n_X} + \dfrac{s_Y^2}{n_Y}}}.$$

Draw a histogram of the result and compare the histogram with the expected
Student t distribution approximation.

3.14 Sample Size and Type II Errors

Experimenters often are interested in determining the size of a random sample. We give
some results in this section. Determining Type II errors is also important, and some
results concerning this idea are also discussed.

Example 3.14.1 An experimenter wishes to determine the value of μ, the mean of a population, by
estimating this by the mean, $\overline{X}$, of a random sample. It is important that this estimate
and μ differ by at most 2 units, with probability 0.95. If it is known that the variance of
the population, σ^2, is 25, how large a sample is necessary?
  We want $P[|\overline{X} - \mu| \le 2] \ge 0.95$, or

$$P\left(-\frac{2}{\sigma/\sqrt{n}} \le \frac{\overline{X} - \mu}{\sigma/\sqrt{n}} \le \frac{2}{\sigma/\sqrt{n}}\right) \ge 0.95.$$

By the central limit theorem it follows that $2/(\sigma/\sqrt{n}) \ge 1.96$ from which we see that
$n \ge [1.96\sigma/2]^2 = 4.9^2 = 24.01$ so a sample of 25 will produce the desired result. Note
that a sample size of 24 will not produce the estimate of μ with the desired probability.
  Example 3.3.4 is similar to this example.   ■

Example 3.14.2 A test is to be conducted on the hypothesis

$$H_0: \mu = 100$$

against the alternative

$$H_1: \mu = 105.$$

The experimenter is willing to have a Type I error of 5% and a Type II error of 20%. If σ
is known to be 10, how large a sample is necessary? Note that α and β are both specified.
  Here the test is one-sided. The sizes of the errors give

$$P\left(\frac{\overline{X} - 100}{10/\sqrt{n}} \le 1.645\right) = 0.95,$$

so the acceptance region for the test is the region for which
$\overline{X} \le 100 + 1.645 \cdot 10/\sqrt{n} = a$, say. Since we also know that

$$P\left(\frac{\overline{X} - 105}{10/\sqrt{n}} \le -0.8416\right) = 0.20,$$

the acceptance region, when the alternative is true, is the region
$\overline{X} \le 105 - 0.8416 \cdot 10/\sqrt{n}$. It follows that $a = 105 - 0.8416 \cdot 10/\sqrt{n}$. The situation
is shown in Figure 3.16.

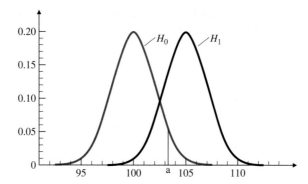

Figure 3.16 Normal Curves for Example 3.14.2

Since the acceptance region must be determined by the same values, whether the hypothesis is true or not, $100 + 1.645 \cdot 10/\sqrt{n} = 105 - 0.8416 \cdot 10/\sqrt{n}$. This is solved to give

$$n = \left[\frac{(1.645 + 0.8416) \cdot 10}{(105 - 100)} \right]^2 = 24.733,$$

so a sample of 25 is sufficient. ∎

Example 3.14.3 Suppose now that we wish to select two samples from normal populations having, possibly, different means. It is important to detect a difference between the means of 3 units with a high probability, say 90%. Suppose that $\alpha = 5\%$. We investigate the sample sizes necessary.

Let the hypotheses be

$$H_0 \colon \mu_X - \mu_Y = 0 \quad \text{and} \quad H_1 \colon \mu_X - \mu_Y = d > 0.$$

Now the following is a normal random variable:

$$Z = \frac{(\overline{X} - \overline{Y}) - (\mu_X - \mu_Y)}{\sqrt{\dfrac{\sigma_X^2}{n_X} + \dfrac{\sigma_Y^2}{n_Y}}}.$$

If $\alpha = 0.05$, then the acceptance region for the null hypothesis is the interval

$$\overline{X} - \overline{Y} \leq 1.645 \cdot \sqrt{\frac{\sigma_X^2}{n_X} + \frac{\sigma_Y^2}{n_Y}}.$$

We wish to detect a difference $d = 3$ between the means with probability 90%. So we want $P(H_0 \text{ is rejected if } d = 3) = 0.90$ and $P(H_0 \text{ is accepted if } d = 3) = 0.10$. This means that $\beta = 10\%$. The upper limit for the acceptance region if the alternative is true must then be a point on the normal curve with mean 3 that has 10% of the normal curve to its left; that is, its z score must be -1.28155 or

$$\frac{(\overline{X} - \overline{Y}) - 3}{\sqrt{\dfrac{\sigma_X^2}{n_X} + \dfrac{\sigma_Y^2}{n_Y}}} = -1.28155.$$

It follows that

$$1.645 \cdot \sqrt{\frac{\sigma_X^2}{n_X} + \frac{\sigma_Y^2}{n_Y}} = 3 - 1.28155 \cdot \sqrt{\frac{\sigma_X^2}{n_X} + \frac{\sigma_Y^2}{n_Y}}.$$

This can be written as

$$\frac{\sigma_X^2}{n_X} + \frac{\sigma_Y^2}{n_Y} = \frac{3^2}{(1.645 + 1.28155)^2}.$$

There are a number of possibilities now. We first note that if the population variances are unknown, there is no exact solution to the problem. (This was discussed in Section 3.13.) If the variances are known and if the sample sizes are equal so that $n_X = n_Y = n$, then we find that

$$n = \frac{(1.645 + 1.28155)^2 \cdot \left(\sigma_X^2 + \sigma_Y^2\right)}{3^2}.$$

If, for example, $\sigma_X^2 = 50$, $\sigma_Y^2 = 75$, we find that

$$n = \frac{(1.645 + 1.28155)^2 \cdot (50 + 75)}{3^2} = 118.954,$$

so samples of size 119 must be selected.

Since the random variable Y has a larger variance, we might want to select a larger sample from the Y population than from the X population. If we can select a sample of say 10 more items from the Y population, then we find that $n_X = 114$ and $n_Y = 124$.   ∎

Unknown Population Variance

We now want to explain, for purposes of completeness, why sample size problems and determining the size of the Type II error, β, when the population variance is not known, are not elementary tasks. Fortunately, computer algebra systems allow us to do calculations that previously had to be approximated. Here we give one example of calculating β and one example of determining sample size.

To test the hypothesis

$$H_0: \mu = \mu_0$$

against the alternative

$$H_1: \mu = \mu_1 = \mu_0 + \delta$$

where the population standard deviation is unknown, the Student t test with $n - 1$ degrees of freedom is used to determine critical values for rejection of the null hypothesis or to determine p-values.

However, the distribution of $(\overline{X} - \mu_1)/(s/\sqrt{n}) = t'$ does not have a Student t distribution. It has been shown that the distribution of this statistic follows what is known as a **noncentral Student t distribution** with $n - 1$ degrees of freedom and noncentrality parameter $\delta\sqrt{n}/\sigma$.

The noncentral Student t distribution is far from elementary, so until recently calculations that used it were usually approximated. With the advent of computer algebra systems, such as *Mathematica*, we have become able to do exact calculations easily, but the noncentrality parameter, $\delta\sqrt{n}/\sigma$, is dependent upon the unknown population standard deviation, σ.

If it is possible to write the quantity δ as a multiple of σ, then the noncentrality parameter can be specified. This is shown in the following example.

Example 3.14.4   Consider the hypothesis

$$H_0: \mu = 150$$

against the alternative

$$H_1: \mu = 180.$$

We have a sample of 20 observations and set the size of the Type I error to be 5%. If the null hypothesis is true, a one-sided rejection region is $t_{19} > 1.72913$.

We want to calculate β. We must make some supposition about the size of the deviation of the alternative, $\mu = 180$, from the hypothesis. Let us proceed supposing that the deviation is about $0.5\,\sigma$ units. When the standard deviation is unknown, experimenters are often interested in detecting deviations that are some multiple of the standard deviation from the value specified by the null hypothesis. The noncentrality parameter in this case is then

$$\frac{0.5\sigma\sqrt{n}}{\sigma} = 0.5\sqrt{20} = 2.2361.$$

Then β is the area to the left of 1.72913 under a noncentral t distribution with 19 degrees of freedom and noncentrality parameter 2.2361. *Mathematica* gives this area as 0.304839. The situation is shown in Figure 3.17.

Sample size problems can also be solved using the noncentral t distribution.   ∎

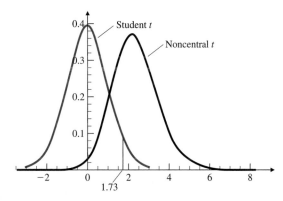

Figure 3.17 Student t and Noncentral Student t Distributions

Example 3.14.5   Consider the hypotheses in the previous example,

$$H_0: \mu = 150$$

against the alternative

$$H_1: \mu = 180.$$

Determine the sample size so that $\alpha = 0.05$ and $\beta = 0.02$, using a one-sided test. We first assume that σ is known and then we will contrast this solution to that in which σ is unknown.

Values of $\overline{X}$ that give $Z > 1.64485$ lead to the rejection of the null hypothesis. These are values of $\overline{X} > 150 + 1.64485 \cdot \sigma/\sqrt{n}$. The value of β tells us that

$$\frac{150 + 1.64485 \cdot \dfrac{\sigma}{\sqrt{n}} - 180}{\dfrac{\sigma}{\sqrt{n}}} = -2.05375.$$

If we assume for the time being that $\sigma = 30$, this gives $n = 13.6796$ so a sample size of 14 is necessary.

Now we proceed to the case in which σ is unknown. We must, of course, use the noncentral t distribution with $n - 1$ degrees of freedom. We suppose here that $\delta/\sigma = 1$ so that the noncentrality parameter is $\sqrt{n}$.

If $n = 16$, the critical Student t value is 1.75305. The probability that a noncentral Student t variable with 15 degrees of freedom and noncentrality parameter $\sqrt{16} = 4$ is less than 1.75305 is 0.0151523. This is the size of β. A similar calculation using a sample of size 15 and so a noncentrality parameter of $\sqrt{15}$ and a critical t value of 1.76131, gives $\beta = 0.0210838$, so we would probably choose the larger sample. Interestingly enough, the unknown standard deviation increases the sample size, but not by all that much. This calculation is highly dependent upon the assumption that $\delta/\sigma = 1$; as this ratio decreases, the required sample size increases greatly. ∎

EXERCISES 3.14

1. An automatic machine fills bottles of soft drink. Past experience indicates that the true standard deviation is $\sigma = 0.5$ oz. How large a sample must be taken so that the sample mean and the true population mean differ by at most 0.15 oz with probability 0.95?

2. The weights of bags of cement are being studied. We want to test $H_0: \mu = 150$ against the alternative $H_1: \mu > 150$. Assume that $\sigma = 4$. How large a sample is necessary if $\alpha = 0.05$ and $\beta = 0.10$ for the alternative $\mu = 152$?

3. The breaking strength of wire for a construction site is known to have $\sigma = 100$ lb. How large a sample must be chosen so that the probability the sample mean and the true mean differ by at most 5 lb is 0.90?

4. In testing $H_0: \mu = 22$ against the alternative $H_1: \mu > 22$ we want to be able to detect the fact that $\mu = 24$ with probability 0.95. If $\sigma = 4$ and $\alpha = 0.05$, how large a sample is necessary?

5. For two populations it is known that $\sigma_1^2 = 25$ and $\sigma_2^2 = 45$. The null hypothesis $H_0: \mu_1 - \mu_2 = 0$ is to be tested against $H_1: \mu_1 - \mu_2 \neq 0$. The experimenter wishes to have $\alpha = 0.05$ and the probability that H_0 is accepted if, in fact $\mu_1 - \mu_2 = -3$, equal to 1%.

   a. If equal sample sizes are to be drawn, what should the sample size be?

   b. If it is possible to select five more observations from population 2 than from population 1, what should the sample sizes be?

6. A random sample of households from a large community is to be selected to estimate the true average residential electricity usage per household during a winter month. How many households should be sampled if the sample mean is to be within 20 kilowatt hours of the true mean with probability 0.95? Assume that $\sigma = 200$ kilowatt hours.

THE REMAINING PROBLEMS IN THIS SECTION REQUIRE A COMPUTER.

7. A test of H_0: $\mu = 400$ against the alternative H_1: $\mu = 450$ is being conducted by selecting a sample of size 30. Supposing that the difference of 50 units between the hypotheses is about 0.75σ, find β for a Type I error of size 5%.

8. In Exercise 7, suppose that β is 0.07. Find the sample sizes necessary if
   a. $\sigma = 50$
   b. σ is unknown, again assuming that $\delta = 0.75\sigma$.

3.15 Tests for Categorical Data

Data often arise in industrial settings that can be classified into categories that usually cannot be distinguished on a numerical scale. Parts are either acceptable for use or unacceptable; a drug either improves a patient's condition or it doesn't; separate tests on produced items are identified as test 1, test 2, and so on. We consider tests on such categorical data in this section.

Contingency Tables

Example 3.15.1 An industry is investigating the effect of education upon the performance level of some of its production line workers. The workers have either high school educations or some community college education; their performance is categorized as "Poor," "Satisfactory," or "Above Average." The data are as follows:

	Poor	Satisfactory	Above Average
Community College	14	24	22
High School	16	15	19

The table exhibiting the data is often called a **contingency table.**

Although many hypotheses can be tested with data that are categorical, we will consider only one here, namely, that the categories are independent. In this example, we then test to see if education and job performance are independent of each other.

To be specific, let p_{ij} denote the probability that an item is in row i and column j of the data. In this example, $i = 1, 2$ and $j = 1, 2, 3$. Further, let $p_{i.}$ denote the probability an item is in the ith row and $p_{.j}$ denote the probability an item is in the jth column. The hypothesis of independence then becomes

$$H_0: p_{ij} = p_{i.} \cdot p_{.j} \quad \text{for all } i \text{ and } j.$$

We want to find maximum likelihood estimates of the cell probabilities, namely, the p_{ij}.

The probability that n_{ij} items are in cell (i, j) is $p_{ij}^{n_{ij}}$.

The likelihood function, L, is the probability that the observed data appear in the cells of a contingency table having r rows and k columns so

$$L = \prod_{i=1}^{r} \cdot \prod_{j=1}^{k} p_{ij}^{n_{ij}}.$$

We could work with the general case, but, in order to see the results of our mathematics more clearly, we are going to restrict ourselves to a contingency table with only two rows and two columns. The situation is precisely similar to the general one, but let us consider then the contingency table

$$\begin{matrix} a & c \\ b & d \end{matrix}.$$

The likelihood function for this table is

$$L = \prod_{i=1}^{2} \cdot \prod_{j=1}^{2} p_{ij}^{n_{ij}},$$

which becomes, in this case,

$$L = p_{11}^{a} \cdot p_{12}^{c} \cdot p_{21}^{b} \cdot p_{22}^{d}.$$

Under the hypothesis of independence this becomes

$$L = p_{1.}^{a} \cdot p_{.1}^{a} \cdot p_{1.}^{c} \cdot p_{.2}^{c} \cdot p_{2.}^{b} \cdot p_{.1}^{b} \cdot p_{2.}^{d} \cdot p_{.2}^{d}.$$

This, in turn, can be simplified to

$$L = p_{1.}^{a+c} \cdot p_{.1}^{a+b} \cdot p_{2.}^{b+d} \cdot p_{.2}^{c+d}.$$

But since $p_{1.} + p_{2.} = 1$, we may write L as

$$L = (1 - p_{2.})^{a+c} \cdot p_{.1}^{a+b} \cdot p_{2.}^{b+d} \cdot p_{.2}^{c+d}.$$

We can maximize the logarithm of L so it follows that

$$\ln L = (a + c) \ln(1 - p_{2.}) + (a + b) \ln p_{.1} + (b + d) \ln p_{2.} + (c + d) \ln p_{.2}$$

and we find, for example, that

$$\frac{\partial \ln L}{\partial p_{2.}} = -\frac{a + c}{1 - p_{2.}} + \frac{b + d}{p_{2.}}.$$

When this is equated to zero and solved for the maximum likelihood estimator, we find that

$$\widehat{p_{2.}} = \frac{b + d}{a + b + c + d}.$$

It then follows that

$$\widehat{p_{1\cdot}} = 1 - \widehat{p_{2\cdot}} = \frac{a+c}{a+b+c+d}.$$

In an entirely similar way, it is found that

$$\widehat{p_{\cdot 1}} = \frac{a+b}{a+b+c+d}$$

and that

$$\widehat{p_{\cdot 2}} = \frac{c+d}{a+b+c+d}.$$

These maximum likelihood estimators are simply the row or column totals divided by the total number of observations. We denote these as

$$\frac{n_{i\cdot}}{n} \quad \text{and} \quad \frac{n_{\cdot j}}{n}.$$

Now

$$\widehat{p_{ij}} = \widehat{p_{i\cdot}} \cdot \widehat{p_{\cdot j}} = \frac{n_{i\cdot}}{n} \cdot \frac{n_{\cdot j}}{n}.$$

If we wanted to predict the number of observations in the (i, j) cell, we would estimate this as

$$n \cdot \frac{n_{i\cdot}}{n} \cdot \frac{n_{\cdot j}}{n}.$$

These are called **cell expectations.**

If we do this with the numerical example with which we began, the table, with expectations (rounded to the nearest integer) in parentheses, becomes

	Poor	Satisfactory	Above Average
Community College	14 (16)	24 (21)	22 (22)
High School	16 (14)	15 (18)	19 (19)

Denote the observation and the expectation in a typical cell, say the mth cell, as O_m and E_m. In this case, the range of m is 1 to 6, but in general it will vary from 1 to $r \cdot k$.

Now it can be shown, although not easily, that

$$\sum_{m=1}^{r\cdot k} \frac{(O_m - E_m)^2}{E_m} = \chi^2_{(r-1)\cdot(k-1)}.$$

The chi-squared variable compares the observed values in the cells to the expected values in the cells, the expectations being calculated using the hypothesis of independence. If the expectations are close to the observations, then χ^2 will be small. This leads us to reject the hypothesis of independence only if χ^2 is too large.

In our example, then, we find that

$$\chi^2_{(2-1)\cdot(3-1)} = \frac{(14-16)^2}{16} + \frac{(16-14)^2}{14} + \frac{(24-21)^2}{21} + \frac{(15-18)^2}{18}$$

$$+ \frac{(22-22)^2}{22} + \frac{(19-19)^2}{19} = 1.4643.$$

The upper 5% value of χ^2_2 is 5.991, so the hypothesis of independence is accepted. ∎

We pause here to present two facts, which can be established by doing the requisite algebra.

FACT 1

For the contingency table with two rows and two columns,

$$\begin{matrix} a & c \\ b & d \end{matrix},$$

then

$$\chi^2_1 = \frac{(a+b+c+d)(ad-bc)^2}{(a+b)(a+c)(b+d)(c+d)}.$$

FACT 2

A test for the difference between two proportions was given in Section 3.11. If we construct a 2-by-2 contingency table with a the number of successes and b the number of failures, for $a + b$ trials, and if c and d are the number of successes and failures, for $c + d$ trials, then the test on the hypothesis that the two proportions are equal is equivalent to the chi-squared test given above. The estimate for the common value of p must be

$$\frac{a+c}{a+b+c+d}.$$

Fitting Curves to Data

Experimenters often want to fit a curve of some sort to a set of data. This, in fact, is a common problem in science and engineering, and we will consider it at length in the next chapter when we discuss regression and correlation. For now, we give an example of fitting a curve to a set of data because one test that the curve does in fact fit the data uses the chi-squared distribution.

Example 3.15.2   A Geiger counter is recording emissions from a radioactive substance. In Table 3.3, the random variable X represents the number of observations recorded in a brief time period and f represents the frequency with which the values of X were observed.

x	f
0	40
1	34
2	18
3	5
4	2
5	2

Table 3.3

A graph of these observations is shown in Figure 3.18.

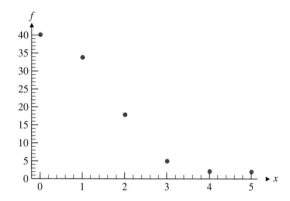

Figure 3.18 Geiger Counter Data

The Poisson probability distribution function defined as

$$f(x) = P(X = x) = \frac{e^{-\lambda}\lambda^x}{x!} \qquad \text{for } x = 0, 1, 2, \ldots$$

is frequently used as a model for this sort of data, so we want to know if it is a good model for this specific data set.

It is not difficult to show that the maximum likelihood estimator for the parameter λ is $\bar{x}$, the mean of the sample. For these data, $\bar{x} = 1.01980$. In Example 3.15.1, we calculated expectations for each cell. We do that here as well, using $\lambda = 1.01980$ in $f(x)$ above. The general idea is to compute expectations and compare them to observations.

Since there are 101 total observations, we find expectations for each value of X by multiplying the probabilities by 101. The results, rounded off, are shown in Table 3.4.

Now the formula $\sum_m (O_m - E_m)^2/E_m$ cannot be applied due to the cell with expectation 0. In fact, small expectations inflate χ^2, so we must take some precautions to

x	Observation	Expectation
0	40	36
1	34	37
2	18	19
3	5	6
4	2	2
5	2	0

Table 3.4

prevent this. The chi-squared distribution is only an approximation to the true distribution of $\sum_m (O_m - E_m)^2 / E_m$ and it has been found that the approximation is best if the cell expectations are at least 5. To achieve this, we combine adjacent cells until the expectations are each at least 5. The result is shown in Table 3.5.

x	Observations	Expectations
0	40	36
1	34	37
2	18	19
3 or more	9	8

Table 3.5

It is also known that the number of degrees of freedom for χ^2 is:

degrees of freedom = number of cells − number of parameters estimated − 1.

Since one parameter, namely λ, has been estimated from the data, χ^2 has $4 - 1 - 1 = 2$ degrees of freedom. We find that

$$\chi_2^2 = \frac{(40-36)^2}{36} + \frac{(34-37)^2}{37} + \frac{(18-19)^2}{19} + \frac{(9-8)^2}{8} = 0.86532.$$

We also find that $P(\chi_2^2 \geq 0.86532) = 0.648781$. Since the curve fits well if the observations and expectations are close, producing a rejection region consisting of large values of χ^2, we would conclude that the Poisson model fits the data well here. ∎

We end this chapter with an example of fitting a continuous distribution to a set of data. We could proceed in a manner similar to that for discrete data. Continuous data can be divided into classes and counts can be made for the number of observations in each class. Then parameters for the entire distribution can be estimated, usually by the method of maximum likelihood, predictions can be made for each class, producing expectations, and, finally, the observations and expectations can be compared using the chi-squared variable used in the previous examples. The result in the continuous case is highly dependent upon the classes chosen.

In the following example we show another procedure that has become popular with the availability of computer software. It is called a **quantile-quantile, or q-q, plot.**

Example 3.15.3   We want to fit a set of data, which has sample mean 5, to an exponential distribution with probability density function

$$f(x) = \frac{1}{5}e^{-\frac{x}{5}}, \qquad x \geq 0.$$

The mean of this distribution is 5 as well.

We first calculate the **quantiles** of this distribution, that is, values x_q so that

$$P(X \leq x_q) = q$$

for $q = 0.1, 0.2, \ldots, 1$. This divides the distribution into 10 parts. These quantiles are shown in the column labeled x_q in Table 3.6.

We expect 10% of the data to lie in each of the intervals defined by the quantiles. We could count the number of data points contained in each of the intervals defined by the quantiles and use the chi-squared variable again.

We show another solution based on a graph. We took a data set of 1000 observations (which is not shown here) and calculated the quantiles for this data set. These are shown in Table 3.6.

q	x_q	Observed Quantile
0.1	0.526803	0.416819
0.2	1.11572	1.02238
0.3	1.78337	1.68019
0.4	2.55413	2.43697
0.5	3.46574	3.47048
0.6	4.58145	4.64408
0.7	6.01986	6.09554
0.8	8.04719	8.10811
0.9	11.5129	11.5406
1.0	∞	32.5989

Table 3.6

A graph of the points (Observed quantile, x_q) produces the quantile-quantile plot. This is shown in Figure 3.19, where the point $(32.5989, \infty)$ is replaced by $(32.5989, 30)$. ∎

The goodness of fit here is judged by how close the points are to a straight line with slope 1 (indicating a perfect match between the observed quantiles and the true quantiles).

We can only observe here that the fit appears to the eye to be remarkably good, but this is a visual observation only. This leads us to consider the question of when a straight

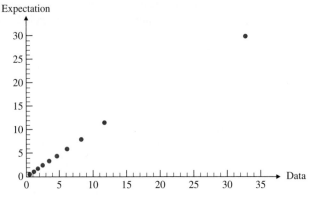

Figure 3.19 A Quantile-Quantile Plot

line fits a data set well. This consideration, as well as many others, is the subject of Chapter 4.

EXERCISES 3.15

1. Two operators of the same machine show the following numbers of good and defective parts produced.

Operator	A	B
Good	21	17
Defective	79	83

a. Test the hypothesis that the categories are independent using the chi-squared test for contingency tables.
b. Test the hypothesis that the proportions of good items are equal using the test for the difference between proportions.

2. The results of a sample survey of voter opinions on a particular issue were as follows.

	For	**Against**	**Undecided**
Male	30	35	10
Female	40	32	8

Are the opinions on this issue independent of gender? Use $\alpha = 5\%$.

3. A random-number generator is being tested to determine if each of 10 digits is being produced with equal probability. A sample of 1000 digits gave the following results.

Digit	0	1	2	3	4	5	6	7	8	9
Frequency	99	106	98	97	104	106	92	98	99	101

Is the random-number generator behaving properly?

4. A study is made of the number of calls a busy switchboard receives during a given hour with the following results.

Number of calls	0	1	2	3	4	5	6	7	8	9
Frequency	0	12	24	33	31	16	14	10	9	1

Test the hypothesis at the 5% level that the data follow a Poisson distribution.

5. Accidents at a busy intersection occur with the following frequencies in a study.

Number of accidents	0	1	2	3
Frequency	7	8	7	3

Do the data conform to a Poisson distribution?

6. A production process uses five machines during each of two work shifts. The frequencies with which these machines need to be repaired are as follows.

	Machine				
	1	2	3	4	5
Shift					
1	12	6	12	14	6
2	23	4	9	16	5

Are the machine breakdowns independent of work shift?

7. Lifetimes (in thousands of hours) of fluorescent lighting fixtures have been studied. The following observed quantiles of the data have been found.

q	Observed Quantile
0.1	0.835
0.2	1.467
0.3	2.626
0.4	4.011
0.5	5.063
0.6	6.421
0.7	8.566
0.8	11.277
0.9	18.904
1.0	26.912

The data are thought to arise from an exponential distribution with mean 7. Show a quantile-quantile plot and decide whether or not this supposition is correct.

8. A sample of observations is thought to come from a normal distribution with mean 23 and standard deviation 4. The following quantiles were observed.

q	Observed Quantile
0.1	17.6048
0.2	20.3750
0.3	20.9024
0.4	22.9865
0.5	23.0566
0.6	25.0123
0.7	25.0988
0.8	27.4563
0.9	29.3567
1.0	34.9924

Show a quantile-quantile plot of the data and decide whether or not the data were likely to arise from a normal distribution.

Chapter Review

This chapter was concerned with *statistical inference,* that is, the ways in which random samples, selected from populations with unknown characteristics, can be used to draw conclusions about those unknown population characteristics. Statistical inference is divided into two parts—*estimation* and *hypothesis testing*. Each of these topics was considered in this chapter.

Key Concepts

- Constructing a point estimate of a population parameter
- Constructing a least squares estimate of a population parameter
- Constructing a maximum likelihood estimate of a population parameter
- Constructing and interpreting confidence intervals for μ and σ^2
- Using the probability distribution of the sample proportion
- Testing a hypothesis
- Finding the power of a test

- Constructing the best test of a hypothesis
- Constructing tests comparing two samples
- Finding the *p*-value of a test
- Interpreting a contingency table
- Fitting theoretical values to a set of sample values

Key Terms

Unbiased Estimator	We call $\widehat{\theta}$ an *unbiased estimator* of the unknown population parameter θ if $E(\widehat{\theta}) = \theta$.
Maximum Likelihood Estimator	The *maximum likelihood estimator* for θ is that value $\widehat{\theta}$ that maximizes the likelihood function $$L(X;\theta) = f(x_1;\theta) \cdot f(x_2;\theta) \cdot f(x_3;\theta) \cdots f(x_n;\theta).$$
Student *t* Distribution	When sampling from a normal distribution with unknown variance, the sample mean, $\overline{X}$, follows a *Student t distribution*.
F Distribution	The *F distribution* is used in this chapter to test hypotheses on two variances.
Hypothesis	Both the null and alternative *hypotheses* are statements concerning the parameters or the form of a probability distribution.
Critical Region	The *critical region* is the set of values of the test statistic where the null hypothesis is rejected.
Type I Error	A *Type I error,* whose size is denoted by α, occurs when the null hypothesis is rejected if it is true.
Type II Error	A *Type II error* occurs, whose size is denoted by β, when the null hypothesis is accepted when it is false.
Power of a Test	The *power of a test* is the probability the null hypothesis is rejected when it is false.
p-Value for a Test	The *p-value for a test* is the probability that the test statistic assumes a value equal to or more extreme than the observed value.
Non-Central *t* Distribution	The *non-central t distribution* is used to determine the size of the Type II error when the population variance is unknown.
Contingency Table	A display of data in a *table* with rows and columns.

Key Theorems and Facts

Student t Random Variable

The *Student t* distribution with n degrees of freedom is defined as the ratio of two independent *random variables*
$\frac{N(0,1)}{\sqrt{\chi^2/n}} = t_n$ We used this to determine that
$\frac{\overline{X}-\mu}{s/\sqrt{n}} = t_{n-1}$.

Confidence Interval for a Proportion

A 95% *confidence interval for* an unknown *proportion, p,* based on a sample proportion, p_s, is of the form

$$P\left[-2\sqrt{\frac{p \cdot q}{n}} \le p_s - p \le 2\sqrt{\frac{p \cdot q}{n}}\right] \ge 0.95.$$

The end points of the confidence interval can be found by
1) solving the inequalities
2) replacing p and q by $1/2$
3) replacing p and q by p_s and q_s

Difference Between Proportions

Confidence intervals for the *difference between two proportions* are of the form

$$P\left[\begin{array}{c} (p_{s_1} - p_{s_2}) - z \cdot \sqrt{\frac{p_{s_1} \cdot q_{s_1}}{n_1} + \frac{p_{s_2} \cdot q_{s_2}}{n_2}} \le p_1 - p_2 \\ \le (p_{s_1} - p_{s_2}) + z \cdot \sqrt{\frac{p_{s_1} \cdot q_{s_1}}{n_1} + \frac{p_{s_2} \cdot q_{s_2}}{n_2}} \end{array}\right] = p.$$

Distribution of the Sample Variance

If the sampling is from a normal population, then $\frac{(n-1) \cdot s^2}{\sigma^2}$ follows a χ^2_{n-1} distribution.

Distribution of the Ratio of Two Variances

If the samples arise from normal populations, then $\frac{\chi^2_n/n}{\chi^2_m/m} = F(n, m)$.
We used this fact to find that

$$\frac{\dfrac{s_1^2}{\sigma_1^2}}{\dfrac{s_2^2}{\sigma_2^2}} = F(n_1 - 1, n_2 - 1).$$

Constructing Best Tests

The Neyman-Pearson Lemma can sometimes be used to *construct tests* with minimal Type II error for a given Type I error.

In testing the simple hypothesis H_0: $\theta = \theta_0$ against the simple alternative hypothesis H_1: $\theta = \theta_1$, if there exists a region R and a constant k such that $\dfrac{L(X,\theta_0)}{L(X,\theta_1)} \leq k$ inside R, and $\dfrac{L(X,\theta_0)}{L(X,\theta_1)} \geq k$ outside R, where $L(X, \theta)$ is the likelihood function, then R is the best critical region of size α.

Tests on Two Samples

There are several cases.

1) If the population variances are known,

$$z = \frac{(\overline{X}-\overline{Y})-(\mu_X-\mu_Y)}{\sqrt{\dfrac{\sigma_X^2}{n_x}+\dfrac{\sigma_Y^2}{n_y}}} \quad \text{is a } N(0, 1) \text{ variable,}$$

2) If the population variances are unknown but can be presumed to be equal,

$$t_{n_A+n_B-2} = \frac{(\overline{X}-\overline{Y})-(\mu_X-\mu_Y)}{s_p\sqrt{\dfrac{1}{n_X}+\dfrac{1}{n_Y}}} \quad \text{where}$$

$$s_p^2 = \frac{(n_x-1)s_X^2+(n_y-1)s_Y^2}{n_x+n_y-2}.$$

3) If the population variances are unknown and cannot be presumed to be equal, then T is approximately a Student t variable with ν degrees of freedom where

$$T = \frac{(\overline{X}-\overline{Y})-(\mu_X-\mu_Y)}{\sqrt{\dfrac{s_X^2}{n_x}+\dfrac{s_Y^2}{n_y}}} \quad \text{and}$$

$$\nu = \frac{\left(\dfrac{s_X^2}{n_X}+\dfrac{s_Y^2}{n_Y}\right)^2}{\dfrac{\left(\dfrac{s_X^2}{n_X}\right)^2}{n_X-1}+\dfrac{\left(\dfrac{s_Y^2}{n_Y}\right)^2}{n_Y-1}}.$$

Paired Observations

When an experiment is designed with *pairs of observations* from each of two populations, the difference, d, between the pairs of observations is approximately a Student t variable with $n_d - 1$ degrees of freedom where

$$t_{n_d-1} = \frac{\overline{d}-\mu}{s_d/\sqrt{n_d}}.$$

A Test for Contingency Tables	A *test for* the independence of the categories in an *r* by *k contingency table* with observations O_m is based on the statistic $$\chi^2_{(r-1)\cdot(k-1)} = \sum_{m=1}^{r\cdot k} \frac{(O_m - E_m)^2}{E_m}$$ where the cell expectations, the E_m, are found using the hypothesis of independence.
Fitting Observations to a Theoretical Distribution	If E_m is the expectation, based on a theoretical distribution for a category with observation O_m then a test that the *observations* arose from the *theoretical distribution* is based on $$\chi^2 = \sum_m \frac{(O_m - E_m)^2}{E_m}$$ where the degrees of freedom is determined by $$df = number\ of\ cells$$ $$- number\ of\ parameters\ estimated - 1$$
Q-Q Plots	A graphic test for the goodness of fit of observations to a theoretical distribution is based on a graph of the set of points $(x_{q,th}, x_{q,obs})$ where $x_{q,th}$ is a quantile from the theoretical distribution and $x_{q,obs}$ is the corresponding quantile from the set of observations. If the points are well-approximated by a straight line then the observations probably arose from the theoretical distribution. The plot is called a *q-q*, or *quantile-quantile plot*.

4 Simple Linear Regression—Summarizing Data with Equations

4.1 Introduction

W E NOW COME TO A FUNDAMENTAL PROBLEM in science and engineering, that of *model building.* In this chapter, we seek to explain in what manner one random variable is dependent upon another random variable. In Chapter 3, we considered more than one random variable and considered tests, such as the equality of means, but we did not consider that one of the variables might be a function of the other variable.

Science and engineering are often concerned with discovering models that explain parts of the physical world. Relations such as $I = E/R$ or $E = mc^2$ are familiar examples of models in which one variable is a linear function of one or more other variables. In this chapter, we will explore the case where a variable is dependent upon a single independent variable, a situation called **simple linear regression.** (The case where the dependent variable is a function of more than one independent variable is the subject of Chapter 5.)

Unlike the models above, which arise from theoretical considerations, the models we seek are established through sampling and so are based on only some of the possible observations. Our models are then established with some uncertainty. Simple linear regression is, in reality, not so simple, although that is the manner in which it is characterized in the statistical literature.

We begin with an example.

Example 4.1.1   A manufacturer of coils for portable heaters is testing the quality of the coil. If the coils are used under normal operating temperatures to determine how long they last, normally a period of thousands of hours of use over a period of several years, the coils might well become obsolete while they are being tested because of the speed of technical advances.

Manufacturers therefore often perform *accelerated life testing* in which the product is purposely tested under severe conditions. The following data from a test on heater coils show the length of their useful life in hours at various very high temperatures (in degrees Fahrenheit).

Temperature (X)	500	600	700	800	900
Length of Life (Y)	804	791	658	599	562

The data raise several questions. Is it reasonable to assume that X and Y are linearly related, say by the linear equation $Y = \alpha + \beta X$? Of course, we mean that the linear equation represents a reasonable, not an exact, relationship between the variables. If this is so, what are α and β? What is an estimate for the length of life had the temperature $625°F$ been observed? Can we give a reasonable estimate in the form of an interval for this length of life? (Other questions arise as well, many of which we will answer in this chapter.) ∎

This is an example of data collected in the form of points, or pairs of data, such as in the data set $\{(x_1, y_1), (x_2, y_2), (x_3, y_3), \ldots, (x_n, y_n)\}$. The variables here, X and Y, might be summarized individually, but it is usually important to decide if they behave in some detectable fashion together. For example, it might be important to determine if Y is dependent on X according to the linear function above or according to some more general function $Y = f(X)$ for which the data set is a good fit. So the analysis of data of this form takes quite a different character from the analysis of the types of data we have considered up to this point. X is often called a **predictor variable,** and Y is called a **response variable.**

We will begin in this chapter by seeking a linear function to summarize data; in the next chapter we will expand our horizons considerably and allow for nonlinear functions as well as functions of more than one variable to perform the data summarization.

The topics considered in this and the next three chapters are among the most important applications of statistics in science and engineering since they arise in many experimental settings.

4.2 Least Squares

Let us suppose that we have two variables that are in reality linearly related; that is, their true relationship is of the form $Y = \alpha + \beta X$. This relationship might arise from some theoretical considerations or it might be entirely hypothetical on the part of the experimenter. Its exact form could be verified only by the entire population, but of course we have only a finite data set, or sample,

$$\{(x_1, y_1), (x_2, y_2), (x_3, y_3), \ldots, (x_n, y_n)\}$$

of pairs of data for the random variables X and Y.

Nature does not allow us to observe the function $\alpha + \beta X$ directly; if we could do so, two data points would be sufficient to find both α and β. Instead, nature adds a random variable, which we denote by ε, every time we observe a value for Y at a particular value for the independent variable X. So the values of the random variable Y we observe arise from the model $Y = \alpha + \beta X + \varepsilon$.

We begin by asking, how can the data set be used to estimate, in some optimal way, the unknown parameters α and β? In this next example we continue with the problems raised in Example 4.1.1.

Example 4.2.1 The data should be plotted first in order to see that there is no obvious non-linear relationship between the variables. The data, shown in Figure 4.1, appear to be fairly linear, and no periodic or curvilinear relationship is evident. In fact, since the data appear to be closely variable about a straight line, we seek the best approximation to this straight line. However, we must give a caution at this point. Drawing conclusions from a data set on the basis of a picture is quite often dangerous. See, for example, Figure 4.2, which shows that the data are quite variable, although it too shows that Y decreases as X increases.

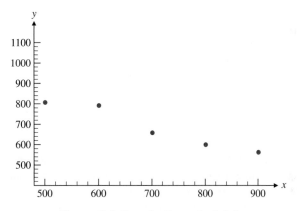

Figure 4.1 Data for Example 4.1.1

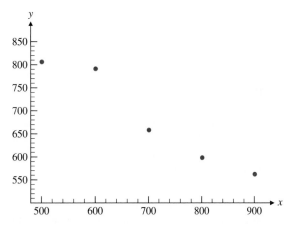

Figure 4.2 Data for Example 4.1.1

The point is that Figures 4.1 and 4.2 show exactly the *same* data! The vertical scale used in Figure 4.1 has simply been stretched to a different vertical scale to produce the different picture shown in Figure 4.2.

As this example shows, the usefulness of visualization is limited when it comes to drawing conclusions concerning linearity. While crude estimates can be made, it would be very difficult to make accurate estimates of α, the vertical intercept of the line, and β, the slope of the line, based on visualization alone.

One might guess, for example, that $\alpha = 1000$ and that $\beta = -0.70$ or that $\alpha = 1150$ and that $\beta = -0.85$. How can one measure which set of estimates is better? Consider the first set of estimates, $\alpha = 1000$, $\beta = -0.70$. We could use the resulting line, $Y = 1000 - 0.70X$, to find how close the predicted values of Y from this line are to the actual observed values of Y. The results are shown in Table 4.1.

X	Y	Predicted Y	Error	Squared Error
500	804	650	154	23,716
600	791	580	211	44,521
700	658	510	148	21,904
800	599	440	159	25,281
900	562	370	192	36,864
Total				152,286

Table 4.1

In Table 4.1, the errors are the differences between the observed and the predicted values of Y. We have also squared these errors and added them.

The predicted values as a rule are somewhat close to the observed values, so, on that basis alone, the model $Y = 1000 - 0.70X$ is not a bad one, but this is not to say that there is not a better model to fit the data. The errors or deviations themselves are of interest, but their sum might not be very meaningful since it is possible that in some situations negative deviations could offset positive deviations. In order to remove the signs and avoid this possibility, it would appear sensible to square the deviations and add them up, which is what we have done. In this case, the sum is 152,286. The details are shown in Table 4.1.

We will not show the details, but if we were to make similar calculations for the model $Y = 1150 - 0.85X$, the sum of the squares of the deviations would be 87,621. On this basis alone, then, the model $Y = 1150 - 0.85X$ is a far better one. As we alter the model, the deviations change in size and sign; it may not be obvious, in the absence of numerical calculations, that the sum of squares we have calculated varies at all. This example shows that it does. In fact, Figure 4.3 shows a graph of the sum of squares as a function of various values of α and β. ∎

The function shown as the quadratic surface in Figure 4.3 attains a minimum value. We define the values of α and β for which the sum of squares is a minimum as the "best" values for α and β. We denote these, consistent with our work in estimation, as $\widehat{\alpha}$ and

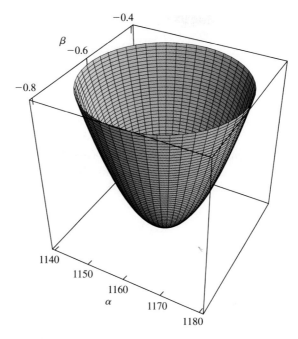

Figure 4.3 Sum of Squares for Example 4.2.1

$\widehat{\beta}$. Now our estimates $\widehat{\alpha}$ and $\widehat{\beta}$ are highly dependent upon the random sample selected and so they are random variables. Before discovering some of the properties of these estimators as random variables, we generalize the procedure we have just followed.

Principle of Least Squares

For the model $Y = \alpha + \beta X$, the best estimators of α and β based on the data $\{(x_1, y_1), (x_2, y_2), (x_3, y_3), \ldots, (x_n, y_n)\}$ are $\widehat{\alpha}$ and $\widehat{\beta}$ where $\widehat{\alpha}$ and $\widehat{\beta}$ minimize the sum of the squares of the deviations we have considered previously. So we wish to minimize

$$S = \sum_{i=1}^{n} (y_i - \alpha - \beta x_i)^2.$$

The estimators $\widehat{\alpha}$ and $\widehat{\beta}$ making S a minimum are called **least squares estimators.**

It is not difficult to determine formulas for $\widehat{\alpha}$ and $\widehat{\beta}$. The minimum values of S are attained where

$$\frac{\partial S}{\partial \alpha} = 0 \quad \text{and} \quad \frac{\partial S}{\partial \beta} = 0.$$

We see that

$$\frac{\partial S}{\partial \alpha} = \frac{\partial}{\partial \alpha} \sum_{i=1}^{n} (y_i - \alpha - \beta x_i)^2 = 2 \sum_{i=1}^{n} (y_i - \alpha - \beta x_i)(-1)$$

and

$$\frac{\partial S}{\partial \beta} = \frac{\partial}{\partial \beta} \sum_{i=1}^{n} (y_i - \alpha - \beta x_i)^2 = 2 \sum_{i=1}^{n} (y_i - \alpha - \beta x_i)(-x_i).$$

Ignoring the factors of -2 in each equation, upon expanding each sum, simplifying, and equating each to 0, the following equations in $\widehat{\alpha}$ and $\widehat{\beta}$ (the specific values that satisfy the two equations) result:

$$n\widehat{\alpha} + \widehat{\beta} \sum_{i=1}^{n} x_i = \sum_{i=1}^{n} y_i \quad \text{and} \quad \widehat{\alpha} \sum_{i=1}^{n} x_i + \widehat{\beta} \sum_{i=1}^{n} x_i^2 = \sum_{i=1}^{n} x_i y_i. \tag{1}$$

The line of the form $y_i = \widehat{\alpha} + \widehat{\beta} x_i$ is called the **least squares regression line** for the data set $\{(x_1, y_1), (x_2, y_2), (x_3, y_3), \ldots, (x_n, y_n)\}$.

Dividing the first least squares equation through by n gives $\widehat{\alpha} + \widehat{\beta}\overline{x} = \overline{y}$, showing that the least squares regression line passes through the point $(\overline{x}, \overline{y})$.

Equations (1) can be solved for both $\widehat{\alpha}$ and $\widehat{\beta}$. We find that

$$\widehat{\beta} = \frac{n \cdot \sum_{i=1}^{n} x_i y_i - \left(\sum_{i=1}^{n} x_i\right) \cdot \left(\sum_{i=1}^{n} y_i\right)}{n \sum_{i=1}^{n} x_i^2 - \left(\sum_{i=1}^{n} x_i\right)^2}.$$

In this example, $\sum_{i=1}^{5} x_i = 3500$, $\sum_{i=1}^{5} y_i = 3414$, $\sum_{i=1}^{5} x_i^2 = 2{,}550{,}000$, and $\sum_{i=1}^{5} x_i y_i = 2{,}322{,}200$ so

$$\widehat{\beta} = \frac{5 \cdot 2{,}322{,}200 - (3500) \cdot (3414)}{5 \cdot 2{,}550{,}000 - (3500)^2} = -\frac{169}{250} = -0.676.$$

The formula for $\widehat{\beta}$ can be rewritten as

$$\widehat{\beta} = \frac{\sum_{i=1}^{n} (x_i - \overline{x})(y_i - \overline{y})}{\sum_{i=1}^{n} (x_i - \overline{x})^2}$$

and we have previously seen that

$$\widehat{\alpha} = \overline{y} - \widehat{\beta}\overline{x}.$$

In this example, then,

$$\widehat{\alpha} = \frac{3414}{5} + 0.676 \cdot \frac{3500}{5} = 1156.$$

These computations are shown in order to illustrate the use of the formulas. The actual computation of $\widehat{\alpha}$ and $\widehat{\beta}$ is usually done by computer, especially for large data sets. However, we will have considerable theoretical use for the formulas; they are more important in this sense than for computation.

Minitab gives the least squares regression line as $y_i = 1156 - 0.676x_i$ for the data in this example, as we have calculated. It can also be found that the sum of squares minimized by the principle of least squares is $\sum_{i=1}^{5}(y_i - 1156 + 0.676x_i)^2 = 2929.2$.

A Vector Formulation

How can we explain the fact that the observations do not lie exactly on a straight line? It is useful to think that the observations arise from two sources: a deterministic part of the model, $\widehat{\alpha} + \widehat{\beta}x_i$, and a random part, ε_i. We could show each observed data point as being composed of the predicted part from the least squares line plus a random error. Each of these errors could actually be calculated for each observed point and then added to the prediction from the least squares line, producing the observation.

For example, for the first observation, the point (500, 804), we find that the predicted value at $x = 500$ is $1156 - 0.676 \cdot 500 = 818$. So, expressing the observation as the sum of the predicted value and an error, we have that

$$804 = (1156 - 0.676 \cdot 500) - 14$$

or

$$804 = 818 - 14.$$

All of the data points could be expressed in this way at the same time most efficiently by using vectors.

If we write the vector of observations as

$$\begin{bmatrix} 804 \\ 791 \\ 658 \\ 599 \\ 562 \end{bmatrix},$$

we can decompose this into the sum of the vectors

$$\begin{bmatrix} 818.0 \\ 750.4 \\ 682.8 \\ 615.2 \\ 547.6 \end{bmatrix} \quad \text{and} \quad \begin{bmatrix} -14 \\ 40.6 \\ -24.8 \\ -16.2 \\ 14.4 \end{bmatrix}$$

representing the predictions from the model and the random errors, respectively. So

$$\begin{bmatrix} 804 \\ 791 \\ 658 \\ 599 \\ 562 \end{bmatrix} = \begin{bmatrix} 818.0 \\ 750.4 \\ 682.8 \\ 615.2 \\ 547.6 \end{bmatrix} + \begin{bmatrix} -14 \\ 40.6 \\ -24.8 \\ -16.2 \\ 14.4 \end{bmatrix}.$$

Using the intercept and the slope, we could further decompose the prediction vector as

$$\begin{bmatrix} 818 \\ 750.4 \\ 682.8 \\ 615.2 \\ 547.6 \end{bmatrix} = \begin{bmatrix} 1156 \\ 1156 \\ 1156 \\ 1156 \\ 1156 \end{bmatrix} - 0.676 \begin{bmatrix} 500 \\ 600 \\ 700 \\ 800 \\ 900 \end{bmatrix}$$

so we have

$$\begin{bmatrix} 804 \\ 791 \\ 658 \\ 599 \\ 562 \end{bmatrix} = \begin{bmatrix} 1156 \\ 1156 \\ 1156 \\ 1156 \\ 1156 \end{bmatrix} - 0.676 \begin{bmatrix} 500 \\ 600 \\ 700 \\ 800 \\ 900 \end{bmatrix} + \begin{bmatrix} -14 \\ 40.6 \\ -24.8 \\ -16.2 \\ 14.4 \end{bmatrix}.$$

This can also be written as

$$\begin{bmatrix} 804 \\ 791 \\ 658 \\ 599 \\ 562 \end{bmatrix} = \begin{bmatrix} 1 & 500 \\ 1 & 600 \\ 1 & 700 \\ 1 & 800 \\ 1 & 900 \end{bmatrix} \begin{bmatrix} 1156 \\ -0.676 \end{bmatrix} + \begin{bmatrix} -14 \\ 40.6 \\ -24.8 \\ -16.2 \\ 14.4 \end{bmatrix}.$$

The 5×2 matrix

$$\begin{bmatrix} 1 & 500 \\ 1 & 600 \\ 1 & 700 \\ 1 & 800 \\ 1 & 900 \end{bmatrix},$$

which we generally denote by **X,** has a column of $1's$ corresponding to the constant value in the equation as well as a second column corresponding to the values of X. The vector

$$\begin{bmatrix} 1156 \\ -0.676 \end{bmatrix}$$

is

$$\begin{bmatrix} \widehat{\alpha} \\ \widehat{\beta} \end{bmatrix}$$

and is generally denoted by $\widehat{\beta}$, while the vector

$$\begin{bmatrix} -14.0 \\ 40.6 \\ -24.8 \\ -16.2 \\ 14.4 \end{bmatrix}$$

is the vector of residuals or the vector of differences between the vector of observations and the vector of predictions. We denote this by ε and the vector of observations by $\mathbf{Y}$, so our vector equation becomes $\mathbf{Y} = \mathbf{X}\beta + \varepsilon$.

EXERCISES 4.2

1. An engineer is measuring the strength (Y) of a material as a function of the amount of an ingredient (X). The data, which have been coded, are as follows.

x	1	2	3	4	5	6
y	4	7	12	14	18	17

a. Make a scatter plot of the data.
b. By examining the scatter plot, estimate the values of α and β in the model $y_i = \alpha + \beta x_i + \varepsilon_i$, $i = 1, 2, \ldots, 6$.
c. Calculate the predicted values, $\widehat{y}_i = \widehat{\alpha} + \widehat{\beta} x_i$, $i = 1, 2, \ldots, 6$ using your estimates for $\widehat{\alpha}$ and $\widehat{\beta}$ from part (b).
d. Calculate $\sum_{i=1}^{6}(y_i - \widehat{y}_i)^2$.
e. Find $\widehat{\alpha}$ and $\widehat{\beta}$ using the principle of least squares and calculate $\sum_{i=1}^{6}(y_i - \widehat{y}_i)^2$ using the least squares estimators.
f. Show the data in the form $\mathbf{Y} = \mathbf{X}\beta + \varepsilon$.

2. In order to illustrate some simple ideas in physics, an instructor has five groups of students measure the velocity of a freely falling object at five different times. The experiment is arranged so that the initial velocity of the object is unknown. The data follow where x is in minutes and y is in inches/minute.

x	0.5	1.0	1.5	2.0	2.5
y	1.4	3.2	4.4	5.7	8.0

a. Make a scatter plot of the data.
b. By examining the scatter plot, estimate the values of α and β in the model $y_i = \alpha + \beta x_i + \varepsilon_i$, $i = 1, 2, \ldots, 5$.

c. Calculate the predicted values, $\widehat{y}_i = \widehat{\alpha} + \widehat{\beta} x_i$, $i = 1, 2, \ldots, 5$ using your estimates for $\widehat{\alpha}$ and $\widehat{\beta}$ from part (b).
d. Calculate $\sum_{i=1}^{5}(y_i - \widehat{y}_i)^2$.
e. Find $\widehat{\alpha}$ and $\widehat{\beta}$ using the principle of least squares and calculate $\sum_{i=1}^{5}(y_i - \widehat{y}_i)^2$ using the least squares estimators.
f. Show the data in the form $\mathbf{Y} = \mathbf{X}\beta + \varepsilon$.
g. Estimate the initial velocity of the object.

3. Raw material used in the production of a synthetic fiber is stored in a place that has no humidity control. Measurements of the relative humidity in the storage place and the moisture content of a sample of the raw material, both measured as percentages, on 12 days were as follows.

Humidity (X)	46	53	37	42	34	29	60	44	41	48	33	40
Moisture (Y)	12	14	11	13	10	8	17	12	10	15	9	13

a. Make a scatter plot of the data.
b. Estimate the values of α and β in the model $y_i = \alpha + \beta x_i + \varepsilon_i$, $i = 1, 2, \ldots, 12$.
c. Calculate the predicted values, $\widehat{y}_i = \widehat{\alpha} + \widehat{\beta} x_i$, $i = 1, 2, \ldots, 12$ using your estimates for $\widehat{\alpha}$ and $\widehat{\beta}$ from part (b).
d. Calculate $\sum_{i=1}^{12}(y_i - \widehat{y}_i)^2$.
e. Find $\widehat{\alpha}$ and $\widehat{\beta}$ using the principle of least squares and calculate $\sum_{i=1}^{12}(y_i - \widehat{y}_i)^2$ using the least squares estimators.
f. Show the data in the form $\mathbf{Y} = \mathbf{X}\beta + \varepsilon$.

4. Use the principle of least squares to derive the least squares equations for fitting the model $y_i = ae^{bx_i}$ to the data set $\{(x_i, y_i)| i = 1, 2, \ldots, n\}$, $i = 1, 2, \ldots, n$. Simplify the resulting equations as much as possible, but do not attempt to solve for $\widehat{a}$ and $\widehat{b}$.

COMPUTER EXPERIMENT

    a. Generate 300 observations from the model $y_i = 3 + 5x_i + \varepsilon_i$ where $x_i = 1,2,\ldots,10$ and $\varepsilon_i \sim N(0,12)$.
    b. Make a scatter plot of the data.
    c. Find the least squares regression line and compare the results to the expected results.
    d. Find the value of $\sum_{i=1}^{300}(y_i - \widehat{y}_i)^2$.

4.3   A Linear Model

Let us now express the considerations discussed in the previous section in a more formal way. This will allow us to determine some properties of $\widehat{\alpha}$ and $\widehat{\beta}$ and, more importantly, allow us to determine their probability distributions as random variables.

Recall Figure 4.1, which shows that the data points for Example 4.1.1 do not lie exactly along a straight line. Consistent with our observations in Section 4.2, it is useful to explain this by supposing that nature indicates that for a fixed value, say x_i, we cannot observe $\alpha + \beta x_i$ directly. If we could observe $\alpha + \beta x_i$, then two observations, for different values of x, would be sufficient to determine both α and β. There must be some factor that is causing a disturbance in the observations. We suppose that the observed values, y_i, arise both from the true value, $\alpha + \beta x_i$, and some added random variable, say ε_i, so we hypothesize a model as

$$y_i = \alpha + \beta x_i + \varepsilon_i \quad \text{for } i = 1, 2, \ldots, n.$$

What shall we suppose about the random variables, ε_i? It is common to suppose that these are independently distributed normal random variables with mean 0 and common variance σ^2, that is,

$$\varepsilon_i \sim NID(0, \sigma) \quad \text{for } i = 1, 2, \ldots, n.$$

Now, to generalize the linear model in the previous section, we let

$$\mathbf{Y} = \begin{bmatrix} y_1 \\ y_2 \\ \vdots \\ y_n \end{bmatrix}, \quad \mathbf{X} = \begin{bmatrix} 1 & x_1 \\ 1 & x_2 \\ \vdots & \vdots \\ 1 & x_n \end{bmatrix}, \quad \beta = \begin{bmatrix} \alpha \\ \beta \end{bmatrix}, \quad \text{and } \varepsilon = \begin{bmatrix} \varepsilon_1 \\ \varepsilon_2 \\ \vdots \\ \varepsilon_n \end{bmatrix}$$

so the model is

$$\mathbf{Y} = \mathbf{X}\beta + \varepsilon.$$

This is our first example of a **linear model.** The word *linear* here refers to the linearity of the parameters α and β, not to the data $\{x_i, y_i\}$, which are known.

Maximum Likelihood

The linear model uses the assumption that $\varepsilon_i \sim NID(0, \sigma)$, so, since $y_i = \alpha + \beta x_i + \varepsilon_i$, and since $\alpha + \beta x_i$ is constant for a fixed value x_i, each value of y_i is composed of a constant plus a normal random variable. Since $E(\varepsilon_i) = 0$, this means that the y_i's are normally distributed with mean $\alpha + \beta x_i$ and variance σ^2.

We can then use the principle of maximum likelihood to estimate the regression parameters. The likelihood function is

$$L = \frac{1}{\sigma\sqrt{2\pi}} e^{-\frac{1}{2}(y_1-\alpha-\beta x_1)^2} \cdot \frac{1}{\sigma\sqrt{2\pi}} e^{-\frac{1}{2}(y_2-\alpha-\beta x_2)^2} \cdots \frac{1}{\sigma\sqrt{2\pi}} e^{-\frac{1}{2}(y_n-\alpha-\beta x_n)^2},$$

which can be written as

$$L = \frac{1}{(\sigma\sqrt{2\pi})^n} e^{-\frac{1}{2}\sum_{i=1}^{n}(y_i-\alpha-\beta x_i)^2},$$

showing that the likelihood function is maximized when $\sum_{i=1}^{n}(y_i - \alpha - \beta x_i)^2$ is minimized. This, of course, is exactly the principle of least squares. Thus we see that the least squares estimators for regression parameters are also maximum likelihood estimators provided the errors are normally distributed.

4.4 Properties of Least Squares Estimators

Having used the principle of least squares to find estimators, $\widehat{\alpha}$ and $\widehat{\beta}$, for the parameters α and β, we now determine some properties arising from the fact that $S = \sum_{i=1}^{n}(y_i - \alpha - \beta x_i)^2$ attains a minimum value for these estimators.

We have seen that

$$\widehat{\beta} = \frac{\sum\limits_{i=1}^{n}(x_i - \overline{x})(y_i - \overline{y})}{\sum\limits_{i=1}^{n}(x_i - \overline{x})^2}.$$

Consider the x-values to be fixed and let

$$a_i = \frac{x_i - \overline{x}}{\sum\limits_{j=1}^{n}(x_j - \overline{x})^2}.$$

For a fixed value of i, a_i is a constant. So we can write $\widehat{\beta} = \sum_{i=1}^{n} a_i(y_i - \overline{y})$.
But

$$\sum_{i=1}^{n} a_i = \sum_{i=1}^{n} \frac{x_i - \overline{x}}{\sum\limits_{j=1}^{n}(x_j - \overline{x})^2} = \frac{1}{\sum\limits_{j=1}^{n}(x_j - \overline{x})^2} \sum_{i=1}^{n}(x_i - \overline{x}) = 0$$

since $\sum_{i=1}^{n}(x_i - \overline{x}) = 0$.

We can then write

$$\widehat{\beta} = \sum_{i=1}^{n} a_i(y_i - \overline{y}) = \sum_{i=1}^{n} a_i y_i - \sum_{i=1}^{n} a_i \overline{y} = \sum_{i=1}^{n} a_i y_i - \overline{y} \sum_{i=1}^{n} a_i$$

$$\widehat{\beta} = \sum_{i=1}^{n} a_i y_i.$$

The estimator $\widehat{\beta}$ is then a weighted average of the observed y-values. The average is a weighted one since the a_i's are not constant from one value of i to another. It is interesting to look more closely at $\widehat{\beta}$ as a weighted average. Consider the data in Example 4.1.1. Since $\overline{x} = 700$, and since it is easy to calculate that $\sum_{i=1}^{5}(x_i - \overline{x})^2 = 100{,}000$, we have that $a_i = (x_i - 700)/100{,}000$ so we find that

$$a_1 = \frac{x_1 - 700}{100{,}000} = \frac{500 - 700}{100{,}000} = -\frac{1}{500},$$

$$a_2 = \frac{x_2 - 700}{100{,}000} = \frac{600 - 700}{100{,}000} = -\frac{1}{1000},$$

$$a_3 = \frac{x_3 - 700}{100{,}000} = \frac{700 - 700}{100{,}000} = 0,$$

$$a_4 = \frac{x_4 - 700}{100{,}000} = \frac{800 - 700}{100{,}000} = \frac{1}{1000},$$

and

$$a_5 = \frac{x_5 - 700}{100{,}000} = \frac{900 - 700}{100{,}000} = \frac{1}{500}.$$

This gives

$$\widehat{\beta} = \sum_{i=1}^{n} a_i y_i = -\frac{1}{500}(804) - \frac{1}{1000}(791) + 0 \cdot (658) + \frac{1}{1000}(599) + \frac{1}{500}(562)$$

$$= -0.676$$

as before.

While this procedure cannot be recommended for purposes of calculation, it does illustrate some important points. For one, since $a_i = 0$ when $x_i = \overline{x}$, the value of y when $x_i = \overline{x}$ plays absolutely no role whatsoever in the calculation of $\widehat{\beta}$. So the value of y when $x_i = \overline{x}$ can be altered at will without having the least effect upon the slope of the regression line, but we will see that the value of $\widehat{\alpha}$ will be affected because that quantity depends upon $\overline{y}$.

More importantly, we see that the farther x_i is from $\overline{x}$, the larger a_i becomes and hence the greater influence the corresponding value of y has upon $\widehat{\beta}$. For example, if we were to change the value of y when $x = 900$ to 662 from 562, the value of $\widehat{\beta}$ would increase by 0.2. Different data points then exercise entirely different influences on the slope of the regression line.

This observation has some important consequences. Occasionally, a data point appears to be unusual and may perhaps be an outlier. Such points can arise from random variation, but are much more likely to arise from some experimental anomaly. In any event, we can say here that we can measure the influence every data point has upon the entire regression line and decide whether or not to regard a point as an outlier. Since we do not have enough theoretical material to work with at this point, we cannot pursue this idea further here, but we will return to it in Section 4.9.

We now develop some additional theoretical ideas.

Expected Values and Variances

The model for simple linear regression is

$$y_i = \alpha + \beta x_i + \varepsilon_i \quad \text{for } i = 1, 2, \ldots, n$$

where $\varepsilon_i \sim NID(0, \sigma)$.

Now we also know that $\widehat{\beta} = \sum_{i=1}^{n} a_i y_i$, so

$$E(\widehat{\beta}) = E \left(\sum_{i=1}^{n} a_i y_i \right) = \sum_{i=1}^{n} a_i E(y_i)$$

since a_i is a constant for each fixed i. But $E(y_i) = \alpha + \beta x_i$, so

$$E(\widehat{\beta}) = \sum_{i=1}^{n} a_i (\alpha + \beta x_i) = \alpha \sum_{i=1}^{n} a_i + \beta \sum_{i=1}^{n} a_i x_i$$

$$= 0 + \beta \sum_{i=1}^{n} a_i (x_i - \overline{x}) = \beta \frac{\sum_{i=1}^{n} (x_i - \overline{x})^2}{\sum_{j=1}^{n} (x_j - \overline{x})^2} = \beta,$$

showing that $\widehat{\beta}$ is an unbiased estimator for β.

From the first least squares equation, $\widehat{\alpha} + \widehat{\beta}\overline{x} = \overline{y}$, we see that $\widehat{\alpha} = \overline{y} - \widehat{\beta}\overline{x}$, so

$$E(\widehat{\alpha}) = E(\overline{y} - \widehat{\beta}\overline{x}) = E(\overline{y}) - E(\widehat{\beta}\overline{x}) = \alpha + \beta\overline{x} - \beta\overline{x} = \alpha.$$

So both $\widehat{\alpha}$ and $\widehat{\beta}$ are unbiased estimators. This is not a peculiarity of the simple linear regression model, but is a property of least squares estimators in general. The principle of least squares has many remarkable properties, some of which we will discover here and in subsequent chapters.

We now calculate a variance.

$$\text{Var}(\widehat{\beta}) = \text{Var} \left(\sum_{i=1}^{n} a_i y_i \right) = \sum_{i=1}^{n} a_i^2 \, \text{Var}(y_i) = \sigma^2 \sum_{i=1}^{n} a_i^2 = \frac{\sigma^2}{\sum_{i=1}^{n} (x_i - \overline{x})^2}.$$

We would like to be able to calculate $\text{Var}(\widehat{\alpha})$, but first we must consider variances of sums and differences.

Variances of Sums and Differences

Suppose X and Y are random variables and consider

$$\text{Var}(X+Y) = E[(X+Y) - \mu_{X+Y}]^2 = E[(X+Y) - (\mu_X + \mu_Y)]^2$$

$$= E[(X - \mu_X) + (Y - \mu_Y)]^2$$

$$= E(X - \mu_X)^2 + 2E[(X - \mu_X)(Y - \mu_Y)] + E(Y - \mu_Y)^2.$$

We recognize the first and third terms as $\text{Var}(X)$ and $\text{Var}(Y)$, respectively. The term in the middle is twice the **covariance** of X and Y, a quantity first encountered in Section 2.12. We denote this by $\text{Cov}(X, Y)$. Had we considered $\text{Var}(X - Y)$, the only change above would be in the sign of the covariance, so we conclude that

$$\text{Var}(X \pm Y) = \text{Var}(X) \pm 2\,\text{Cov}(X, Y) + \text{Var}(Y).$$

It is easy to show that

$$\text{Cov}(X, Y) = E(X \cdot Y) - E(X) \cdot E(Y) = E(X \cdot Y) - \mu_X \cdot \mu_Y.$$

In the special case where X and Y are independent, we have $E(X \cdot Y) = E(X) \cdot E(Y)$ and then $\text{Cov}(X, Y) = 0$. So if X and Y are independent, then their covariance is 0. The converse of this is not true, however. If $\text{Cov}(X, Y) = 0$, then X and Y are called **uncorrelated;** they are not necessarily independent.

Now we consider

$$\text{Cov}(aX + bY, cX + dY) = E\left\{[aX + bY - E(aX + bY)] \cdot [cX + dY - E(cX + dY)]\right\}$$

$$= E\left\{[a(X - \mu_X) + b(Y - \mu_Y)] \cdot [c(X - \mu_X) + d(Y - \mu_Y)]\right\}$$

$$= ac E(X - \mu_X)^2 + (ad + bc)E[(X - \mu_X) \cdot (Y - \mu_Y)]$$

$$+ bd E(Y - \mu_Y)^2.$$

So we conclude that

$$\text{Cov}(aX + bY, cX + dY) = ac\,\text{Var}(X) + (ad + bc)\,\text{Cov}(X, Y) + bd\,\text{Var}(Y).$$

Before considering $\text{Var}(\widehat{\alpha})$, consider

$$\text{Cov}(\bar{y}, \widehat{\beta}) = \text{Cov}\left(\frac{1}{n}y_1 + \frac{1}{n}y_2 + \cdots + \frac{1}{n}y_n, a_1 y_1 + a_2 y_2 + \cdots + a_n y_n\right).$$

Since the y_i's are independent, and thus have zero covariances,

$$\text{Cov}(\bar{y}, \widehat{\beta}) = \sum_{i=1}^{n} \frac{1}{n} a_i \,\text{Var}(y_i) = \sigma^2 \cdot \frac{1}{n} \cdot \sum_{i=1}^{n} a_i = 0.$$

Now we can consider $\text{Var}(\widehat{\alpha}) = \text{Var}(\bar{y} - \widehat{\beta}\bar{x})$, which can be written as $\text{Var}(\bar{y}) - 2\bar{x}\,\text{Cov}(\bar{y}, \widehat{\beta}) + \bar{x}^2\,\text{Var}(\widehat{\beta})$ since $\bar{x}$ is a constant.

So $\text{Var}(\widehat{\alpha}) = \text{Var}(\bar{y}) + \bar{x}^2\,\text{Var}(\widehat{\beta})$, and so

$$\text{Var}(\widehat{\alpha}) = \frac{\sigma^2}{n} + \frac{\sigma^2 \bar{x}^2}{\displaystyle\sum_{i=1}^{n}(x_i - \bar{x})^2}$$

or

$$\text{Var}(\widehat{\alpha}) = \sigma^2 \left[\frac{1}{n} + \frac{\overline{x}^2}{\sum\limits_{i=1}^{n} (x_i - \overline{x})^2} \right],$$

and it is easy to simplify this to

$$\text{Var}(\widehat{\alpha}) = \sigma^2 \frac{\sum\limits_{i=1}^{n} x_i^2}{n \sum\limits_{i=1}^{n} (x_i - \overline{x})^2}.$$

We have developed point estimators and have found some of their properties. We could establish some confidence intervals for these estimators, but their calculation is dependent on our knowledge of σ^2, which is not usually known. An estimate of σ^2 is dependent upon some considerations regarding the analysis of variance, so for this reason we postpone a discussion of confidence intervals until Section 4.7. We now turn to a test for the linearity of the data.

4.5 A Test for Linearity

The least squares regression line shown in Section 4.2 can be calculated for any set of data, whether it is linear or not. To cite an extreme example, one could sample points from a semicircle and fit a straight line to them. The fit, obviously, will not be a very good one. The calculation itself will not reveal whether or not the fit is satisfactory or if one could reasonably decide that the data set is in reality linear. These are fundamental questions, and they cannot be answered easily.

Experimenters frequently compare observed values with values predicted by the least squares regression line. If the discrepancies are not serious for the experimenter, then the fitted linear function is an acceptable summarization of the data. This procedure, while practical, lacks theoretical justification and leaves the decision-making process to the judgment of the individual experimenter.

We will now develop a systematic approach for answering the question of the adequacy of the fit of the least squares regression line. We begin with our first encounter with a remarkable partition of numbers associated with the sample and the least squares regression line.

Consider a set of observed values, $\{y_i \,|\, i = 1, 2, \ldots, n\}$, and the set of predicted values, $\{\widehat{y_i} \,|\, i = 1, 2, \ldots, n\}$, these predicted values arising from the least squares regression line, so

$$\widehat{y_i} = \widehat{\alpha} + \widehat{\beta} x_i, \qquad i = 1, 2, \ldots, n.$$

Now consider the sum of the squared deviations of the observed values from their mean,

$$\sum_{i=1}^{n}(y_i - \bar{y})^2.$$

This is the numerator of the variance of the y_i values. It can be rewritten, by adding and subtracting $\widehat{y}_i$, as

$$\sum_{i=1}^{n}(y_i - \bar{y})^2 = \sum_{i=1}^{n}(y_i - \widehat{y}_i + \widehat{y}_i - \bar{y})^2.$$

Squaring and expanding, this can be written as

$$\sum_{i=1}^{n}(y_i - \bar{y})^2 = \sum_{i=1}^{n}(y_i - \widehat{y}_i)^2 + 2\sum_{i=1}^{n}(y_i - \widehat{y}_i)(\widehat{y}_i - \bar{y}) + \sum_{i=1}^{n}(\widehat{y}_i - \bar{y})^2.$$

Since $\widehat{y}_i = \widehat{\alpha} + \widehat{\beta}x_i = \bar{y} + \widehat{\beta}(x_i - \bar{x})$, the middle term (ignoring the factor of 2) becomes

$$\sum_{i=1}^{n}(y_i - \widehat{y}_i)(\widehat{y}_i - \bar{y}) = \sum_{i=1}^{n}[(y_i - \bar{y}) - \widehat{\beta}(x_i - \bar{x})]\widehat{\beta}(x_i - \bar{x})$$

$$= \widehat{\beta}\sum_{i=1}^{n}(y_i - \bar{y})(x_i - \bar{x}) - \widehat{\beta}^2\sum_{i=1}^{n}(x_i - \bar{x})^2.$$

But

$$\widehat{\beta} = \frac{\sum\limits_{i=1}^{n}(y_i - \bar{y})(x_i - \bar{x})}{\sum\limits_{i=1}^{n}(x_i - \bar{x})^2}$$

so

$$\sum_{i=1}^{n}(y_i - \widehat{y}_i)(\widehat{y}_i - \bar{y}) = 0.$$

So we have

$$\sum_{i=1}^{n}(y_i - \bar{y})^2 = \sum_{i=1}^{n}(y_i - \widehat{y}_i)^2 + \sum_{i=1}^{n}(\widehat{y}_i - \bar{y})^2.$$

This is called an **analysis of variance** partition, although it is more accurately described as an analysis of a sum of squares rather than a partition of a variance.

We show these calculations for the data in Example 4.1.1 in Table 4.2.

y_i	$\widehat{y}_i$	$y_i - \overline{y}$	$(y_i - \overline{y})^2$	$\widehat{y}_i - \overline{y}$	$(\widehat{y}_i - \overline{y})^2$	$y_i - \widehat{y}_i$	$(y_i - \widehat{y}_i)^2$
804	818	121.2	14,689.44	135.2	18,279.04	-14	196
791	750.4	108.2	11,707.24	67.6	4,569.76	40.6	1648.36
658	682.8	-24.8	615.04	0	0	-24.8	615.04
599	615.2	-83.8	7,022.44	-67.6	4,569.76	-16.2	262.44
562	547.6	-120.8	14,592.64	-135.2	18,279.04	14.4	207.36
Total			48,626.8	0	45,697.6	0	2929.2

Table 4.2

Here the partition is

$$\sum_{i=1}^{n}(y_i - \overline{y})^2 = 48{,}626.8 = \sum_{i=1}^{n}(y_i - \widehat{y}_i)^2 + \sum_{i=1}^{n}(\widehat{y}_i - \overline{y})^2 = 2929.2 + 45{,}697.6.$$

If the regression line fits well, we expect the observations and predictions to be reasonably close together and hence we expect $\sum_{i=1}^{n}(y_i - \widehat{y}_i)^2$ to be a small proportion of $\sum_{i=1}^{n}(y_i - \overline{y})^2$. It is here. Similarly, if the regression line is a good fit for the data, we expect $\sum_{i=1}^{n}(\widehat{y}_i - \overline{y})^2$ to be a large proportion of $\sum_{i=1}^{n}(y_i - \overline{y})^2$, as it is here. This is evidence that our least squares regression line is a good fit for the data. However, we still lack a mathematical basis for our conclusion. We need to establish this so that we can reach decisions in cases that are much less clear cut than this one.

The facts are as follows. (Establishing the facts claimed here is beyond the scope of this book. Several recent books on regression are listed in the Bibliography, many of which contain the relevant mathematical derivations.)

Recall that our model for this situation is

$$y_i = \alpha + \beta x_i + \varepsilon_i \quad \text{for } i = 1, 2, \ldots, n,$$

where $\varepsilon_i \sim NID(0, \sigma)$.

The assumption that the errors, ε_i, are independent and normally distributed is crucial to the establishment of the following facts.

1. The **total sum of squares,**

$$\sum_{i=1}^{n}(y_i - \overline{y})^2,$$

is a chi-squared random variable with $n - 1$ degrees of freedom.
2. The **sum of squares due to regression,**

$$\sum_{i=1}^{n}(\widehat{y}_i - \overline{y})^2,$$

is a chi-squared random variable with 1 degree of freedom.

3. The **error sum of squares,**

$$\sum_{i=1}^{n}(y_i - \widehat{y}_i)^2,$$

is a chi-squared random variable with $n - 2$ degrees of freedom.

4. The **sums of squares**

$$\sum_{i=1}^{n}(\widehat{y}_i - \overline{y})^2 \quad \text{and} \quad \sum_{i=1}^{n}(y_i - \widehat{y}_i)^2$$

are independently distributed so

5. The **ratio**

$$\frac{\displaystyle\sum_{i=1}^{n}(\widehat{y}_i - \overline{y})^2/1}{\displaystyle\sum_{i=1}^{n}(y_i - \widehat{y}_i)^2/(n-2)}$$

follows the $F(1, n-2)$ distribution.

The numerator in the fraction in fact (5) is known as the **mean square for regression,** and the denominator is known as the **mean square for error.**

These facts are commonly exhibited in an **analysis of variance table.** Minitab gives the analysis of variance table shown in Table 4.3 for the data in Example 4.1.1.

Source	df	Sum of Squares	Mean Square	F	p
Regression	1	45,697.6	45,697.6	46.8021	0.00639199
Error	3	2,929.2	976.4		
Total	4	48,626.8			

Table 4.3

The analysis of variance table for simple linear regression in general is shown in Table 4.4.

Source	df	Sum of Squares	Mean Square	F	p
Regression	1	$\displaystyle\sum_{i=1}^{n}(\widehat{y}_i - \overline{y})^2$	$\displaystyle\sum_{i=1}^{n}(\widehat{y}_i - \overline{y})^2/1$	$\dfrac{\displaystyle\sum_{i=1}^{n}(\widehat{y}_i - \overline{y})^2/1}{\displaystyle\sum_{i=1}^{n}(y_i - \widehat{y}_i)^2/(n-2)}$	
Error	$n-2$	$\displaystyle\sum_{i=1}^{n}(y_i - \widehat{y}_i)^2$	$\displaystyle\sum_{i=1}^{n}(y_i - \widehat{y}_i)^2/(n-2)$		
Total	$n-1$	$\displaystyle\sum_{i=1}^{n}(y_i - \overline{y})^2$			

Table 4.4

The column headed **p** is the p-value for the F value and in this case is the probability that $F(1, n - 2)$ exceeds the calculated value.

The Hypothesis

What hypothesis is being tested by the analysis of variance table and how is the p-value to be interpreted?

If it should happen that $\sum_{i=1}^{n}(\widehat{y}_i - \overline{y})^2$ is small, we must conclude that each of the predicted values, the $\widehat{y}_i$, must be close to $\overline{y}$. This means that $\widehat{y}_i$ is not dependent upon x_i, so the slope of the regression line would be 0 or very close to it. In this case, $\sum_{i=1}^{n}(\widehat{y}_i - \overline{y})^2$ is a small proportion of $\sum_{i=1}^{n}(y_i - \overline{y})^2$ and the F value is small.

On the other hand, if $\sum_{i=1}^{n}(\widehat{y}_i - \overline{y})^2$ is large, then the predicted values tend to be dissimilar to the constant $\overline{y}$ and the F value is large. In this case, $\widehat{y}_i$ is highly dependent upon x_i and the true slope of the regression line is not 0.

Figure 4.4 shows an instance where β, the true slope of the regression line, is 0.

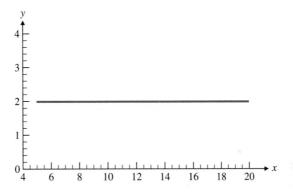

Figure 4.4 A Line with Slope 0

It is obvious in the case where $\beta = 0$ that the observed values y_i are not in fact dependent upon the value of x that is chosen, since the regression line has exactly the same height for every value of x. This is a trivial case, and one for which a regression line is unnecessary. We are interested then in the case where $\beta \neq 0$. The F ratio in the analysis of variance table then tests

$$H_0: \beta = 0$$

against

$$H_1: \beta \neq 0.$$

Since

$$F(1, n - 2) = \frac{\sum\limits_{i=1}^{n}(\widehat{y}_i - \overline{y})^2/1}{\sum\limits_{i=1}^{n}(y_i - \widehat{y}_i)^2/(n - 2)},$$

the null hypothesis should be rejected only if F is too large. The p-value in the analysis of variance table in (Table 4.3) indicates that $P[F(1, 3) > 46.8021] = 0.00639199$. No doubt H_0 would be rejected in favor of H_1.

EXERCISES 4.5

1. Consider again the data given in Exercise 1 of Exercises 4.2:

x	1	2	3	4	5	6
y	4	7	12	14	18	17

a. Use the data and the least squares line found in Exercise 1 of Exercises 4.2 to find:

1. $\sum_{i=1}^{n}(\widehat{y}_i - \overline{y})^2$,

2. $\sum_{i=1}^{n}(y_i - \widehat{y}_i)^2$, and

3. $\sum_{i=1}^{n}(y_i - \overline{y})^2$

and show that the analysis of variance partition holds.

b. Complete the analysis of variance table and draw a conclusion about H_0: $\beta = 0$ against H_1: $\beta \neq 0$.

2. a. In the previous problem, calculate the weight,

$$a_i = \frac{x_i - \overline{x}}{\sum_{i=1}^{n}(x_i - \overline{x})^2},$$

with which each y_i is weighted in the calculation of $\widehat{\beta}$.

b. Use the a_i to calculate $\widehat{\beta}$.

c. Change the last data point to (6, 14.4) and determine the new value of $\widehat{\beta}$.

3. Consider again the data given in Exercise 2 of Exercises 4.2:

x	0.5	1.0	1.5	2.0	2.5
y	1.4	3.2	4.4	5.7	8.0

a. Use the data and the least squares line found in Exercise 2 of Exercises 4.2 to find:

1. $\sum_{i=1}^{n}(\widehat{y}_i - \overline{y})^2$,

2. $\sum_{i=1}^{n}(y_i - \widehat{y}_i)^2$, and

3. $\sum_{i=1}^{n}(y_i - \overline{y})^2$.

and show that the analysis of variance partition holds.

b. Complete the analysis of variance table and draw a conclusion about H_0: $\beta = 0$ against H_1: $\beta \neq 0$.

4. a. In the previous problem, calculate the weight,

$$a_i = \frac{x_i - \overline{x}}{\sum_{i=1}^{n}(x_i - \overline{x})^2},$$

with which each y_i is weighted in the calculation of $\widehat{\beta}$.

b. Use the a_i to calculate $\widehat{\beta}$.

c. Change the last data point to (2.5, 6.3) and determine the new value of $\widehat{\beta}$.

5. The viscosity of motor oil decreases with increasing temperature. Viscosity measurements of SAE 30 motor oil are taken at 5°F increments starting at 165° and ending at 200°. The data follow where x denotes temperature and y is a measure of viscosity.

x	165	170	175	180	185	190	195	200
y	28.5	26.1	23.9	22.0	20.4	18.5	17.1	15.8

a. Find the equation of the least squares regression line.

b. Find the analysis of variance table and use it to test H_0: $\beta = 0$ against H_1: $\beta \neq 0$.

6. The following data represent chemistry grades (y) for a random sample of 12 college freshmen and their scaled scores on an intelligence test (x).

x	65	50	55	65	55	70	65	70	55	70	50	55
y	85	74	76	90	85	87	94	98	81	91	76	74

a. Find the equation of the least squares regression line.

b. Find the analysis of variance table and use it to test H_0: $\beta = 0$ against H_1: $\beta \neq 0$.

7. A study of the amount of rainfall (x) in an area and the quantity of air pollution removed (y) gave the following data.

x	4.3	4.5	5.9	5.6	6.1	5.2	3.8	2.1	7.5
y	126	121	116	118	114	118	132	141	108

a. Find the equation of the least squares regression line.
b. Find the analysis of variance table and use it to test H_0: $\beta = 0$ against H_1: $\beta \neq 0$.

8. The data shown below represent the weld diameter (x) and the shear strength (y) of spot welds on a certain type of steel.

x	140	155	160	165	175	165	195	185	195	210
y	350	380	385	450	465	485	535	555	590	605

a. Find the equation of the least squares regression line.
b. Find the analysis of variance table and use it to test H_0: $\beta = 0$ against H_1: $\beta \neq 0$.

9. A set of 12 points (x_i, y_i) gave $y_i = 10.20833 + 1.925x_i$ as the least squares regression line.
a. Complete the following analysis of variance table.

Source	df	SS	MS	F	p
Regression					
Error		241.7917			
Total		982.9167			

b. What conclusion can be drawn from the analysis of variance table?

10. The yield of a chemical process is thought to be a linear function of the amount of catalyst added to the reaction. An experiment gave the following data where x is the amount of catalyst (in pounds) and y is the yield (in percent).

x	0.9	1.4	1.6	1.7	1.8	2.0	2.1	2.3
y	60.54	63.86	63.76	60.15	66.66	71.66	70.81	65.72

a. Find the equation of the least squares regression line.
b. Test H_0: $\beta = 0$ against H_1: $\beta \neq 0$.

11. Show that

$$\text{Var}(\widehat{\alpha}) = \sigma^2 \left[\frac{1}{n} + \frac{\overline{x}^2}{\sum_{i=1}^{n}(x_i - \overline{x})^2} \right]$$

can be simplified to

$$\text{Var}(\widehat{\alpha}) = \sigma^2 \frac{\sum_{i=1}^{n} x_i^2}{n \sum_{i=1}^{n}(x_i - \overline{x})^2}.$$

4.6 Correlation

Simple linear regression involves two variables, X and Y. If Y is in fact a linear function of X, then Y is completely determined by X. Otherwise Y may be partially determined by X. One way to measure the size of the dependence of Y on X is to calculate the *covariance* of X and Y, which was introduced in Sections 2.12, 4.4, and 4.5. It is defined as follows.

DEFINITION

The **covariance** of random variables X and Y is

$$\text{Cov}(X, Y) = E[(X - \mu_X)(Y - \mu_Y)].$$

It is obvious that $\text{Cov}(X, X) = \text{Var}(X)$, but, unlike the variance, the covariance can be, and often is, negative. The covariance can also be used to determine the *correlation coefficient*.

DEFINITION

The **correlation coefficient** of random variables X and Y is

$$\rho_{X,Y} = \frac{\text{Cov}(X, Y)}{\sigma_X \cdot \sigma_Y}.$$

We encountered $\rho_{X,Y}$ in Section 2.12 when we considered the bivariate normal probability distribution. Our estimate of $\rho_{X,Y}$ based on a sample is

$$r_{X,Y} = \frac{\sum\limits_{i=1}^{n}(x_i - \bar{x})(y_i - \bar{y})/(n-1)}{s_x \cdot s_y}$$

where s_x and s_y are the estimates of σ_X and σ_Y, the true standard deviations of X and Y, respectively.

Example 4.6.1   For the data in Example 4.1.1, we find that $\sum_{i=1}^{5} y_i^2 = 2{,}379{,}706$. The numerator of $r_{X,Y}$ can be written as

$$\frac{\sum\limits_{i=1}^{n}(x_i - \bar{x})(y_i - \bar{y})}{n - 1} = \frac{\sum\limits_{i=1}^{n} x_i y_i - n \cdot \bar{x} \cdot \bar{y}}{n - 1}.$$

So we have that

$$r_{X,Y} = \frac{\left(\sum\limits_{i=1}^{n} x_i y_i - n \cdot \bar{x} \cdot \bar{y}\right)/(n-1)}{\sqrt{\left[\dfrac{n \sum\limits_{i=1}^{n} x_i^2 - \left(\sum\limits_{i=1}^{n} x_i\right)^2}{n \cdot (n-1)}\right] \cdot \left[\dfrac{n \sum\limits_{i=1}^{n} y_i^2 - \left(\sum\limits_{i=1}^{n} y_i\right)^2}{n \cdot (n-1)}\right]}}$$

$$r_{X,Y} = \frac{\left(2{,}322{,}200 - 5 \cdot \dfrac{3500}{5} \cdot \dfrac{3414}{5}\right)/4}{\sqrt{\left[\dfrac{5(2{,}550{,}000) - (3500)^2}{5 \cdot 4}\right]\left[\dfrac{5(2{,}379{,}706) - (3414)^2}{5 \cdot 4}\right]}} = -0.96941.$$

The negative sign indicates that Y decreases as X increases. The calculation above is for a very small data set. This calculation is almost always done by computer; the formula shows the extent of the calculation and its difficulty for a large data set.   ■

Interpreting the Correlation Coefficient

There is an interesting relationship between the correlation coefficient, $r_{X,Y}$, and $\widehat{\beta}$, the least squares estimator for β. We know that

$$\widehat{\beta} = \frac{\sum_{i=1}^{n}(x_i - \bar{x})(y_i - \bar{y})}{\sum_{i=1}^{n}(x_i - \bar{x})^2}$$

so we can write

$$r_{X,Y} = \widehat{\beta}\sqrt{\frac{\sum_{i=1}^{n}(x_i - \bar{x})^2}{\sum_{i=1}^{n}(y_i - \bar{y})^2}}.$$

So $\widehat{\beta} = r \cdot s_y/s_x$, showing that the least squares regression line could be written as

$$\frac{\widehat{y_i} - \bar{y}}{s_y} = r \cdot \frac{x_i - \bar{x}}{s_x}.$$

If r is positive, then Y tends to increase with increasing values of X; if r is negative, then Y tends to decrease with increasing values of X. There is one more interesting and useful fact we can establish concerning r.

Recall the analysis of variance partition considered in Section 4.5 and consider the ratio of the sum of squares due to regression to the total sum of squares:

$$\frac{\sum_{i=1}^{n}(\widehat{y_i} - \bar{y})^2}{\sum_{i=1}^{n}(y_i - \bar{y})^2}.$$

This could be written as

$$\widehat{\beta}^2 \cdot \frac{\sum_{i=1}^{n}(x_i - \bar{x})^2}{\sum_{i=1}^{n}(y_i - \bar{y})^2}$$

but this is r^2. The ratio $\sum_{i=1}^{n}(\widehat{y_i} - \bar{y})^2/\sum_{i=1}^{n}(y_i - \bar{y})^2$ is often called the **coefficient of determination.** It is the proportion of the total sum of squares that is due to regression. Since $\sum_{i=1}^{n}(\widehat{y_i} - \bar{y})^2$ is at most $\sum_{i=1}^{n}(y_i - \bar{y})^2$, it follows that $0 \le r^2 \le 1$ and so

$$-1 \le r \le 1.$$

If Y is exactly a linear function of X, then the sum of squares due to error, $\sum_{i=1}^{n}(y_i - \widehat{y_i})^2 = 0$, and then $\sum_{i=1}^{n}(\widehat{y_i} - \bar{y})^2 = \sum_{i=1}^{n}(y_i - \bar{y})^2$ and $r = 1$ if Y is an increasing function of X. If Y is a decreasing function of X, then $r = -1$, presuming again that Y is exactly a linear function of X.

However, the correlation coefficient alone is not sufficient to establish a linear relationship between two variables. The square of the correlation coefficient indicates the proportion of the total variation in Y that is due to regression and is generally regarded as more informative than the value of r itself. For example, if $r = 0.7$, a fairly substantial correlation coefficient provided the sample size is large, then $r^2 = 0.49$ showing that 49% of the variation in Y can be explained from regression, while 51% of the variation in Y is random.

In Example 4.1.1, we find that

$$\frac{\sum\limits_{i=1}^{n}(\widehat{y}_i - \overline{y})^2}{\sum\limits_{i=1}^{n}(y_i - \overline{y})^2} = \frac{45{,}697.6}{48{,}626.8} = 0.93976 = (-0.96941)^2.$$

(We previously calculated $r = -0.96941$.)

Testing Hypotheses on the Correlation Coefficient

The following theorem can be proved.

Theorem

If r is a sample correlation coefficient from a bivariate normal population with true correlation coefficient ρ, then

$$\tanh^{-1} r = \frac{1}{2}\ln\frac{1+r}{1-r} \sim N\left(\frac{1}{2}\ln\frac{1+\rho}{1-\rho}, \frac{1}{\sqrt{n-3}}\right).$$

The theorem, whose proof exceeds the scope of this book, can be used to test hypotheses concerning ρ and to establish confidence intervals for ρ.

Example 4.6.2   Suppose that a sample of size $n = 5$ shows a sample correlation coefficient $r = 0.40$. Use this data to test H_0: $\rho = 0$ against H_1: $\rho \neq 0$.

The theorem above indicates that the variable z' (to distinguish this from a standard normal variable) where

$$z' = \frac{1}{2}\ln\frac{1+r}{1-r} = \frac{1}{2}\ln\frac{1+0.40}{1-0.40} = 0.42365$$

is an $N(0, 1/\sqrt{2} = 0.70711)$ variable. The standard z score for this value of z' is then $(0.42365 - 0)/0.70711 = 0.59913$, giving a p-value of 0.274543, most probably far too large to reject the null hypothesis.

We might also determine the smallest value of r that would give a p-value of, say, 0.01. The corresponding z score is then 2.57583, so the corresponding z' value is

$2.57583 = (z' - 0)/0.70711$ giving $z' = 1.8214$. But

$$z' = \frac{1}{2} \ln \frac{1 + r}{1 - r}.$$

Solving this for r gives $r = 0.948978$. So it requires an extremely large value of r before we are likely to conclude that $\rho \neq 0$. We should extend a caution here, namely that the sample size is extremely small. The graph in Figure 4.5 shows how the critical value of r varies with the sample size, n. We have used a p-value of 0.01 in constructing the graph. ∎

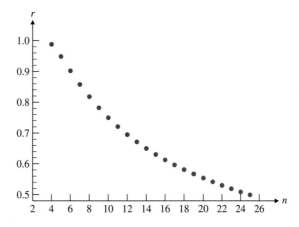

Figure 4.5 Critical Values of r as a Function of the Sample Size, n

Example 4.6.3   Find a 95% confidence interval for ρ if a sample of size 28 from a bivariate normal distribution gave $r = 0.80$.

Since

$$z' = \frac{1}{2} \ln \frac{1 + r}{1 - r} \sim N\left(\frac{1}{2} \ln \frac{1 + \rho}{1 - \rho}, \frac{1}{\sqrt{n - 3}} \right),$$

we have that

$$z' = \frac{1}{2} \ln \frac{1 + 0.8}{1 - 0.8} = 1.09861$$

so a 95% confidence interval is

$$-1.96 \leq \frac{z' - \frac{1}{2} \ln \frac{1 + \rho}{1 - \rho}}{1/5} \leq 1.96$$

or

$$z' - \frac{1.96}{5} \leq \frac{1}{2} \ln \frac{1 + \rho}{1 - \rho} \leq z' + \frac{1.96}{5}$$

or

$$0.706612 \leq \tanh^{-1} \rho \leq 1.49061.$$

Solving this for ρ, we find

$$0.608548 \leq \rho \leq 0.903437$$

is a 95% confidence interval. The range for the confidence interval is still quite large, a function of the sample size. ∎

Example 4.6.4   The confidence interval in Example 4.6.3 is quite wide, ranging from 0.608548 to 0.903437. Increasing the sample size, n, should decrease this range so we are tempted to calculate the endpoints of the confidence interval for ρ as a function of the sample size. Unfortunately, this turns out to be very difficult. It would be useful if we could calculate the sample size giving a confidence interval of a given length for a given confidence coefficient, but this can only be done approximately.

We find that a sample size of 205 is necessary to produce a 95% confidence interval for ρ of length 0.10, if the sample correlation coefficient is $r = 0.80$. We used Mathematica to calculate the sample size giving a 95% confidence interval of length ≤ 0.10 for various values of r, the sample correlation coefficient. The results are shown in Table 4.5.

r	n
0.2	1417
0.3	1274
0.4	1086
0.5	867
0.6	633
0.7	404
0.8	205

Table 4.5

The sample sizes given in Table 4.5 indicate that the endpoints of the confidence intervals, as functions of the sample size, are quite variable. The values in Table 4.5 are shown in Figure 4.6. This indicates that the data may well be linear. The least squares

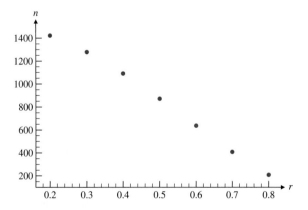

Figure 4.6 Sample Size to Produce Confidence Intervals for ρ

regression line fitted to the data in Table 4.5 gives

$$f(r) = 1881.75 - 2081.79r.$$

The fit is quite good, yielding a p value of $4 \cdot 10^{-7}$.   ■

EXERCISES 4.6

In each of the following exercises, assume that the sample arises from a bivariate normal population.

1. A sample of size 50 gave $r = 0.20$. Find the p-value for testing H_0: $\rho = 0$ against H_1: $\rho \neq 0$.

2. Find a 95% confidence interval for ρ if a sample of size 28 gave $r = 0.62$.

3. Find a 95% confidence interval for ρ if a sample of size 50 gave $r = -0.35$.

4. A sample of size 42 gave $r = 0.29$. Find the p-value for testing H_0: $\rho = 0$ against H_1: $\rho \neq 0$.

5. A sample of size 10 gave $r = -0.87$. Find the p-value for testing H_0: $\rho = 0$ against H_1: $\rho \neq 0$.

6. Find a 90% confidence interval for ρ if a sample of size 16 gave $r = 0.48$.

7. A sample of size 27 gave $r = -0.44$. Find the p-value for testing H_0: $\rho = -0.50$ against H_1: $\rho \neq -0.50$.

8. Show that the analysis of variance F test can be expressed as

$$F(1, n - 2) = \frac{r^2(n - 2)}{1 - r^2}$$

where r is the sample correlation coefficient. Conclude that an equivalent test uses

$$t_{n-2} = \frac{r\sqrt{n - 2}}{\sqrt{1 - r^2}}.$$

9. Show that

$$\frac{\sum_{i=1}^{n}(\widehat{y_i} - \overline{y})^2}{\sum_{i=1}^{n}(y_i - \overline{y})^2} = \widehat{\beta}^2 \cdot \frac{\sum_{i=1}^{n}(x_i - \overline{x})^2}{\sum_{i=1}^{n}(y_i - \overline{y})^2}.$$

COMPUTER EXPERIMENTS

1. Find the sample size necessary to produce a 95% confidence interval of length 0.10 if the sample correlation coefficient is $r = 0.57$.

2. Find the sample size necessary to produce a 95% confidence interval of length 0.10 if the sample correlation coefficient is $r = 0.90$.

3. Find the sample size necessary to produce a 95% confidence interval of length ≤ 0.10 if the sample correlation coefficient is $r = 0.10$.

4.7   Estimation and Prediction

Experimenters often use the regression line to make predictions for values of X that are close to the observed set of values or to make predictions for values of X that are *within* the range of the observed values of X but for some reason were not actually observed themselves. In Example 4.1.1, we asked, for the temperature and life-length data, what life-length should be predicted for temperature 625°F? Our answer uses the least squares equation and it appears natural to use $1156 - 0.676 \cdot 625 = 733.5$, since the least squares regression line is $y_i = 1156 - 0.676 \cdot x_i$. But it is not clear what properties this point estimate might have, nor have we found a way to find a confidence interval.

We must establish some theoretical results before determining the properties of our estimates and constructing confidence intervals. Recall, from Section 4.4, that

$$\text{Var}(\widehat{\beta}) = \frac{\sigma^2}{\displaystyle\sum_{i=1}^{n}(x_i - \overline{x})^2}$$

and

$$\text{Var}(\widehat{\alpha}) = \sigma^2 \left[\frac{1}{n} + \frac{\overline{x}^2}{\displaystyle\sum_{i=1}^{n}(x_i - \overline{x})^2} \right].$$

The model for simple linear regression is $y_i = \alpha + \beta x_i + \varepsilon_i$ for $i = 1, 2, \ldots, n$ where $\varepsilon_i \sim NID(0, \sigma)$.

Consider now making repeated observations of Y for the same value of X, say x_i. The result will be a variety of values of Y. From the model we see that $E(y_i) = \alpha + \beta x_i$ since $E(\varepsilon_i) = 0$. This is also $E(\widehat{y}_i) = E(\widehat{\alpha} + \widehat{\beta} x_i) = \alpha + \beta x_i$ since the estimators $\widehat{\alpha}$ and $\widehat{\beta}$ are unbiased estimators of α and β.

So $\widehat{y}_i$ is in fact an estimate of the mean value of the y's that could be observed for a fixed value of x, say x_i. The model tells us that the expected value of y_i, at $x = x_i$, is $\alpha + \beta x_i$; these observations are normally distributed with common variance σ^2.

We want now to determine $\text{Var}(\widehat{y}_i)$. It happens that $\widehat{\alpha}$ and $\widehat{\beta}$ are correlated; we can determine their covariance. From the first least squares equation we know that $\overline{y} = \widehat{\alpha} + \widehat{\beta}\overline{x}$. So

$$\text{Var}(\overline{y}) = \text{Var}(\widehat{\alpha} + \widehat{\beta}\overline{x}) = \text{Var}(\widehat{\alpha}) + 2\overline{x}\,\text{Cov}(\widehat{\alpha}, \widehat{\beta}) + \overline{x}^2\,\text{Var}(\widehat{\beta})$$

Using the fact that $\text{Var}(\overline{y}) = \sigma^2/n$ and the formulas for $\text{Var}(\widehat{\alpha})$ and $\text{Var}(\widehat{\beta})$, we have that

$$\frac{\sigma^2}{n} = \sigma^2 \left[\frac{1}{n} + \frac{\overline{x}^2}{\displaystyle\sum_{i=1}^{n}(x_i - \overline{x})^2} \right] + 2\overline{x}\,\text{Cov}(\widehat{\alpha}, \widehat{\beta}) + \overline{x}^2 \frac{\sigma^2}{\displaystyle\sum_{i=1}^{n}(x_i - \overline{x})^2}.$$

Solving this gives

$$\text{Cov}(\widehat{\alpha}, \widehat{\beta}) = \frac{-\sigma^2 \displaystyle\sum_{i=1}^{n} x_i}{n \displaystyle\sum_{i=1}^{n}(x_i - \overline{x})^2}.$$

The Variance-Covariance Matrix

Recall the matrix

$$\mathbf{X} = \begin{bmatrix} 1 & x_1 \\ 1 & x_2 \\ \vdots & \vdots \\ 1 & x_n \end{bmatrix}$$

and consider the matrix $\mathbf{X}'\mathbf{X}$ where $\mathbf{X}'$ denotes the transpose of the matrix $\mathbf{X}$. This is

$$\mathbf{X}'\mathbf{X} = \begin{bmatrix} n & \sum\limits_{i=1}^{n} x_i \\ \sum\limits_{i=1}^{n} x_i & \sum\limits_{i=1}^{n} x_i^2 \end{bmatrix}.$$

Now we find that

$$(\mathbf{X}'\mathbf{X})^{-1} = \frac{1}{n \sum\limits_{i=1}^{n} x_i^2 - \left(\sum\limits_{i=1}^{n} x_i\right)^2} \begin{bmatrix} \sum\limits_{i=1}^{n} x_i^2 & -\sum\limits_{i=1}^{n} x_i \\ -\sum\limits_{i=1}^{n} x_i & n \end{bmatrix}$$

$$(\mathbf{X}'\mathbf{X})^{-1} = \frac{1}{n \sum\limits_{i=1}^{n} (x_i - \bar{x})^2} \begin{bmatrix} \sum\limits_{i=1}^{n} x_i^2 & -\sum\limits_{i=1}^{n} x_i \\ -\sum\limits_{i=1}^{n} x_i & n \end{bmatrix}.$$

The matrix $\sigma^2(\mathbf{X}'\mathbf{X})^{-1}$ can now be recognized as

$$\sigma^2(\mathbf{X}'\mathbf{X})^{-1} = \begin{bmatrix} \text{Var}(\widehat{\alpha}) & \text{Cov}(\widehat{\alpha}, \widehat{\beta}) \\ \text{Cov}(\widehat{\alpha}, \widehat{\beta}) & \text{Var}(\widehat{\beta}) \end{bmatrix}.$$

For this reason, the matrix $\sigma^2(\mathbf{X}'\mathbf{X})^{-1}$ is called the **variance-covariance** matrix. The only instance in which $\mathbf{X}'\mathbf{X}$ is singular (that is, when $\mathbf{X}'\mathbf{X}$ is not invertible) occurs when $\sum_{i=1}^{n}(x_i - \bar{x})^2 = 0$, that is, when all the values of x_i are equal, a clearly trivial instance when considering simple linear regression.

We can now determine the variance of the expected or mean predicted value, $\widehat{y}_i$, at $x = x_i$.

$$\text{Var}(\widehat{y}_i) = \text{Var}(\widehat{\alpha} + \widehat{\beta}x_i) = \text{Var}(\widehat{\alpha}) + 2x_i\,\text{Cov}(\widehat{\alpha}, \widehat{\beta}) + x_i^2\,\text{Var}(\widehat{\beta})$$

$$= \sigma^2 \begin{bmatrix} \dfrac{1}{n} + \dfrac{\bar{x}^2}{\sum\limits_{j=1}^{n}(x_j - \bar{x})^2} \end{bmatrix} - \frac{2x_i\sigma^2 \sum\limits_{j=1}^{n} x_j}{n \sum\limits_{j=1}^{n}(x_j - \bar{x})^2} + \frac{\sigma^2 x_i^2}{\sum\limits_{j=1}^{n}(x_j - \bar{x})^2}.$$

This can be simplified to

$$\text{Var}(\widehat{y_i}) = \sigma^2 \left[\frac{1}{n} + \frac{(x_i - \overline{x})^2}{\sum\limits_{j=1}^{n} (x_j - \overline{x})^2} \right].$$

$\widehat{y_i}$ is our estimate of the mean value of y when repeated observations are made at $x = x_i$. Note that this variance increases as the distance of x_i from $\overline{x}$ increases.

Now we can calculate the variance of an individual value of y when $x = x_i$. The individual values can be thought of as the mean or expected value plus the random error, ε_i, as an added term. Since these errors are independent, it follows that

$$\text{Var}(y_i) = \text{Var}(\widehat{y_i} + \varepsilon_i),$$

and since the errors are uncorrelated with the y-values,

$$\text{Var}(y_i) = \text{Var}(\widehat{y_i}) + \text{Var}(\varepsilon_i)$$

$$= \sigma^2 \left[\frac{1}{n} + \frac{(x_i - \overline{x})^2}{\sum\limits_{j=1}^{n} (x_j - \overline{x})^2} \right] + \sigma^2$$

so

$$\text{Var}(y_i) = \sigma^2 \left[1 + \frac{1}{n} + \frac{(x_i - \overline{x})^2}{\sum\limits_{j=1}^{n} (x_j - \overline{x})^2} \right].$$

Since these variances increase as the distance of x_i from $\overline{x}$ increases, the confidence intervals for both individual predictions and mean values can be shown as curves around the least squares regression line. These are shown in Figure 4.7 on page 204.

How can the formulas derived here be put to practical use? The problem, of course, is that while all the other quantities in the formulas are known from the data, σ^2 is unknown. Fortunately, it can be shown that

$$E \left[\frac{\sum\limits_{i=1}^{n} (y_i - \widehat{y_i})^2}{n - 2} \right] = \sigma^2.$$

We recognize

$$\frac{\sum\limits_{i=1}^{n} (y_i - \widehat{y_i})^2}{n - 2}$$

as the mean square for error in the analysis of variance. Let this be denoted by s^2.

One final crucial fact, whose proof is beyond the scope of this book, is that since the errors in the linear model are normal and the sums of squares (and hence the variances) of the quantities we have been considering follow chi-squared distributions, the random variables involved follow the Student t distribution with $n - 2$ degrees of freedom. We summarize all this information below.

For the mean value of Y at a fixed value of X, which we denote by $\mu_{Y|X}$,

$$t_{n-2} = \frac{\widehat{y}_i - \mu_{Y|X}}{\sqrt{s^2 \left[\dfrac{1}{n} + \dfrac{(x_i - \overline{x})^2}{\displaystyle\sum_{j=1}^{n} (x_j - \overline{x})^2} \right]}}.$$

For an individual value of Y at $X = x_i$,

$$t_{n-2} = \frac{\widehat{y}_i - y_i}{\sqrt{s^2 \left[1 + \dfrac{1}{n} + \dfrac{(x_i - \overline{x})^2}{\displaystyle\sum_{j=1}^{n} (x_j - \overline{x})^2} \right]}}.$$

We now show uses for these formulas.

Example 4.7.1   Consider the mean value predicted from the regression line found in Example 4.1.1 when $x = 625$.

The variance of this mean value is estimated as

$$s^2 \left[\frac{1}{n} + \frac{(x_i - \overline{x})^2}{\displaystyle\sum_{j=1}^{n} (x_j - \overline{x})^2} \right] = 976.4 \cdot \left[\frac{1}{5} + \frac{(625 - 700)^2}{100{,}000} \right] = 250.2025,$$

giving the two-sided 95% confidence interval as

$$(733.5 - 3.182 \cdot \sqrt{250.2025}, \ 733.5 + 3.182 \cdot \sqrt{250.2025}) = (683.17, \ 783.83).$$

Finally, a 95% confidence interval for a single predicted value at $x = 625$ has estimated variance as

$$s^2 \left[1 + \frac{1}{n} + \frac{(x_i - \overline{x})^2}{\displaystyle\sum_{j=1}^{n} (x_j - \overline{x})^2} \right] = 976.4 \cdot \left[1 + \frac{1}{5} + \frac{(625 - 700)^2}{100{,}000} \right] = 1226.6025$$

giving a 95% confidence interval as

$$(733.5 - 3.182 \cdot \sqrt{1226.6025}, \ 733.5 + 3.182 \cdot \sqrt{1226.6025}) = (622.06, \ 844.94).$$

The confidence intervals for both mean and predicted values determine confidence bands around the least squares regression line. These are shown for the data in Example 4.1.1 in Figure 4.7.   ■

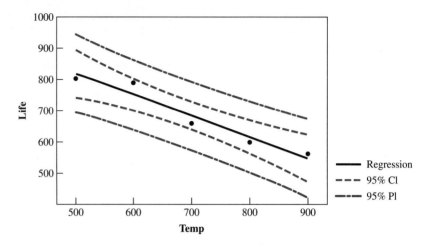

Figure 4.7 Confidence Bands

Now we consider hypothesis tests on the slope and the intercept of the least squares regression line.

Tests on Individual Regression Parameters

We consider first a test on the slope of the least squares regression line, β. We estimate the variance of $\widehat{\beta}$ by

$$\widehat{\operatorname{Var}(\beta)} = \frac{s^2}{\displaystyle\sum_{i=1}^{n}(x_i - \overline{x})^2}.$$

Then

$$t_{n-2} = \frac{\widehat{\beta} - \beta}{\sqrt{\dfrac{s^2}{\displaystyle\sum_{i=1}^{n}(x_i - \overline{x})^2}}}.$$

In a similar way, we find that

$$t_{n-2} = \frac{\widehat{\alpha} - \alpha}{\sqrt{s^2\left[\dfrac{1}{n} + \dfrac{\overline{x}^2}{\displaystyle\sum_{i=1}^{n}(x_i - \overline{x})^2}\right]}}.$$

These facts can be used to fashion tests and confidence intervals on the regression parameters.

Example 4.7.2   We illustrate these formulas using the data in Example 4.1.1.

Note, from the analysis of variance table in Section 4.5, that our estimate of σ^2 is the mean square for error, $2929.2/3 = 976.4$. We used this value for s^2 in the formulas above.

Our estimate for $\text{Var}(\widehat{\beta})$ is then

$$\frac{976.4}{\sum_{i=1}^{n}(x_i - \bar{x})^2} = \frac{976.4}{100,000} = 0.009764,$$

producing the standard deviation for $\widehat{\beta} = \sqrt{0.009764} = 0.098813$, so a 95% confidence interval for β is

$$(-0.676 - 3.182 \cdot 0.098813, -0.676 + 3.182 \cdot 0.098813) = (-0.99042, -0.36158).$$

Our estimate of $\text{Var}(\widehat{\alpha})$ is

$$s^2\left[\frac{1}{n} + \frac{\bar{x}^2}{\sum_{i=1}^{n}(x_i - \bar{x})^2}\right] = 976.4\left[\frac{1}{5} + \frac{700^2}{100,000}\right] = 4979.64.$$

A 95% confidence interval for α is then

$$(1156 - 3.182\sqrt{4979.64}, 1156 + 3.182\sqrt{4979.64} = (931.46, 1380.5). \quad \blacksquare$$

Hypotheses can also be tested, as we will see in the next example.

Example 4.7.3   For the data in Example 4.1.1, test $H_0: \beta = 0$ against the alternative $H_1: \beta \neq 0$.

The t test above gives

$$t_3 = \frac{\widehat{\beta} - \beta}{\sqrt{\dfrac{s^2}{\sum_{i=1}^{n}(x_i - \bar{x})^2}}} = \frac{-0.676 - 0}{\sqrt{\dfrac{976.4}{100,000}}} = -6.8412,$$

leading without doubt to the rejection of the null hypothesis. $\quad \blacksquare$

We previously tested this hypothesis in Section 4.5 where we found, from the analysis of variance table, that $F(1, 3) = 46.8021$. We apparently have two different tests for the same hypothesis. However, it can be shown that $F(1, k) = t_k^2$, so the tests are in fact equivalent. We note here that $-\sqrt{46.8021} = -6.8412$, showing the tests to be equivalent. Notice that the F test is one-sided while the t test is two-sided.

One advantage the t test has over the F test is that the t test can be used to test the hypothesis that β is some nonzero value.

Example 4.7.4   We test $H_0: \beta = -1$ against the alternative $H_1: \beta \neq -1$ for the data in Example 4.1.1. We expect this hypothesis to be rejected at the 5% level since it is not contained in the

95% confidence interval calculated in Example 4.7.2. We find the t value to be

$$t_3 = \frac{-0.676 + 1}{\sqrt{\dfrac{976.4}{100,000}}} = 3.2789,$$

corresponding to a p-value of 0.0232322.   ∎

Computer Results

Minitab provides a table indicating t values for the regression parameters in Example 4.2.1 as shown in Table 4.6.

Predictor	Coef	StDev	t	p
Constant	1156	70.57	16.38	0.000
Age	−0.676	0.09881	−6.84	0.006

Table 4.6

Mathematica provides a table of confidence intervals for individual predicted values, which appears in Table 4.7.

Observed	Predicted	Standard Error	Confidence Interval
804	818	39.5252	(692.213,943.787)
791	750.4	35.6275	(637.017,863.783)
658	682.8	34.2298	(573.865,791.735)
599	615.2	35.6275	(501.817,728.583)
562	547.6	39.5252	(421.813,673.387)

Table 4.7

Example 4.7.5   We can estimate the variance-covariance matrix for the data in Example 4.1.1. We use s^2 as an estimate of σ^2 and calculate the following.

$$\sigma^2(\widehat{X'X})^{-1} = 976.4 \begin{bmatrix} 5.1 & -0.007 \\ -0.007 & 0.00001 \end{bmatrix}$$

$$\sigma^2(\widehat{X'X})^{-1} = \begin{bmatrix} 4979.64 & -6.8348 \\ -6.8348 & 0.009764 \end{bmatrix}$$

∎

EXERCISES 4.7

1. In Section 4.1, we showed the data for the temperature and life-length experiment in Example 4.1.1. Find the predicted value for temperature at 715°F. Also find a 90% confidence interval for the predicted value at $x = 715$°F.

2. The data from Exercise 8 in Section 4.5 are as follows.

x	140	155	160	165	175	165	195	185	195	210
y	350	380	385	450	465	485	535	555	590	605

a. Refer to the analysis of variance tests done in Exercise 8 in Section 4.5. Use the t test to test $H_0: \beta = 0$ and show that the two tests are equivalent.
b. Find a 95% confidence interval for β.

3. The data from Exercise 8 in Section 4.5 are as follows.

x	140	155	160	165	175	165	195	185	195	210
y	350	380	385	450	465	485	535	555	590	605

a. Find a 95% confidence interval for a single predicted value at $x = 170$.
b. Find a 95% confidence interval for the mean value of the predicted values at $x = 170$.

4. The data from Exercise 7 in Section 4.5 are as follows.

x	4.3	4.5	5.9	5.6	6.1	5.2	3.8	2.1	7.5
y	126	121	116	118	114	118	132	141	108

a. Refer to the analysis of variance tests done in Exercise 7 in Section 4.5. Use the t test to test $H_0: \beta = 0$ and show that the two tests are equivalent.
b. Find a 95% confidence interval for β.

5. The data from Exercise 7 in Section 4.5 are as follows.

x	4.3	4.5	5.9	5.6	6.1	5.2	3.8	2.1	7.5
y	126	121	116	118	114	118	132	141	108

a. Find a 95% confidence interval for a single predicted value at $x = 6.0$.
b. Find a 95% confidence interval for the mean value of the predicted values at $x = 6.0$.

6. In a study comparing engine size (x), measured in cubic inches of displacement, and miles per gallon estimates (y) for eight compact automobiles, the data are as follows.

x	121	120	97	98	122	97	85	122
y	30	31	34	27	29	34	38	32

a. Show two tests for $H_0: \beta = 0$ and show that they are equivalent.
b. Find a 90% confidence interval for the mean predicted value at $x = 100$.
c. Find a 90% confidence interval for a single predicted value at $x = 100$.
d. Find a 90% confidence interval for β.
e. Test $H_0: \beta = 0.1$ against $H_1: \beta \neq 0.1$ and find a p-value for the test.

7. A recent paper reported a study of the relationship between applied stress (x), measured in kg/mm, and the time to fracture (y), measured in hours, for a type of stainless steel under uniaxial stress in a solution maintained at a constant temperature. The data are as follows.

x	2.5	5.0	10.0	15.0	17.5	20.0	25.0	30.0	35.0	45.0
y	63	58	55	61	62	37	38	45	46	19

a. Show two tests for $H_0: \beta = 0$ and show that they are equivalent.
b. Find a 95% confidence interval for the mean predicted value at $x = 16$.
c. Find a 95% confidence interval for a single predicted value at $x = 16$.
d. Find a 95% confidence interval for β.
e. Test $H_0: \beta = -1$ against $H_1: \beta \neq -1$ and find a p-value for the test.

COMPUTER EXPERIMENTS

1. Consider the model $y_i = 3 + 5x_i + \varepsilon_i$ where $\varepsilon_i \sim N(0, 15)$. Find 500 observations from this model when $x = 5$. Show a 95% confidence interval for the computed values and compare the number of observations in the confidence interval to the expected number of observations in the confidence interval.
2. Using the model in Exercise 1, draw 10 samples of size 100 each. For each sample calculate the slope of the least squares regression line. Then find a confidence interval for β, the true slope of the regression line. What proportion of the sample $\widehat{\beta}$'s is contained in the confidence interval?

4.8 Regression Through the Origin

Occasionally it is known that the true regression line passes through the origin. In this case we know that $\alpha = 0$, so the model becomes

$$y_i = \beta x_i + \varepsilon_i \quad \text{for } i = 1, 2, \ldots, n$$

where $\varepsilon_i \sim NID(0, \sigma)$.

It is very important to realize that the *true* value of α is 0. This does not mean that we can make our *estimate* of α, $\widehat{\alpha}$, zero in the least squares equations previously derived. In fact, doing that leads to incorrect results, so we must apply the principle of least squares using the new model. We must then minimize

$$S = \sum_{i=1}^{n} (y_i - \beta x_i)^2$$

for the data set $\{(x_1, y_1), (x_2, y_2), (x_3, y_3), \ldots, (x_n, y_n)\}$. We find that

$$\frac{dS}{d\beta} = 2 \sum_{i=1}^{n} (y_i - \beta x_i)(-x_i).$$

Equating this to 0 and solving for $\widehat{\beta}$ gives

$$\widehat{\beta} = \frac{\sum_{i=1}^{n} x_i y_i}{\sum_{i=1}^{n} x_i^2}.$$

The least squares regression line is then $\widehat{y}_i = \widehat{\beta} x_i$.

Analysis of Variance

An analysis of variance partition of the total sum of squares, which is $\sum_{i=1}^{n} y_i^2$ in this case since the mean value is not estimated from the data, can be found as follows:

$$\sum_{i=1}^{n} y_i^2 = \sum_{i=1}^{n} (y_i - \widehat{y}_i + \widehat{y}_i)^2 = \sum_{i=1}^{n} (y_i - \widehat{y}_i)^2 + 2 \sum_{i=1}^{n} (y_i - \widehat{y}_i)\widehat{y}_i + \sum_{i=1}^{n} \widehat{y}_i^2.$$

Now, ignoring the factor of 2, the middle term becomes

$$\sum_{i=1}^{n} (y_i - \widehat{y}_i)\widehat{y}_i = \sum_{i=1}^{n} (y_i - \widehat{\beta} x_i)\widehat{\beta} x_i = \widehat{\beta} \sum_{i=1}^{n} y_i x_i - \widehat{\beta}^2 \sum_{i=1}^{n} x_i^2 = 0$$

due to the definition of $\widehat{\beta}$. The analysis of variance is then shown in Table 4.8 for testing $H_0: \beta = 0$ against the alternative $H_1: \beta \neq 0$.

Note that the total sum of squares, $\sum_{i=1}^{n} y_i^2$, is the sum of n independent chi-squared random variables and so has n degrees of freedom.

Source	df	Sum of Squares	Mean Square	F	p
Regression	1	$\sum_{i=1}^{n} \widehat{y_i}^2$	$\sum_{i=1}^{n} \widehat{y_i}^2/1$		
Error	$n-1$	$\sum_{i=1}^{n} (y_i - \widehat{y_i})^2$	$\sum_{i=1}^{n} (y_i - \widehat{y_i})^2/(n-1)$	$\dfrac{\sum_{i=1}^{n} \widehat{y_i}^2/1}{\sum_{i=1}^{n} (y_i - \widehat{y_i})^2/(n-1)}$	
Total	n	$\sum_{i=1}^{n} y_i^2$			

Table 4.8

Example 4.8.1   A study of age (X) and systolic blood pressure (Y) reported by Snedecor and Cochran [30] gave the following data.

X	35	45	55	65	75
Y	114	124	143	158	166

The analysis of variance table for the data given in this example, assuming a regression line that passes through the origin, is shown in Table 4.9. Minitab gives this regression line as $Y = 2.49X$.

Source	df	Sum of Squares	Mean Square	F	p
Regression	1	99,995.3	99,995.3	297.226	0.0000664198
Error	4	1345.71	336.428		
Total	5	101,341			

Table 4.9

The hypothesis that the true slope of the regression line is 0 is most probably rejected. ∎

EXERCISES 4.8

1. Exercise 2 in Exercises 4.2 referred to an experiment concerning the velocity of a freely falling object at five different times. The experiment was arranged so that the initial velocity of the object is unknown. Now suppose that the initial velocity is 0, so that the regression line passes through the origin.

In the following data x is in minutes and y is in inches/minute.

x	0.5	1.0	1.5	2.0	2.5
y	1.4	3.2	4.4	5.7	8.0

a. Fit the least squares regression line that passes through the origin to this data.

b. Show the analysis of variance table and state your conclusions.

2. An engineer is seeking a relationship between the current in amps and the voltage in volts across a diode. The following data are found, which have been coded for convenience.

V	0.5	0.6	0.8	0.9	1.0
A	0.02	0.05	0.10	0.25	0.75

a. Assuming the regression line passes through the origin, find the equation of the line.

b. Compare the line found in part (a) to a least squares regression line that is not restricted to pass through the origin.

c. Using analysis of variance tables, state your conclusions.

3. The following data show the true amount of molybdenum (x) in a substance and the measured amount of molybdenum (y).

x	1	1	2	2	3	3	4	4	5	5
y	1.8	1.6	3.1	2.6	3.6	3.4	4.9	4.2	6.0	5.9

x	6	6	7	7	8	8	9	9	10	10
y	6.8	6.9	8.2	7.3	8.8	8.5	9.5	9.5	10.6	10.6

a. Compare, using analysis of variance tables, a regression line that passes through the origin to one that is not so restricted.

b. What conclusions can be drawn from the results in part (a)?

4.9 Influential Observations

We noted in Section 4.4 that the least squares estimate of β can be written as $\widehat{\beta} = \sum_{i=1}^{n} a_i y_i$ where

$$a_i = \frac{x_i - \overline{x}}{\sum_{j=1}^{n} (x_j - \overline{x})^2}.$$

This shows that $\widehat{\beta}$ is a weighted average of the observed y_i values. We also noted in Section 4.4 that the weights, the values of a_i, vary with the distance of x_i from $\overline{x}$, producing different influences of a data point on $\widehat{\beta}$.

There are several different ways to measure the size of the influence a particular data point has on the resulting regression parameters. We discuss only one of these here, from Cook [24]. All of the methods are computer intensive and would be very difficult to carry out by hand calculation.

Cook defines a *distance, d,* for each data point. In brief, d measures the effect of each data point by omitting each data point from the data set, calculating the resulting regression equation, and measuring the change in the regression parameters brought about by the omission of the data point.

The values of Cook's d are calculated as follows in the case of simple linear regression: For the ith data point, let ε_i denote the residual at that point. Let

$$h_i = \frac{1}{n} + \frac{(x_i - \overline{x})^2}{\sum_{j=1}^{n} (x_j - \overline{x})^2},$$

and denote the mean square for regression in the analysis of variance by *MSE*. Then Cook's *d* for the *i*th data point is

$$d_i = \frac{\varepsilon_i^2}{2 \cdot MSE} \left(\frac{h_i}{(1 - h_i)^2} \right).$$

We illustrate the use of this for the fifth data point in Example 4.1.1, (900, 562). The predicted value for this data point from the least squares regression line is 547.6, so $\varepsilon_i = 14.4$ and

$$h_5 = \frac{1}{n} + \frac{(x_5 - \bar{x})^2}{\sum\limits_{j=1}^{n} (x_j - \bar{x})^2} = \frac{1}{5} + \frac{(900 - 700)^2}{100,000} = \frac{3}{5}.$$

This produces

$$d_5 = \frac{\varepsilon_5^2}{2 \cdot MSE} \left(\frac{h_5}{(1 - h_5)^2} \right) = \frac{(14.4)^2}{2 \cdot 976.4} \left(\frac{3/5}{(1 - 3/5)^2} \right) = 0.398197.$$

The complete set of values of *d* for the data set in Example 4.1.1 is as follows:

$$\{0.376383, 0.516796, 0.0984228, 0.0822806, 0.398197\}.$$

A general rule is that a data point should be investigated as a possible outlier if *d* exceeds the 50th percentile of the $F[2, n - 2]$ distribution. In this example, $F_{0.50}[2, 3] = 0.881102$.

None of the data points in this example is then unduly influential upon the regression line. It is often found that the data points at the ends of the range of the values of the independent variable are influential. (Minitab and other computer statistical packages can be used to calculate the values of *d*.)

The values of h_i are often used alone to detect influential points. Now the sum of the h_i values is

$$\sum_{i=1}^{n} h_i = \sum_{i=1}^{n} \left[\frac{1}{n} + \frac{(x_i - \bar{x})^2}{\sum\limits_{j=1}^{n} (x_j - \bar{x})^2} \right] = \frac{1}{n} \cdot n + \sum_{i=1}^{n} \left[\frac{(x_i - \bar{x})^2}{\sum\limits_{j=1}^{n} (x_j - \bar{x})^2} \right] = 1 + 1 = 2,$$

so the mean value of h_i is $2/n$. Generally, any value of h_i more than twice the average, namely $4/n$, we regard as unusual. In this example, then, we would look for values of h_i that exceed $4/5$, but none of the data points exceeds this criterion.

Recalling from Section 4.3 the definition of the matrix

$$\mathbf{X} = \begin{bmatrix} 1 & x_1 \\ 1 & x_2 \\ \vdots & \vdots \\ 1 & x_n \end{bmatrix},$$

the values of h_i can be shown to be the diagonal elements of the matrix $\mathbf{X}(\mathbf{X}'\mathbf{X})^{-1}\mathbf{X}'$. The matrix $\mathbf{X}(\mathbf{X}'\mathbf{X})^{-1}\mathbf{X}'$ is often called the **hat matrix**.

We will return to the detection of influential points in Chapter 5 when we consider multiple regression.

4.10 Testing for Normality

In Section 3.15 we considered how to decide whether or not a sample arose from a given population. We used a chi-squared test for a data set arising from a discrete distribution. In the case of a continuous distribution, we relied on a graph that we called a **quantile-quantile, or q-q plot.** We did not have a way, other than visual observation, of deciding whether or not the data arose from a given population. We can now improve on that situation considerably and offer a regression test.

The most common presumption among statistical tests is that the data came from a normal distribution. We have seen that t tests rely upon this assumption; in regression, we have presumed that the random errors are normally distributed. So the presumption of normality is a frequent one, and it is a very important one to test.

Recall that x_q is called the **qth quantile** if $P(X < x_q) = q$.

Example 4.10.1   A sample of 1000 observations was selected from a population. Some of the values are:

$$\{14.7622, 10.513, 13.0708, 27.3855, 17.865, 11.7839, 14.2884, 4.58916,$$
$$13.3286, 15.9219, 13.5237, 12.9351, 16.4753, 18.3969, 15.191, 18.2475,$$
$$15.4953, 10.1976, 9.99946, 13.7601, 5.19656, 9.62611, 5.2793, 3.14178,$$
$$2.60383, 0.298653, 6.45915, 7.6787, 8.4658, 9.20346, 17.9811, \ldots\}$$

We want to test the hypothesis that the sample came from a normal population with mean 10 and standard deviation 5.

To do this, we will calculate the quantiles of the data set and compare them to the quantiles of the $N(10, 5)$ distribution. We elect here to calculate nine quantiles for each data set, but more quantiles could just as easily be used.

- The quantiles for the data set are: $\{3.52598, 5.59546, 7.28802, 8.52198, 9.81575, 10.934, 12.2336, 13.7601, 16.1476\}$.
- The quantiles for the $N(10, 5)$ distribution are: $\{3.59224, 5.79189, 7.378, 8.73326, 10, 11.2667, 12.622, 14.2081, 16.4078\}$.

If the data set came from the $N(10, 5)$ distribution, the points whose coordinates are the corresponding quantiles should lie on a straight line with slope 1, indicating that the quantiles are in fact approximately equal. So we fit a least squares regression line to the set of points

$$(3.59224, 3.52598), (5.79189, 5.59546), \ldots, (16.4078, 16.1476).$$

The graph is shown in Figure 4.8. To the eye, the data appear to fit a straight line very closely. Now, however, we can do better than a visual test: We can calculate the least squares regression line and do the analysis of variance test.

We find that the least squares regression line is $y = -0.00121977 + 0.975928x$, so the slope is very close to 1. The analysis of variance test shows a p-value for the hypothesis that $\beta = 0$ to be $4.72289 \cdot 10^{-13}$, a value so small that the hypothesis that $\beta = 0$ is certainly rejected.   ■

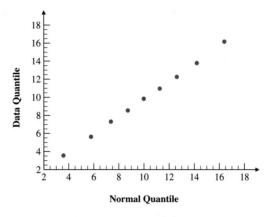

Figure 4.8 Quantile-Quantile Plot for a Normal Distribution

4.11 Checking the Assumptions

Our linear model for simple linear regression is

$$y_i = \alpha + \beta x_i + \varepsilon_i \quad \text{for } i = 1, 2, \ldots, n$$

where we have presumed that $\varepsilon_i \sim NID(0, \sigma)$. It is possible to test the normality of the ε_i, the residuals, by using the q-q plot described in the previous section or by graphical analysis.

To illustrate these procedures, it is best to have a large data set, so we use a data set available in Minitab, called **Grades.mtw.** We will not show the data set here. The data we use here predict graduating grade-point average as a function of the student's Mathematics SAT score for a group of 200 college students. A least squares regression line was calculated, then each of the fitted values was calculated. From these and the observed values, the residuals were calculated. Figure 4.9 shows a graph of these residuals as a function of the order of the observation.

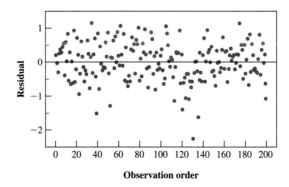

Figure 4.9 Residuals Versus the Order of the Data

In order to assess the randomness of the errors, we first observe that there is no discernible pattern in the residuals. Since the residuals do not appear to become more or less variable as we increase the order of the observation, it would appear that the variance is approximately constant. There are tests for this hypothesis, but they can be used only if we have repeated observations for each value of the independent variable.

But we can test to see if the slope of a regression line for the residuals is 0. The least squares regression line is

$$\text{GPA} = 2.69 - 0.000580 \cdot \text{Order}.$$

The test of H_0: $\beta = 0$ gives a p-value of 0.416, so the hypothesis is not rejected.

A normal probability plot is shown in Figure 4.10, indicating that the residuals can be considered normal. A final graph of the residuals in a histogram is shown in Figure 4.11, which also exhibits a normal state. These graphs are easily produced with most statistical computer packages.

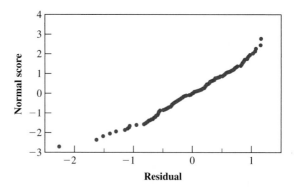

Figure 4.10 Normal Probability Plot of the Residuals

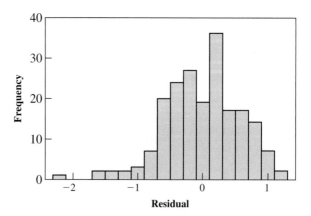

Figure 4.11 Histogram of the Residuals

EXERCISES 4.11

1. Use the following data from Exercise 3 of Section 4.2.

Humidity (X)	46	53	37	42	34	29	60	44	41	48	33	40
Moisture (Y)	12	14	11	13	10	8	17	12	10	15	9	13

Make a plot of the residuals from the least squares regression line and decide whether or not they show random variation.

2. Use the following data from Exercise 5 of Section 4.5.

x	165	170	175	180	185	190	195	200
y	28.5	26.1	23.9	22.0	20.4	18.5	17.1	15.8

a. Make a plot of the residuals from the least squares regression line and decide whether or not they show random variation.

b. Calculate the value of Cook's d for the first data point.

3. Use the following data from Exercise 6 of Section 4.5.

x	65	50	55	65	55	70	65	70	55	70	50	55
y	85	74	76	90	85	87	94	98	81	91	76	74

a. Make a plot of the residuals from the least squares regression line and decide whether or not they show random variation.

b. Calculate the value of Cook's d for the first data point.

4. Use the following data from Exercise 7 of Section 4.5.

x	4.3	4.5	5.9	5.6	6.1	5.2	3.8	2.1	7.5
y	126	121	116	118	114	118	132	141	108

Make a plot of the residuals from the least squares regression line and decide whether or not they show random variation.

5. Use the following data from Exercise 8 of Section 4.5.

x	140	155	160	165	175	165	195	185	195	210
y	350	380	385	450	465	485	535	555	590	605

Make a plot of the residuals from the least squares regression line and decide whether or not they show random variation.

COMPUTER EXPERIMENTS

1. Choose a sample of 1000 observations from the $N(5, 15)$ distribution.
a. Make a q-q plot of the data, using ten quartiles. The fit should be very good.
b. Find the equation of the least squares regression line for the points used in the q-q plot.
c. Draw conclusions from the slope and the intercept for the line in part (b).
2. Consider the data in Problem 8 in Section 4.5,

x	140	155	160	165	175	165	195	185	195	210
y	350	380	385	450	465	485	535	555	590	605

a. Calculate Cook's d for each data point.
b. Alter the value of y at $x = 175$ and show that this has no effect on the influence of that data point.
c. Alter the value of y at $x = 210$ until the point has significant influence.
d. Alter the value of y at $x = 140$ until the point has significant influence.

4.12   Fitting Other Linear Models

The model $y_i = \alpha + \beta x_i + \varepsilon_i$, with the assumption that $\varepsilon_i \sim NID(0, \sigma)$, is often used to cover a multitude of models in addition to straight-line models. Note again that the data set $\{(x_i, y_i) \mid i = 1, 2, \ldots, n\}$ is known, so the linearity of the model refers to the linearity in the parameters α and β. Often the observations can be transformed using a mathematical function and these transformations of the data set then are often used to fit models other than straight lines. We show a series of examples.

Example 4.12.1   The following data represent the population growth of a city over a period of years.

x (year)	1900	1910	1920	1930	1940	1950
y (population)	10,331	14,204	19,372	26,543	36,320	49,920

x (year)	1960	1970	1980	1990
y (population)	68,589	94,275	129,358	178,472

The rate of growth is not linear, as Figure 4.12 shows.

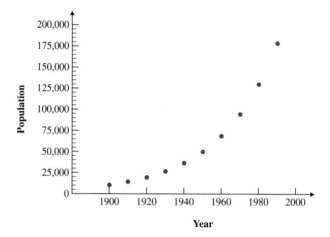

Figure 4.12 Data for Example 4.12.1

We might consider the model $y_i = a \cdot e^{bx_i} + \varepsilon_i$. We could work out the least squares equations for this situation, but this is not commonly done because the resulting equations are very difficult to solve numerically. We will consider a simpler solution, although one that has some limitations. First, let $x = \text{Year} - 1900$. We will transform this model into the straight-line model we have been using by ignoring the error term and taking logarithms

$$\ln y_i = \ln a + b x_i$$

and then use the ordinary least squares equations fitting the logarithm of y to the observed x-values. Then to estimate a, we take the antilog of $\widehat{\alpha}$ for the fitted straight-line model. The result, given by Mathematica, is $\widehat{\ln a} = 9.24007$ and $\widehat{b} = 0.0316273$, showing that $\widehat{a} = 10{,}301.8$ and giving the exponential model

$$y_i = 10{,}301.8 \cdot e^{0.0316273 x_i}.$$

Actually, the data were generated by the model $y_i = 10{,}000 \cdot e^{0.032x_i}$ to which a normal random variable was added.

Now a word of caution: If the true model is $y_i = a \cdot e^{bx_i} + \varepsilon_i$, we note that the errors are additive and assumed to be independent. This in general will no longer be true after the transformation by logarithms is done. The point is an important one because the least squares equations were derived assuming the errors were additive. The model here, as well as the ones discussed below, is acceptable only if the assumption about the errors is correct. ∎

Other Models

We show here a succession of models that can be transformed into the straight-line model.

1. $y_i = a \cdot e^{b/x_i}$. Taking logarithms, we have $\ln y_i = \ln a + b/x_i$, so we fit our straight-line model to the data set $\{1/x_i, \ln y_i\}$. Again, the antilogarithm of the estimated intercept must be calculated.

2. $y_i = a x_i^b$. Taking logarithms gives $\ln y_i = \ln a + b \ln x_i$, so the linear model is fitted to the data set $\{\ln x_i, \ln y_i\}$.

3. $y_i = a + b \ln x_i$ uses the data set $\{\ln x_i, y_i\}$.

4. $y_i = 1/(a + be^{-x_i})$ can be transformed to $1/y_i = a + be^{-x_i}$.

5. The model $y_i = x_i/(ax_i - b)$ can be transformed into the model $1/y_i = a - b/x_i$.

Example 4.12.2 An experimenter has gathered the data

$$\{(0, 0.208869), (0.5, 0.285723), (1, 0.384298), (1.5, 0.418224),$$
$$(2, 0.496861), (2.5, 0.541015), (3, 0.519434), (3.5, 0.53667),$$
$$(4, 0.556184), (4.5, 0.518767), (5, 0.548126)\}.$$

A graph of the data appears as Figure 4.13.

Theory indicates that the model $y = 1/(a + be^{-x})$ is appropriate. To write this in a form useful for simple linear regression, we find that $1/y = a + be^{-x}$, so it appears that we should fit the data set $\{e^{-x_i}, 1/y_i\}$. When that is done, the resulting simple linear regression line is

$$\frac{1}{y_i} = 1.72797 + 2.95235 e^{-x_i}.$$

The p-value in the analysis of variance is of the order 10^{-9}, so we accept the model. ∎

We have studied simple linear regression in this chapter where we have a response, y, which is presumed to be an unknown linear function of a single predictor variable, x. The conclusion reached in this situation is either that the single variable selected as

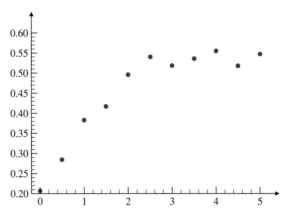

Figure 4.13 Data for Example 4.12.2

the predictor variable has predictive value or that it does not have predictive value, so the experiment must have some theoretical reasons for choosing x or be exceedingly fortunate in its selection. The model built in this case is a very restrictive one, containing only a single independent variable. In Chapter 5 we will shift our focus to models containing two or more independent variables.

EXERCISES 4.12

1. Fit a curve of the form $y = ax^b$ to the data set

{(0.5, 0.801004), (1, 3.07252), (1.5, 6.79272), (2, 12.0825), (2.5, 18.7921), (3, 27.0486), (3.5, 36.8306), (4, 48.0389)}.

Be sure to state your estimates for a and b.

2. Fit a curve of the form $y = x/(ax + b)$ to the data set

{(0.7, 7.00609), (1.1, 0.851431), (1.5, 0.602644), (1.9, 0.521569), (2.3, 0.478522), (2.7, 0.445204), (3.1, 0.431989), (3.5, 0.421423), (3.9, 0.409697), (4.3, 0.400418), (4.7, 0.395242), (5.1, 0.38508), (5.5, 0.385427), (5.9, 0.384514)}.

Be sure to state estimates for a and b.

3. The following data show the compressive strength of an alloy (y) and the concentration of an additive to the alloy (x).

x	10	10	10	15	15	15	20	20
y	25.2	27.3	28.7	29.8	31.1	27.8	31.2	32.6

x	20	25	25	25	30	30	30
y	32.6	31.7	30.1	31.3	29.4	40.8	32.8

Fit a curve of the form $y = ax^2$ to the data and determine whether or not the fit is a good one.

4. A test on the strength of packaging materials containing cans of soft drink involves varying the speed of impact (x) and measuring the numbers of cans damaged (y). The data follow.

x	3	3	3	5	5	5	6	7	7	8
y	54	62	65	94	122	84	142	139	184	254

a. Fit a curve of the form $y = 27 + ax^2$ to the data and decide whether the model is a good one or not.

b. Suggest a procedure for fitting a curve of the form $y = b + ax^2$ to the data.

COMPUTER EXPERIMENT

Calculate a table of values for the function $f(x) = 4x^3$ for $x = 0, 0.5, 1, \ldots, 6$. To each of these values add a random observation from a uniform probability distribution on the interval $[0, 0.01]$. Then fit this data to an appropriate linear function and state your conclusions.

Chapter Review

Simple linear regression refers to approximating a data set $\{\{x_1, y_1\}, \{x_2, y_2\}, \{x_3, y_3\} \ldots, \{x_n, y_n\}\}$ by a linear relationship, say $y = \alpha + \beta x$. Since the data are not exactly linear, an analysis of variance test for the quality of the linear fit is used. We identified several random variables in the course of our investigation along with their probability distributions.

Key Concepts

- Summarizing data sets such as $\{\{x_1, y_1\}, \{x_2, y_2\}, \{x_3, y_3\} \ldots, \{x_n, y_n\}\}$ by linear equations such as $y = \alpha + \beta x$.
- Establishing α and β as random variables in the equation $y = \alpha + \beta x$.
- Estimating α and β in the equation $y = \alpha + \beta x$ by using the principle of least squares.
- Using the linear model $y_i = \alpha + \beta x_i + \varepsilon_i$ for $i = 1, 2, \ldots, n$ where we presume the random errors, ε_i, are $NID(0, \sigma)$.
- Testing H_0: $\beta = 0$ in the linear model using the analysis of variance.
- Using the sample correlation coefficient, r, as a random variable and testing it for significance.
- Finding confidence intervals for the estimates of α and β, $\hat{\alpha}$ and $\hat{\beta}$.
- Finding a regression line that passes through the origin.
- Identifying influential observations.
- Testing data for normality.
- Fitting models other than straight lines.

Key Terms

Residual	The *residual* for a data point $\{x_i, y_i\}$ is $y_i - (\hat{\alpha} + \hat{\beta} x_i)$.
Least Squares Estimators	*Estimators* of α and β, $\hat{\alpha}$ and $\hat{\beta}$, are found using the principle of *least squares* that determines $\hat{\alpha}$ and $\hat{\beta}$ as those values that minimize $S = \sum_{i=1}^{n}(y_i - \alpha - \beta x_i)^2$.

Estimators for α and β

Estimators for α and β based on the principle of least squares are

$$\hat{\beta} = \frac{\sum\limits_{i=1}^{n}(x_i - \bar{x})(y_i - \bar{y})}{\sum\limits_{i=1}^{n}(x_i - \bar{x})^2}$$

$$= \frac{n\sum\limits_{i=1}^{n}x_i y_i - \left(\sum\limits_{i=1}^{n}x_i\right)\left(\sum\limits_{i=1}^{n}y_i\right)}{n\sum\limits_{i=1}^{n}x_i^2 - \left(\sum\limits_{i=1}^{n}x_i\right)^2}$$

and

$$\hat{\alpha} = \bar{y} - \hat{\beta}\bar{x}.$$

Analysis of Variance Identity

If $\hat{y}_i = \hat{\alpha} + \hat{\beta}x_i$, then

$$\sum_{i=1}^{n}(y_i - \bar{y})^2 = \sum_{i=1}^{n}(y_i - \hat{y}_i)^2 + \sum_{i=1}^{n}(\hat{y}_i - \bar{y})^2.$$

Sample Correlation Coefficient

The *sample correlation coefficient* is

$$r_{X,Y} = \frac{\sum\limits_{i=1}^{n}(x_i - \bar{x})(y_i - \bar{y})/(n-1)}{s_x \cdot s_y}.$$

Influential Observations

Influential observations are those with values of Cook's d exceeding a critical value, usually $F_{0.50}[2, n-2]$.

Cook's d

Cook's d is defined as

$$d_i = \frac{\varepsilon_i^2}{2 \cdot MSE}\left(\frac{h_i}{(1-h_i)^2}\right),$$

where

$$h_i = \frac{1}{n} + \frac{(x_i - \bar{x})^2}{\sum\limits_{j=1}^{n}(x_j - \bar{x})^2}.$$

Matrix Formulation of the Model

The model can be written as $\mathbf{Y} = \mathbf{X}\boldsymbol{\beta} + \boldsymbol{\varepsilon}$ where

$$\mathbf{Y} = \begin{bmatrix} y_1 \\ y_2 \\ \vdots \\ y_n \end{bmatrix}, \mathbf{X} = \begin{bmatrix} 1 & x_1 \\ 1 & x_2 \\ \vdots & \vdots \\ 1 & x_n \end{bmatrix}, \boldsymbol{\beta} = \begin{bmatrix} \alpha \\ \beta \end{bmatrix}, \boldsymbol{\varepsilon} = \begin{bmatrix} \varepsilon_1 \\ \varepsilon_2 \\ \vdots \\ \varepsilon_n \end{bmatrix}.$$

Variance-Covariance Matrix

The *variance-covariance matrix* is

$$\sigma^2(\mathbf{X'X})^{-1} = \begin{bmatrix} \mathrm{Var}(\hat{\alpha}) & \mathrm{Cov}(\hat{\alpha}, \hat{\beta}) \\ \mathrm{Cov}(\hat{\alpha}, \hat{\beta}) & \mathrm{Var}(\hat{\beta}) \end{bmatrix}.$$

Normal Probability Plot

A *normal probability plot* is a plot of the quartiles of a data set (here the residuals from a linear model) and corresponding quartiles from a normal distribution.

Key Theorems and Facts

Analysis of Variance Table

The *analysis of variance table* for testing $H_0\colon \beta = 0$ can be found on page 190.

Then $H_0\colon \beta = 0$ is rejected for large values of F.

Probability Distribution of r

The *probability distribution* useful for the sample correlation coefficient r is $\tanh^{-1} r = \frac{1}{2} \ln \frac{1+r}{1-r}$ is approximately $N\left(\frac{1}{2} \ln \frac{1+\rho}{1-\rho}, \sqrt{\frac{1}{n-3}}\right)$ where ρ is the true population correlation coefficient.

Coefficient of Determination

The *coefficient of determination* is

$$r^2 = \frac{\sum_{i=1}^{n} (\hat{y}_i - \bar{y})^2}{\sum_{i=1}^{n} (y_i - \bar{y})^2}.$$

Probability Distribution of $\hat{\alpha}$

The *probability distribution of* the random variable $\hat{\alpha}$ is

$$N\left[\alpha, \sigma \sqrt{\frac{1}{n} + \frac{\bar{x}^2}{\sum_{i=1}^{n} (x_i - \bar{x})^2}}\right].$$

Probability Distribution of $\hat{\beta}$

The *probability distribution of* the random variable $\hat{\beta}$ is

$$N\left[\beta, \sigma \sqrt{\frac{1}{\sum_{i=1}^{n} (x_i - \bar{x})^2}}\right].$$

Unbiased Estimator for σ^2

It can be shown that

$$E\left[\frac{\sum_{i=1}^{n}(y_i - \hat{y}_i)^2}{n-2}\right] = E(s^2) = \sigma^2$$

where $\dfrac{\sum_{i=1}^{n}(y_i - \hat{y}_i)^2}{n-2} = s^2$ is the mean square for error in the analysis of variance.

Confidence Interval for a Predicted Value

A *confidence interval for* the *predicted value* at $x = x_i$ is based upon

$$t_{n-2} = \frac{\hat{y}_i - y_i}{\sqrt{s^2\left[1 + \dfrac{1}{n} + \dfrac{(x_i - \overline{x})^2}{\sum\limits_{j=1}^{n}(x_j - \overline{x})^2}\right]}}.$$

Confidence Interval for a Mean Predicted Value

A *confidence interval for the mean predicted value* at $x = x_i$ is based upon

$$t_{n-2} = \frac{\hat{y}_i - \mu_{Y|x}}{\sqrt{s^2\left[\dfrac{1}{n} + \dfrac{(x_i - \overline{x})^2}{\sum\limits_{j=1}^{n}(x_j - \overline{x})^2}\right]}}.$$

Regression Through the Origin

An analysis of variance table for the model $y_i = \beta x_i, i = 1, 2 \ldots, n$ can be found on page 209.

5 Multiple Linear Regression

5.1 Introduction

WE KNOW THAT EXPERIMENTERS IN SCIENCE AND ENGINEERING focus much of their activity on discovering models. In this chapter we continue our study of model building by considering **multiple linear regression** models.

Experimenters often must deal with cases that include a number of possible candidates for the predictor variable. When only one of these variables will be used, they must determine which one to study. When more than one variable can be studied, the model becomes much more complex. Including all the candidate predictor variables would likely prove wasteful and costly. Choosing some but not all of the variables would probably be the best course. The experimenters then would have to decide which subset to study. If more than one predictor variable is selected, the variables may interact with each other and the individual variables may influence the predicted value in different ways. Multiple linear regression models help experimenters perform complex analyses in very practical situations.

Here are some examples:

- In the production of steel the hardness of the product is a function of the temperature of the production process as well as the percentage of carbon in the steel. (There may be many other variables involved in this example as well.)
- Rainfall at a certain time is a function of humidity and atmospheric pressure.
- A college student's grade-point average might be predicted by her rank in her high school class, her score on the Scholastic Aptitude Test, and the quality of the high school.

We now turn our attention to multiple regression models in which some response variable, y, is a function of two or more predictor variables, say $x_1, x_2, \ldots, x_{k-1}$ where $k \geq 3$. We illustrate the situation with an example.

Example 5.1.1 A university admissions officer would like to be able to predict the college grade-point average for a prospective student currently in high school. A multiple regression model could be helpful. Such prediction equations, if they can be found, might be important not only in determining admission to a college or university, but in determining financial aid as well. In order to build the model, the admissions office takes a random sample of 25 graduating college seniors. For each of the graduates, the office has the following data: Rank in Class, Class Size, Scholastic Aptitude English Score (SAT Eng), Scholastic Aptitude Mathematics Score (SAT Math), and a measure of the high school's quality, School Rank (Sch Rank). The data are shown in Table 5.1.

GPA	Rank	Class Size	SAT Eng	SAT Math	Sch Rank
1.73	8	50	354	608	7
1.92	7	49	705	435	8
3.64	4	51	502	568	15
2.97	9	55	491	489	12
0.58	7	53	441	419	3
1.35	9	47	481	509	6
3.68	11	49	375	367	15
3.81	9	51	525	348	15
3.61	8	46	436	391	15
5.31	12	46	359	570	21
3.56	6	49	582	587	14
2.62	10	51	518	634	11
3.84	3	51	560	651	16
1.79	4	48	618	593	7
0.61	11	48	544	407	3
3.90	8	50	506	409	16
2.22	6	47	484	445	9
2.20	9	52	516	392	9
1.65	10	49	481	485	7
3.76	1	49	578	733	15
2.84	5	51	417	549	12
1.91	9	49	381	495	8
3.28	6	52	461	757	13
2.48	3	52	538	540	10
3.06	9	52	538	513	12

Table 5.1

Trends in the data are not easily seen because there are many predictor variables, making the situation much more complex than what we saw in simple linear regression. In this example we will want to be able to predict graduation grade-point average (GPA) as a function of the predictor variables Rank, Class Size, SAT Eng, SAT Math, and Sch Rank. ∎

Some specific questions may now arise, such as: If we fit a linear equation to these data, how well does the equation predict graduating grade-point average? Do we need to use all the variables? Would the predictions deteriorate seriously if we did not use all the variables? Are the variables related to each other in any way? Are any of the data points unusually influential in determining the linear equation?

As in the case of simple linear regression, it will be helpful if we develop some theory before we attempt to answer the questions raised here. We will return to this example later in this chapter.

Some Terminology

In the case of simple linear regression, we referred to the predictor variable, x, as the **independent variable** since that is the familiar mathematical terminology for x. To describe x as the independent variable is also completely accurate in a statistical sense since there is only one variable.

When two or more x variables are used, it is common to speak of them as *mathematically* independent since that refers only to the fact that they are used as predictor variables. However—and here is where the difficulty arises—the x variables may not be *statistically* independent since, although they are used as predictor variables, they may in fact be correlated. If the correlation coefficient is large, we probably will not use all the variables, although there are situations in which they are all used. When the correlation coefficient is small, then the variables come closer to being independent in both the mathematical and the statistical senses. So we refer to the x variables as **predictor variables.**

The simplest case of multiple regression is that in which there are only two predictor variables. Since many of the difficulties encountered in the general case, where there are more than two predictors, will be encountered here, we will begin our study by considering the case with two predictor variables.

5.2 Two Predictors

In Chapter 4 the principle of least squares was used to find what we called the best straight line to fit a set of data. Recall that we have a set of data $\{(x_1, y_1), (x_2, y_2), \ldots, (x_n, y_n)\}$ in that case and that the model is

$$y_i = \alpha + \beta x_i + \varepsilon_i \quad \text{for } i = 1, 2, \ldots, n.$$

The random errors, the ε_i, are assumed to be independently distributed according to a normal distribution with mean 0 and variance σ^2, which we denote as $NID(0, \sigma)$.

We defined the design matrix X as

$$X = \begin{bmatrix} 1 & x_1 \\ 1 & x_2 \\ \vdots & \vdots \\ 1 & x_n \end{bmatrix}.$$

Let the vector

$$\beta = \begin{bmatrix} \alpha \\ \beta \end{bmatrix},$$

the vector

$$Y = \begin{bmatrix} y_1 \\ y_2 \\ \vdots \\ y_n \end{bmatrix},$$

so that the model is $Y = X\beta + \varepsilon$ where

$$\varepsilon = \begin{bmatrix} \varepsilon_1 \\ \varepsilon_2 \\ \vdots \\ \varepsilon_n \end{bmatrix}.$$

The principle of least squares shows that the best estimates of the parameters α and β are the components of $\widehat{\beta}$ where

$$\widehat{\beta} = (X'X)^{-1}X'Y.$$

The word "best" refers to the fact that the estimators are unbiased and have the smallest possible variance.

We also found that the variance-covariance matrix of the parameters $\widehat{\alpha}$ and $\widehat{\beta}$ was $\sigma^2(X'X)^{-1}$. Now we wish to extend the model; we begin with its simplest possible generalization.

Suppose we have a response variable, y, and we want to find an equation expressing y as a function of two predictor variables, x_1 and x_2, based on n sets of observations on these variables. Let x_{ij} denote the jth observation of the variable x_i. Our data are then

$$(x_{11}, x_{21}, y_1), (x_{12}, x_{22}, y_2), \ldots, (x_{1n}, x_{2n}, y_n).$$

We presume the model is

$$y_i = \beta_0 + \beta_1 x_{1i} + \beta_2 x_{2i} + \varepsilon_i \quad \text{for } i = 1, 2, \ldots, n$$

where again the $\varepsilon_i \sim NID(0, \sigma)$. That is, we seek a plane in three-dimensional space that approximates the data. Figure 5.1 shows some points in three-dimensional space and an approximating plane.

The principle of least squares in this instance, as in the case of simple linear regression, seeks to find those estimators, $\widehat{\beta}_0$, $\widehat{\beta}_1$, and $\widehat{\beta}_2$, that minimize $\sum_{i=1}^{n} \varepsilon_i^2$. If we let the

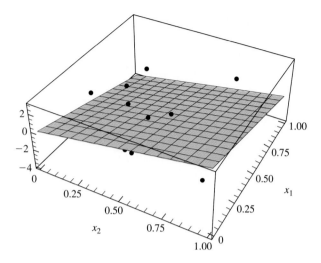

Figure 5.1 Some Points and Approximating Plane

design matrix

$$X = \begin{bmatrix} 1 & x_{11} & x_{21} \\ 1 & x_{12} & x_{22} \\ \vdots & \vdots & \vdots \\ 1 & x_{1n} & x_{2n} \end{bmatrix},$$

the vector

$$Y = \begin{bmatrix} y_1 \\ y_2 \\ \vdots \\ y_n \end{bmatrix},$$

the vector of parameters

$$\beta = \begin{bmatrix} \beta_0 \\ \beta_1 \\ \beta_2 \end{bmatrix},$$

and the vector

$$\varepsilon = \begin{bmatrix} \varepsilon_1 \\ \varepsilon_2 \\ \vdots \\ \varepsilon_n \end{bmatrix},$$

then the model becomes $Y = X\beta + \varepsilon$. It can be shown that the best linear unbiased estimator of the vector β, found using the principle of least squares, is

$$\widehat{\beta} = (X'X)^{-1}X'Y.$$

It can also be shown that the variance-covariance matrix is $\sigma^2(X'X)^{-1}$, so simple regression is a special case of multiple regression where we have only one predictor. As we shall subsequently see, similar results follow for a set of k predictors.

In the case of two predictors, it is fairly simple to calculate that

$$X'X = \begin{bmatrix} n & \sum_{i=1}^{n} x_{1i} & \sum_{i=1}^{n} x_{2i} \\ \sum_{i=1}^{n} x_{1i} & \sum_{i=1}^{n} x_{1i}^2 & \sum_{i=1}^{n} x_{1i} x_{2i} \\ \sum_{i=1}^{n} x_{2i} & \sum_{i=1}^{n} x_{1i} x_{2i} & \sum_{i=1}^{n} x_{2i}^2 \end{bmatrix}$$

and so the least squares equations, $X'X\widehat{\beta} = X'Y$, are

$$n\widehat{\beta}_0 + \widehat{\beta}_1 \sum_{i=1}^{n} x_{1i} + \widehat{\beta}_2 \sum_{i=1}^{n} x_{2i} = \sum_{i=1}^{n} y_i,$$

$$\widehat{\beta}_0 \sum_{i=1}^{n} x_{1i} + \widehat{\beta}_1 \sum_{i=1}^{n} x_{1i}^2 + \widehat{\beta}_2 \sum_{i=1}^{n} x_{1i} x_{2i} = \sum_{i=1}^{n} x_{1i} y_i,$$

and

$$\widehat{\beta}_0 \sum_{i=1}^{n} x_{2i} + \widehat{\beta}_1 \sum_{i=1}^{n} x_{1i} x_{2i} + \widehat{\beta}_2 \sum_{i=1}^{n} x_{2i}^2 = \sum_{i=1}^{n} x_{2i} y_i.$$

These least squares equations can be solved uniquely if the matrix $X'X$ is nonsingular. It is an amazing fact that the least squares equations always have a solution, although if the $X'X$ matrix is singular—that is, is not invertible—the solution may not be unique. We will not consider this case. We give a simple example now.

Example 5.2.1   Suppose that our data set is

y	16	19	22	41	37
x_1	1	2	3	4	5
x_2	0	4	2	0	3

A graph of the data is shown in Figure 5.2.

Then the matrix

$$X = \begin{bmatrix} 1 & 1 & 0 \\ 1 & 2 & 4 \\ 1 & 3 & 2 \\ 1 & 4 & 0 \\ 1 & 5 & 3 \end{bmatrix}$$

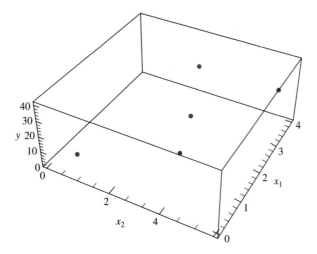

Figure 5.2 Data for Example 5.2.1

and so

$$X'X = \begin{bmatrix} 5 & 15 & 9 \\ 15 & 55 & 29 \\ 9 & 29 & 29 \end{bmatrix}$$

and the matrix

$$X'Y = \begin{bmatrix} 1 & 1 & 1 & 1 & 1 \\ 1 & 2 & 3 & 4 & 5 \\ 0 & 4 & 2 & 0 & 3 \end{bmatrix} \begin{bmatrix} 16 \\ 19 \\ 22 \\ 41 \\ 37 \end{bmatrix} = \begin{bmatrix} 135 \\ 469 \\ 231 \end{bmatrix}$$

showing that the least squares equations, $(X'X)\widehat{\beta} = X'Y$, are

$$5\widehat{\beta}_0 + 15\widehat{\beta}_1 + 9\widehat{\beta}_2 = 135,$$
$$15\widehat{\beta}_0 + 55\widehat{\beta}_1 + 29\widehat{\beta}_2 = 469,$$

and

$$9\widehat{\beta}_0 + 29\widehat{\beta}_1 + 29\widehat{\beta}_2 = 231.$$

Solving these equations gives the least squares estimators for the components of $\widehat{\beta} = (X'X)^{-1}X'Y$ as

$$\widehat{\beta} = \begin{bmatrix} 10.200 \\ 6.800 \\ -2.00 \end{bmatrix}$$

so the least squares plane is $\widehat{y} = 10.200 + 6.800x_1 - 2.000x_2$, which is the equation used to find predicted values. The predicted values from the model, as well as the residuals, are shown in Table 5.2.

x_1	x_2	y	$\widehat{y} = Fit$	Residual
1	0	16.00	17.00	−1.00
2	4	19.00	15.80	3.20
3	2	22.00	26.60	−4.60
4	0	41.00	37.40	3.60
5	3	37.00	38.20	−1.20

Table 5.2

The principle of least squares assures us that the sum of the squares of the residuals,

$$\sum_{i=1}^{5}(y_i - \widehat{y}_i)^2 = (-1.00)^2 + (3.20)^2 + (-4.60)^2 + (3.60)^2 + (-1.20)^2 = 46.80,$$

is the smallest among any linear combination of the predictor variables. We give an informal verification of this. We fix the intercept at 10.2 (so that we can see a graph of a function of two variables later) and vary the parameters β_1 and β_2. Then the sum of the squares of the residuals becomes a function of β_1 and β_2 so

$$\sum_{i=1}^{5}(y_i - \widehat{y}_i)^2 = f(\beta_1, \beta_2) = (16 - 10.2 - \beta_1 \cdot 1 - \beta_2 \cdot 0)^2$$
$$+ (19 - 10.2 - \beta_1 \cdot 2 - \beta_2 \cdot 4)^2 + (22 - 10.2 - \beta_1 \cdot 3 - \beta_2 \cdot 2)^2$$
$$+ (41 - 10.2 - \beta_1 \cdot 4 - \beta_2 \cdot 0)^2 + (37 - 10.2 - \beta_1 \cdot 5 - \beta_2 \cdot 3)^2.$$

Mathematica simplifies this to

$$f(\beta_1, \beta_2) = 1917.2 + 55\beta_1^2 - 278.4\beta_2 + 29\beta_2^2 - 632\beta_1 + 58\beta_1\beta_2.$$

With one of the parameters fixed, the sum of the squares of the residuals will always be a general quadratic function of the two remaining variables.

Figure 5.3 shows the surface $f(\beta_1, \beta_2)$. It is not at all unreasonable to suppose that the minimum value of the function is near $(6.8, -2)$.

One can also solve the simultaneous equations

$$\frac{\partial f(\beta_1, \beta_2)}{\partial \beta_1} = 0 \quad \text{and} \quad \frac{\partial f(\beta_1, \beta_2)}{\partial \beta_2} = 0$$

to find that $\widehat{\beta}_1 = 6.80$ and $\widehat{\beta}_2 = -2.00$. Of course, least squares without restriction treats the residual sum of squares as a function of three variables, β_0, β_1, and β_2, but we have specified one of them here so that we could see some of the geometry involved. ∎

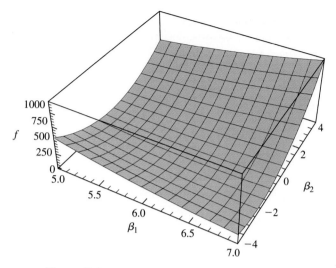

Figure 5.3 Quadratic Surface for Example 5.2.1

Our next task will be to establish some test of the significance, or lack of it, for the regression equation.

EXERCISES 5.2

Exercises 1–4 can be done with a hand-held calculator. The remaining exercises are most easily done using a computer statistical program.

1. Consider the following data set.

y	x_1	x_2
18	1	1
6	0	0
−12	−1	1

a. Show a graph of the data set.
b. Show the data in the form $Y = X\beta + \varepsilon$.
c. Show the least squares equations and solve them for $\widehat{\beta}$.
d. Calculate all the predicted values from the model. Explain why the predicted values equal the observed values.

2. Consider the following data set.

y	x_1	x_2
8	−1	0
10	−1	1
16	1	0
6	1	−1

a. Show a graph of the data set.
b. Show the data in the form $Y = X\beta + \varepsilon$.
c. Show the least squares equations and solve them for $\widehat{\beta}$.
d. Calculate all the predicted values from the model.
e. Calculate the sum of the squares of the residuals.
f. Assuming that $\widehat{\beta}_0 = 10$, express the sum of the squares of the residuals as a function, f, of β_1 and β_2, and then solve the equations

$$\frac{\partial f(\beta_1, \beta_2)}{\partial \beta_1} = 0 \quad \text{and} \quad \frac{\partial f(\beta_1, \beta_2)}{\partial \beta_2} = 0$$

confirming the solution to the least squares equations.

3. Consider the following data set.

y	x_1	x_2
2	-2	1
1	-1	1
3	0	-1
9	1	5
3	2	-1

a. Show a graph of the data set.
b. Show the data in the form $Y = X\beta + \varepsilon$.
c. Show the least squares equations and solve them for $\widehat{\beta}$.
d. Calculate all the predicted values from the model.
e. Calculate the sum of the squares of the residuals.

4. Consider the following data set.

y	x_1	x_2
1	-2	1
1	-1	1
6	0	-1
1	1	5
1	2	-1

a. Show a graph of the data set.
b. Show the data in the form $Y = X\beta + \varepsilon$.
c. Show the least squares equations and solve them for $\widehat{\beta}$.
d. Calculate all the predicted values from the model.
e. Explain why the variable x_1 has no predictive value whatsoever. What characteristic must the y values possess for this to be true?

5. Consider the following data set.

y	x_1	x_2
-2	1	1
-1	1	1
0	-1	6
1	5	1
2	-1	1

(The data set is a variant of the data set in the previous problem.)
a. Show a graph of the data set.
b. Show the data in the form $Y = X\beta + \varepsilon$.
c. Show the least squares equations and solve them for $\widehat{\beta}$.

d. Calculate all the predicted values from the model.
e. Explain why neither of the predictor variables has any predictive value. What characteristic must the design matrix possess for this to be true?

6. A manufacturer of plastic washers used in the manufacture of roofing nails studies the hardness of the washer, y, as a function of the time the plastic is allowed to cure, x_1, and the percentage of a chemical component in the plastic, x_2. The data are shown below.

Hardness	Time	Component (%)
276	22	36
297	34	53
264	12	44
290	40	22
276	34	17
293	44	25
323	60	45
293	40	25
276	32	15
291	27	46

a. Write the model in the form $Y = X\beta + \varepsilon$.
b. Write the least squares equations in numerical form.
c. Solve the least squares equations for $\widehat{\beta}$.

7. The data below are from a study of two different types of light bulbs (coded as -1 and 1) operating under five typical household voltages (108, 114, 120, 126, and 132 volts, coded as 1, 2, 3, 4, and 5). The data give the number of bulbs that blew out during the study.

Blowouts	Type	Voltage
2	-1	1
6	-1	2
11	-1	3
27	-1	4
45	-1	5
3	1	1
8	1	2
13	1	3
31	1	4
46	1	5

Source: Minitab, Lightbul.mtw.

a. Write the model in the form $Y = X\beta + \varepsilon$.
b. Write the least squares equations in numerical form.
c. Solve the least squares equations for $\widehat{\beta}$.

COMPUTER EXPERIMENTS

1. Using the data in Example 5.1.1, suppose we want to predict graduating GPA as a function of the predictor variables SAT Eng and SAT Math only.
   a. Show the X matrix.
   b. Show the least squares equations.
   c. Solve the least squares equations.
   d. Use the least squares equations to calculate each of the predicted values and each of the residual values.
   e. Show a graph of the fitted values and the residuals.
2. Again use the data in Example 5.1.1. Suppose we want to predict graduating GPA as a function of the predictor variables SAT Eng and SAT Math only.
   a. The result of Problem 1 above is that graduating GPA is roughly a constant plus 0.002 times the difference between the SAT Math and SAT Eng scores. This suggests the model GPA $= \beta_0 + \beta_1$ Diff where *Diff* is the difference between an individual's SAT Math and SAT Eng scores. Express the sum of the squares of the residuals using this model.
   b. Graph the sum of the squares of the residuals as found in part (a) and show that the minimum is in fact near the result given by the principle of least squares.

5.3 Analysis of Variance

An analysis of variance, partitioning a total sum of squares, was derived in Chapter 4 for simple linear regression. One can also be derived for the multiple regression model, although the derivation is more complex than the derivation in the case of simple regression. We give a very informal explanation of the partition of the total sum of squares here. For more details, see Saville and Wood [11].

As usual, we let y_i denote the ith observation and $\widehat{y}_i$ denote the estimate of y_i found using the principle of least squares. Let $y_i - \overline{y}$ denote the deviation of y_i from its mean. The first least squares equation,

$$n\widehat{\beta}_0 + \widehat{\beta}_1 \sum_{i=1}^{n} x_{1i} + \widehat{\beta}_2 \sum_{i=1}^{n} x_{2i} = \sum_{i=1}^{n} y_i,$$

can be written, by dividing through by n, as

$$\widehat{\beta}_0 + \widehat{\beta}_1 \overline{x}_1 + \widehat{\beta}_2 \overline{x}_2 = \overline{y}.$$

But the mean of the predicted values is $\overline{\widehat{y}_i} = \widehat{\beta}_0 + \widehat{\beta}_1 \overline{x}_1 + \widehat{\beta}_1 \overline{x}_2$. This shows that the mean of the predicted values, the values of $\widehat{y}_i$, is $\overline{y}$. So $\widehat{y}_i - \overline{y}$ denotes the deviation of $\widehat{y}_i$ from its mean.

Now consider the vectors $\overrightarrow{y_i - \overline{y}}$, $\overrightarrow{y_i - \widehat{y}_i}$ and $\overrightarrow{\widehat{y}_i - \overline{y}}$, whose ith components are $y_i - \overline{y}$, $y_i - \widehat{y}_i$, and $\widehat{y}_i - \overline{y}$, respectively. Then

$$\overrightarrow{y_i - \overline{y}} = \overrightarrow{y_i - \widehat{y}_i} + \overrightarrow{\widehat{y}_i - \overline{y}}.$$

The vectors cannot be accurately depicted for higher dimensions, but Figure 5.4 may be useful.

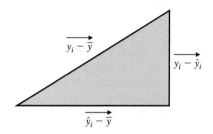

Figure 5.4 Vectors for an Analysis of Variance

Figure 5.4 depicts the vectors $\overrightarrow{y_i - \hat{y}_i}$ and $\overrightarrow{\hat{y}_i - \bar{y}}$ as perpendicular, which indeed they are. This is not easy to show and we will not do it here. Assuming that this is the case, then, since each sum is the squared length of the appropriate vector, the Pythagorean theorem shows that

$$\sum_{i=1}^{n}(y_i - \bar{y})^2 = \sum_{i=1}^{n}(\hat{y}_i - \bar{y})^2 + \sum_{i=1}^{n}(y_i - \hat{y}_i)^2.$$

This is the basic analysis of variance partition that we will use in multiple regression. We pause here to show that this is the case in our example. Table 5.3 shows the relevant calculations. The sample mean is $\bar{y} = 27$ for this example.

x_1	x_2	y_i	$y_i - \bar{y}$	$\hat{y}_i$	$y_i - \hat{y}_i$	$\hat{y}_i - \bar{y}$
1	0	16.00	−11	17.00	−1.00	−10.0
2	4	19.00	−8	15.80	3.20	−11.2
3	2	22.00	−5	26.60	−4.60	−0.4
4	0	41.00	14	37.40	3.60	10.4
5	3	37.00	10	38.20	−1.20	11.2

Table 5.3

In this case,

$$\sum_{i=1}^{n}(y_i - \bar{y})^2 = (16 - 27)^2 + (19 - 27)^2 + (22 - 27)^2$$
$$+ (41 - 27)^2 + (37 - 27)^2 = 506.00,$$

$$\sum_{i=1}^{n}(\hat{y}_i - \bar{y})^2 = (17 - 27)^2 + (15.8 - 27)^2 + (26.6 - 27)^2$$
$$+ (37.4 - 27)^2 + (38.2 - 27)^2 = 459.20,$$

$$\sum_{i=1}^{n}(y_i - \hat{y}_i)^2 = (16 - 17)^2 + (19 - 15.8)^2 + (22 - 26.6)^2$$
$$+ (41 - 37.4)^2 + (37 - 38.2)^2 = 46.80,$$

and

$$459.20 + 46.80 = 506.00.$$

Note that the dot product of the vectors $\overrightarrow{y_i - \hat{y}_i}$ and $\overrightarrow{\hat{y}_i - \bar{y}}$ is

$$(-10)(-1) + (-11.2)(3.2) + (-0.4)(-4.6) + (10.4)(3.6) + (11.2)(-1.2) = 0,$$

showing that these vectors are, in this case, perpendicular. This of course does not establish this central fact in our argument, but is an indication that it may, in actuality, be true.

Suffice it to say for our purposes that the analysis of variance partition is

$$\sum_{i=1}^{n}(y_i - \bar{y})^2 = \sum_{i=1}^{n}(\hat{y}_i - \bar{y})^2 + \sum_{i=1}^{n}(y_i - \hat{y}_i)^2.$$

In this equation,

1. $\sum_{i=1}^{n}(y_i - \bar{y})^2$ is the total sum of squares of the observations around their mean value. It is denoted by *SST*.
2. $\sum_{i=1}^{n}(\hat{y}_i - \bar{y})^2$ is the sum of squares of the deviations of the predicted values and their mean value. This is expected to be large if the regression is significant. It is denoted by *SSR*, or the sum of squares due to regression.
3. $\sum_{i=1}^{n}(y_i - \hat{y}_i)^2$ is the sum of squares of the deviations of the observed values and the predicted values. This is denoted by *SSE*, or the sum of squares due to error. If the regression is significant, this is expected to be small.

These sums of squares can be shown to be independently distributed chi-squared random variables whose ratios are F random variables.

Minitab gives the analysis of variance table for the data in the example as shown in Table 5.4. (Details for using Minitab for multiple linear regression may be found in Appendix B). The p-value in the table indicates that the regression is of marginal significance. It is most probable that we have selected predictor variables that are not of much value and so we should search for variables that have more predictive value. This analysis of variance test is one for the entire regression, although tests and confidence intervals can be constructed for individual regression parameters; these will be shown in subsequent sections.

Source	df	SS	MS	F	p
Regression	2	459.20	229.60	9.81	0.092
Error	2	46.80	23.40		
Total	4	506.00			

Table 5.4

EXERCISES 5.3

Exercises 1–4 can be done with a hand-held calculator. The remaining exercises are most easily done using a computer statistical program.

1. Consider again the data from Exercise 2 of Section 5.2.

y	x_1	x_2
8	−1	0
10	−1	1
16	1	0
6	1	−1

a. Using the results of Exercise 2 of Section 5.2, calculate $\sum_{i=1}^{n}(y_i - \bar{y})^2$, $\sum_{i=1}^{n}(\hat{y}_i - \bar{y})^2$, and $\sum_{i=1}^{n}(y_i - \hat{y}_i)^2$ and verify that these satisfy the analysis of variance partition.
b. Show the complete analysis of variance table. Write a paragraph describing the conclusions that can be drawn from it.
c. Determine each of the vectors $\overrightarrow{y_i - \hat{y}_i}$ and $\overrightarrow{\hat{y}_i - \bar{y}}$ and show that they are perpendicular to each other.

2. Consider again the data from Exercise 3 of Section 5.2.

y	x_1	x_2
2	−2	1
1	−1	1
3	0	−1
9	1	5
3	2	−1

a. Using the results of Exercise 3 of Section 5.2, calculate $\sum_{i=1}^{n}(y_i - \bar{y})^2$, $\sum_{i=1}^{n}(\hat{y}_i - \bar{y})^2$, and $\sum_{i=1}^{n}(y_i - \hat{y}_i)^2$ and verify that these satisfy the analysis of variance partition.
b. Show the complete analysis of variance table. Write a paragraph describing the conclusions that can be drawn.
c. Determine each of the vectors $\overrightarrow{y_i - \hat{y}_i}$ and $\overrightarrow{\hat{y}_i - \bar{y}}$ and show that they are perpendicular to each other.

3. Consider again the data from Exercise 4 of Section 5.2.

y	x_1	x_2
1	−2	1
1	−1	1
6	0	−1
1	1	5
1	2	−1

a. Using the results of Exercise 4 of Section 5.2, calculate $\sum_{i=1}^{n}(y_i - \bar{y})^2$, $\sum_{i=1}^{n}(\hat{y}_i - \bar{y})^2$, and $\sum_{i=1}^{n}(y_i - \hat{y}_i)^2$ and verify that these satisfy the analysis of variance partition.
b. Show the complete analysis of variance table. Write a paragraph describing the conclusions that can be drawn.
c. Determine each of the vectors $\overrightarrow{y_i - \hat{y}_i}$ and $\overrightarrow{\hat{y}_i - \bar{y}}$ and show that they are perpendicular to each other.

4. Consider again the data from Exercise 5 of Section 5.2.

y	x_1	x_2
−2	1	1
−1	1	1
0	−1	6
1	5	1
2	−1	1

a. Using the results of Exercise 5 of Section 5.2, calculate $\sum_{i=1}^{n}(y_i - \bar{y})^2$, $\sum_{i=1}^{n}(\hat{y}_i - \bar{y})^2$, and $\sum_{i=1}^{n}(y_i - \hat{y}_i)^2$ and verify that these satisfy the analysis of variance partition.
b. Show the complete analysis of variance table. Write a paragraph describing the conclusions that can be drawn.
c. Determine each of the vectors $\overrightarrow{y_i - \hat{y}_i}$ and $\overrightarrow{\hat{y}_i - \bar{y}}$ and show that they are perpendicular to each other.

5. Consider again the data from Exercise 6 of Section 5.2.

Hardness	Time	Component (%)
276	22	36
297	34	53
264	12	44
290	40	22
276	34	17
293	44	25
323	60	45
293	40	25
276	32	15
291	27	46

a. Show the analysis of variance table.
b. Explain, using formulas, how each of the quantities in the analysis of variance table is calculated.
c. Write a paragraph describing the conclusions that can be drawn from the analysis of variance table.

COMPUTER EXPERIMENTS

1. Refer to the computer experiments at the end of Section 5.2. Calculate each of the sums of squares used in the analysis of variance table directly from the data, verifying the analysis of variance partition of the total sum of squares.
2. Use a computer algebra system to verify the p-values given in the analysis of variance table.

5.4 Variances, Covariances, Confidence Intervals, and Tests

In order to find tests and confidence intervals for the individual components of β, we continue to develop Example 5.2.1. The $X'X$ matrix is

$$X'X = \begin{bmatrix} 5 & 15 & 9 \\ 15 & 55 & 29 \\ 9 & 29 & 29 \end{bmatrix}$$

and its inverse is

$$(X'X)^{-1} = \begin{bmatrix} 1.21613 & -0.280645 & -0.0967742 \\ -0.280645 & 0.103226 & -0.016129 \\ -0.0967742 & -0.016129 & 0.0806452 \end{bmatrix}.$$

It can be shown that an unbiased estimator for σ^2 is the mean square for error in the analysis of variance, which is 23.40 in this case. We will subsequently use the fact that the mean square for error has 2 degrees of freedom. So our estimated variance-covariance matrix is

$$\widehat{\sigma^2}(X'X)^{-1} = \begin{bmatrix} 28.457 & -6.5671 & -2.2645 \\ -6.5671 & 2.4155 & -0.37742 \\ -2.2645 & -0.37742 & 1.8871 \end{bmatrix}.$$

Note that the estimates of β_0, β_1, and β_2 are correlated. The variance-covariance matrix also tells us that

$$\text{Var}(\widehat{\beta_0}) = 28.457 \text{ so StDev}(\widehat{\beta_0}) = \sqrt{28.457} = 5.3345,$$

$$\text{Var}(\widehat{\beta_1}) = 2.4155 \text{ so StDev}(\widehat{\beta_1}) = \sqrt{2.4155} = 1.5542,$$

$$\text{Var}(\widehat{\beta_2}) = 1.8871 \text{ so StDev}(\widehat{\beta_2}) = \sqrt{1.8871} = 1.3737.$$

It can be shown, under the assumptions of the model, that the individual components of $\widehat{\beta}$ are Student t variables. This fact allows us to test each of the individual coefficients for significance. For example, $\widehat{\beta_0} = 10.2$ and its estimated standard deviation is 5.3345 so

$$t_2 = \frac{10.2 - 0}{5.3345} = 1.9121$$

is the appropriate Student t value with 2 degrees of freedom for testing the hypothesis $H_0\colon \beta_0 = 0$. The p-value is 0.196, so the hypothesis would probably not be rejected. Note that the Student t tests are two-sided, while the F test in the analysis of variance is one-sided.

Predictor	Coeff	StDev	t	p
Constant	10.200	5.335	1.91	0.196
x_1	6.800	1.554	4.38	0.048
x_2	-2.000	1.374	-1.46	0.283

Table 5.5

Minitab produces Table 5.5, which shows the tests on each of the regression coefficients. This suggests that x_1 is the only coefficient of significance and that a model of the form $y_i = \beta_1 x_{1i}$ should be fit. Minitab gives the analysis of variance for this model in Table 5.6. (The analysis of variance given there is a simple regression through the origin, which was discussed in Chapter 4). This suggests that the values predicted by this model are very close to the original data.

Source	df	SS	MS	F	p
Regression	1	3999.3	3999.3	105.45	0.001
Error	4	151.7	37.9		
Total	5	4151.0			

Table 5.6

Confidence Intervals

Confidence intervals can be found for both the regression coefficients and predicted values from the regression model.

Confidence Intervals for Regression Coefficients Consider a two-sided 95% confidence interval for the coefficient β_1 in the preceding example. Since the Student t value at the 97.5% level is 4.30265, $\widehat{\beta}_1 = 6.80$, and the standard deviation of the estimate is 1.554, a two-sided 95% confidence interval is

$$(6.80 - 4.30265(1.554), \ 6.80 + 4.30265(1.554)) = (0.113682, 13.4863).$$

The confidence interval is quite wide, a consequence of the small sample size in this example.

Confidence Intervals for Predicted Values Since the predicted value from the regression model could be written as

$$\widehat{y} = x_0\widehat{\beta}_0 + x_1\widehat{\beta}_1 + x_2\widehat{\beta}_2$$

where $x_0 \equiv 1$, it follows that

$$\text{Var}(\widehat{y}) = \sum_{i=0}^{2} x_i^2 \text{Var}(\widehat{\beta}_i) + 2 \sum_{i>j=0}^{2} x_i x_j \text{Cov}(\widehat{\beta}_i, \widehat{\beta}_j)$$

where $x_0 \equiv 1$, which can be expanded to

$$\text{Var}(\widehat{y}) = \text{Var}(\widehat{\beta}_0) + x_1^2 \text{Var}(\widehat{\beta}_1) + x_2^2 \text{Var}(\widehat{\beta}_2)$$
$$+ 2x_1 \text{Cov}(\widehat{\beta}_1, \widehat{\beta}_0) + 2x_2 \text{Cov}(\widehat{\beta}_2, \widehat{\beta}_0) + 2x_2 x_1 \text{Cov}(\widehat{\beta}_2, \widehat{\beta}_1).$$

In the example we have been following in this section, consider the predicted value when $x_1 = 5$ and $x_2 = 3$. We find that

$$\text{Var}(\widehat{y}) = 28.457 + 5^2 \cdot (2.4155) + 3^2 \cdot (1.8871) + 2 \cdot 5 \cdot (-6.5671)$$
$$+ 2 \cdot 3 \cdot (-2.2645) + 2 \cdot 5 \cdot 3 \cdot (-0.37742)$$
$$= 15.2478.$$

The regression plane gives the predicted value as 38.20. Then a two-sided 95% confidence interval for this predicted value is

$$38.20 \pm 4.30265\sqrt{15.2478} = (21.3988, 55.0012).$$

Minitab and other computer statistical programs provide confidence intervals for multiple regression.

EXERCISES 5.4

Exercises 1–4 can be completed using a hand-held cal-culator. A computer statistical program will be useful for the remaining exercises.

1. We repeat the data considered in Exercise 2 of Section 5.2 and Exercise 1 of Section 5.3.

y	x_1	x_2
8	−1	0
10	−1	1
16	1	0
6	1	−1

a. Show the variance-covariance matrix.
b. Conduct tests on the hypotheses
   (i) H_0: $\beta_0 = 0$ against H_1: $\beta_0 \neq 0$,
   (ii) H_0: $\beta_1 = 0$ against H_1: $\beta_1 \neq 0$, and
   (iii) H_0: $\beta_2 = 0$ against H_1: $\beta_2 \neq 0$
   and state conclusions in a written paragraph.

2. We repeat the data considered in Exercise 3 of Section 5.2 and Exercise 2 of Section 5.3.

y	x_1	x_2
2	−2	1
1	−1	1
3	0	−1
9	1	5
3	2	−1

a. Show the variance-covariance matrix.
b. Conduct tests on the hypotheses
   (i) H_0: $\beta_0 = 0$ against H_1: $\beta_0 \neq 0$,
   (ii) H_0: $\beta_1 = 0$ against H_1: $\beta_1 \neq 0$, and
   (iii) H_0: $\beta_2 = 0$ against H_1: $\beta_2 \neq 0$
   and state conclusions in a written paragraph.

3. We repeat the data considered in Exercise 4 of Section 5.2 and Exercise 3 of Section 5.3.

y	x_1	x_2
1	−2	1
1	−1	1
6	0	−1
1	1	5
1	2	−1

a. Show the variance-covariance matrix.
b. Conduct tests on the hypotheses
   (i) H_0: $\beta_0 = 0$ against H_1: $\beta_0 \neq 0$,
   (ii) H_0: $\beta_1 = 0$ against H_1: $\beta_1 \neq 0$, and
   (iii) H_0: $\beta_2 = 0$ against H_1: $\beta_2 \neq 0$
   and state conclusions in a written paragraph.

4. We repeat the data considered in Exercise 5 of Section 5.2 and Exercise 4 of Section 5.3.

y	x_1	x_2
−2	1	1
−1	1	1
0	−1	6
1	5	1
2	−1	1

a. Show the variance-covariance matrix.
b. Conduct tests on the hypotheses
   (i) H_0: $\beta_0 = 0$ against H_1: $\beta_0 \neq 0$,
   (ii) H_0: $\beta_1 = 0$ against H_1: $\beta_1 \neq 0$, and
   (iii) H_0: $\beta_2 = 0$ against H_1: $\beta_2 \neq 0$,
   and state conclusions in a written paragraph.

5. We repeat the data considered in Exercise 6 of Section 5.2 and Exercise 5 of Section 5.3.

Hardness	Time	Component (%)
276	22	36
297	34	53
264	12	44
290	40	22
276	34	17
293	44	25
323	60	45
293	40	25
276	32	15
291	27	46

a. Show the variance-covariance matrix.
b. Conduct tests on the hypotheses
   (i) H_0: $\beta_0 = 0$ against H_1: $\beta_0 \neq 0$,
   (ii) H_0: $\beta_1 = 0$ against H_1: $\beta_1 \neq 0$, and
   (iii) H_0: $\beta_2 = 0$ against H_1: $\beta_2 \neq 0$
   and state conclusions in a written paragraph.

6. Using the data given in Exercise 1,
   a. find a 95% confidence interval for β_1.
   b. find a 95% confidence interval for the predicted value when $x_1 = -1$ and $x_2 = 1$.

7. Using the data given in Exercise 2,
   a. find a 95% confidence interval for β_2,
   b. find a 95% confidence interval for the predicted value when $x_1 = 1$ and $x_2 = 5$.

8. Using the data given in Exercise 3,
   a. find a 95% confidence interval for β_1.

   b. find a 95% confidence interval for the predicted value when $x_1 = -2$ and $x_2 = 1$.

9. Using the data given in Exercise 4,
   a. find a 95% confidence interval for β_2.
   b. find a 95% confidence interval for the predicted value when $x_1 = -1$ and $x_2 = 1$.

10. Using the data given in Exercise 5,
   a. find a 95% confidence interval for β_0.
   b. find a 95% confidence interval for the predicted value when Time $= 40$ and Component (%) $= 22$.

5.5 Correlation and the Coefficient of Determination

When we discussed the use of the term *predictor variable* rather than *independent variable* earlier in this chapter, we noted that the difference rested in the distinction between variables that are mathematically independent and those that are statistically independent.

With this in mind, the correlations between predictor variables should always be investigated. Minitab makes this easy to do. For the variables in Example 5.2, Minitab produces the correlations found in Table 5.7. (Some instructions on using Minitab can be found in Appendix B).

	y	**x₁**
x₁	0.900	
	0.038	
x₂	−0.149	0.177
	0.811	0.776

Table 5.7

The entries in Table 5.7 are, first, the correlation coefficient and, second, the p-value for the test of the hypothesis that the true correlation coefficient is 0. We conclude that x_2 is not highly correlated with either y or x_1 and that x_1 is probably the only predictor that should be used in the regression equation. However, it is very important to investigate the correlation structure since it may be unnecessary, and perhaps counterproductive, to include variables that are highly correlated with each other. Minitab often automatically excludes such variables in performing a regression analysis, but the correlation structure should be known nonetheless.

Coefficient of Determination In the case of simple linear regression, we saw that the ratio of the sum of squares due to regression and the total sum of squares is the square of the correlation coefficient $r_{x,y}$, so

$$r_{x,y}^2 = \frac{\sum\limits_{i=1}^{n} (\widehat{y}_i - \overline{y})^2}{\sum\limits_{i=1}^{n} (y_i - \overline{y})^2}.$$

In the case of multiple regression, the ratio above is called the **coefficient of determination** and is denoted by R^2, so

$$R^2 = \frac{\sum\limits_{i=1}^{n} (\widehat{y}_i - \overline{y})^2}{\sum\limits_{i=1}^{n} (y_i - \overline{y})^2} = 1 - \frac{\sum\limits_{i=1}^{n} (y_i - \widehat{y}_i)^2}{\sum\limits_{i=1}^{n} (y_i - \overline{y})^2}.$$

If $\widehat{y}_i = y_i$, then $R^2 = 1$ and if $\widehat{y}_i = \overline{y}$, then $R^2 = 0$ so $0 \leq R^2 \leq 1$. It is also known that if the predictor variables are orthogonal, then

$$R^2 = r_{y,x_1}^2 + r_{y,x_2}^2 + \cdots + r_{y,x_n}^2.$$

If the predictor variables are highly correlated, then R^2 is usually considerably less than the sum above. In any event, R^2 is the proportion of the total sum of squares that is due to regression. For small samples, it is common to calculate an adjusted coefficient of determination,

$$R_{\text{Adj}}^2 = 1 - \frac{n-1}{n-k}\left(1 - R^2\right),$$

where $k - 1$ is the number of independent predictor variables.

Minitab produces both R_{Adj}^2 and R^2. Disparity between R_{Adj}^2 and R^2 is often a sign that the model is overdetermined; that is, the model contains variables that do not contribute to the regression. That is certainly the case with our example since we have concluded that x_2 is not needed in the regression equation. For the analysis of variance given in Table 5.4, $R^2 = 459.20/506.00 = 0.9075$ while

$$R_{\text{Adj}}^2 = 1 - \frac{5-1}{5-3}\left(1 - \frac{459.20}{506.00}\right) = 0.8150.$$

A Warning   One might think that a model that has a value of R^2 larger than the value of R^2 for another model might be the better choice. This is not necessarily the case. The problem is that *any* variable, even a variable totally irrelevant to the study at hand, has a positive sum of squares, which will reduce the error sum of squares and hence inflate R^2. In our GPA data, we could add a variable denoting the color of the student's algebra textbook. This would increase R^2! Common sense must prevail. If a model is predicting the response variable adequately in the eyes of the experimenter, then it is a good and sufficient model.

Minitab gives both R^2 and R_{Adj}^2. If the sample size, n, is large, then clearly R^2 and R_{Adj}^2 are approximately equal and

$$R_{\text{Adj}} \rightarrow R^2 \quad \text{as} \quad n \rightarrow \infty.$$

We now turn to a special case where the predictors are necessarily correlated, that of polynomial regression.

5.6 Polynomial and Other Regression Models

Consider the data in Example 5.2.1, but suppose that we use only the variable x_1 and that we want to fit a parabola to the data. We can use the theory developed here for two predictor variables, using the variable $x_2 = x_1^2$. The model is then of the form $y_i = \beta_0 + \beta_1 x_1 + \beta_2 x_1^2$. Minitab produces the parabola

$$y = 8.80 + 5.54x_1 + 0.143x_1^2.$$

Table 5.8 shows the analysis of variance. The fit here is worse than the straight line using the variable x_1 alone and is worse than the original fit using both x_1 and x_2. Of the four regressions we have tried, the best uses a model with no intercept and the variable x_1 alone.

Source	df	SS	MS	F	p
Regression	2	409.89	204.94	4.26	0.190
Error	2	96.11	48.06		
Total	4	506.00			

Table 5.8

It is easy with a computer statistical package to let x_2 be virtually any function of x_1 that we wish to investigate. Minitab has a large variety of transformations that can be used easily, or the transformation can be defined by the user. Some common functions are the following.

$$x_2 = e^{x_1}$$
$$x_2 = \sin(x_1)$$
$$x_2 = x_1^2$$
$$x_2 = \ln(x_1)$$
$$x_2 = \sqrt{x_1}$$
$$x_2 = |x_1|$$

Many other functions are possible. Note, however, that whatever function is selected, the model $y_i = \beta_0 + \beta_1 x_{1i} + \beta_2 x_{2i} + \varepsilon_i$, for $i = 1, 2, \ldots, n$, remains linear in the transformed variables since the word *linear* refers to the linearity in the βs, the unknown coefficients. We will not consider any models that are not linear.

We have shown an example of fitting various models to the same data set. This is an elementary illustration of the search for a model that best fits the data, a topic we will consider more fully later in this chapter in Section 5.9. This ends our discussion of the multiple regression model with two predictors. Now we move on to the general multiple regression model.

EXERCISES 5.6

1. The following data show the number of students per computer in American schools for academic years 1983–1984 through 1997–1998.

Academic Year	Students per Computer
1983–1984	125
1984–1985	75
1985–1986	50
1986–1987	37
1987–1988	32
1988–1989	25
1989–1990	22
1990–1991	20
1991–1992	18
1992–1993	16
1993–1994	14
1994–1995	10.5
1995–1996	10
1996–1997	7.8
1997–1998	6.1

Source: World Almanac and Book of Facts, 2000

   a. Show a plot of the data.
   b. Fit a parabola to the data predicting Students per Computer as a quadratic function of the variable Year.
   c. Show all the residuals for the model produced in part (b).
   d. Fit a straight line to the data, showing all the residuals.
   e. Write a paragraph comparing the models found in parts (b) and (d).

2. Using the data given in Exercise 1,
   a. fit an exponential model to the data.
   b. write a paragraph comparing the model found in part (a) with those found in Exercise 1.

3. Calculate the value of R^2 and R^2_{Adj} for each of the three models found in Exercises 1 and 2. Interpret each of these values. Which model is best? Why?

4. The following table shows the debt of the United States (in billions of dollars) for the years 1975–1998.

Year	Debt
1975	533.2
1976	620.4
1977	698.8
1978	771.5
1979	826.5
1980	907.7
1981	997.9
1982	1142.0
1983	1377.2
1984	1572.3
1985	1823.1
1986	2125.3
1987	2350.3
1988	2602.3
1989	2857.4
1990	3233.3
1991	3665.3
1992	4064.6
1993	4411.5
1994	4692.8
1995	4974.0
1996	5224.8
1997	5413.1
1998	5526.2

Source: World Almanac and Book of Facts, 2000

   a. Show a plot of the data.
   b. Fit an exponential curve to the data, coding the years as 0 through 23. Show all the residuals and the analysis of variance.
   c. Fit a quadratic function to the data. Show all the residuals and the analysis of variance.
   d. Write a paragraph explaining the values of R^2 for the two models found in parts (b) and (c).
   e. Which model is best? Why?

5. Logarithmic models are often used in predicting the growth of populations. In the following data, x represents time and y is a coded measure of the size of a population.

x	y
0.1	−2.810
0.5	−1.170
1.0	0.022
1.5	0.550
2.0	0.905
2.5	1.05
3.0	2.03
3.5	1.77
4.0	1.37
4.5	2.06
5.0	2.25
5.5	2.34
6.0	2.18
6.5	2.11
7.0	2.72
7.5	2.46
8.0	3.24
8.5	2.18
9.0	3.19
9.5	2.45
10.0	3.13

and y is a measure of the subject's ability to perform the task.

x	y
0.0	7.06
0.5	4.64
1.0	3.03
1.5	4.22
2.0	5.02
2.5	3.43
3.0	4.37
3.5	3.33
4.0	5.99
4.5	8.60
5.0	11.78
5.5	17.07
6.0	19.78
6.5	22.60
7.0	27.22
7.5	33.63
8.0	39.27
8.5	46.37
9.0	53.78
9.5	59.38
10.0	65.47

a. Show a graph of the data.

b. The graph in part (a) indicates that a logarithmic function may fit the data well. Fit a logarithmic function to the data.

c. Write a paragraph describing the results of the statistical analysis of the data.

6. A psychologist is measuring the ability of a subject to perform a repetitive task under the influence of a stimulant. In the following data, x is a measure of time

a. Plot a graph of the data.

b. The graph in part (a) suggests that a quadratic function may fit the data well. Fit a least squares quadratic function to the data.

c. Show the analysis of variance table for the fit in part (b).

d. Write a paragraph describing the conclusions that can be drawn from the statistical analysis.

COMPUTER EXPERIMENTS

1. Make a table of values of the function $y = 125 + 3x + 4x^2 + 2\log(x)$ for $x = 1$, $2, \ldots, 20$. Now add a random observation from the distribution $N(100, 100)$ to each value of y. Show the analysis of variance and conclude that the normal errors produce a result quite distinct from the original function. Show graphs of the original y and the values of y with the normal errors added on the same set of axes.

2. Continuing Problem 1, add observations from $N(100, 1000)$ to the original values of y. Show the analysis of variance and conclude that the result now is largely random. Show graphs of the original y and the values of y with the normal errors added on the same set of axes.

5.7 The General Case

Consider now the case where there are $k - 1$ predictors, say $x_1, x_2, \ldots, x_{k-1}$, where $k \geq 3$, and a constant term so that the model is

$$y_i = \beta_0 + \beta_1 x_{1i} + \beta_2 x_{2i} + \cdots + \beta_{k-1} x_{k-1,i} + \varepsilon_i \quad \text{for } i = 1, 2, \ldots, n.$$

Let

$$\beta = \begin{bmatrix} \beta_0 \\ \beta_1 \\ \vdots \\ \beta_{k-1} \end{bmatrix}.$$

The X matrix is then

$$X = \begin{bmatrix} 1 & x_{11} & x_{21} & \cdots & x_{k-1,1} \\ 1 & x_{12} & x_{22} & \cdots & x_{k-1,2} \\ \vdots & \vdots & \vdots & & \vdots \\ 1 & x_{1n} & x_{2n} & \cdots & x_{k-1,n} \end{bmatrix},$$

so that the model is $Y = X\beta + \varepsilon$ in matrix terms. It can be shown that the least squares equations are $(X'X)\widehat{\beta} = X'Y$, with solution, if $(X'X)$ is nonsingular, as

$$\widehat{\beta} = (X'X)^{-1} X'Y.$$

The analysis of variance partition of the total sum of squares is the same as that found in the case for two predictor variables, namely,

$$\sum_{i=1}^{n}(y_i - \bar{y})^2 = \sum_{i=1}^{n}(\widehat{y}_i - \bar{y})^2 + \sum_{i=1}^{n}(y_i - \widehat{y}_i)^2.$$

Predicted values are now found from

$$\widehat{y}_i = \widehat{\beta}_0 + \widehat{\beta}_1 x_{1i} + \widehat{\beta}_2 x_{2i} + \cdots + \widehat{\beta}_{k-1} x_{k-1,i}, \quad \text{for } i = 1, 2, \ldots, n.$$

The individual sums of squares are again independent chi-squared random variables whose ratio is a F random variable. The analysis of variance table is shown in Table 5.9.

Source	df	SS	MS	F	p
Regression	$k-1$	SSR	SSR/$(k-1)$	MSR/MSE	
Error	$n-k$	SSE	SSE/$(n-k)$		
Total	$n-1$	SST			

Table 5.9

The p-value is now $\Pr[\,MSR/MSE > F(k-1, n-k)\,]$.

As in the case of simple regression, or that of multiple regression with two predictors, it can be shown that the variance-covariance matrix is $\sigma^2(X'X)^{-1}$.

The p-value now tests the hypothesis

$$H_0: \beta_0 = \beta_1 = \cdots = \beta_{k-1} = 0$$

against the alternative

$$H_1: \text{at least one of the } \beta_i's \text{ is not zero.}$$

Confidence Intervals for Predicted Values The predicted value from a general regression model is

$$\widehat{y} = \widehat{\beta_0}x_0 + \widehat{\beta_1}x_1 + \widehat{\beta_2}x_2 + \cdots + \widehat{\beta_{k-1}x_{k-1}}$$

where $x_0 \equiv 1$. Then it follows that

$$\text{Var}(\widehat{y}) = \sum_{i=0}^{k-1} x_i^2 \text{Var}(\widehat{\beta_i}) + 2 \sum_{i>j=0}^{k-1} x_i x_j \text{Cov}(\widehat{\beta_i}, \widehat{\beta_j})$$

where $x_0 \equiv 1$.

These facts can be used to construct confidence intervals for predicted values of the form

$$\widehat{y} \pm t_{n-k} \cdot \sqrt{\text{Var}(\widehat{y})}.$$

An example is shown below.

Example 5.7.1 We return to the data given for Example 5.1.1 in Table 5.1. Recall that the data are an attempt to predict a college student's graduating grade-point average as a function of rank in high school class, class size, SAT English score, SAT Mathematics score, and a measure of the secondary school's quality. This prediction equation might be very important in determining admission to an honors program as well as determining general admission to a college or university.

Table 5.10 shows the data for the original predictor variables as well as two new ones. A regression equation could be found using the original variables, but it is thought best to create other variables from the original ones. Therefore, we let

$$\text{SAT Total} = \text{SAT Eng} + \text{SAT Math}$$

and we create the variable Class Size/Rank in Class. The variable SAT Total reflects the sum of the SAT scores, although there are instances where the individual scores might prove to be useful, while the variable Class Size/Rank in Class gives greatest weight to students who rank high in large classes. Table 5.10 shows the resulting data.

The least squares equation (from Mathematica) is

$$\text{GPA} = -0.288906 + 0.256052 \text{ Sch Rank}$$
$$+ 0.000172585 \text{ SAT Total} - 0.00113076 \text{ Size/Rank}$$

GPA	Rank	Class Size	SAT Eng	SAT Math	Sch Rank	SAT Total	Size/Rank
1.73	8	50	354	608	7	962	6.2500
1.92	7	49	705	435	8	1140	7.0000
3.64	4	51	502	568	15	1070	12.7500
2.97	9	55	491	489	12	980	6.1111
0.58	7	53	441	419	3	860	7.5714
1.35	9	47	481	509	6	990	5.2222
3.68	11	49	375	367	15	742	4.4545
3.81	9	51	525	348	15	873	5.6667
3.61	8	46	436	391	15	827	5.7500
5.31	12	46	359	570	21	929	3.8333
3.56	9	49	582	587	14	1169	5.4444
2.62	10	51	518	634	11	1152	5.1000
3.84	3	51	560	651	16	1211	17.0000
1.79	4	48	618	593	7	1211	12.0000
0.61	11	48	544	407	3	951	4.3636
3.90	8	50	506	409	16	915	6.2500
2.22	6	47	484	445	9	929	7.8333
2.20	9	52	516	392	9	908	5.7778
1.65	10	49	481	485	7	966	4.9000
3.76	1	49	578	733	15	1311	49.0000
2.84	5	51	417	549	12	966	10.2000
1.91	9	49	381	495	8	876	5.4444
3.28	6	52	461	757	13	1218	8.6667
2.48	3	52	538	540	10	1078	17.3333
3.06	9	52	538	513	12	1051	5.7778

Table 5.10

A least squares regression program in Mathematica produces the analysis of variance shown in Table 5.11. (Both Mathematica and Minitab were used for this example in order to use the numerical accuracy in Mathematica. Some instructions on using Minitab can be found in Appendix B.)

Source	df	SS	MS	F	p
Regression	3	31.0521	10.3507	1769.43	$1 \cdot 10^{-28}$
Error	21	0.122844	0.00584973		
Total	24	31.1749			

Table 5.11

So the model is a very good one.

It is very valuable now to examine the actual predictions from the regression equation. They are shown in Table 5.12.

GPA	Fit	Residual	Cook's d
1.73	1.66242	0.0675807	0.01831
1.92	1.94834	−0.0283436	0.00572
3.64	3.72213	−0.0821262	0.02562
2.97	2.94594	0.024056	0.00130
0.58	0.619113	−0.0391126	0.02821
1.35	1.41236	−0.0623617	0.02015
3.68	3.67490	0.00510155	0.00047
3.81	3.69614	0.113864	0.08265
3.61	3.68810	−0.0781033	0.05509
5.31	5.24419	0.0658124	0.10746
3.56	3.49142	0.0685792	0.05657
2.62	2.72072	−0.100720	0.08709
3.84	3.99771	−0.157707	0.24222
1.79	1.69889	0.0911088	0.09131
0.61	0.63845	−0.0284451	0.00953
3.90	3.95878	−0.0587776	0.02160
2.22	2.16704	0.0529619	0.00903
2.22	2.16574	0.034262	0.00405
1.65	1.66464	−0.0146362	0.00087
3.76	3.72273	0.0372707	2.54506
2.84	2.93890	−0.0989042	0.02352
1.91	1.90454	0.0054599	0.00015
3.28	3.24018	0.0398183	0.01922
2.48	2.43806	0.0419367	0.00734
3.06	2.95857	0.101426	0.03261

Table 5.12

The residuals should be examined graphically; Figure 5.5 shows a normal plot of the residuals. The residuals appear to fit a straight line very well but one residual, −0.157707, appears to be somewhat extreme. This residual arises from the GPA score of 3.84. The predicted score is 3.99771, so the point may be influential in determining the regression equation. Figure 5.5 shows a normal probability plot of the residuals.

Figure 5.6 shows a plot of the residuals against the order of the data, which reveals a random pattern with no discernible trends. Such trends in the residuals, when they exist, are often signs of a lack of independence among the residuals.

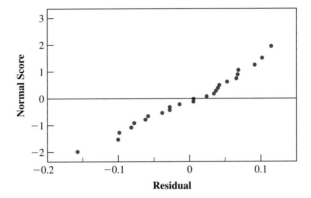

Figure 5.5 Normal Probability Plot of the Residuals

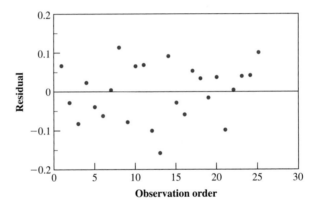

Figure 5.6 Residuals Versus the Order of the Data

Minitab also produces the correlations and p-values shown in Table 5.13. The figures in Table 5.13 give the correlation coefficients and the p-values for testing the hypothesis that the true correlation coefficient is zero. Certainly, the variable Sch Rank is highly correlated with GPA, so we might conclude that this variable is the cause of at least some

	GPA	Sch Rank	SAT Total
Sch Rank	0.998		
	0.000		
SAT Total	0.102	0.086	
	0.627	0.684	
Size/Rank	0.198	0.195	0.583
	0.343	0.351	0.002

Table 5.13

of the significance. We will pursue this point in the next section. We also notice that the variables Size/Rank and SAT Total are correlated, so perhaps we don't need to include both in the regression equation. We will return to this point later in this chapter as well.

This example concludes with the calculation of a confidence interval for a predicted value. Consider the data point where

Sch Rank = 16, SAT Total = 1211, and Size/Rank = 17. The observed GPA = 3.84.

Now

$$\text{Var}(\widehat{y}) = \sum_{i=0}^{k-1} x_i^2 \text{Var}(\widehat{\beta_i}) + 2 \sum_{i>j=0}^{k-1} x_i x_j \text{Cov}(\widehat{\beta_i}, \widehat{\beta_j})$$

where $x_0 \equiv 1$, and we need the variance-covariance matrix. Mathematica produces the $(X'X)^{-1}$ matrix. When multiplied by the mean square for regression in the analysis of variance, we find the estimate of the variance-covariance matrix given in Table 5.14.

$$\widehat{\sigma^2}(X'X)^{-1} =$$

	Constant	Sch Rank	SAT Total	Size/Rank
Constant	$1.75571 \cdot 10^{-2}$	$-1.4775 \cdot 10^{-4}$	$-1.67872 \cdot 10^{-5}$	$1.4196 \cdot 10^{-4}$
Sch Rank	$-1.4775 \cdot 10^{-4}$	$1.286 \cdot 10^{-5}$	$1.67319 \cdot 10^{-8}$	$-1.38126 \cdot 10^{-6}$
SAT Total	$-1.67872 \cdot 10^{-5}$	$1.67319 \cdot 10^{-8}$	$1.79334 \cdot 10^{-8}$	$-1.67327 \cdot 10^{-7}$
Size/Rank	$1.4196 \cdot 10^{-4}$	$-1.38126 \cdot 10^{-6}$	$-1.67327 \cdot 10^{-7}$	$4.64612 \cdot 10^{-6}$

Table 5.14

It follows, for this data point, that

$$\begin{aligned}\text{Var}(\widehat{y}) &= 0.0175571 + 16^2 \cdot 1.286 \cdot 10^{-5} + 1211^2 \cdot 1.79334 \cdot 10^{-8} \\ &+ 17^2 \cdot 4.64612 \cdot 10^{-6} + 2 \cdot 16 \cdot (-1.4775) \cdot 10^{-4} \\ &+ 2 \cdot 17 \cdot 1.4196 \cdot 10^{-4} + 2 \cdot 1211 \cdot (-1.67872) \cdot 10^{-5} \\ &+ 2 \cdot 16 \cdot 17 \cdot (-1.38126) \cdot 10^{-6} + 2 \cdot 16 \cdot 1211 \cdot 1.67319 \cdot 10^{-8} \\ &+ 2 \cdot 17 \cdot 1211 \cdot (-1.67327) \cdot 10^{-7} \\ &= 0.000939205.\end{aligned}$$

The Student t value has 21 degrees of freedom. The critical t value is 2.07961 at the 97.5% level, so the confidence interval is

$$\left(3.84 - 2.07961 \cdot \sqrt{0.000939205}, \; 3.84 + 2.07961 \cdot \sqrt{0.000939205}\right) = (3.77627, 3.90373).$$

This confidence interval is very short due to the small value of Var $(\widehat{y})$. ∎

We have addressed several questions regarding multiple regression raised in the Introduction in this chapter and we are yet to answer others. In the next section we will consider the fact that often some variables are more important than others and show how these important variables can be determined. As in simple linear regression, some data points are more influential upon the resulting regression equation than other data points; we will discuss how these points are found in the next section as well.

5.8 Significant Variables and Influential Observations

The analysis of variance in Table 5.11 indicates that the regression equation fits the data quite well. It is possible to detect particularly significant variables, as we did in Chapter 4.

Each of the $\hat{\beta}_i$ can be shown to be a Student t variable under the model where $\varepsilon_i \sim NID(0, \sigma)$, so tests of significance on these coefficients can be constructed. We used Minitab to produce Table 5.15, which shows tests of significance on each of the coefficients of the predictor variables for the data in Example 5.1.1. We used Mathematica to find the p-values.

Predictor	Coeff	StDev	t	p
Constant	−0.288906	0.132503	−2.18037	0.0407501
Sch Rank	0.256052	0.00358609	71.4015	$1 \cdot 10^{-26}$
SAT Total	0.000172585	0.000133916	1.28876	0.211498
Size/Rank	−0.00113076	0.00215549	−0.524596	0.605358

Table 5.15

The analysis of variance in Table 5.11 indicates that the regression is quite significant, but we cannot tell from the table which of the predictor variables, or which combination of predictor variables, is the cause of the significance. The analysis of variance in Table 5.11 shows a sum of squares due to the total regression of 31.052. It is possible to calculate a sum of squares for each of the individual predictor variables. These are shown in Table 5.16.

Source	df	SS	MS	F	p
Sch Rank	1	31.042	31.042	5306.57	$1 \cdot 10^{-26}$
SAT Total	1	0.009	0.009	1.54	0.228303
Size/Rank	1	0.002	0.002	0.341896	0.564967
Sub Total	3	31.052			

Table 5.16

In Table 5.16, the individual factors have been entered into the model sequentially, accounting for the differences in the p-values from those found in Table 5.15. It is important to note here that the sums of squares for the individual predictors are usually highly dependent on the order in which the variables are entered into the equation. This point is explored in the Computer Experiment at the end of the section.

We would probably conclude that GPA is adequately predicted by Sch Rank alone and that the remaining two variables can be dropped from the regression equation. If the regression is calculated using the predictor Sch Rank alone, Mathematica produces the regression equation

$$\text{GPA} = -0.125071 + 0.256082 \text{ Sch Rank}$$

and the analysis of variance as shown in Table 5.17.

Source	df	SS	MS	F	p
Regression	1	31.0419	31.0419	5368.30	$1 \cdot 10^{-28}$
Error	23	0.132996	0.00578244		
Total	24	31.1749			

Table 5.17

We now compare the predictions using the full model and those using the predictor variable Sch Rank alone. The results are shown in Table 5.18. The reduced model appears to be adequate for prediction purposes, since the discrepancies in the predicted grade-point averages from the full model and the predictions from the reduced model are fairly small. The reduced model is the best we can do with the data available. If we want to predict grade-point average with greater accuracy, then more variables than those available in this study should be measured.

Influential Observations

We used Cook's Distance in Chapter 4 as a measure of the influence of a data point in simple linear regression. Just as in simple linear regression, data points in multiple linear regression can have great influence upon the resulting regression equation. Since many variables are involved, extreme values are often difficult to detect by examining their coordinates alone, so a measure is required. We again use Cook's Distance, as we did in simple linear regression. We gave some formulas for calculating Cook's Distance in the case of simple linear regression in Section 4.9.

It is more difficult to calculate Cook's d for multiple regression. For the ith data point, the value of Cook's d is given by

$$d_i = \frac{1}{ks^2} \sum_{j=1}^{n} (\widehat{Y}_{j(i)} - \widehat{Y}_j)^2$$

where $\widehat{Y}_{j(i)}$ is the predicted value for the jth data point when the ith data point is not used in the computation of the least squares regression equation. The quantity s^2 is the estimate of σ^2 from the analysis of variance and k is the total number of regression coefficients, including the constant. Unlike the case of simple linear regression, the computation is clearly very difficult without a computer.

Minitab produces the values of Cook's d. Table 5.12 shows the values of Cook's d for the data set in our example using the full model.

GPA	Fit (Full Model)	Fit (Reduced Model)
1.73	1.66242	1.6675
1.92	1.94834	1.92358
3.64	3.72213	3.71615
2.97	2.94594	2.94791
0.58	0.61911	0.64317
1.35	1.41236	1.41142
3.68	3.67490	3.71615
3.81	3.69614	3.71615
3.61	3.68810	3.71615
5.31	5.24419	5.25264
3.56	3.49142	3.46007
2.62	2.72072	2.69183
3.84	3.99771	3.97224
1.79	1.69889	1.6675
0.61	0.63845	0.64317
3.90	3.95878	3.97224
2.22	2.16704	2.17966
2.20	2.16574	2.17966
1.65	1.66464	1.6675
3.76	3.72273	3.71615
2.84	2.93890	2.94791
1.91	1.90454	1.92358
3.28	3.24018	3.20399
2.48	2.43806	2.43575
3.06	2.95857	2.94791

Table 5.18

It is a generally accepted rule that the values of Cook's d should be compared with the 50th percentile for the $F[k, n-k]$ distribution (although Cook's d is known not to be exactly an F variable). Here $F_{0.50}[4, 21] = 0.866875$. Only one value of Cook's d, namely 2.54506, exceeds this critical value. This arises from the data point where Sch Rank = 15, SAT Total = 1311, and Size/Rank = 49, all rather large values for these variables. The next largest value of Cook's d is 0.24222, which arises from the data point where Sch Rank = 16, SAT Total = 1211, and Size/Rank = 17, which again are all relatively large values for these variables. Cook [24] indicates that it is useful to investigate values of Cook's d exceeding 0.5, while values of Cook's d exceeding 1 should always be investigated.

It is very important to note that neither of these comparisons constitutes a test of significance and that they are useful for investigative purposes only.

Since the regression is very significant, no great advantage will be achieved by deleting these data points, but that might be done in other data sets.

Multiple regression is a central tool in determining which variables are important in predicting a measured response. We have seen some techniques for determining variables that are important in the model as well as those that are not important to include in the model. In the next section we will consider some very important and interesting procedures, which are of great importance in creating or building a model. These are called *stepwise* procedures.

EXERCISES 5.8

1. Several variables are important in the production of quality plywood. A chuck is inserted at each end of a log of wood and then a thin layer of wood is cut off by a saw blade. A laboratory measured the torque that could be applied before the chuck became ineffective by spinning out. The variables measured were the temperature of the log, the diameter of the log, and the chuck penetration on the log. The data are given below.

Diameter	Penetration	Temperature	Torque
4.5	1.00	60	17.30
4.5	1.50	60	18.05
4.5	2.25	60	17.40
4.5	3.25	60	17.40
4.5	1.00	120	16.70
4.5	1.50	120	17.95
4.5	2.25	120	18.60
4.5	3.25	120	18.55
4.5	1.00	150	15.75
4.5	1.50	150	16.65
4.5	2.25	150	15.25
4.5	3.25	150	15.85
7.5	1.00	60	29.55
7.5	1.50	60	31.50
7.5	2.25	60	36.75
7.5	3.25	60	41.20
7.5	1.00	120	23.20
7.5	1.50	120	25.90
7.5	2.25	120	35.65
7.5	3.25	120	37.60
7.5	1.00	150	22.55
7.5	1.50	150	22.90
7.5	2.25	150	28.90
7.5	3.25	150	35.20

Source: Minitab, Plywood.mtw

a. Find the least squares regression equation predicting Torque from the variables Diameter, Penetration, and Temperature.
b. Show the analysis of variance table. Write a paragraph describing the results and conclusions that can be drawn.
c. Show all the fits and residuals from the least squares equation. Show a graph of the residuals and state the conclusions that can be drawn from the graph.
d. Show all the values of Cook's d and state any conclusions that can be drawn.
e. Test each of the regression coefficients for significance and state any conclusions that can be drawn.
f. Calculate a two-sided 95% confidence interval for the predicted value where Diameter $= 7.5$, Penetration $= 1.00$, and Temperature $= 60$.

2. The following data show the average census (in thousands), the occupancy rate (as a percent), the number of personnel (in thousands), and the number of beds (in thousands) for hospitals in nine northeastern states in the United States.

State	Average Census	Occupancy Rate	Personnel	Beds
ME	3.1	68.2	18.6	4.5
NH	2.3	66.1	14.2	3.4
VT	1.1	65.0	6.6	1.7
MA	15.8	72.3	105.0	21.8
RI	2.4	76.9	15.3	3.1
NY	64.2	84.2	319.1	76.3
NJ	24.9	79.5	118.9	31.3
PA	37.7	72.6	221.5	52.0
CT	6.9	75.6	45.1	9.1

Source: Statistical Abstract of the United States, 1994

a. Find the least squares regression equation predicting Average Census from the variables Occupancy Rate, Personnel, and Beds.

b. Show the analysis of variance table. Write a paragraph describing the results and conclusions that can be drawn.

c. Show all the fits and residuals from the least squares equation. Show a graph of the residuals and state the conclusions that can be drawn from the graph.

d. Show all the values of Cook's d and state any conclusions that can be drawn.

e. Test each of the regression coefficients for significance and state any conclusions that can be drawn.

3. The following data give information about full-page ads in two magazines for the years 1989, 1991, and 1993.

Year	FullAds	Magazine	Pages
1	60	1	76
1	61	1	76
1	45	1	84
1	72	1	100
2	36	1	74
2	63	1	74
2	34	1	68
2	60	1	90
3	27	1	76
3	53	1	82
3	36	1	76
3	61	1	90
1	53	2	88
1	56	2	122
1	33	2	88
1	44	2	106
2	32	2	82
2	58	2	104
2	55	2	96
2	64	2	140
3	75	2	202
3	58	2	90
3	45	2	86
3	35	2	86

Source: Minitab, Ads.mtw

a. Find the least squares regression equation predicting Pages from the variables Magazine, FullAds, and Year.

b. Show the analysis of variance table. Write a paragraph describing the results and conclusions that can be drawn.

c. Show all the fits and residuals from the least squares equation. Show a graph of the residuals and state the conclusions that can be drawn from the graph.

d. Show all the values of Cook's d and state any conclusions that can be drawn.

e. Test each of the regression coefficients for significance and state any conclusions that can be drawn.

f. Calculate a two-sided 95% confidence interval for the predicted value where Year $= 2$, FullAds $= 64$, and Magazine $= 2$.

4. The following data concern trade in petroleum products. The response variable is the total number of Petroleum Products as a function of imports from the Persian Gulf, Total Imports, and Total Exports.

Year	Persian Gulf Imports	Total Imports	Total Exports	Petroleum Products
1975	1165	6056	209	16322
1976	1840	7313	223	17461
1977	2448	8807	243	18431
1978	2219	8363	362	18847
1979	2069	8456	471	18513
1980	1519	6909	544	17056
1981	1219	5996	595	16058
1982	696	5113	815	15296
1983	442	5051	739	15231
1984	506	5437	722	15726
1985	311	5067	781	15726
1986	912	6224	785	16281
1987	1077	6678	764	16665
1988	1541	7402	815	17283
1989	1861	8061	859	17325
1990	1966	8018	857	16988
1991	1845	7627	1001	16714
1992	1778	7888	950	17033
1993	1782	8620	1003	17237
1994	1728	8996	942	17718
1995	1573	8835	949	17725
1996	1604	9399	981	18234
1997	1755	10162	1003	18620
1998	2091	10708	945	18680

Source: World Almanac and Book of Facts, 2000

a. Find the least squares regression equation predicting Petroleum Products from the variables Persian Gulf Imports, Total Imports, and Total Exports.

b. Show the analysis of variance table. Write a paragraph describing the results and conclusions that can be drawn.

c. Show all the fits and residuals from the least squares equation. Show a graph of the residuals and state the conclusions that can be drawn from the graph.

d. Show all the values of Cook's d and state any conclusions that can be drawn.

e. Test each of the regression coefficients for significance and state any conclusions that can be drawn.

5. The following data concern a study made on the cost of a computer as a function of the number of floppy drives, the size of the hard drive, the CPU speed (coded), and the CPU Size (coded).

Price	HD	Floppy	CPUCode	SpeedCode
2599	80	2	1	−1
6730	330	1	1	1
2849	80	1	−1	−1
4075	200	2	−1	1
3995	80	1	1	−1
7675	200	2	1	1
2495	80	2	−1	1
3795	213	2	1	1
2098	80	1	−1	−1
2498	100	2	−1	1
2595	80	2	−1	−1
2995	200	2	−1	1
3495	200	2	1	−1
3995	200	2	1	−1
3399	100	2	−1	−1
3999	200	2	−1	1
5399	200	2	1	−1
6299	200	2	1	1
2854	89	2	−1	1
3404	125	2	1	−1
4924	125	2	1	1
2695	100	2	−1	1
3595	200	2	1	−1
3995	200	2	1	1
3195	200	2	−1	1
3295	100	2	1	−1
3395	110	1	−1	1
2999	110	1	1	−1
7299	200	1	1	1
3755	210	2	1	−1
3995	210	2	1	1
3395	200	2	−1	1
3995	200	2	1	−1
5495	200	2	1	1
3995	85	1	1	−1

Source: Minitab, Computer.mtw

a. Find the least squares regression equation predicting Price from the variables HD, Floppy, CPUCode, and CPU Speed Code.

b. Show the analysis of variance table. Write a paragraph describing the results and conclusions that can be drawn.

c. Show all the fits and residuals from the least squares equation. Show a graph of the residuals and state the conclusions that can be drawn from the graph.

d. Show all the values of Cook's d and state any conclusions that can be drawn.

e. Test each of the regression coefficients for significance and state any conclusions that can be drawn.

f. Calculate a two-sided 95% confidence interval for the predicted price where HD = 200, Floppy = 2, CPUCode = 1 and SpeedCode = 1.

6. The following data concern the salary of an employee as a function of the following variables: number of years employed in the firm, number of years of experience, years of post–high school education, employee ID, gender, employee's department, and number of employees supervised by the employee.

Salary	Yrs	Exp	Educ	ID	Gender	Dept	Super
38985	18	7	9	412	0	1	5
32920	15	3	9	458	1	1	4
29548	5	6	1	604	0	1	0
24749	6	2	0	598	1	1	1
41889	22	16	7	351	0	1	7
31528	3	11	3	674	0	1	6
38791	21	4	5	356	0	1	9
39828	18	6	5	415	1	1	5
28985	0	1	4	693	1	1	4
32782	0	1	7	694	0	1	0
43674	6	9	4	625	0	2	2
35467	3	6	6	354	1	2	3
29876	2	0	3	268	1	2	5
36431	9	4	4	984	1	2	2
56326	12	3	8	651	0	2	6
36571	6	1	4	359	0	2	2
35468	9	5	4	647	1	2	5
26578	0	6	2	845	1	2	2
47536	15	5	6	972	0	3	4
23654	0	0	0	539	1	3	2
37548	19	9	4	649	0	3	6
36578	4	4	8	824	0	3	8
54679	20	3	6	649	1	3	4

(*continued*)

Salary	Yrs	Exp	Educ	ID	Gender	Dept	Super
53234	25	0	6	624	0	3	3
31425	7	6	5	891	1	3	6
39743	9	6	5	974	1	4	1
26452	1	3	2	648	1	4	0
34632	5	4	4	321	0	4	0
35631	6	4	4	264	0	4	2
46211	14	5	6	291	1	4	5
34231	6	2	6	267	0	4	3
26548	5	1	0	548	0	4	2
36512	6	6	4	555	1	4	2
34869	7	5	4	366	1	4	1
41255	9	4	6	246	0	4	4
39331	9	3	6	215	1	4	1
35487	8	2	2	814	1	4	2
36487	6	5	2	212	0	4	3
68425	25	2	12	526	0	4	1
69246	22	3	10	778	0	4	45
65487	27	0	12	486	1	4	44
48695	6	9	8	913	0	4	40
51698	18	6	6	679	0	4	1
46184	20	3	4	602	0	4	1
34987	9	6	2	359	1	4	3
54899	12	5	8	675	1	4	0

Source: Minitab, Techn.mtw

a. Find the least squares regression equation predicting Salary from the variables Yrs, Exp, Educ, ID, Gender, Dept, and Super.

b. Show the analysis of variance table. Write a paragraph describing the results and conclusions that can be drawn.

c. Show all the fits and residuals from the least squares equation. Show a graph of the residuals and state the conclusions that can be drawn from the graph.

d. Show all the values of Cook's d and state any conclusions that can be drawn.

e. Test each of the regression coefficients for significance and state any conclusions that can be drawn.

7. The data below are gathered from the 2001 NCAA Women's Collegiate Basketball Championship Tournament. Sixty-four teams are ranked, 16 in each of four national regions. Each of the 16 teams is ranked, or seeded, from 1 to 16, 1 being the highest seed and 16 the lowest seed. The teams then play the first round of games by the first seeded team playing the 16th, the second playing the 15th, and so on. In the second round, the winners of the first rounds play in a given

order. The data that follow give the difference, or gap, in the seedings and the gap in the score in the game. For example, the gap between the first and the 16th seeded teams is 15. The way in which the teams are divided into sub-brackets makes a seed gap of 7 or 8 in the second round most likely. The gap in the score is then the difference between the lower seeded team's score and the higher seeded team's score. For example, in the first round, Notre Dame, seeded sixth, beat 13th-seeded California State Northbridge by a score of 83 to 75, giving a seed gap of 7 and a score gap of 8. Negative gaps then signify upsets, that is, the higher seeded team is beaten by the lower seeded team. Data are given for both the first and second rounds of play.

First Round

Seed Gap	Score Gap			
Region	East	Midwest	Mideast	West
15	72	49	42	33
13	29	39	43	6
11	36	26	13	32
9	19	7	28	37
7	8	22	2	18
5	−2	20	17	2
3	−3	2	20	−25
1	−14	10	−4	−9

Second Round

Seed Gap	Score Gap			
Region	East	Midwest	Mideast	West
8	45		17	21
8	21			17
8	−13			
7		34		
5		15	21	
3		6	3	−11
1	4	−9	15	−7

a. For the first-round data, fit a regression line to the data, using the score gap as the response and the seed gap as the predictor. Show the analysis of variance.

b. For the first-round data, in each region separately, fit a regression line to the data using score gap as the response and the seed gap as the predictor. Show the analysis of variance.

c. For the second round, predict score gap as a function of seed gap.

d. Show the values of Cook's d for each of the models in parts (a) and (c).

e. Write a paragraph stating the conclusions that can be drawn from the analysis of these data. Of the five upsets in the first round, four of the winning teams were defeated in the second round.

8. The following data are from a study of home furnaces. The study measured several variables, including the energy used when the furnace damper was in, BTU In, as a function of Furnace Type, House Type, and Age of House.

BTU In	Furnace	House	Age
7.87	1	3	8
9.43	2	2	75
7.16	1	2	44
8.67	2	2	75
12.31	3	2	30
9.84	3	3	4
16.90	1	2	45
10.04	1	1	16
12.62	3	5	45
7.62	3	5	40
11.12	1	1	22
13.43	2	2	40
9.07	1	1	13
6.94	1	2	99
10.28	1	1	19
9.37	1	2	30
7.93	2	2	60
13.96	1	2	30
6.80	1	2	10
4.00	1	2	60
8.58	1	1	24
8.00	1	2	70
5.98	1	1	12
15.24	3	2	60
8.54	1	2	40
11.09	1	1	17
11.70	1	4	15
12.71	1	1	18
6.78	1	1	4
9.82	1	1	5
12.91	1	2	75
10.35	1	1	14

(continued)

BTU In	Furnace	House	Age
9.60	1	1	8
9.58	1	2	99
9.83	1	1	99
9.52	1	1	34
18.26	1	2	80
10.64	1	2	99
6.62	1	2	99
5.20	1	4	6

Source: Minitab, Furnace.mtw

a. Find the least squares regression equation predicting BTU In from the variables Furnace Type, House Type, and Age.

b. Show the analysis of variance table. Write a paragraph describing the results and conclusions that can be drawn.

c. Show all the fits and residuals from the least squares equation. Show a graph of the residuals and state the conclusions that can be drawn from the graph.

d. Show all the values of Cook's d and state any conclusions that can be drawn.

e. Test each of the regression coefficients for significance and state any conclusions that can be drawn.

9. The file Salary.mtw in Minitab contains data on 171 college faculty members' salaries several years ago. Some of the factors measured were Department, Beginning Salary, Experience, Gender, and Rank. Some of the data follow.

Salary	Dept	Gender	Exp.	BegSal	Rank
35000	8	0	1.00	8900	3
43000	8	1	4.00	7500	4
26000	5	0	8.00	17550	2
51100	1	1	4.00	9100	4
49200	8	1	19.50	22200	4
44900	8	1	3.50	14000	4
34400	3	1	5.00	22500	2
40600	7	0	5.00	17655	4
35200	5	1	7.50	19000	3
40600	3	1	3.00	9500	4
29400	8	1	6.00	19000	2

If you can retrieve the entire file, do so; otherwise, use the reduced data set given here.

a. Find the least squares regression equation predicting Salary from the variables Gender, Rank, Department, and Beginning Salary.

b. Show the analysis of variance table. Write a paragraph describing the results and conclusions that can be drawn.

c. Show all the fits and residuals from the least squares equation. Show a graph of the residuals and state the conclusions that can be drawn from the graph.

d. Show all the values of Cook's d and state any conclusions that can be drawn.

e. Test each of the regression coefficients for significance and state any conclusions that can be drawn.

10. Minitab contains the file Assess.mtw which contains data on 81 assessments of homes. In addition to the assessed value of the home, assessors measured several variables including Acreage, Height (a measure of the number of stories in the home), First Floor Area, Exterior Condition, Type of Fuel, Number of Rooms, Number of Bedrooms, Number of Full Baths, Number of Half Baths, Number of Fireplaces, and the existence of a Garage. Some of the data follow.

state the conclusions that can be drawn from the graph.

d. Show all the values of Cook's d and state any conclusions that can be drawn.

e. Test each of the regression coefficients for significance and state any conclusions that can be drawn.

11. Use the following data set to solve parts (a)–(d).

x	y	x	y
1.0	0.5834	3.6	−0.5445
1.2	0.8728	3.8	−0.3350
1.4	0.9483	4.0	−0.7563
1.6	0.8516	4.2	−0.7942
1.8	0.7251	4.4	−0.9923
2.0	0.9806	4.6	−1.1188
2.2	0.7832	4.8	−1.1577
2.4	0.5075	5.0	−0.4756
2.6	0.4833	5.2	−0.7080
2.8	0.5020	5.4	−0.6902
3.0	0.0385	5.6	−0.4651
3.2	−0.1763	5.8	−0.4690
3.4	−0.1095	6.0	−0.1047

Value	Acre	Ht	1stFl	Ext	Fuel	Rms	BdRms	Bth	HalBath	Fpl	Gar
199657	1.630	3	1726	2	1	8	4	2	1	2	1
78482	0.495	1	1184	2	1	6	2	1	0	0	0
119962	0.375	4	1014	2	2	7	3	2	0	1	1
116492	0.981	1	1260	3	2	6	3	2	0	1	1
131263	1.140	4	1314	3	1	8	4	2	1	2	0
244128	3.240	8	1040	2	1	10	5	3	0	1	1
240777	1.942	4	1752	2	1	9	4	3	2	3	1
381984	1.377	4	2504	2	1	9	4	3	2	2	1
137275	3.650	4	708	2	1	8	4	1	0	0	0
131625	1.200	1	1776	2	2	7	3	7	1	2	1

If you can retrieve the entire file, do so; otherwise, use the reduced data set given here.

a. Find the least squares regression equation predicting Value from all the remaining variables.

b. Show the analysis of variance table. Write a paragraph describing the results and conclusions that can be drawn.

c. Show all the fits and residuals from the least squares equation. Show a graph of the residuals and

a. Show a graph of the data.

b. The graph in part (a) suggests that a sinusoidal function may fit the data well. Use a Taylor approximation to the sine function to fit a sinusoidal function to the data.

c. Show the statistical analysis of the data.

d. Write a paragraph describing the conclusions that can be drawn from the statistical analysis of the data.

COMPUTER EXPERIMENT

We observed in Section 5.8 that the sums of squares associated with the individual predictor variables depends on the order in which the variables are entered into the regression equation. Consider once again the data in Example 5.1.1 and suppose that we wish to predict graduating GPA from the variables SAT Eng and SAT Math. Show the regression for each of the two orders in which the predictor variables may be entered into the equation and observe that the sequential sums of squares differ in each case. Calculate F values for each sequential sum of squares and then verify the t values given when tests are made on the individual predictor variables.

5.9 Model Building—Stepwise Regression

An experimenter frequently has several variables and wishes to discover an acceptable model containing only a few of these variables. We indicated one solution to this problem in the previous sections: We found a regression model with all the variables included and then eliminated two of them since the F values for the sum of squares associated with each of them was small, in the order in which the variables were entered. This is a slow procedure if the number of predictor variables is large. Fortunately, there are computer programs that provide an objective procedure for selecting some predictors rather than others. These are called **stepwise** procedures.

There are two primary procedures, known as **forward** and **backward** selections. We will describe each of these procedures as they are implemented in Minitab. Some details about using the stepwise procedures as they are implemented in Minitab are given in Appendix B.

We begin with the forward procedure. In brief, the forward procedure starts with the constant and none of the predictor variables and adds predictor variables one at a time as long as the p-value associated with that variable does not exceed some critical value, called **Alpha to Enter.**

The backward procedure starts with all the predictor variables and deletes variables one at a time until all of the predictor variables not included in the equation have p-values exceeding a critical value, called **Alpha to Remove.** The p-values then for the variables included in the equation all have p-values less than the value of *Alpha to Remove.*

We use the example concerning graduating grade-point averages that we have used in previous sections and apply each of the stepwise procedures.

Forward Selection

We choose *Alpha to Enter* as 0.75. Minitab produces the results shown in Table 5.19. The results in Table 5.19 can be read as follows. In step 1, the predictor variable with the smallest p-value that does not exceed *Alpha to Enter*, Sch Rank, is added to the constant producing the regression equation

$$GPA_1 = -0.1251 + 0.2561 \text{ Sch Rank}.$$

Step	1	2	3
Constant	−0.1251	−0.2544	−0.2889
Sch Rank	0.2561	0.2557	0.2561
t-Value	73.27	73.70	71.40
p-Value	0.000	0.000	0.000
SAT Total		0.00013	0.00017
t-Value		1.23	1.29
p-Value		0.232	0.211
Size/Rank			−0.0011
t-Value			−0.52
p-Value			0.605
s	0.0760	0.0752	0.0765
R^2	99.57	99.60	99.61
R^2(adj)	99.55	99.56	99.55

Table 5.19

The t value is 73.27, but since $t_n^2 = F[1, n]$, we find that the corresponding F value is $(73.27)^2 = 5368.5$. This is equivalent to the analysis of variance given in Table 5.17.

In step 2, the next most significant predictor variable, SAT Total, is entered producing the regression equation

$$\text{GPA}_2 = -0.2544 + 0.2557 \text{ Sch Rank} + 0.00013 \text{ SAT Total}.$$

Appropriate t values are shown for each of the predictors. The t value for SAT Total is not significant, so the procedure would normally stop with step 1. However, in order to continue and to illustrate the process, we have set *Alpha to Enter* artificially high.

Finally, step 3 adds the least significant predictor variable whose p-value does not exceed 0.75, Size/Rank, giving the regression equation

$$\text{GPA}_3 = -0.2889 + 0.2561 \text{ SchRank} + 0.00017 \text{ SAT Total} - 0.0011 \text{ Size/Rank}.$$

The t value for Size/Rank is not significant.

The process may of course stop before all of the predictor variables are entered. Had we used *Alpha to Enter* as 0.25, the process would stop at the second step since the p-value for the remaining predictor variable, Size/Rank, is 0.605, which exceeds 0.25.

One can see in this procedure that the coefficients change as we proceed through the steps, but they do not change very much. Also notice that the values of R^2 increase as variables, significant or insignificant, are added. This is not necessarily true for R^2(adj) whose values are also given in the Minitab output.

Backward Selection

Now we begin with all the predictor variables. We have set the value of *Alpha to Remove* as 0.10. From the forward procedure, we know that the *p*-values for the variables SAT Total and Size/Rank exceed 0.10, while the *p*-value for the variable Sch Rank is less than 0.10. The process then should eliminate the variable with the largest *p*-value, Size/Rank, first, followed by the elimination of the variable SAT Total, and stopping with the variable Sch Rank still included in the regression equation. Minitab produces the results shown in Table 5.20.

Step	1	2	3
Constant	−0.2889	−0.2544	−0.1251
Sch Rank	0.2561	0.2557	0.2561
t-Value	71.40	73.70	73.27
p-Value	0.000	0.000	0.000
SAT Total	0.00017	0.00013	
t-Value	1.29	1.23	
p-Value	0.211	0.232	
Size/Rank	−0.0011		
t-Value	−0.52		
p-Value	0.605		
s	0.0765	0.0752	0.0760
R^2	99.61	99.60	99.57
R^2(adj)	99.55	99.56	99.55

Table 5.20

The backward procedure then gives the following regression equations.

$$GPA_1 = -0.2889 + 0.2561 \text{ Sch Rank} + 0.00017 \text{ SAT Total} - 0.0011 \text{ Size/Rank}$$

$$GPA_2 = -0.2544 + 0.2557 \text{ Sch Rank} + 0.00013 \text{ SAT Total}$$

$$GPA_3 = -0.1251 + 0.2561 \text{ Sch Rank}$$

In this instance, the results of both the forward and backward procedures are the same at every step. This is not always the case in examples with large numbers of variables, some of which are highly correlated with each other.

Since these stepwise procedures are very useful in determining easily the most important variables to be included in a model they have become widely used in science and engineering. This is not to say, however, that equations with more variables than other equations are more accurate or more useful. The usual regression analysis should be performed on each of the models and the p-values for the total regression compared. In this case, Minitab will return a p-value of 0 for each of the three regression equations. More accurate computation of these p-values by Mathematica shows that they are, for the equations with three, two, and one predictor variables, respectively, $3 \cdot 10^{-25}$, $4 \cdot 10^{-27}$, and $2.6 \cdot 10^{-30}$. So any of these equations summarize the data very adequately, and an investigator might well proceed with other studies using the variable Sch Rank alone. In any event, the results of any stepwise procedure should be investigated thoroughly. In some cases varying the order in which the possible predictor variables are specified to the stepwise regression program may produce quite different results.

Statistical programs other than Minitab may request F or other values as a criterion for entering or leaving the equation.

This chapter concludes with a study of **response surfaces** where the problem is to determine values of predictor variables that maximize or minimize a response. This is an important industrial application of multiple regression and statistical methods in instances when variables in a production process must be selected so as to maximize or minimize a characteristic of a product.

EXERCISES 5.9

1. Use the data for the plywood experiment shown in Exercise 1 of Section 5.8.
   a. Show a forward stepwise regression using *Alpha to Enter* as 0.25.
   b. Show a backward stepwise regression using *Alpha to Remove* as 0.005.
   c. Show that the results of parts (a) and (b) are identical after two steps.
   d. Show that the results of parts (a) and (b) using all the variables are identical.
   e. Which regression equation is better? Why?

2. Use the hospital data given in Exercise 2 of Section 5.8.
   a. Show a forward stepwise regression using *Alpha to Enter* as 1.
   b. Show a backward stepwise regression using *Alpha to Remove* as 0.005.
   c. Which of the several regression equations produced is best? Why?

3. Use the magazine data given in Exercise 3 of Section 5.8.
   a. Show a forward stepwise regression using *Alpha to Enter* as 0.25.
   b. Show a backward stepwise regression using *Alpha to Remove* as 0.0019.
   c. Compare the results of the procedures in parts (a) and (b).

4. Use the data on petroleum products given in Exercise 4 of Section 5.8.
   a. Show a forward stepwise regression using *Alpha to Enter* as 0.25.
   b. Show a backward stepwise regression using *Alpha to Remove* as 0.1.
   c. Show that the results of parts (a) and (b) are identical.
   d. Which regression equation is best? Why?

5. Use the data on computer configurations given in Exercise 5 of Section 5.8.
   a. Show a forward stepwise regression using *Alpha to Enter* as 0.25.
   b. Show a backward stepwise regression using *Alpha to Remove* as 0.05.
   c. Compare the results of parts (a) and (b).
   d. Which regression equation is best? Why?

6. Use the data given in Exercise 6 of Section 5.8.
   a. Show a forward stepwise regression using *Alpha to Enter* as 0.50.
   b. Show a backward stepwise regression using *Alpha to Remove* as 0.1.
   c. Show that the results of parts (a) and (b) are identical after four steps.
   d. Which regression equation is best? Why?

5.10 Response Surfaces

Experiments are frequently devised in order to determine values of variables that maximize or minimize a measured characteristic. This is usually easy to do if an exact model is available, but it is not so straightforward when the model is based on sampling and so is subject to some error. We begin our discussion of response surfaces with an example.

Example 5.10.1 A manufacturer of plastic parts performs an experiment to see how varying two major factors in the manufacture of the parts, namely, temperature and pressure, affects the quality of the produced parts. The response is the percentage of unacceptable parts produced under the experimental conditions, which is denoted by "quality." We wish to minimize this quantity and seek the values of temperature and pressure that will do this. The data are shown in Table 5.21.

Pressure	Temperature	Quality
10	40	15
10	50	14
10	60	7
10	70	8
10	80	21
20	40	13
20	50	4
20	60	6
20	70	6
20	80	12
30	40	16
30	50	6
30	60	5
30	70	3
30	80	14
40	40	14
40	50	5
40	60	4
40	70	9
40	80	18
50	40	32
50	50	13
50	60	2
50	70	1
50	80	30

Table 5.21

It is not uncommon for an experimenter to measure only one factor at a time. If we graph the data in Table 5.21 using only one of the factors at a time, we find the graphs shown in Figures 5.7 and 5.8. We might conclude from Figures 5.7 and 5.8 that the

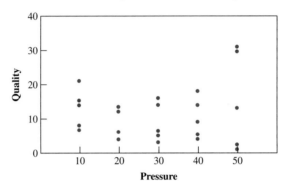

Figure 5.7 Quality as a Function of Pressure in Example 5.10.1

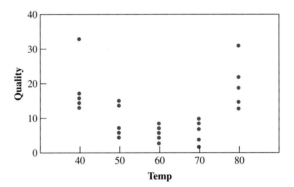

Figure 5.8 Quality as a Function of Temperature in Example 5.10.1

minimum quality occurs at a temperature of 60 and a pressure of about 20. We shall see that while one of these conclusions is fairly accurate, the other is quite far from the true minimum. This leads us to conclude that it is not always possible to make accurate conclusions about factors unless they are considered together rather than separately.

The data in Table 5.21 are also fairly puzzling to analyze simply by examining the table. Responses from a combination of several factors are often difficult, if not impossible, to compare by simply inspecting the data.

The manufacturer in this case is not so much interested in a regression equation as she is in determining values of temperature and pressure that will minimize the percentage of unacceptable product. If an exact model predicting quality as a function of our two predictor variables were available, determining these values would be fairly simple. Because we lack an exact model, we settle for an approximation to this relationship; a least squares regression equation of a general quadratic function is used, this being the most common function to fit to response data. Using the letters T for temperature and P for pressure, Minitab shows this equation to be

$$\text{Quality} = 137 - 4.25\,T - 0.655\,P + 0.0361\,T^2 + 0.0153\,P^2 - 0.00320\,P \cdot T.$$

The analysis of variance gives a p-value of $9.65 \cdot 10^{-5}$ so the fit is quite good; a close fit for the data is important for an accurate location of the minimum point. A graph of the residuals shows no significant departure from normality.

The interesting graph of the surface in Figure 5.9 was produced using Mathematica. There appears to be a minimum setting near Temperature = 60 and Pressure = 30. The surface shown in Figure 5.9 is called a **response surface.**

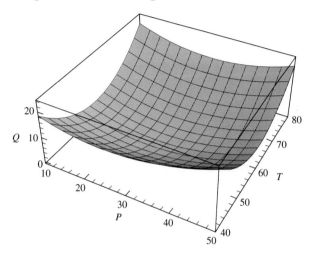

Figure 5.9 Least Squares Surface for Example 5.10.1

To see that there is a minimum point on the response surface, consider a **contour plot** of the surface, shown in Figure 5.10. The contours in Figure 5.10 are the intersections of the response surface with planes where the value of the response quality is kept constant.

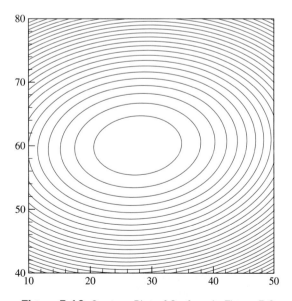

Figure 5.10 Contour Plot of Surface in Figure 5.9

P	T	Q
27.6	59.7	0.242873
27.6	59.8	0.240436
27.6	59.9	0.238721
27.6	60.0	0.237728
27.6	60.1	0.237457
27.6	60.2	0.237908
27.6	60.3	0.239081
27.6	60.4	0.240976
27.6	60.5	0.243593

Table 5.22

P	T	Q
27.2	60.1	0.241009
27.3	60.1	0.239662
27.4	60.1	0.238621
27.5	60.1	0.237886
27.6	60.1	0.237457
27.7	60.1	0.237334
27.8	60.1	0.237517
27.9	60.1	0.238006
28.0	60.1	0.238801

Table 5.23

The easiest way to find the minimum point on the response surface is to set the necessary partial derivatives to zero and solve the resulting equations. A computer algebra system is again of great utility. Mathematica produces $T = 60.0915$ and $P = 27.6893$ as the coordinates of the minimum point.

Another good use of a computer algebra system is to do calculations in any region of the surface. Tables 5.22 and 5.23 show some of the calculations done with Mathematica. The points on the quadratic surface were calculated allowing T to range from 59 to 61 in increments of 0.1 while P ranged from 26 to 28 in increments of 0.1. The resulting calculations, shown in Table 5.22, are easy to do and indicate that one might examine the responses at various pressures near a temperature of 60.1. Table 5.23 displays these results, which give an absolute minimum near $P = 27.7$ and $T = 60.1$ and produce a measure of the resulting quality as well. ∎

Method of Steepest Descent

The complete ranges of both temperatures and pressures were available to the experimenter in Example 5.10.1. That is, we measured data points in the range of 40 to 60 for temperature and in the range of 10 to 50 for pressure assuming that these are the only sensible ranges of values of factors to use. Since the complete ranges of data were available, the experimenter gathered data at all the possible combinations of the values, increasing each by 10 units at a time, and then examined the quadratic surface that results, producing the desired minimum. Often, however, the ranges of values for the factors involved are unknown, so the experimenter is unable to follow the procedure in Example 5.10.1.

Here is a way to proceed that is frequently used when the complete range of values for either of the factors is unknown. We will use the data in Example 5.10.1 to illustrate the procedure, while ignoring the fact that we already know the answer.

Since the full range of values of the factors is unknown, we start with values of temperature in the range 40 to 60 and of the pressure in the range 10 to 30. In Table 5.21, we are starting in what proves to be one of the corners of the data. This process will work for any of the corners in the example. That being the case, suppose then that we have only the data points given in Table 5.24.

Pressure	Temperature	Quality
10	40	15
20	40	13
30	40	16
10	50	14
20	50	4
30	50	6
10	60	7
20	60	6
30	60	5

Table 5.24

We now fit a regression plane to these data. Minitab produces the plane

$$Q = 34.2 - 0.433T - 0.150P,$$

which is shown in Figure 5.11.

It is obvious from Figure 5.9 that the experimenter should proceed toward the point where Pressure $= 30$ and Temperature $= 60$. From our previous calculations, we know that the minimum is actually in this area, but the experimenter at this point would not have that knowledge. It is also not clear that a minimum on a quadratic surface might appear before we reach that point, nor is it clear how we might approach the minimum most rapidly. It is now useful to examine a contour plot of the plane in Figure 5.11. This is shown in Figure 5.12.

The contours are parallel lines since the surface is a plane. To proceed toward the minimum most rapidly, we should move in a direction perpendicular to the contour lines. These contour lines have equations of the form $P = c - (0.433/0.150)T$, so the equations of the perpendicular lines have slope $0.150/0.433 = 0.3464$. That is, for every 0.433 units in T we should move 0.150 units in P, or 3.464 units in P for every 10 units

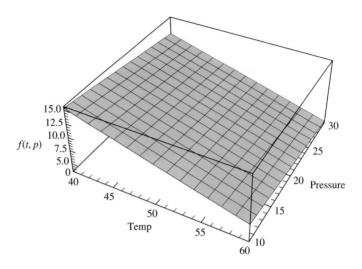

Figure 5.11 Plane Used to Illustrate the Method of Steepest Descent

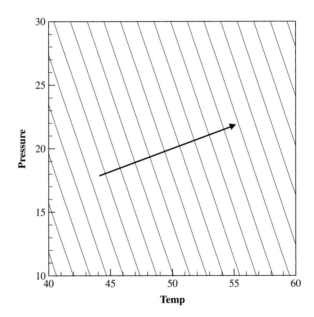

Figure 5.12 Contour Plot of the Plane in Figure 5.11

in T. This would produce the following sequence of observations:

$$
\begin{array}{ll}
T = 40 & P = 10.000 \\
T = 50 & P = 13.464 \\
T = 60 & P = 16.928 \\
T = 70 & P = 20.392
\end{array}
$$

In this case, the experimenter would find that the measurements of quality would decrease and then increase. Since we do not have actual measurements in this case, we use the least squares quadratic surface to find the following values for quality for the values of T and P above: 18.46, 6.55041, 2.00631, and 4.82769. These are representative of the results the experimenter would find. This shows that the minimum is near $T = 60$ and $P = 16.928$. One can then experiment further in this region for the location of the minimum. This process, which is usually called the **method of steepest descent** (or **ascent** if a maximum is being located), often results in fewer data points than an experiment run on the full range of values of the factors (if these ranges are indeed known).

EXERCISES 5.10

1. The quality, Q, of a manufactured product is known to be a function of the temperature, X, at which the product is manufactured and a measure of the amount of a catalyst, Y, used in the product. The following data are gathered.

X	Y	Q
10	10	112
10	20	400
10	30	503
10	40	407
10	50	97
20	10	409
20	20	692
20	30	808
20	40	713
20	50	408
30	10	512
30	20	783
30	30	907
30	40	796
30	50	506
40	10	398
40	20	698
40	30	776
40	40	706
40	50	393
50	10	113
50	20	393
50	30	511
50	40	409
50	50	107

a. Plot the data as a function of X alone. Where does the maximum appear to be?

b. Plot the data as a function of Y alone. Where does the maximum appear to be?

c. Fit the least squares quadratic function to the data, showing the analysis of variance table and writing your conclusions in a paragraph.

d. Determine the maximum point on the surface in part (c). Calculate values of the function near the maximum point in order to verify numerically that the maximum point is correct. Is the result consistent with those in parts (a) and (b)?

e. Now suppose that only the following observations were available.

X	Y	Q
10	10	86
10	15	279
10	20	397
15	10	289
15	15	458
15	20	583
20	10	387
20	15	589
20	20	708

Fit the least squares plane to the data, show the analysis of variance table, and state conclusions that can be drawn in a paragraph.

f. Show a contour plot of the plane found in part (e). Determine the slope of the perpendicular lines to the contours and use this to determine the maximum point on the quadratic surface.

2. The following data represent coded variables X and Y with a response, Z.

X	Y	Z
-6	-6	197
-6	-4	172
-6	-2	136
-6	0	137
-6	2	126
-6	4	145
-6	6	131
-4	-6	90
-4	-4	90
-4	-2	74
-4	0	100
-4	2	42
-4	4	82
-4	6	71
-2	-6	46
-2	-4	64
-2	-2	20
-2	0	-5
-2	2	21
-2	4	32
-2	6	72
0	-6	27
0	-4	27
0	-2	-4
0	0	-35
0	2	1
0	4	54
0	6	27
2	-6	57
2	-4	43
2	-2	48
2	0	2
2	2	34
2	4	34
2	6	62
4	-6	63
4	-4	48
4	-2	79
4	0	60
4	2	83
4	4	85
4	6	126
6	-6	189

Table continued		
X	Y	Z
6	-4	140
6	-2	150
6	0	139
6	2	122
6	4	160
6	6	213

a. Plot the data as a function of X alone. Where does the minimum appear to be?

b. Plot the data as a function of Y alone. Where does the minimum appear to be?

c. Fit the least squares quadratic function to the data, showing the analysis of variance table and writing your conclusions in a paragraph.

d. Determine the minimum point on the surface in part (c). Calculate values of the function near the minimum point in order to verify numerically that the minimum point is correct. Is this conclusion consistent with those found in parts (a) and (b) ?

e. Now suppose that only the following observations were available.

X	Y	Z
-6	-6	180
-6	-4	160
-6	-4	160
-6	-2	148
-6	0	144
-6	2	148
-6	6	180
-4	-6	100
-4	-4	80
-4	-2	68
-4	0	64

Fit the least squares plane to the data, show the analysis of variance table, and state conclusions that can be drawn in a paragraph.

f. Show a contour plot of the plane found in part (e). Determine the slope of the perpendicular lines to the contours and use this to determine the minimum point on the quadratic surface.

COMPUTER EXPERIMENT

Consider the surface $f(x, y) = 4x^2 + 3y^2 + 8x - 6y + 100$.

a. Plot the surface and determine the minimum point.

b. Using both x and y in the range from -5 to 5, calculate a table of values of $f(x, y)$. To each of these values, add a random observation from the normal distribution with mean 0 and variance 20. Fit the least squares quadratic surface to these values, show the analysis of variance table and state conclusions that can be drawn in a paragraph.

c. Plot the data in part (b) as a function of x alone. Where does the minimum value appear to be?

d. Plot the data in part (b) as a function of y alone. Where does the minimum value appear to be?

e. Now suppose that the values of the function in part (b) are available only for combinations of x and y where $x = -5$ and -4 and $y = -5, -4, -3$, and -2. Fit the least squares plane to these data. Show a contour plot of the plane and determine the slope of the perpendiculars to the contours, determining the direction in which to proceed in order to determine the minimum point. Does the direction in fact lead to the minimum point?

Chapter Review

The topic of this chapter was **multiple regression,** in which some response variable, y, is a function of two or more predictor variables, say $x_1, x_2, \ldots, x_{k-1}$, where $k \geq 3$. The situation here is much more complex than that of simple regression in which there is only one predictor variable for the given response. In multiple regression, the predictor variables may be correlated and the variables may have different influences upon the predicted response; both of these facts may cause difficulties in assessing the quality of the fitted response.

Key Concepts

- Using a model with more than one predictor variable to predict a response.
- Using the principle of least squares to estimate the coefficients in the model.
- Assessing the quality of the model using the analysis of variance.
- Identifying significant variables for a model.
- Building a model by inserting the variables one at a time or by starting with all the variables and eliminating them one at a time until a satisfactory model is obtained.
- Determining observations that may be unduly influential in determining the model.
- Fitting a response surface to data in order to maximize or minimize a response.

Key Terms	
Predictor Variables	*Predictor variables* are variables used in the model to predict a single response. A variable is usually denoted by x and the response by y.
Multiple Regression Model	The general *multiple regression model* is $$y_i = \beta_0 + \beta_1 x_{1i} + \beta_2 x_{2i} + \cdots + \beta_{k-1} x_{k-1,i} + \varepsilon_i$$ $$\text{for } i = 1, 2, \ldots, n, k \geq 3.$$
Principle of Least Squares	The *principle of least squares* produces estimators, $\widehat{\beta}_i$ for the coefficients β_i in the multiple regression model that minimize $$\sum_{i=1}^{n} \varepsilon_i^2.$$
Predicted Value	The *value* for *y predicted* by the least squares model is $$\hat{y}_i = \widehat{\beta}_0 + \widehat{\beta}_1 x_{1i} + \widehat{\beta}_2 x_{2i} + \cdots + \widehat{\beta_{k-1}} x_{k-1,i}.$$
Residual	The *residual* is the difference between the observed and the predicted values of y_i, namely $y_i - \hat{y}_i$.
Influential Observation	An observation is *influential* if its value of Cook's d exceeds $F_{0.50}[k, n-k]$.
Significant Variable	A variable is regarded as *significant* if its p-value is not greater than a specified value.
Forward Selection	The *forward* stepwise regression procedure adds variables to the model one at a time if the variable's p-value is less than a specified value known as **Alpha to Enter.**
Backward Selection	The *backward* stepwise regression procedure begins will all the candidate variables and removes them one at a time if the variable's p-value exceeds a value known as **Alpha to Remove.**

Key Theorems/Facts	
Perpendicular Vectors	The *vectors* $\overrightarrow{y_i - \overline{y}}$, $\overrightarrow{\hat{y}_i - \overline{y}}$, and $\overrightarrow{y_i - \hat{y}_i}$ are *perpendicular.*

Analysis of Variance Identity	The *analysis of variance identity* for multiple regression is

$$\sum_{i=1}^{n}(y_i - \overline{y})^2 = \sum_{i=1}^{n}(\hat{y}_i - \overline{y})^2 + \sum_{i=1}^{n}(y_i - \hat{y}_i)^2$$

Analysis of Variance Table — The *analysis of variance table* for multiple regression can be found on page 246.

Multiple Regression Model — The *multiple regression model* can be written in matrix form as

$$Y = X\beta + \varepsilon \text{ where}$$

$$\beta = \begin{bmatrix} \beta_0 \\ \beta_1 \\ \vdots \\ \beta_{k-1} \end{bmatrix}, \quad X = \begin{bmatrix} 1 & x_{11} & x_{21} & \cdots & x_{k-1,1} \\ 1 & x_{12} & x_{22} & \cdots & x_{k-1,2} \\ \vdots & \vdots & \vdots & & \vdots \\ 1 & x_{1n} & x_{2n} & \cdots & x_{k-1,n} \end{bmatrix},$$

$$Y = \begin{bmatrix} y_1 \\ y_2 \\ \vdots \\ y_n \end{bmatrix} \quad \text{and} \quad \varepsilon = \begin{bmatrix} \varepsilon_1 \\ \varepsilon_2 \\ \vdots \\ \varepsilon_n \end{bmatrix}$$

Least Squares Equations — The *least squares equations* can be written as $\hat{\beta} = (X'X)^{-1}X'Y$.

Variance-Covariance Matrix — The *variance-covariance matrix* is $\sigma^2(X'X)^{-1}$.

Estimator for σ^2 — An unbiased *estimator for σ^2* is the mean square for error in the analysis of variance.

Coefficient of Determination — The *coefficient of determination* is

$$R^2 = \frac{\sum_{i=1}^{n}(\hat{y}_i - \overline{y})^2}{\sum_{i=1}^{n}(y_i - \overline{y})^2}.$$

Confidence Intervals for Regression Coefficients — *Confidence intervals for regression coefficients* can be found using the fact that

$$\frac{\hat{\beta}_i - \beta_i}{\text{StDev}(\hat{\beta}_i)} = t_{n-k}.$$

Variance of a Predicted Value

The *variance of a predicted value* can be found using the fact that

$$\text{Var}(\hat{y}) = \sum_{i=0}^{k-1} x_i^2 \text{Var}(\widehat{\beta_i})$$

$$+ 2 \sum_{i>j=0}^{k-1} x_i x_j \text{Cov}(\widehat{\beta_i}, \widehat{\beta_j}).$$

where $x_0 \equiv 1$.

Influential Observations

Observations having a large *influence* on the regression equation are those with values of $d_i = 1/ks^2 \sum_{j=1}^{n} (\hat{Y}_{j(i)} - \hat{Y}_j)^2$ exceeding $F_{0.50}[k, n-k]$ where $\hat{Y}_{j(i)}$ is the predicted value for the jth data point when the ith data point is not used in the computation of the least squares regression equation. The quantity s^2 is the estimate of σ^2 from the analysis of variance and k is the number of predictor variables including the constant.

Response Surface

A *response surface* is used to determine the values of x and y for which the response, z, is a maximum or minimum when only a data set

$$\{(x_i, y_i, z_i) \mid i = 1, 2, \ldots, n.\}$$

is available. Generally a quadratic function in two variables such as

$$z = f(x, y) \\ = ax^2 + by^2 + cx + dy + exy + g$$

is fit to the data set.

Method of Steepest Descent

The *method of steepest descent* (or *ascent*) is used to proceed most rapidly toward the minimum (or maximum) on a response surface. A plane is fit to some of the points, the contours of the plane are determined and then, if one proceeds in the direction of the perpendiculars to these contours, the maximum or minimum point on the surface can be reached most rapidly.

Polynomial Models

By using a suitable transformation, *polynomial models* can be fit using a multiple regression model.

6 Design of Science and Engineering Experiments

6.1 Introduction

SCIENTISTS AND ENGINEERS OFTEN CONDUCT EXPERIMENTS and investigations in order to determine laws and models governing situations. For example, a scientist might want to determine the effect of sodium in a person's diet. To do so, an experiment is planned in which responses from patients with varying amounts of sodium in their diets would be measured. It is clear, however, that there are many other possible influences on the results. A person's weight, general health condition, general dietary conditions, as well as many other factors can, and generally will, influence the results. Is it possible to separate out the effect of sodium itself? The answer to this is "yes" if the experiment is planned properly. It is also possible, as we shall see, to determine not only the effect of sodium but the effects of other factors as well. The way in which sodium interacts with the other factors in the experiment can also be found and measured.

To accomplish these goals, the experiment must be planned and conducted in a manner that allows the scientist to determine both the size and the statistical significance of each of the **factors** (one of which is sodium in this case) and both the size and the statistical significance of the **interactions** (the factors' behavior with each other). The **design** of the experiment refers to the planning of the experiment, including the sampling process that must occur. We must know how many and what observations to take in order to answer the questions arising in the investigation. One carefully planned and executed experiment can measure the effects of the factors and the interactions as well as of several different experiments, to determine each of the factors and interactions, had been performed.

In our example, for instance, one might select a group of people, give them first a diet with a low level of sodium, and then measure the effect on, say, metabolism. Then

277

one might increase the level of sodium and measure the results. This might be done for a variety of levels, or amounts, of the sodium with the results recorded. This is an example of a **one-factor-at-a-time** experiment. As we shall see, this kind of experiment is generally very inefficient. While we might learn something about the effect of sodium level in the diet, we would not be able to determine what effect the sodium had with other factors, such as magnesium. For this reason, one-factor-at-a-time experiments are rarely performed today. In fact, the interactions are frequently more interesting and important than the individual factors themselves. It is obvious that in order to determine these interactions the factors must be studied together. It might appear that this would hopelessly pollute the data, but in fact, that is not the case.

We will consider several types of experimental designs and the conclusions that can be drawn from data that are selected from them. We will find that some of the results are quite surprising and perhaps run counter to intuition. The statistical design of experiments has become an important tool in science and engineering because of the efficiencies it provides to experimenters in reaching conclusions from experimental data. In exploring this subject, we will also point out some commonalities it shares mathematically with multiple linear regression. We begin with a fairly simple example.

6.2 An Example

Example 6.2.1 A group of investigators are concerned about the time that it takes for a computer to perform a complex mathematical process. They presume that two factors, processor speed (Sp) and random access memory size (RAM), are important. Two processor speeds, 133 MHz and 400 MHz, are studied together with two sizes of RAM, namely, 128 MB and 256 MB. The mathematical program is run with each combination of processor speed and RAM size and the time taken to perform the program is measured in thousandths of a second. Three observations are taken with each of the four possible combinations of processor speed and RAM size, so the observations are **replicated.** The data are given in Table 6.1.

	Sp_1		Sp_2	
	30		16	
RAM_1	26		9	
	16	(24)	11	(12)
	22		6	
RAM_2	12		10	
	14	(16)	8	(8)
				[15]

Table 6.1

The mean of each treatment combination is shown in parentheses and the overall mean is shown in square brackets. It appears that increasing the processor speed decreases the calculation time and increasing the random access memory size also decreases the calculation time, but the amount of the decrease in calculation time is not at all clear from an examination of the data. A graph allows us to interpret these data visually. See Figure 6.1.

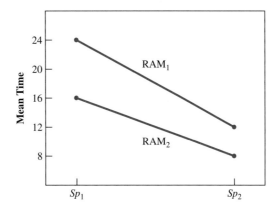

Figure 6.1 Data from Table 6.1

Figure 6.1 shows the two **levels** (or values) of the factor speed, Sp_1 and Sp_2, on the horizontal axis. Mean performance times are shown on the vertical axis. The straight lines show the mean performance times for the two levels of random access memory size, RAM_1 and RAM_2, for speeds Sp_1 and Sp_2. The graph indicates that calculation times decrease with an increase in processor speed, but the amount of decrease in calculation times differs with the two different sizes of RAM. The lines on the graph then are not parallel, showing that the influence on performance time differs for the two sizes of RAM. If the decrease in calculation time were constant for the different values of RAM, then these lines would be parallel. When the lines are not parallel, this is an indication that the two factors, Sp and RAM, **interact.** In other words, the two factors, processor speed and the size of the random access memory, influence performance time by acting together in some fashion.

It appears that there are three influences on the data: the processor speed, Sp, the size of the random access memory, RAM, and the interaction of these two influences, which we denote by $Sp \cdot RAM$.

Two important questions now arise: What is the size of the influence of each of the factors? and, are the factors Sp and RAM or their interaction statistically significant? We now seek some way in which to answer these questions, and we will begin by measuring the size of these influences, or effects. To simplify the data somewhat we calculate the means of the observations for each of the factor combinations. Figure 6.2 shows the mean performance times arranged at the corners of a square.

Consider first the influence of the processor speed, Sp. How can this be measured? In Figure 6.2 we have arbitrarily labeled the lower level of Sp (133 MHz) as -1 and the

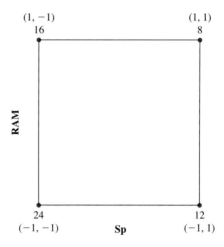

$(1, -1)$ $(1, 1)$
16   8

RAM

24   12
$(-1, -1)$ $(-1, 1)$

Sp

Figure 6.2 Data from Table 6.1 at the Corners of a Square

higher level of Sp (400 MHz) as $+1$. These numbers have nothing whatever to do with the actual values of the factor. Similarly, we have labeled the lower level (128 MB) of the random access memory, RAM, as -1 and the higher level of the random access memory (256 MB) as $+1$. The corners are then labeled as ordered pairs in the order (RAM, Sp).

The means of the observations at the right side of the square are where the factor Sp is at the $+1$ level and the means of the observations along the left side of the square are where the factor Sp is at the -1 level. One might consider the difference between these means in order to calculate the effect of Sp for the two values of RAM and calculate

$$\frac{8 + 12}{2} - \frac{24 + 16}{2} = -10.$$

It is customary to divide this by 2 (for the difference between the high level of Sp, $+1$ and the low level of Sp, -1) to find that

$$\text{Effect of Sp} = \frac{1}{2}\left[\frac{8 + 12}{2} - \frac{24 + 16}{2}\right] = -5.$$

This means that as we move from the lower level of Sp to the higher level of Sp, the performance time decreases 5 thousandths of a second on average. This amount may or may not be significant to the experimenter and it may or may not be statistically significant.

What is the effect of the random access memory, RAM? In a similar manner, by comparing the means along the top side of the square, where the factor RAM is at the $+1$ level with the means along the bottom side of the square, where the factor RAM is at the -1 level, and then dividing by 2 we find that

$$\text{Effect of RAM} = \frac{1}{2}\left[\frac{16 + 8}{2} - \frac{24 + 12}{2}\right] = -3.$$

Again the divisor of 2 is used since this is the distance between the $+1$ and -1 values of the factor RAM. This shows that the increase in RAM also tends to decrease this particular performance time; this may or may not be of practical or statistical significance.

Now how do we calculate how the two main factors, Sp and RAM, behave, or *interact*, with each other? It would appear to be sensible to measure this interaction, which we have denoted by Sp · RAM, by taking the mean where the *product* of the levels of the individual factors is positive and subtracting the mean where the product of the levels of the two factors is negative, and again dividing by 2, to find

$$\text{Effect of Sp} \cdot \text{RAM} = \frac{1}{2}\left[\frac{24+8}{2} - \frac{16+12}{2}\right] = +1.$$

We conclude that the interaction tends to increase performance times.

Now what do these numbers mean? As it happens, these numbers can be used in the following linear model:

$$\text{Mean Time} = 15 - 5\,\text{Sp} - 3\,\text{RAM} + 1 \cdot \text{Sp} \cdot \text{RAM}$$

provided that the variables Sp, RAM, and Sp · RAM are regarded as either $+1$ or -1 depending upon the corner of the square in Figure 6.2 being considered.

For example, consider the corner where Sp $= +1$ and RAM $= -1$. The mean time at that corner is 12. The linear model is then used with these values to obtain

$$12 = 15 - 5 + 3 - 1$$

so the model exactly explains the corner observation. The model explains the mean times in the remaining corners in a similar way.

Note that in calculating the main effects and the interaction, we used each of the sides and the diagonals of the square in Figure 6.2.   ∎

A Linear Model

The linear model above is certainly reminiscent of the linear models encountered when we considered multiple regression. It appears that this model is a special case of the model

$$y = \mu + \beta_1 x_1 + \beta_2 x_2 + \beta_3 x_3$$

where y represents the mean calculation time, the x's are either -1 or $+1$, and the β's represent the effects Sp, RAM, or Sp · RAM. This model, as we have seen, explains the means at the corners exactly. If we were to consider the individual observations at the corners, these could be explained by the model

$$y = \mu + \beta_1 x_1 + \beta_2 x_2 + \beta_3 x_3 + \varepsilon$$

where ε represents a random error.

These results appear to be identical to those considered when we considered multiple regression and indeed they are exactly equivalent. In fact, if a multiple regression analysis is done using this data and the x's take on the values -1 and $+1$ as above, the following regression equation is found:

$$\text{Mean Time} = 15 - 5\,\text{Sp} - 3\,\text{RAM} + 1 \cdot \text{Sp} \cdot \text{RAM}.$$

A multiple regression analysis may then be used to calculate all the effects for this data set, but it is interesting and useful to see the geometry involved.

While we can calculate the size of the effects geometrically, we have no way so far of judging whether they are of statistical significance or not unless we employ a multiple regression model. It is possible to do this, and we will consider this computer calculation time data again when we analyze the data in a somewhat different way in Section 6.6.

We abandon this example for the time being and turn to an entirely different context in which we will see that the observations taken have a large effect upon the accuracy and precision of the conclusions that can be drawn from them, illustrating the importance of the design of experiments. We again will begin with a fairly simple example.

EXERCISES 6.2

1. Another experiment involving processor speed (Sp) and size of random access memory (RAM) for computers produced the following data.

	Sp₁	**Sp₂**
RAM₁	48	35
	37	37
	38	39
RAM₂	30	22
	35	26
	31	30

a. Find all the main effects and the interaction geometrically.
b. Find all the main effects and interaction using a regression model.
c. Show the linear model and use it to explain each of the means at the corners of a square.

2. A study of the gasoline consumption of two different models (M) of a car involves two separate terrains (T) over which the cars are tested. The data follow.

	M₁	**M₂**
T₁	20.4	16.2
	19.3	22.1
	18.6	17.9
	20.5	18.8
T₂	19.6	22.5
	19.6	23.6
	20.4	24.1
	18.4	20.2

a. Find all the main effects and the interaction geometrically.
b. Find all the main effects and interaction using a regression model.
c. Show the linear model and use it to explain each of the means at the corners of a square.

3. A chain grocery has two stores (S) in a small city. The owners think that the volume of sales of soft drinks varies with the time of day (T). The following data are gathered where the measure is the number of cases of soft drink sold.

	S₁	**S₂**
T₁	106	145
	89	83
	74	94
	99	88
T₂	74	91
	87	94
	63	98
	88	101

a. Find all the main effects and the interaction geometrically.
b. Find all the main effects and interaction using a regression model.
c. Show the linear model and use it to explain each of the means at the corners of a square.
d. What implications does this study have for the owners of the stores?

6.3 **Some Weighing Designs**

Two Weights

Suppose we have two objects whose weights are to be determined and a pan balance with a scale. If two, and only two, weighings are allowed, what is the best way in which to determine the unknown weights?

The word "best" of course could be defined in many ways. In accordance with our previous definition, we would like our estimates of the unknown weights from the weighings to be unbiased and to have the smallest variance possible.

First suppose the true weights are W_1 and W_2. If the pan balance were exactly accurate, then weighing the objects separately would determine their true weights exactly. However, the pan balance adds a random error each time it is used. Suppose that the pan balance in fact gives readings w_1 and w_2 for the true weights W_1 and W_2 so that the model we have in this case is

$$w_i = W_i + \varepsilon_i \quad \text{for } i = 1, 2$$

where the errors, ε_i, are independently distributed with mean zero and variance σ^2 so we write $\varepsilon_i \sim ID(0, \sigma)$.

It follows that $E(w_i) = E(W_i + \varepsilon_i) = E(W_i) + E(\varepsilon_i) = W_i + 0 = W_i$ for $i = 1, 2$, showing that the w_i are unbiased estimators for the true weights, W_i. We can also determine the variance of these estimators. We see that

$$\text{Var}(w_i) = \text{Var}(W_i + \varepsilon_i) = \text{Var}(\varepsilon_i) = \sigma^2 \quad \text{for } i = 1, 2$$

since the true weights, W_i, are constants.

These estimates are also uncorrelated since

$$\text{Cov}(w_1, w_2) = \text{Cov}(W_1 + \varepsilon_1, W_2 + \varepsilon_2)$$
$$= \text{Cov}(\varepsilon_1, \varepsilon_2) = 0.$$

We know then that the estimators, w_i, are unbiased and uncorrelated estimators of the true weights, W_i, and each has variance σ^2.

However, by considering other alternatives, we may be able to do much better. Suppose we weigh the objects together, both on the same side of the pan balance, and then, for a second weighing, we put one weight on one side of the balance and the other weight on the other side of the balance. We then observe estimates for the sum of the weights and for the difference between the weights. Denote the observations by O_1 and O_2. Then

$$O_1 = W_1 + W_2 + \varepsilon_1 \quad \text{and} \quad O_2 = W_1 - W_2 + \varepsilon_2.$$

Now consider the random variables

$$\frac{O_1 + O_2}{2} \quad \text{and} \quad \frac{O_1 - O_2}{2}.$$

We see that

$$E\left(\frac{O_1 + O_2}{2}\right) = \frac{E(O_1) + E(O_2)}{2} = \frac{(W_1 + W_2) + (W_1 - W_2)}{2} = W_1$$

and that

$$E\left(\frac{O_1 - O_2}{2}\right) = \frac{E(O_1) - E(O_2)}{2} = \frac{(W_1 + W_2) - (W_1 - W_2)}{2} = W_2,$$

showing that our observations are again unbiased estimators of the true weights. However,

$$\text{Var}\left(\frac{O_1 + O_2}{2}\right) = \frac{\text{Var}(O_1 + O_2)}{4} = \frac{\text{Var}(O_1) + \text{Var}(O_2)}{4} = \frac{\sigma^2 + \sigma^2}{4} = \frac{\sigma^2}{2}$$

and

$$\text{Var}\left(\frac{O_1 - O_2}{2}\right) = \frac{\text{Var}(O_1 - O_2)}{4} = \frac{\text{Var}(O_1) + \text{Var}(O_2)}{4} = \frac{\sigma^2 + \sigma^2}{4} = \frac{\sigma^2}{2},$$

which is half the variance of the estimators obtained by weighing the weights one at a time. This, then, is equivalent to weighing each object individually twice and taking the mean of the observations. Clearly, the experimental design has greatly increased efficiency in the experiment.

These estimates are also uncorrelated since, omitting the divisor of 4,

$$\text{Cov}\left(O_1 + O_2, O_1 - O_2\right) = \text{Var}(O_1) - \text{Cov}(O_1, O_2) + \text{Cov}(O_2, O_1) - \text{Var}(O_2) = 0$$

since $\text{Cov}(O_1, O_2) = \text{Cov}(O_2, O_1)$ and $\text{Var}(O_1) = \text{Var}(O_2) = \sigma^2$.

So judicious use of the two weighings produces unbiased and uncorrelated estimators with smaller variances. The **design** of the experiment here then produces results that might not otherwise be anticipated. We will see that this is the case generally in situations much more complex than this one.

The Linear Model

The first design considered above could be shown in a matrix formulation. Recall that $w_i = W_i + \varepsilon_i$ for $i = 1, 2$ so that

$$\begin{bmatrix} w_1 \\ w_2 \end{bmatrix} = \begin{bmatrix} 1 & 0 \\ 0 & 1 \end{bmatrix} \cdot \begin{bmatrix} W_1 \\ W_2 \end{bmatrix} + \begin{bmatrix} \varepsilon_1 \\ \varepsilon_2 \end{bmatrix}.$$

The second, more efficient, design can also be shown in a matrix formulation. Recall that

$$O_1 = W_1 + W_2 + \varepsilon_1 \quad \text{and} \quad O_2 = W_1 - W_2 + \varepsilon_2$$

so

$$\begin{bmatrix} O_1 \\ O_2 \end{bmatrix} = \begin{bmatrix} 1 & 1 \\ 1 & -1 \end{bmatrix} \cdot \begin{bmatrix} W_1 \\ W_2 \end{bmatrix} + \begin{bmatrix} \varepsilon_1 \\ \varepsilon_2 \end{bmatrix}.$$

Putting this in our familiar $Y = X\beta + \varepsilon$ form, we see that

$$Y = \begin{bmatrix} O_1 \\ O_2 \end{bmatrix}, \qquad X = \begin{bmatrix} 1 & 1 \\ 1 & -1 \end{bmatrix}, \qquad \beta = \begin{bmatrix} W_1 \\ W_2 \end{bmatrix}, \quad \text{and} \quad \varepsilon = \begin{bmatrix} \varepsilon_1 \\ \varepsilon_2 \end{bmatrix}.$$

The least squares estimator for β, namely $\hat{\beta}$, is then $\left(X'X\right)^{-1}X'Y$ or

$$\begin{bmatrix} \widehat{W_1} \\ \widehat{W_2} \end{bmatrix} = \begin{bmatrix} \dfrac{1}{2} & \dfrac{1}{2} \\ \dfrac{1}{2} & -\dfrac{1}{2} \end{bmatrix} \cdot \begin{bmatrix} O_1 \\ O_2 \end{bmatrix}$$

$$= \begin{bmatrix} \dfrac{1}{2}(O_1 + O_2) \\ \dfrac{1}{2}(O_1 - O_2) \end{bmatrix},$$

giving the results obtained previously. The variance-covariance matrix is

$$\sigma^2(X'X)^{-1} = \sigma^2 \begin{bmatrix} \dfrac{1}{2} & 0 \\ 0 & \dfrac{1}{2} \end{bmatrix},$$

showing the variances as previously calculated and showing that the estimators are uncorrelated.

For the first model,

$$\begin{bmatrix} w_1 \\ w_2 \end{bmatrix} = \begin{bmatrix} 1 & 0 \\ 0 & 1 \end{bmatrix} \cdot \begin{bmatrix} W_1 \\ W_2 \end{bmatrix} + \begin{bmatrix} \varepsilon_1 \\ \varepsilon_2 \end{bmatrix},$$

the least squares estimators are

$$\begin{bmatrix} \widehat{W_1} \\ \widehat{W_2} \end{bmatrix} = \begin{bmatrix} 1 & 0 \\ 0 & 1 \end{bmatrix} \cdot \begin{bmatrix} w_1 \\ w_2 \end{bmatrix}$$

and the variance-covariance matrix is

$$\sigma^2 \begin{bmatrix} 1 & 0 \\ 0 & 1 \end{bmatrix}.$$

The X matrix is called the **design matrix** and in this case is the identity matrix.

Other Models for Two Weights

Given two weighings, the plan we have suggested is by no means the only plan. For example, the model

$$\begin{bmatrix} O_1 \\ O_2 \end{bmatrix} = \begin{bmatrix} 1 & 0 \\ 1 & -1 \end{bmatrix} \cdot \begin{bmatrix} W_1 \\ W_2 \end{bmatrix} + \begin{bmatrix} \varepsilon_1 \\ \varepsilon_2 \end{bmatrix}$$

is an alternative possibility.

In this case, we can calculate that

$$\begin{bmatrix} \widehat{W_1} \\ \widehat{W_2} \end{bmatrix} = \begin{bmatrix} 1 & 0 \\ 1 & -1 \end{bmatrix} \begin{bmatrix} O_1 \\ O_2 \end{bmatrix}.$$

We also find that the variance-covariance matrix is

$$\sigma^2 \begin{bmatrix} 1 & 1 \\ 1 & 2 \end{bmatrix}.$$

Now

$$E(\widehat{W_1}) = E(O_1) = E(W_1 + \varepsilon_1) = W_1$$

and

$$E(\widehat{W_2}) = E(O_1 - O_2) = E(O_1) - E(O_2) = E(W_1 + \varepsilon_1) - E(W_1 - W_2 + \varepsilon_2)$$
$$= W_1 - (W_1 - W_2) = W_2$$

so we see that the estimates are unbiased (which we need not check since this is a consequence of least squares), and, while they have minimum variance (also a consequence of least squares), since the off diagonal entries are not 0, they are correlated.

Orthogonal Matrices

An **orthogonal matrix** X is one for which $X^{-1} = X'$ so that different rows of the matrix have dot product 0 with each other giving the product $X'X$ as the identity matrix.

We will be interested in uncorrelated estimators and, since the variance-covariance matrix is $\sigma^2(X'X)^{-1}$, we will consider constant multiples of such design matrices here, because these will produce uncorrelated estimators. The matrix

$$X = \begin{bmatrix} 1 & 1 \\ 1 & -1 \end{bmatrix}$$

used above has

$$X'X = \begin{bmatrix} 2 & 0 \\ 0 & 2 \end{bmatrix},$$

showing that our estimators are uncorrelated. Clearly, only multiples of orthogonal matrices will produce uncorrelated estimators.

Both our models give uncorrelated estimates, but the variance of the second model is less than that of the first model.

Three Weights

A variety of models is possible here, supposing that we have three objects to weigh and three weighings. One possibility is the model

$$\begin{bmatrix} O_1 \\ O_2 \\ O_3 \end{bmatrix} = \begin{bmatrix} 1 & 1 & 1 \\ 1 & -1 & -1 \\ -1 & 1 & -1 \end{bmatrix} \cdot \begin{bmatrix} W_1 \\ W_2 \\ W_3 \end{bmatrix} + \begin{bmatrix} \varepsilon_1 \\ \varepsilon_2 \\ \varepsilon_3 \end{bmatrix}.$$

Calculation shows that the least squares estimates are

$$
\begin{bmatrix} \widehat{W_1} \\ W_2 \\ W_3 \end{bmatrix} = \begin{bmatrix} \dfrac{1}{2} & \dfrac{1}{2} & 0 \\[2mm] \dfrac{1}{2} & 0 & \dfrac{1}{2} \\[2mm] 0 & -\dfrac{1}{2} & -\dfrac{1}{2} \end{bmatrix} \cdot \begin{bmatrix} O_1 \\ O_2 \\ O_3 \end{bmatrix}
$$

and the variance-covariance matrix is

$$
\sigma^2 \begin{bmatrix} \dfrac{1}{2} & \dfrac{1}{4} & -\dfrac{1}{4} \\[2mm] \dfrac{1}{4} & \dfrac{1}{2} & -\dfrac{1}{4} \\[2mm] -\dfrac{1}{4} & -\dfrac{1}{4} & \dfrac{1}{2} \end{bmatrix},
$$

showing that the estimates are correlated.

The matrix X is called the **design matrix** since it determines the observations to be made. If the design matrix is a multiple of an orthogonal matrix, so that $X^{-1} = X'$, then the least squares estimators are given by

$$
\widehat{\beta} = (X'X)^{-1} X'Y.
$$

We omit the possible multiple of the design matrix and conclude that this becomes $\widehat{\beta} = X'Y$. The variance-covariance matrix is then $\sigma^2 (X'X)^{-1} = \sigma^2 I$, so the estimates are uncorrelated. The design matrix in our example using two weights is

$$
\begin{bmatrix} 1 & 1 \\ 1 & -1 \end{bmatrix},
$$

which is a multiple of an orthogonal matrix. In this case, the variance-covariance matrix will still be a multiple of the identity matrix, so the estimates remain uncorrelated.

It is possible to discover uncorrelated estimators for the three weights. One possibility is the design matrix

$$
\begin{bmatrix} 0 & -1 & 1 \\ 0 & 1 & 1 \\ 1 & 0 & 0 \end{bmatrix},
$$

which gives

$$
\sigma^2 \begin{bmatrix} 1 & 0 & 0 \\ 0 & \dfrac{1}{2} & 0 \\[2mm] 0 & 0 & \dfrac{1}{2} \end{bmatrix}
$$

for the variance-covariance matrix. Many other design matrices are possible in this case.

Four Weights

One possible orthogonal design matrix in this case is the matrix

$$\begin{bmatrix} 1 & 1 \\ 1 & -1 \end{bmatrix} \otimes \begin{bmatrix} 1 & 1 \\ 1 & -1 \end{bmatrix}$$

where the symbol $\otimes$ represents the direct product of the matrices. Each of the component matrices in the direct product is orthogonal and the direct product matrix is orthogonal as well. This gives the design matrix as

$$X = \begin{bmatrix} 1 & 1 & 1 & 1 \\ 1 & -1 & 1 & -1 \\ 1 & 1 & -1 & -1 \\ 1 & -1 & -1 & 1 \end{bmatrix}.$$

The variance-covariance matrix is $(\sigma^2/4)I$.

The weighing designs considered here are interesting and establish some of the basic principles of the design of experiments. Some interesting applications of design matrices to spectroscopy and other fields can be found in Sloane [47] and Banerjee [33]. We now turn to other models where the design and analysis of the data give some unexpected results.

EXERCISES 6.3

1. Suppose in weighing two objects that the sum of the weights is found in the first weighing and then the first weight is weighed alone for the second weighing.
   a. Show that the design matrix is

   $$\begin{bmatrix} 1 & 1 \\ 1 & 0 \end{bmatrix}.$$

   b. Write the model in the form $Y = X\beta + \varepsilon$.
   c. Find the variance-covariance matrix, assuming $\varepsilon \sim ID(0, \sigma)$ and show that the least squares estimates are correlated.
   d. Find the least squares estimates of $\widehat{\beta}$ and show that these estimates are correlated.

2. Suppose in weighing two objects that the sum of the weights is found in the first weighing and then the second weight is weighed alone for the second weighing.
   a. Show that the design matrix is

   $$\begin{bmatrix} 1 & 1 \\ 0 & 1 \end{bmatrix}.$$

   b. Write the model in the form $Y = X\beta + \varepsilon$.
   c. Find the variance-covariance matrix, assuming $\varepsilon \sim ID(0, \sigma)$ and show that the least squares estimates are correlated.

   d. Find the least squares estimates of $\widehat{\beta}$ and show that these estimates are correlated.

3. In weighing three objects, show that the design matrices

   $$X_1 = \begin{bmatrix} 1 & -1 & -1 \\ -1 & 1 & -1 \\ -1 & -1 & 1 \end{bmatrix} \quad \text{and} \quad X_2 = \begin{bmatrix} 1 & 1 & -1 \\ 1 & -1 & 1 \\ -1 & 1 & 1 \end{bmatrix}$$

   give the same least squares estimates for $\widehat{\beta}$ and produce the same variance-covariance structure.

4. Suppose in weighing three weights that two weights are selected at a time and the difference between these weights is observed in each of three weighings.
   a. Find the design matrix.
   b. Write the model in the form $Y = X\beta + \varepsilon$.
   c. Show that the matrix $X'X$ is singular.

5. Show how to determine eight weights in eight weighings with uncorrelated estimates for $\widehat{\beta}$.

6. Pan balances have a single pan for a weight and a scale. An entry of 1 in the design matrix indicates that the weight is in the pan while a 0 indicates that the weight is not in the pan.

a. For the design matrix

$$X = \begin{bmatrix} 1 & 0 & 1 \\ 0 & 1 & 1 \\ 1 & 0 & 0 \end{bmatrix},$$

find the least squares estimates of the weights and find the variance-covariance matrix.

b. For the design matrix

$$X = \begin{bmatrix} 0 & 1 & 1 & 1 \\ 1 & 1 & 0 & 0 \\ 1 & 0 & 1 & 0 \\ 1 & 0 & 0 & 1 \end{bmatrix},$$

find the least squares estimates of the weights and find the variance-covariance matrix.

c. For the design matrix

$$X = \begin{bmatrix} 1 & 0 & 1 & 0 & 1 & 0 & 1 \\ 0 & 1 & 1 & 0 & 0 & 1 & 1 \\ 1 & 1 & 0 & 0 & 1 & 1 & 0 \\ 0 & 0 & 0 & 1 & 1 & 1 & 1 \\ 1 & 0 & 1 & 1 & 0 & 1 & 0 \\ 0 & 1 & 1 & 1 & 1 & 0 & 0 \\ 1 & 1 & 0 & 1 & 0 & 0 & 1 \end{bmatrix},$$

find the least squares estimates of the weights and find the variance-covariance matrix. Note the relationship between this matrix and the design matrix in part (a).

6.4 One-Way Classification

In Chapter 3 we considered comparing two means from data that arose from two different samples. This is a very important situation since experimenters often compare established processes or control groups with new processes or experimental groups. Our test involved the Student t distribution, provided we could assume that the populations producing the two samples had equal variances, which may not be the case. We showed an approximate test when the variances were unequal. One might now wonder how three or four samples could be compared, so we now want to extend the comparison of the means of two samples to the situation where we have, say, k different samples.

One might think that the Student t test could be done on all the possible pairs of samples. However, supposing that each individual test on a pair of samples is run at the 95% level, the probability then of choosing samples suggesting at least one significant difference when in reality there is no such difference among the k samples is

$$1 - (0.95)^{\binom{k}{2}}.$$

We find that this is 0.142625 for $k = 3$ samples and 0.401263 for $k = 5$ samples, so it is apparent that with this procedure control is very rapidly lost on the size of the Type I error. It is interesting to note that for $k = 10$ samples, the probability of a Type I error is about 0.90 and for $k = 14$ samples, the probability becomes greater than 0.99.

However, it is possible to compare the k samples and retain control of the size of the Type I error. We will generalize the two-sample Student t test in two ways, showing both a regression analysis and an analysis of variance approach. We would caution, however, that this procedure is in fact a one-factor-at-a-time analysis and should be used only when the experimenter is absolutely sure that only one factor is of importance. When more than one factor is involved, it is unfortunately common to run one-at-a-time experiments on the factors individually. Practice often calls for these experiments, which may show, say, the maximum response for the factors individually. However, if the factors interact with each other, maximizing the response using the individual factors may not produce the maximum response when the factors are considered together. We have seen this situation in multiple regression. As we shall see, there are much more efficient experimental designs than the one-factor-at-a-time experiments when more than one factor is involved.

Example 6.4.1   A chemist is studying the effect of the level of a catalyst on the concentration of a component in a liquid solution.

Experimental conditions as well as spoiled samples produce unequal numbers of observations as the data in Table 6.2 show. Although in most subsequent cases we will be concerned with data that are regarded as "balanced," data with unequal numbers of observations is a fairly common occurrence in experimentation.

Level 1	Level 2	Level 3	Level 4
70.6	70.3	67.7	62.4
68.4	67.6	68.9	63.0
71.8	68.4	63.8	64.3
71.4	69.4		65.1
67.0			

Table 6.2

We look first at some comparative boxplots of the data. These are shown in Figure 6.3. It would appear that there is a downward trend as the catalyst changes level, so we look first at a regression analysis.

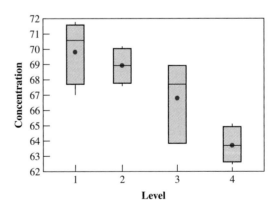

Figure 6.3 Boxplots of Concentration by Levels
(Means are indicated by solid circles.)

Regression Analysis

In order to see the linear trend in the data, the catalyst level values were coded as 1, 2, 3, and 4 and the data were entered as a regression problem. Minitab gives the least squares regression line as Concentration $= 72.3 - 2.03$ Catalyst Level. The analysis of variance, given in Table 6.3, strongly rejects the hypothesis that the slope is 0. (The p-value was calculated using Mathematica.)

		ANALYSIS OF VARIANCE			
Source	**df**	**SS**	**MS**	**F**	**p**
Regression	1	89.975	89.975	28.23	0.000109517
Error	14	44.614	3.187		
Total	15	134.589			

Table 6.3

The regression analysis is of interest and produces another response surface. Much more information, however, is contained in the data. We now investigate the data in a different way. We will refer to this approach as the **analysis of variance** although conclusions from the regression analysis also used an analysis of variance. ∎

Analysis of Variance

We first show a generalization of the data set in Table 6.4.

L_1	L_2	$\cdots$	L_k	
x_{11}	x_{21}	$\cdots$	x_{k1}	
x_{12}	x_{22}	$\cdots$	x_{k2}	
$\vdots$	$\vdots$		$\vdots$	
x_{1n_1}	x_{2n_2}	$\cdots$	x_{kn_k}	
T_1	T_2	$\cdots$	T_k	
$\overline{x}_1$	$\overline{x}_2$	$\cdots$	$\overline{x}_k$	$\overline{\overline{x}}$

Table 6.4

Table 6.4 indicates that there are k treatments, or levels, of a single factor; these are denoted by $L_1, L_2, \ldots, L_k$. These are considered to be all the levels of the factor rather than a sampling of all the possible levels. Individual observations are then denoted by x_{ij} where i denotes the treatment, or level, of the factor and j denotes the observation within the ith column. Note that this is the inverse of the usual matrix notation. Here i denotes the column index and j denotes the row index. Regrettably, this inversion is the standard notation used in books and research papers in statistics.

The totals of the columns are denoted by T_k, the means of the columns are denoted by $\bar{x}_i$, and the overall mean of the data is denoted by $\bar{\bar{x}}$.

The Model

As in the case of the weighing designs, we suppose that an individual observation, x_{ij}, arises from the sum of an overall mean, μ, an effect of the ith treatment, τ_i, and a random error, ε_{ij}, so that

$$x_{ij} = \mu + \tau_i + \varepsilon_{ij}.$$

We will assume that $\sum_{i=1}^{k} \tau_i = 0$, since any other sum would not be detectable in the presence of the mean term, μ. We are interested in the null hypothesis that the τ_i are all zero, namely,

$$H_0: \tau_1 = \tau_2 = \cdots = \tau_k = 0.$$

Acceptance of this hypothesis would indicate that there is in fact no difference between the means of the columns or the catalysts. Against this hypothesis is the alternative hypothesis, namely,

$$H_1: \text{at least one of the } \tau_i \text{ is not } 0.$$

The analysis of the data is now approached in a manner similar to that encountered in simple and multiple regression. The variance of the complete data set is of interest. Consider a multiple of this variance, which we call the **total sum of squares.** This quantity involves the square of the deviation of each data point and the overall mean of the data. Denoting this by *SST*, we have

$$SST = \sum_{i=1}^{k} \sum_{j=1}^{n_i} \left(x_{ij} - \bar{\bar{x}}\right)^2.$$

Adding and subtracting $\bar{x}_i$, we see that this can be written as

$$\sum_{i=1}^{k} \sum_{j=1}^{n_i} \left(x_{ij} - \bar{\bar{x}}\right)^2 = \sum_{i=1}^{k} \sum_{j=1}^{n_i} \left[(x_{ij} - \bar{x}_i) + (\bar{x}_i - \bar{\bar{x}})\right]^2.$$

Squaring, we have

$$\sum_{i=1}^{k} \sum_{j=1}^{n_i} \left(x_{ij} - \bar{\bar{x}}\right)^2 = \sum_{i=1}^{k} \sum_{j=1}^{n_i} (x_{ij} - \bar{x}_i)^2 + 2 \sum_{i=1}^{k} \sum_{j=1}^{n_i} (x_{ij} - \bar{x}_i)(\bar{x}_i - \bar{\bar{x}})$$

$$+ \sum_{i=1}^{k} \sum_{j=1}^{n_i} (\bar{x}_i - \bar{\bar{x}})^2.$$

Consider the middle term. This can be written as

$$\sum_{i=1}^{k}\sum_{j=1}^{n_i}(x_{ij}-\bar{x}_i)\,(\bar{x}_i-\bar{\bar{x}}) = \sum_{i=1}^{k}(\bar{x}_i-\bar{\bar{x}})\left\{\sum_{j=1}^{n_i}(x_{ij}-\bar{x}_i)\right\}.$$

But $\sum_{j=1}^{n_i}(x_{ij}-\bar{x}_i)$ represents, within the ith column, the sum of the deviations of the observations and their mean. So, for fixed i, this sum is 0, showing that the middle term above is 0. We have then that

$$\sum_{i=1}^{k}\sum_{j=1}^{n_i}\left(x_{ij}-\bar{\bar{x}}\right)^2 = \sum_{i=1}^{k}\sum_{j=1}^{n_i}(x_{ij}-\bar{x}_i)^2 + \sum_{i=1}^{k}\sum_{j=1}^{n_i}\left(\bar{x}_i-\bar{\bar{x}}\right)^2.$$

This is a fact that is true of any array of numbers; we have made no use of the model here.

It is customary to refer to $\sum_{i=1}^{k}\sum_{j=1}^{n_i}(x_{ij}-\bar{x}_i)^2$ as the **within sum of squares (SSW)** since it is a measure of the variation around the individual column means, summed over all the columns.

We refer to $\sum_{i=1}^{k}\sum_{j=1}^{n_i}(\bar{x}_i-\bar{\bar{x}})^2$ as the **between sum of squares (SSB)** since it is a measure of the variation of the individual column means around their grand mean, $\bar{\bar{x}}$. The identity,

$$\sum_{i=1}^{k}\sum_{j=1}^{n_i}\left(x_{ij}-\bar{\bar{x}}\right)^2 = \sum_{i=1}^{k}\sum_{j=1}^{n_i}(x_{ij}-\bar{x}_i)^2 + \sum_{i=1}^{k}\sum_{j=1}^{n_i}\left(\bar{x}_i-\bar{\bar{x}}\right)^2,$$

which, of course, is called an **analysis of variance identity,** is then abbreviated as $SST = SSW + SSB$.

To speak loosely, if the column means differed significantly from one another, we would expect SSB to be a large percentage of SST. If the column means do not differ significantly from one another, then we expect a large percentage of SST to be found in SSW. The real situation is not quite that simple. To analyze the data accurately, first suppose in the model

$$x_{ij} = \mu + \tau_i + \varepsilon_{ij}$$

that the ε_{ij} are independently and normally distributed with mean 0 and variance σ^2. Note the presumption of equal variances here. It can then be shown, although not easily, that the sums of squares divided by their degrees of freedom (called **mean squares**) are independently distributed chi-squared random variables and that their ratio is distributed according to the F distribution. The null hypothesis, H_0: $\tau_1 = \tau_2 = \cdots = \tau_k = 0$ is rejected only if the F value is too large for a given Type I error. Proofs of these facts can be found in most books on mathematical statistics. Many computer statistics packages, as well as Mathematica, summarize them in an analysis of variance table, which we show in general terms in Table 6.5. We refer to the columns as **treatments.**

ANALYSIS OF VARIANCE

Source	df	SS	MS	F	p
Treatments	$k-1$	SSB	$\dfrac{SSB}{k-1} = MSB$	$\dfrac{MSB}{MSW}$	
Error	$\displaystyle\sum_{i=1}^{k}(n_i-1)$	SSW	$\dfrac{SSW}{\displaystyle\sum_{i=1}^{k}(n_i-1)} = MSW$		
Total	$\displaystyle\sum_{i=1}^{k}n_i-1$	SST			

Table 6.5

The quantity p here is then

$$p = Pr\left[\left(\frac{MSB}{MSW}\right) > F\left(k-1, \sum_{i=1}^{k}(n_i-1)\right)\right].$$

Table 6.6 then gives values for the data set we are considering. Details for producing Table 6.6 using Minitab can be found in Appendix B.

ANALYSIS OF VARIANCE

Source	df	SS	MS	F	p
Catalyst Level	3	94.73	31.58	9.51	0.002
Error	12	39.86	3.32		
Total	15	134.59			

Table 6.6

The p-value here is of significance, so we conclude that the catalyst levels do in fact produce different results. The analysis of variance table, however, does not indicate what differences among the levels actually produce the significance, a fact that might well interest an experimenter. The boxplots in Figure 6.3 also indicate that there may be significant differences between the responses due to specific catalyst level. We now investigate the data further in an attempt to detect some significant differences.

The graphs indicate that there may be some value in comparing catalyst levels 3 and 4. This could be done using the Student t test, but if one also wanted to compare catalyst levels 1 and 2 or, say, catalyst levels 1 and 2 together with catalyst levels 3 and 4, control of the Type I error is quickly lost.

We suggest two procedures here. The first involves calculating confidence intervals on the individual catalyst level means.

Confidence Intervals

In the model,

$$x_{ij} = \mu + \tau_i + \varepsilon_{ij},$$

we now assume that the ε_{ij}, the random errors, are independently distributed from a normal distribution with mean 0 and variance σ^2, which we abbreviate as $\varepsilon_{ij} \sim NID(0, \sigma)$.

Then it follows that the individual column means have Student t distributions so that

$$t_{n-1} = \frac{\bar{x}_i - \mu_i}{s_i / \sqrt{n}}$$

where $\bar{x}_i$ is the observed mean of column i, s_i is the sample standard deviation for column i, and μ_i is the true mean of column i.

Using the data in catalyst level 1, and noting that the critical Student t value is 2.77645, we have a two-sided 95% confidence interval for μ_1 as

$$69.84 \pm 2.77645 \cdot \frac{2.061}{\sqrt{5}}$$

giving the interval (67.2809, 72.3991).

In a similar way, confidence intervals for catalyst levels 2, 3, and 4 can be found to be (67.0539, 70.7963), (60.1674, 73.4326), and (61.7507, 65.6493). Since the confidence intervals for catalysts 2 and 4 do not overlap, this is an indication of a source of the significance in the analysis of variance. Minitab and other statistical computer packages frequently use a pooled variance (from all of the k columns) to compute confidence intervals for each of the means. We have used the variances of each of the columns in the above calculations, so our results differ somewhat from those given by Minitab.

It is also often of interest to look beyond the difference between the sample means. Fortunately, there are several ways to do this. Here we will only consider what we call *contrasts* or *comparisons*.

Contrasts or Comparisons

DEFINITION

Suppose we have k treatments, and suppose that in the ith treatment there are n_i observations giving a *total* T_i.

A **contrast** or **comparison** among the treatment totals is an expression of the form

$$C_p = \sum_{i=1}^{k} c_{ip} T_i = c_{1p} T_1 + c_{2p} T_2 + \cdots + c_{kp} T_k \quad \text{where}$$

$$\sum_{i=1}^{k} n_i c_{ip} = 0.$$

Two contrasts, C_p and $C_q = \sum_{i=1}^{k} c_{iq} T_i$, are **orthogonal** if $\sum_{i=1}^{k} n_i c_{ip} c_{iq} = 0$.

Contrasts allow, in addition to the comparison of two treatments, a wide variety of comparisons to be made among the treatments.

Let us take the data in the example to show some specific contrasts. The data give

$$T_1 = 349.2 \qquad T_2 = 275.7 \qquad T_3 = 200.4 \qquad T_4 = 254.8$$
$$n_1 = 5 \qquad n_2 = 4 \qquad n_3 = 3 \qquad n_4 = 4.$$

We first compare catalyst levels 1 and 2 with catalyst levels 3 and 4. If we let $C_1 = aT_1 + aT_2 - bT_3 - bT_4$, then the condition that $\sum_{i=1}^{k} n_i c_{ip} = 0$ gives

$$5a + 4a - 3b - 4b = 0$$

so

$$9a = 7b,$$

showing that $a = 7$ and $b = 9$ is one set of possible choices so

$$C_1 = 7T_1 + 7T_2 - 9T_3 - 9T_4.$$

Now suppose we want two other contrasts, so as to form a set of three orthogonal contrasts. If we let $C_2 = dT_1 - eT_2$, then d and e must satisfy the condition that $5d - 4e = 0$, so we can choose $d = 4$ and $e = 5$ producing

$$C_2 = 4T_1 - 5T_2.$$

In a similar way, we find another contrast, orthogonal to both C_1 and C_2, as

$$C_3 = 4T_3 - 3T_4.$$

It is easy to check that the condition $\sum_{i=1}^{k} n_i c_{ip} c_{iq} = 0$ applies to each of the three pairs of contrasts found here, so we have constructed an orthogonal set of contrasts. It is not always this easy with unequal numbers of observations in the treatments. When the data consist of equal numbers of observations for the treatments, then the construction of orthogonal sets (there are many choices) is fairly simple. The sizes of the contrasts may be of some interest, but these are usually difficult to interpret. We find here that

$$C_1 = 277.50$$
$$C_2 = 18.30$$
$$C_3 = 37.20,$$

but it is not clear from these numbers whether or not an individual contrast has statistical significance. It is possible to determine the statistical significance, or lack of it, for the individual contrasts.

It can be shown that a contrast $C_p = \sum_{i=1}^{k} c_{ip} T_i = c_{1p} T_1 + c_{2p} T_2 + \cdots + c_{kp} T_k$ has a sum of squares associated with it, namely,

$$SSC_p = \frac{C_p^2}{\displaystyle\sum_{i=1}^{k} n_i c_{ip}^2}.$$

Furthermore, consider the following fact.

> **FACT**
>
> The factor sum of squares in the analysis of variance is completely partitioned by a set of $(k-1)$ orthogonal contrasts where k is the number of treatments.

In this example,

$$C_1 = 7T_1 + 7T_2 - 9T_3 - 9T_4$$

$$= 7(349.2) + 7(275.7) - 9(200.4) - 9(254.8) = 277.50$$

and

$$SSC_1 = \frac{C_1^2}{\sum\limits_{i=1}^{k} n_i c_{i1}^2} = \frac{(277.50)^2}{5(7)^2 + 4(7)^2 + 3(-9)^2 + 4(-9)^2} = 76.3951.$$

Then,

$$C_2 = 4T_1 - 5T_2$$

so

$$C_2 = 4(349.2) - 5(275.7) = 18.30$$

and

$$SSC_2 = \frac{(18.30)^2}{5(4)^2 + 4(-5)^2} = 1.8605.$$

Also

$$C_3 = 4T_3 - 3T_4$$

so

$$C_3 = 4(200.4) - 3(254.8) = 37.20$$

and

$$SSC_3 = \frac{(37.20)^2}{3(4)^2 + 4(-3)^2} = 16.4743.$$

Now $SSC_1 + SSC_2 + SSC_3 = 76.3951 + 1.8605 + 16.4743 = 94.7299$, which is the sum of squares due to treatments in the analysis of variance.

Each of these contrasts may be tested for significance in the analysis of variance as independent sums of squares. This means that the analysis of variance may be written as shown in Table 6.7. Certainly C_1 would be regarded as significant, and most probably C_3 as well, while C_2 is surely not significant.

ANALYSIS OF VARIANCE					
Source	**df**	**SS**	**MS**	**F**	**p**
Factor					
C_1	1	76.3951	76.3951	23.0106	0.00044
C_2	1	1.8605	1.8605	0.5604	0.46852
C_3	1	16.4743	16.4743	4.9621	0.04581
Subtotal	3	94.7299			
Error	12	39.86	3.32		
Total	15	134.59			

Table 6.7

While we have shown how orthogonal contrasts always completely partition the treatment sum of squares, it may well be that nonorthogonal contrasts are of interest. Tests on such contrasts are, of course, not independent. For instance, in the example above, we might be interested only in investigating catalyst level 3 and the mean of the remaining catalyst levels producing the contrast

$$C_4 = 13T_3 - 3T_1 - 3T_2 - 3T_4$$
$$= 13(200.4) - 3(349.2) - 3(275.7) - 3(254.8) = -33.9.$$

We find that

$$SSC = \frac{(-33.9)^2}{3(13)^2 + 5(-3)^2 + 4(-3)^2 + 4(-3)^2} = 1.8417.$$

The F ratio for this single contrast is $1.8417/3.32 = 0.5547$, giving a p-value of 0.470739, which is not enough to justify a conclusion of significance.

Tests and Confidence Intervals

It is also possible to perform tests of hypotheses and construct confidence intervals for contrasts.

Let

$$\lambda = c_1\mu_1 + c_2\mu_2 + \cdots + c_k\mu_k$$

be a contrast among the treatment means where we suppose that $\sum_{i=i}^{k} n_i c_i = 0$. Let

$$l = c_1\frac{T_1}{n_1} + c_2\frac{T_2}{n_2} + \cdots + c_k\frac{T_k}{n_k}$$

be an estimator for λ. Then it is easy to see that

$$E(l) = \lambda \quad \text{and} \quad \text{Var}(l) = \sum_{i=1}^{k} \frac{c_i^2 \sigma^2}{n_i}$$

and so

$$\widehat{\text{Var}}(l) = \sum_{i=1}^{k} \frac{c_i^2 s^2}{n_i}.$$

This means that

$$\frac{l - \lambda}{\sqrt{\sum_{i=1}^{k} \frac{c_i^2 s^2}{n_i}}} = t_{k(n-1)}$$

where s^2 is the within-groups mean square in the analysis of variance.

For example, consider the contrast $\lambda = 4\mu_3 - 3\mu_4$ for the preceding data. Here

$$l = 4\frac{T_3}{n_3} - 3\frac{T_4}{n_4} = 4\frac{200.4}{3} - 3\frac{254.8}{4} = 76.10$$

and

$$\widehat{\text{Var}}(l) = \sum_{i=1}^{k} \frac{c_i^2 s^2}{n_i} = \left[\frac{4^2}{3} + \frac{(-3)^2}{4}\right](3.32) = 25.1767.$$

It follows that a two-sided 95% confidence interval for λ is

$$76.10 - 2.17881\sqrt{25.1767} < \lambda < 76.10 + 2.17881\sqrt{25.1767}$$

or

$$65.1675 < \lambda < 87.0325.$$

We would very likely reject then the null hypothesis H_0: $\lambda = 0$.

We end this discussion of the one-way classification here, although we should point out that it is possible to find, for a given level of significance, procedures that consider *all* possible contrasts among the means. Relevant references can be found in the Bibliography.

The next logical step in our discussion of the design of experiments is to consider experiments in which both the rows and the columns represent factors in the experiment.

EXERCISES 6.4

1. The following data set is a simple set of data for a one-way classification with factors A, B, C, and D.

A	B	C	D
1	3	1	0
−1	2	3	−2
	6	0	−2
	2	4	−4
	2		

a. Calculate each of the sums of squares for the analysis of variance directly without using a computer.

b. Complete the analysis of variance table and state the conclusions that can be drawn from it.

c. Partition the factor sum of squares into parts comparing the totals for treatments A and B with those for treatments C and D, comparing treatment A with treatment B, and finally comparing treatment C with treatment D.

2. Lin and Stephenson [43] report on the results of a viral assay, which is a procedure used to detect the presence or absence of a virus in a solution. The following data show the results of four separate counts of the plaques formed when the virus is present.

Count 1	Count 2	Count 3	Count 4
24	18	27	21
23	13	24	17
18	20	27	23
25	21	33	28
16	12	21	20
18	8	21	19
20	17	11	15
14	18	14	22
16	17	15	21

a. Show comparative boxplots of the data.
b. Show the analysis of variance and state the conclusions that can be drawn from it.
c. Partition the factor sum of squares due to contrasts comparing count 3 with count 4, comparing count 1 with count 2, and finally comparing the means of counts 1 and 2 with the means of counts 3 and 4.

3. The following data were gathered in an experiment on the length of the storage period (SP) on the number of azalea bulbs blooming out of 50 bulbs.

SP$_1$	SP$_2$	SP$_3$	SP$_4$
35	34	35	24
23	26	35	17
45	29	43	33
33	18		42
	32		42
			35

a. Show comparative boxplots of the data.
b. Show the analysis of variance and state the conclusions that can be drawn from it.
c. Partition the factor sum of squares due to contrasts, comparing SP$_1$ and SP$_3$ with SP$_2$, and SP$_4$, comparing SP$_1$ and SP$_3$, and finally comparing SP$_2$ and SP$_4$.

4. An engineer concerned with product development is interested in making the tensile strength of a new synthetic fiber that will be used in heavy men's shirts. It is thought that the tensile strength is dependent upon the percentage of cotton in the fiber. The engineer measures five pieces of fiber for cotton content ranging from 5% to 25%. The data follow.

		Cotton Percentage		
5%	10%	15%	20%	25%
8	11	19	22	8
15	12	20	25	10
11	18	18	19	7
9	17	16	23	12
7	18	11	23	11

a. Show comparative boxplots of the data.
b. Fit a simple regression line to the data and state the conclusions that can be drawn from it.
c. Show the analysis of variance analyzing the data as a one-way classification and state the conclusions that can be drawn from it.
d. Partition the factor sum of squares into four mutually orthogonal contrasts and state any conclusions that can be drawn from this analysis.

5. A paint manufacturer is studying the reflective properties of three different kinds of paint. The data gathered follow. The reflective indices of the paints have been scaled here.

	Paint		
1	2	3	4
195	45	230	110
150	40	115	55
205	195	235	120
120	65	225	50
160	145		80
	195		

a. Show comparative boxplots of the data.
b. Fit a simple regression line to the data and state the conclusions that can be drawn from it.
c. Show the analysis of variance analyzing the data as a one-way classification and state the conclusions that can be drawn from it.
d. Partition the factor sum of squares into three mutually orthogonal contrasts and state any conclusions that can be drawn from this analysis.

6. Heyl [41] measured the gravitational constant, g, using balls made of gold, glass, and platinum. The data that follow are the units in the second and third decimal places in the determination of g.

Gold	Glass	Platinum
83	78	61
81	71	61
76	75	67
78	72	67
79	74	64
72		

a. Show comparative boxplots of the data.
b. Fit a simple regression line to the data and state the conclusions that can be drawn from it.
c. Show the analysis of variance analyzing the data as a one-way classification and state the conclusions that can be drawn from it.
d. Partition the factor sum of squares into two mutually orthogonal contrasts and state any conclusions that can be drawn from this analysis.

7. The following data are from a study of the failure of household light bulbs operating under different voltages, namely, 108, 114, 120, 126, and 132 volts. The data are the number of light bulbs failing out of 50 bulbs studied at each voltage. (The data were previously used in Exercise 7 of Section 5.2.)

108	114	120	126	132
2	6	13	27	45
3	8	11	31	46

Source: Minitab, Lightbul.mtw

a. Analyze the data as a one-way classification showing the analysis of variance.
b. What conclusions can be drawn from the analysis in part (a)?

8. The following data show the number of hazardous waste sites in each state in various regions of the United States. The first column, for example, shows the number of hazardous waste sites in the six New England States.

NEng	MAtl	ENCen	WNCen	SAtl	ESCen	WSCen	Mtn	Pac
15	108	37	21	19	12	12	10	7
9	83	33	10	0	19	12	17	95
26	98	77	42	55	2	10	9	2
17		33	23	13	14	29	8	9
12		40	10	10			1	50
8			2	23			11	
			4	24			13	
							3	

Source: Minitab, Wastes.mtw

a. Show comparative boxplots of the data.
b. Fit a simple regression line to the data and state the conclusions that can be drawn from it.
c. Show the analysis of variance analyzing the data as a one-way classification and state the conclusions that can be drawn from it.

9. An underwriting laboratory has tested the flammability of fabric, using identical methods in five different laboratories. The measurements are the length of the burned portion of a piece of fabric held over flame for a fixed amount of time.

Lab1	Lab2	Lab3	Lab4	Lab5
2.9	2.7	3.3	3.3	4.1
3.1	3.4	3.3	3.2	4.1
3.1	3.6	3.5	3.4	3.7
3.7	3.2	3.5	2.7	4.2
3.1	4.0	2.8	2.7	3.1
4.2	4.1	2.8	3.3	3.5
3.7	3.8	3.2	2.9	2.8
3.9	3.8	2.8	3.2	3.5
3.1	4.3	3.8	2.9	3.7
3.0	3.4	3.5	2.6	3.5
2.9	3.3	3.8	2.8	3.9

Source: Minitab, Fabric.mtw

a. Do the laboratories differ with respect to their measurement of the amount of burned fabric?
b. Show comparative boxplots of the data.
c. By constructing appropriate contrasts, determine which laboratories, if any, differ from each other.

10. The following data refer to the length of time customers spend in a waiting line at a bank. Data are measured for the first, second, . . . , fifth customer of the day. If you can retrieve the file Accounts.mtw from Minitab, do so; otherwise, use the following abbreviated data set from the same study.

Cust1	Cust2	Cust3	Cust4	Cust5
5.0	5.3	4.9	5.0	5.3
6.0	5.4	6.1	6.0	5.5
5.0	6.0	5.7	5.5	5.4
5.5	5.4	5.4	6.0	6.0
5.6	5.4	5.4	4.9	4.7
5.6	4.7	6.5	5.3	5.6
5.6	4.5	5.0	5.1	5.2
4.1	5.4	6.5	7.0	7.2
6.1	6.1	7.0	6.7	6.8
7.1	6.9	6.2	5.4	5.6

Source: Minitab, Accounts.mtw

a. Do the customers differ with respect to their waiting time for service?

b. Show comparative boxplots of the data.

c. By constructing appropriate contrasts, determine which customer orders, if any, differ from each other.

11. In Chapter 3 we compared two samples using the Student t distribution. Recall that in one case, we presumed that the populations from which the samples were drawn had equal variances. Show that the analysis of variance test given in this section is exactly equivalent to the two-sample test given in Chapter 3.

12. Show how the contrast C_4 was found in the subsection on contrasts and comparisons.

6.5 Blocking

We come now to the simplest case where more than one factor is involved in the response measured by the experiment. Before analyzing that case in general, however, experimenters often find themselves in situations where there are two factors, but one is thought to be of significance before the experiment is planned or performed and it is desired to remove whatever influence this factor may have. So we consider the case where we wish to remove the influence of a single factor, which we call a **block.** For example, experimental data may be recorded over several days and if the experimenters believe that the days may have some effect on the experimental response, they will want to remove this influence on the results. Other block effects may include different experimental materials, such as metals, on which an experiment is run. Different patients are often considered to be blocks when a medical experiment is performed. These factors are in general factors thought to be influential in the results, but whose influence should be removed before the data are analyzed because these influences, if left alone, may significantly alter the conclusions drawn from the statistical analysis.

So we begin with a two-way classification where one of the factors is called a *block.* We start with a specific example.

Example 6.5.1   Reeve and Giesbrecht [45] report on an experiment to measure the rate of dissolution of a drug, in this case into the bloodstream of an animal. The rate at which such drugs are dissolved into the bloodstream is of great interest in measuring part of the efficacy of the drug. Experiments such as this are conducted using separate containers or vessels into which the drug is placed along with some solution, such as water, but frequently enzymes found in the animal's bloodstream are used as well. The drug is allowed to dissolve and measurements of the percentage of the drug that has dissolved in a certain period of time are taken. It is thought that the vessels may differ with respect to the absorption rate of the drug, and this is what we wish to test.

The data are shown in Table 6.8. Both row and column means are included because they play a central role in our analysis of the data. Examination of the columns as time

Time Min.	Vessel 1	Vessel 2	Vessel 3	Vessel 4	Vessel 5	Vessel 6	Time Means
0	0	0	0	1	0	0	0.17
15	2	5	0	17	1	12	6.17
30	20	33	9	23	23	32	23.33
45	65	82	48	81	77	61	69.00
60	95	92	81	94	95	93	91.67
120	98	97	101	100	99	99	99.00
Vessel Means	46.67	51.50	39.83	52.67	49.17	49.50	48.2233

Table 6.8

increases shows that time is a factor of great importance. We look at the data in two different ways. Figure 6.4 compares the vessels ignoring time while Figure 6.5 ignores the vessels and shows the percent dissolved as a function of time. Clearly time is a factor of significance.

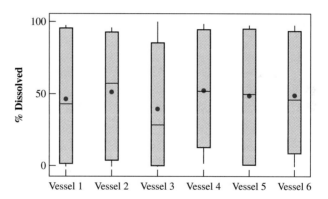

Figure 6.4 Boxplots of Vessels 1–6 (Means are indicated by solid circles.)

A regression analysis shows a definite time trend. It is also clear that this factor may well influence any conclusion from the data regarding the possible difference between the vessels. We will call the time factor in this example a **blocking factor** and show that it is possible to remove this factor from consideration and examine the influence of the vessel alone. ■

To examine the theoretical situation first, we begin with some needed notation. Let x_{ijk} be an observation where i denotes the vessel (coded as 1, 2, ... , 6), j denotes the time or block factor (coded as 1, 2, ... , 6), and k denotes the number of the replication in

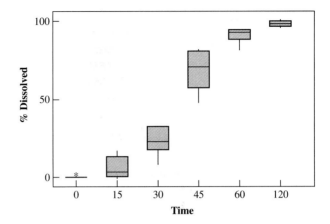

Figure 6.5 Data in Table 6.8 Showing the Influence of Time

the (i, j) cell. The index k is not needed since our specific data have only one observation for each time-vessel combination, but it will be useful to be able to consider the situation where there are multiple observations for each of these combinations. We presume there are t treatments, b blocks, and n observations for each treatment-block combination. Note particularly that here we assume equal sample sizes for each treatment-block combination. For our data, $t = 6$, $b = 6$, and $n = 1$.

The general model for the situation is

$$x_{ijk} = \mu + \tau_i + \beta_j + \varepsilon_{ijk} \quad \text{for} \quad \begin{cases} i = 1, 2, \ldots, t \\ j = 1, 2, \ldots, b \\ k = 1, 2, \ldots, n \end{cases}.$$

The τ_i represent the effects of the column (here the vessel) and the β_j represent the row (here time) effects. We will presume that $\sum_{i=1}^{t} \tau_i = \sum_{j=1}^{b} \beta_j = 0$ since any nonzero sum in either case could not be detected due to the inclusion of the overall mean term, μ, in the model. We denote the block means by $\bar{x}_{.j.}$, the treatment means by $\bar{x}_{i..}$, and the mean of all the observations by $\bar{x}_{...}$. Here the dots indicate the index over which summations have occurred. Dots were not needed before because there is only one index in the one-way classification.

We find an analysis of variance partition of the observations as follows. We let

$$SST = \sum_{i=1}^{t} \sum_{j=1}^{b} \sum_{k=1}^{n} (x_{ijk} - \bar{x}_{...})^2$$

denote the total sum of squares. However,

$$\sum_{i=1}^{t} \sum_{j=1}^{b} \sum_{k=1}^{n} (x_{ijk} - \bar{x}_{...})^2 = \sum_{i=1}^{t} \sum_{j=1}^{b} \sum_{k=1}^{n} [(\bar{x}_{i..} - \bar{x}_{...}) + (\bar{x}_{.j.} - \bar{x}_{...})$$

$$+ (x_{ijk} - \bar{x}_{i..} - \bar{x}_{.j.} + \bar{x}_{...})]^2,$$

which can be simplified as

$$\sum_{i=1}^{t}\sum_{j=1}^{b}\sum_{k=1}^{n}(x_{ijk}-\overline{x}_{...})^2 = \sum_{i=1}^{t}\sum_{j=1}^{b}\sum_{k=1}^{n}(\overline{x}_{i..}-\overline{x}_{...})^2 + \sum_{i=1}^{t}\sum_{j=1}^{b}\sum_{k=1}^{n}(\overline{x}_{.j.}-\overline{x}_{...})^2$$

$$+ \sum_{i=1}^{t}\sum_{j=1}^{b}\sum_{k=1}^{n}(x_{ijk}-\overline{x}_{i..}-\overline{x}_{.j.}+\overline{x}_{...})^2$$

because all of the cross-product terms are zero, a fact that we now establish.

1. $$\sum_{i=1}^{t}\sum_{j=1}^{b}\sum_{k=1}^{n}(\overline{x}_{i..}-\overline{x}_{...})(\overline{x}_{.j.}-\overline{x}_{...}) = n\sum_{i=1}^{t}(\overline{x}_{i..}-\overline{x}_{...})\left[\sum_{j=1}^{b}(\overline{x}_{.j.}-\overline{x}_{...})\right] = 0$$

since $\sum_{j=1}^{b}(\overline{x}_{.j.}-\overline{x}_{...})$ represents the sum of the deviations of the block means around their mean and hence is zero.

2. $$\sum_{i=1}^{t}\sum_{j=1}^{b}\sum_{k=1}^{n}(\overline{x}_{i..}-\overline{x}_{...})(x_{ijk}-\overline{x}_{i..}-\overline{x}_{.j.}+\overline{x}_{...})$$

$$= \sum_{i=1}^{t}(\overline{x}_{i..}-\overline{x}_{...})\left[\sum_{j=1}^{b}\sum_{k=1}^{n}(x_{ijk}-\overline{x}_{i..}-\overline{x}_{.j.}+\overline{x}_{...})\right]$$

but

$$\sum_{j=1}^{b}\sum_{k=1}^{n}(x_{ijk}-\overline{x}_{i..}-\overline{x}_{.j.}+\overline{x}_{...}) = (x_{i..}-nb\overline{x}_{i..}) - n\sum_{j=1}^{b}(\overline{x}_{.j.}-\overline{x}_{...}).$$

Now $(x_{i..}-nb\overline{x}_{i..}) = 0$ and $\sum_{j=1}^{b}(\overline{x}_{.j.}-\overline{x}_{...}) = 0$ so

$$\sum_{j=1}^{b}\sum_{k=1}^{n}(x_{ijk}-\overline{x}_{i..}-\overline{x}_{.j.}+\overline{x}_{...}) = 0$$

and so

$$\sum_{i=1}^{t}\sum_{j=1}^{b}\sum_{k=1}^{n}(\overline{x}_{i..}-\overline{x}_{...})(x_{ijk}-\overline{x}_{i..}-\overline{x}_{.j.}+\overline{x}_{...}) = 0.$$

3. $$\sum_{i=1}^{t}\sum_{j=1}^{b}\sum_{k=1}^{n}(\overline{x}_{.j.}-\overline{x}_{...})(x_{ijk}-\overline{x}_{i..}-\overline{x}_{.j.}+\overline{x}_{...}) = 0$$

by an argument very similar to that given in part 2.

Now if we let

$$SST = \sum_{i=1}^{t}\sum_{j=1}^{b}\sum_{k=1}^{n}(x_{ijk}-\overline{x}_{...})^2,$$

$$SSTmts = \sum_{i=1}^{t}\sum_{j=1}^{b}\sum_{k=1}^{n}(\overline{x}_{i..} - \overline{x}_{...})^2,$$

$$SSBlocks = \sum_{i=1}^{t}\sum_{j=1}^{b}\sum_{k=1}^{n}(\overline{x}_{.j.} - \overline{x}_{...})^2, \text{ and}$$

$$SSError = \sum_{i=1}^{t}\sum_{j=1}^{b}\sum_{k=1}^{n}(x_{ijk} - \overline{x}_{i..} - \overline{x}_{.j.} + \overline{x}_{...})^2,$$

we can then write

$$SST = SSTmts + SSBlocks + SSError.$$

If we presume that the random errors $\varepsilon_i \sim NID(0, \sigma)$, then it can be shown that these sums of squares are independently distributed according to a chi-squared distribution and that their ratios are distributed according to the F distribution. These results are usually exhibited in an analysis of variance table, which we show in general terms in Table 6.9; then the specific analysis of the drug dissolution data is shown in Table 6.10. Blocks are not normally tested for significance; here we simply wish to remove their effect from the analysis of the data.

Source	df	SS	MS	F	p
Blocks	$r-1$	SSR			
Treatments	$c-1$	SSC	$\dfrac{SSC}{c-1} = MSC$	MSC/MSW	
Error	$nrc - r - c + 1$	SSW	$\dfrac{SSW}{nrc - r - c + 1} = MSW$		
Total	$nrc - 1$	SST			

Table 6.9

ANALYSIS OF VARIANCE					
Source	df	SS	MS	F	p
Block (Time)	5	57,569.9			
Vessels	5	634.9	127.0	3.06	0.027
Error	25	1037.4	41.5		
Total	35	59,242.2			

Table 6.10

Table 6.10 indicates that the vessels are in fact different when we remove the block factor of time. In fact, an overwhelming proportion of the total sum of squares is removed in blocking. What would the results be if we had not removed the block factor and analyzed the data as a one-way experiment? The result is shown in Table 6.11, which analyzes the data as a one-way experiment, ignoring the block factor of time. This analysis clearly misleads us entirely by allowing us to conclude that the vessels are not different at all. They are in fact very different.

ANALYSIS OF VARIANCE					
Source	**df**	**SS**	**MS**	**F**	**p**
Vessels	5	634.9	127.0	0.0650	0.997
Error	30	58,607.3	1953.6		
Total	35	59,242.2			

Table 6.11

Blocking is a very special case of the more general situation in which there are two factors, each of which, as well as their interaction, may be of interest and significance. Now we consider a general situation in which there are two factors and we wish to test the significance of each of these factors.

6.6 Two-Way Classifications

We have now an experiment in which there are two factors, say R and C, denoting row and column factors. We also consider the possibility that the factors behave differently when observed in different combinations of their levels. This is called an **interaction** and we denote it by $R \cdot C$. We suppose the factor R has r levels and that the factor C has c levels. We will presume that the experimental data are *balanced* so that there are, say, n observations for each row-column combination. Individual observations are denoted by x_{ijk} where i is the column index, j is the row index, and k is the number of the observation for the (i, j) factor combination. This design is often called a **factorial** design since every level of every factor is observed with every level of every other factor. The general structure is shown in Table 6.12.

We suppose that the individual observations arise from an additive linear model with overall mean, μ, an addition due to the column, α_i, an addition due to the row, β_j, an interaction term, δ_{ij}, and finally an error term, ε_{ijk}, so the model is

$$x_{ijk} = \mu + \alpha_i + \beta_j + \delta_{ij} + \varepsilon_{ijk} \quad \text{for} \quad \begin{cases} i = 1, 2, \ldots, c \\ j = 1, 2, \ldots, r \\ k = 1, 2, \ldots, n \end{cases}.$$

When we consider the levels of the factors to represent all the possible levels of these factors and make no assumptions about the probability distribution of the error terms,

	C_1	C_2	$\ldots$	C_c
	x_{111}	x_{211}	$\cdots$	x_{c11}
R_1	x_{112}	x_{212}	$\cdots$	x_{c12}
	.	.	.	.
	.	.	.	.
	x_{11n}	x_{21n}	$\cdots$	x_{c1n}
R_2	x_{121}	x_{221}	$\cdots$	x_{c21}
	x_{122}	x_{222}	$\cdots$	x_{c22}
	.	.	$\cdots$	.
	.	.	$\cdots$	.
	x_{12n}	x_{22n}	$\cdots$	x_{c2n}
	.	.	$\cdots$	.
	.	.	$\cdots$	.
R_r	x_{1r1}	x_{2r1}	$\cdots$	x_{cr1}
	x_{1r2}	x_{2r2}	$\cdots$	x_{cr2}
	.	.	$\cdots$	.
	x_{1rn}	x_{2rn}	$\cdots$	x_{crn}

Table 6.12

the model is called a **fixed effects** model. Due to the inclusion of the mean term, μ, in the model, we suppose that

$$\sum_{i=1}^{c}\alpha_i = \sum_{j=1}^{r}\beta_j = \sum_{i=1}^{c}\sum_{j=1}^{r}\delta_{ij} = 0,$$

since nonzero sums could not be detected.

Consider

$$\sum_{i=1}^{c}\sum_{j=1}^{r}\sum_{k=1}^{n}(x_{ijk}-\overline{x}_{...})^2.$$

We call this the **total sum of squares** and denote it by *SST*. Now consider

$$\sum_{i=1}^{c}\sum_{j=1}^{r}\sum_{k=1}^{n}(x_{ijk}-\overline{x}_{...})^2 = \sum_{i=1}^{c}\sum_{j=1}^{r}\sum_{k=1}^{n}[(\overline{x}_{i..}-\overline{x}_{...})+(\overline{x}_{.j.}-\overline{x}_{...})$$
$$+(x_{ijk}-\overline{x}_{ij.})+(\overline{x}_{ij.}-\overline{x}_{i..}-\overline{x}_{.j.}+\overline{x}_{...})]^2.$$

Now we square and sum, but all the cross-product terms are zero. We choose two examples:

$$\sum_{i=1}^{c}\sum_{j=1}^{r}\sum_{k=1}^{n}(\overline{x}_{i..}-\overline{x}_{...})\,(\overline{x}_{ij.}-\overline{x}_{i..}-\overline{x}_{.j.}+\overline{x}_{...})$$

$$=\sum_{i=1}^{c}n(\overline{x}_{i..}-\overline{x}_{...})\left[\sum_{j=1}^{r}(\overline{x}_{ij.}-\overline{x}_{i..}-\overline{x}_{.j.}+\overline{x}_{...})\right]$$

$$=\sum_{i=1}^{c}n(\overline{x}_{i..}-\overline{x}_{...})\left[\sum_{j=1}^{r}\left\{(\overline{x}_{ij.}-\overline{x}_{i..})-(\overline{x}_{.j.}-\overline{x}_{...})\right\}\right]$$

$$=\sum_{i=1}^{c}n(\overline{x}_{i..}-\overline{x}_{...})\left[\sum_{j=1}^{r}(\overline{x}_{ij.}-\overline{x}_{i..})-\sum_{j=1}^{r}(\overline{x}_{.j.}-\overline{x}_{...})\right].$$

Now $\sum_{j=1}^{r}(\overline{x}_{ij.}-\overline{x}_{i..})$ and $\sum_{j=1}^{r}(\overline{x}_{.j.}-\overline{x}_{...})$ each represent the sum of deviations of a quantity about its mean and so each is zero. It follows that

$$\sum_{i=1}^{c}\sum_{j=1}^{r}\sum_{k=1}^{n}(\overline{x}_{i..}-\overline{x}_{...})\,(\overline{x}_{ij.}-\overline{x}_{i..}-\overline{x}_{.j.}+\overline{x}_{...})=0.$$

We will show one more of the cross-product terms. Consider

$$\sum_{i=1}^{c}\sum_{j=1}^{r}\sum_{k=1}^{n}(\overline{x}_{i..}-\overline{x}_{...})(x_{ijk}-\overline{x}_{ij.})=\sum_{i=1}^{c}\sum_{j=1}^{r}(\overline{x}_{i..}-\overline{x}_{...})\left[\sum_{k=1}^{n}(x_{ijk}-\overline{x}_{ij.})\right]$$

but again $\sum_{k=1}^{n}(x_{ijk}-\overline{x}_{ij.})$ represents the sum of the deviations of the observations in the (i, j) cell around their mean and hence is zero.

The four remaining cross-product terms can be shown to be zero in a similar way. So we have

$$\sum_{i=1}^{c}\sum_{j=1}^{r}\sum_{k=1}^{n}(x_{ijk}-\overline{x}_{...})^2=\sum_{i=1}^{c}\sum_{j=1}^{r}\sum_{k=1}^{n}[(\overline{x}_{i..}-\overline{x}_{...})^2+(\overline{x}_{.j.}-\overline{x}_{...})^2$$

$$+(x_{ijk}-\overline{x}_{ij.})^2+(\overline{x}_{ij.}-\overline{x}_{i..}-\overline{x}_{.j.}+\overline{x}_{...})^2].$$

In this partition of the total sum of squares,

$$\sum_{i=1}^{c}\sum_{j=1}^{r}\sum_{k=1}^{n}(\overline{x}_{i..}-\overline{x}_{...})^2$$

represents the sum of squares due to the columns and is denoted by *SSC*;

$$\sum_{i=1}^{c}\sum_{j=1}^{r}\sum_{k=1}^{n}(\overline{x}_{.j.}-\overline{x}_{...})^2$$

represents the sum of squares due to the rows and is denoted by SSR;

$$\sum_{i=1}^{c}\sum_{j=1}^{r}\sum_{k=1}^{n}(x_{ijk} - \bar{x}_{ij.})^2$$

represents the sum of squares due to the variation within the cells and is denoted by SSE, giving the error sum of squares, and

$$\sum_{i=1}^{c}\sum_{j=1}^{r}\sum_{k=1}^{n}(\bar{x}_{ij.} - \bar{x}_{i..} - \bar{x}_{.j.} + \bar{x}_{...})^2$$

represents the sum of squares due to the interaction of the two primary factors, and is denoted by $SSR \cdot C$.

So we have

$$SST = SSC + SSR + SSR \cdot C + SSE.$$

This analysis of variance partition is usually exhibited in an analysis of variance table, which we show in Table 6.13.

Source	df	SS	MS	F	p
Rows	$r - 1$	SSR	$\dfrac{SSR}{r-1} = MSR$	$\dfrac{MSR}{MSW}$	
Columns	$c - 1$	SSC	$\dfrac{SSC}{c-1} = MSC$	$\dfrac{MSC}{MSW}$	
Interaction	$(r-1)(c-1)$	$SSR{\cdot}C$	$\dfrac{SSR \cdot C}{(r-1)(c-1)} = MSI$	$\dfrac{MSI}{MSW}$	
Error	$rc(n-1)$	SSW	$\dfrac{SSW}{rc(n-1)} = MSW$		
Total	$nrc - 1$	SST			

Table 6.13

As in the case of the one-way classification, if we assume that the ε_{ijk} are independently normally distributed with mean zero and variance σ^2, the sums of squares can be shown to be independently distributed according to chi-squared distributions and the ratios of the mean squares with the mean square for error can be shown to follow the F distribution. Proofs of these facts can be found in most texts on mathematical statistics.

Example 6.6.1   We return to the computer performance data given in Example 6.2.1 and exhibit them again in Table 6.14. The table shows the cell means, the row and column means, as well as the overall mean.

Statistical packages for computers, as well as Mathematica, compute the sums of squares needed for the analysis of variance table. We show here the arithmetic necessary

	Sp$_1$		**Sp$_2$**		
	30		16		
RAM$_1$	26		9		
	16	(24)	11	(12)	(18)
	22		6		
RAM$_2$	12		10		
	14	(16)	8	(8)	(12)
		(20)		(10)	
					[15]

Table 6.14

to give some idea of the amount of calculation involved.

$$SSE = \sum_{i=1}^{2} \sum_{j=1}^{2} \sum_{k=1}^{3} (x_{ijk} - \overline{x}_{ij.})^2 = (30 - 24)^2 + (26 - 24)^2 + \cdots + (8 - 8)^2 = 194$$

$$SSRAM = \sum_{i=1}^{2} \sum_{j=1}^{2} \sum_{k=1}^{3} (\overline{x}_{i..} - \overline{x}_{...})^2 = \sum_{i=1}^{2} 6(\overline{x}_{i..} - \overline{x}_{...})^2$$
$$= 6[(18 - 15)^2 + (12 - 15)^2] = 108$$

$$SSSp = \sum_{i=1}^{2} \sum_{j=1}^{2} \sum_{k=1}^{3} (\overline{x}_{.j.} - \overline{x}_{...})^2 = 6 \sum_{j=1}^{2} (\overline{x}_{.j.} - \overline{x}_{...})^2$$
$$= 6[(20 - 15)^2 + (10 - 15)^2] = 300$$

$$SSSp \cdot RAM = \sum_{i=1}^{2} \sum_{j=1}^{2} \sum_{k=1}^{3} (\overline{x}_{ij.} - \overline{x}_{i..} - \overline{x}_{.j.} + \overline{x}_{...})^2$$

$$= \sum_{i=1}^{2} \sum_{j=1}^{2} 3(\overline{x}_{ij.} - \overline{x}_{i..} - \overline{x}_{.j.} + \overline{x}_{...})^2$$

$$= 3[(24 - 20 - 18 + 15)^2 + (12 - 10 - 18 + 15)^2$$
$$+ (16 - 20 - 12 + 15)^2$$
$$+ (8 - 10 - 12 + 15)^2] = 12$$

$$SST = \sum_{i=1}^{2} \sum_{j=1}^{2} \sum_{k=1}^{3} (x_{ijk} - \overline{x}_{...})^2 = (30 - 15)^2 + (26 - 15)^2$$
$$+ \cdots + (8 - 15)^2 = 614$$

While the arithmetic in this case is not onerous, it can rapidly become so in the case of a large data set. It is also difficult—if not usually impossible—to calculate the *p*-values without the aid of a computer. We show all the results in the analysis of variance table (Table 6.15).

			ANALYSIS OF VARIANCE		
Source	**df**	**SS**	**MS**	**F**	**p**
RAM	1	108	108	4.45	0.068
Sp	1	300	300	12.37	0.008
Sp·RAM	1	12	12	0.49	0.504
Error	8	194	24.3		
Total	11	614			

Table 6.15

We calculated the size of the effects in Section 6.2, but had no way to make conclusions about their significance. The analysis of variance table now allows us to do this. From Section 6.2, we might conclude that the interaction, Sp · RAM, whose size was measured as +1, had little effect, and that is substantiated by the analysis of variance table. We might judge the size of the Sp effect, −5, as being significant, and the analysis of variance table substantiates this. However, the RAM effect was calculated to be −3, and while one might guess that this is of significance, the analysis of variance table gives its *p*-value as 0.068, of marginal significance. We conclude that the processor speed, Sp, is the only factor of significance and that the random access memory, RAM, can be ignored in further experimentation. ∎

An Added Advantage

One could take the two-way factorial data and analyze them as two one-way classifications, using only the factors RAM and Sp. Minitab gives analysis of variance tables, which we show in Tables 6.16 and 6.17.

Examination of these tables indicates that the sums of squares for the total and for the individual factors are exactly the same as those for the two-way factorial experiment; the one-way classifications add the error sum of squares from the two-way analysis of

			ANALYSIS OF VARIANCE		
Source	**df**	**SS**	**MS**	**F**	**p**
RAM	1	108	108	2.13	0.175
Error	10	506	50.6		
Total	11	614			

Table 6.16

ANALYSIS OF VARIANCE					
Source	df	SS	MS	F	p
Sp	1	300	300	9.55	0.011
Error	10	314	31.4		
Total	11	614			

Table 6.17

variance table to the interaction sum of squares. This produces somewhat different F values and corresponding p-values, but the incredible advantage the two-way factorial analysis has is that it gives all the information found by doing two one-way experiments, and, in addition, we find the interaction as well.

The calculation of interaction effects is of tremendous importance in experimentation because the interaction of factors is often most significant.

We have used a two-way factorial experiment for illustrative purposes, but many experiments involve more than two factors and often these factors occur at more than two levels. This produces a general factorial experiment, which we will consider in the next section.

EXERCISES 6.6

1. An industrial engineer engaged in an eye-focus experiment has five subjects and asks each to focus on an object at distances of 4, 6, 8, and 10 feet. The measurement made is the time for the eye to focus (which has been coded in the following data).

	Subject				
Distance	1	2	3	4	5
4	10	6	6	6	6
6	7	6	6	1	6
8	5	3	3	2	5
10	6	4	4	2	3

a. Since the distances are obviously different, analyze the data as a block experiment with the distances comprising the block. Show the analysis of variance and decide whether the subjects differ with respect to eye-focus time or not.

b. Analyze the data as a one-way experiment ignoring the blocks. What conclusion can be drawn from the data?

2. In a psychological experiment, some grade-school students were asked to stack some blocks in a certain pattern; the time the students took to do this was measured. The students were from grade levels kindergarten to fourth grade, and the number of blocks ranged from one to four. Two students were selected to participate in the experiment from each grade level. The data are shown in the accompanying table.

Number	Grade				
of Blocks	K	1	2	3	4
1	6.4	7.4	5.2	3.6	4.9
	9.1	5.0	4.9	5.2	3.7
2	7.9	8.6	6.3	4.9	3.2
	8.6	8.6	8.9	5.6	4.1
3	10.4	10.9	11.1	9.1	7.9
	12.1	11.4	13.6	8.6	3.4
4	19.2	13.6	14.0	13.6	6.9
	17.3	20.9	9.1	12.0	4.2

a. Analyze the data using the number of blocks to be stacked as the blocking factor. State the conclusions that can be drawn from this analysis.

b. Analyze the data as a one-way experiment ignoring the number of blocks.

3. Researchers studying the effects of cadmium on fish exposed fish to cadmium and recorded the number of aggressive encounters for ten fish (five in the control group and five exposed to cadmium), both before and after exposure. The accompanying data can be found in Aggress.mtw in Minitab.

	Control	Cadmium
	75.57	63.57
	15.57	24.00
Before	20.14	32.57
	5.86	22.29
	13.00	3.43
	90.71	59.00
	22.14	52.86
After	25.57	72.14
	22.00	28.86
	10.14	2.86

a. Analyze the data as a two-way classification. In a paragraph, write the conclusions that may be drawn concerning both the main effects and the interaction.

b. Show that the sums of squares in the analysis of variance table in part (a) for the main effects can be found by performing two one-way analyses of variance.

4. The effect of ozone on two varieties (here labeled *A* and *B*) of Azalea plants over the course of five weeks was the subject of an investigation by plant researchers. The data following, which can be found in azalea.mtw in Minitab, are measurements of leaf damage for the given combination of main factors.

	Week				
	1	2	3	4	5
	1.58	1.09	0.00	2.22	0.20
	1.62	1.03	0.00	2.40	0.40
	2.04	0.00	0.07	2.47	0.34
	1.28	0.46	0.18	1.85	0.00
A	1.43	0.46	0.40	2.50	0.00
	1.93	0.85	0.20	1.20	0.00
	2.20	0.30	0.63	1.33	0.10
	1.96	0.90	0.63	2.40	0.06
	2.23	0.00	0.56	2.23	0.17
	1.54	0.00	0.26	2.57	0.25

	Week				
	1	2	3	4	5
	1.29	0.00	0.78	0.40	0.00
	0.70	0.20	0.64	0.00	0.40
	1.93	0.00	1.00	0.20	0.47
	0.98	0.98	0.42	0.40	0.00
B	0.94	0.00	0.97	0.14	0.00
	1.06	0.62	2.43	0.44	0.00
	0.94	0.67	0.65	1.23	0.00
	1.65	0.00	0.00	0.35	0.00
	0.70	0.00	0.30	0.17	0.00
	0.35	0.00	0.00	0.20	0.00

a. Analyze the data as a two-way classification. In a paragraph, write the conclusions that may be drawn concerning both the main effects and the interaction.

b. Show that the sums of squares in the analysis of variance table in part (a) for the main effects can be found by performing two one-way analyses of variance.

5. The diameter of a product is of concern to the consumer. A manufacturer thinks that the workers' shift as well as the plant may influence the quality of the product. The following data show the results of two measurements taken during each of three different shifts at two different plants.

	Shift		
	1	2	3
Plant 1	64.6	65.1	64.7
	63.9	65.2	64.2
Plant 2	64.7	64.3	64.2
	64.2	64.2	64.1

Source: Minitab, Diameter.mtw

a. Analyze the data as a two-way classification. In a paragraph, write the conclusions that may be drawn concerning both the main effects and the interaction.

b. Show that the sums of squares in the analysis of variance table in part (a) for the main effects can be found by performing two one-way analyses of variance.

6. An investigation of a manufacturing operation involves data gathered on five different machines run on five different days. The entries are the number of units produced by each machine on a given day.

		Machine			
		I	II	III	IV	V

Day		I	II	III	IV	V
	1	173	243	302	312	218
	2	178	249	312	319	226
	3	160	306	291	341	291
	4	168	200	286	286	204
	5	140	236	266	309	282

a. Analyze the data using the factor Day as the blocking factor.
b. What conclusions can be drawn about the different machines? Are they producing equal numbers of units?

7. Three alloys are being tested for tensile strength at six different mold temperatures. The alloys have the same composition except for the amount of a single component.

Temp.		Alloy		
		A	B	C
	1	192	206	172
	2	199	219	182
	3	163	221	194
	4	184	199	173
	5	201	236	189
	6	182	200	194

a. Since the alloys are known to differ, use that factor as a blocking factor and analyze the data.
b. What conclusions can be drawn concerning the temperatures at which the alloys are studied? Do they produce different tensile strengths?

8. A manufacturing plant has three different shifts of workers. The management thinks that the shifts may be producing differing numbers of units. The following data show the number of units produced by each shift on the given day.

Day		Shift		
		A	B	C
	1	78	82	89
	2	62	81	72
	3	84	79	68
	4	79	72	73
	5	80	68	75

a. Analyze the data as a two-way classification.
b. What conclusions can be drawn from the analysis of variance?

9. An automobile testing service tests four different automobiles for gasoline mileage. The cars are run over different terrains and the gasoline mileage is recorded. The data are shown in the accompanying table.

Terrain		Car			
		I	II	III	IV
	A	20.6	28.1	19.2	21.1
	B	19.2	20.6	21.4	21.2
	C	21.9	23.0	22.0	29.3

a. Do the cars differ with respect to gasoline mileage?
b. Do the terrains produce different gasoline mileages?

10. A large discount chain is testing three brands of paint to determine which has the best wear characteristics. The test involves three wear methods. The accompanying data record millimeters of wear, appropriately coded.

Wear Method		Paint		
		A	B	C
	I	23	41	18
		19	26	19
		14	50	22
	II	51	38	31
		48	41	19
		53	52	24
	III	19	22	32
		16	16	21
		21	18	26

a. Analyze the data, using the wear method as a blocking variable.
b. Do the paints differ with respect to their wearing quality?

6.7 Factorial Experiments

We now turn to experiments where there are more than two factors. We illustrate the analysis through an example with three factors. Similar procedures can be used with experiments employing four or more factors.

Example 6.7.1 Czitrom, Sniegowski, and Haugh [40] report on Improving Integrated Circuit Manufacture Using a Designed Experiment. The experiment involved improving the etching process used in the manufacture of integrated circuits. Three variables, Bulk Gas Flow, CF_4 Flow, and Power (measured in watts) were varied simultaneously. We will be interested here in the response variable called Selectivity. The data are shown in Table 6.18. (Two of the observations have been omitted here.)

Selectivity	Bulk Gas Flow	CF_4 Flow	Power
10.93	60	5	550
19.61	180	5	550
7.17	60	15	550
12.46	180	15	550
10.19	60	5	700
17.50	180	5	700
6.94	60	15	700
11.77	180	15	700

Table 6.18

Note that the three factors, Bulk Gas Flow, CF_4 Flow, and Power, each occur at two levels. Note also that we have observed each of the possible $2^3 = 8$ combinations of these levels. When each level of each factor is observed with each level of each of the other factors, the experiment is called a **factorial experiment.** In particular, we will refer to this experiment as a 2^3 factorial experiment, meaning that there are three factors, each at two levels and that each possible combination of factor levels have been observed. Example 6.6.1 is a 2^2 factorial experiment.

We will first consider the data geometrically, then give a regression analysis, and finally provide an analysis of variance. The data are shown at the corners of a cube in Figure 6.6.

Like the analysis given for the data in Section 6.2, we can calculate three main effects, $\binom{3}{2} = 3$ interactions of the factors chosen two at a time, and $\binom{3}{3} = 1$ three-factor interaction. The overall mean of the data is found by adding up all the observations and dividing by 8. The result is 12.07.

Each of the corners of the cube has an attached sequence of plus or minus signs. These indicate the levels of the factors in the order (Bulk Gas, CF_4 Flow, Power) where for Bulk Gas Flow, 60 has been coded as -1 and 180 has been coded as $+1$, 5 for CF_4 Flow becomes -1 and 15 becomes $+1$, and finally 550 becomes -1 and 700 becomes $+1$ for the coding for Power. The coding is completely arbitrary, except for the fact that one level must be coded as -1 and the other level as $+1$. Interchanging some or all of these

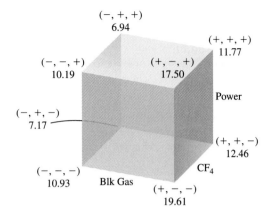

Figure 6.6 Data for a 2^3 Factorial Experiment

from the coding we have chosen will not alter the final result. As in Example 6.2.1, we seek to explain the corner observations as functions of the main effects and interactions. We proceed in a manner very similar to that in Example 6.2.1.

Consider now the main effect for Bulk Gas Flow. We take the mean of the observations for which Bulk Gas Flow has a positive sign and subtract the mean of the observations for which Bulk Gas Flow has a negative sign, and then divide the result by 2. This compares the mean of the observations in the right plane of the cube with the mean of those on the left plane of the cube. The result is

$$\text{Bulk Gas Flow} = \frac{1}{2}\left[\frac{17.5 + 11.77 + 12.46 + 19.61}{4} - \frac{10.19 + 6.94 + 7.17 + 10.93}{4}\right]$$

$$= 3.2638$$

In a similar way, we find that the main effect for Power is -0.471, while the main effect for CF_4 Flow is -2.4863. These compare the means of the observations in the top plane of the cube with those on the bottom plane and the means of the observations on the back plane of the cube with those on the front plane of the cube. Figure 6.7 shows these planes.

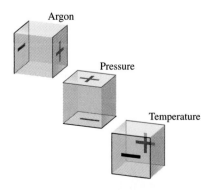

Figure 6.7 Planes Determining Main Effects

The two-factor interactions are also found by comparing means on planes, but these planes go from one edge of the cube to another edge. For example, to find the interaction of Bulk Gas Flow and Power, we compare the mean of the observations for which the product of the signs of Bulk Gas Flow and Power is positive with those for which the product is negative. The result is

Bulk Gas Flow · Power

$$= \frac{1}{2}\left[\frac{10.93 + 7.17 + 17.5 + 11.77}{4} - \frac{19.61 + 12.46 + 10.19 + 6.94}{4}\right] = -0.2288.$$

Similarly, one finds that the two-factor interaction Bulk Gas Flow · CF$_4$ Flow is -0.7338 and that of Power · CF$_4$ Flow is 0.2413.

Figure 6.8 shows the planes passing through the edges of the cube used for calculating the two-factor interactions.

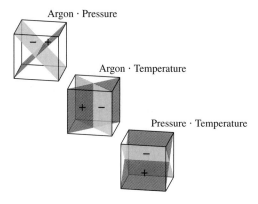

Figure 6.8 Planes for Two-Factor Interactions

Calculation of the main effects and the two-factor interactions then uses up all the planes it is possible to pass through the cube. Where then is the three-factor interaction Bulk Gas Flow · CF$_4$ Flow · Power? This interaction is found by comparing the observations where the product of all the signs at the corner is positive with those for which the product is negative. The result is

Bulk Gas Flow · CF$_4$ Flow · Power

$$= \frac{1}{2}\left[\frac{19.61 + 7.17 + 10.19 + 11.77}{4} - \frac{10.93 + 12.46 + 17.5 + 6.94}{4}\right] = 0.1138.$$

The corners for which the three-factor interaction is positive form a tetrahedron inside the cube as do the corners for which the three-factor interaction is negative. These are shown in Figure 6.9.

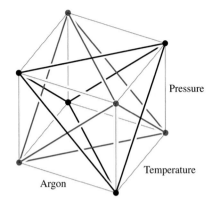

: Argon · Pressure · Temperature (+)
: Argon · Pressure · Temperature (−)

Figure 6.9 Tetrahedra Determining Three-Factor Interactions

These calculations tell us that the eight corner observations can be explained by the linear model

$$y = \mu + \beta_1 x_1 + \beta_2 x_2 + \beta_3 x_3 + \beta_{12} x_1 x_2 + \beta_{13} x_1 x_3 + \beta_{23} x_2 x_3 + \beta_{123} x_1 x_2 x_3$$

where the β's represent the main effects and interactions and the x's are either $+1$ or -1.

Specifically, then, this linear model becomes

$$y = 12.07 + 3.2638 x_1 - 2.4863 x_2 - 0.4713 x_3 - 0.7338 x_1 x_2$$
$$- 0.2288 x_1 x_3 + 0.2413 x_2 x_3 + 0.1138 x_1 x_2 x_3.$$

The model uses Bulk Gas Flow as factor 1, CF_4 Flow as factor 2 and Power as factor 3. The corner observations are explained exactly by the linear model. Consider, for example, the observation 10.19, which occurs where the signs are $(-, -, +)$. With $x_1 = -1$, $x_2 = -1$, and $x_3 = 1$, it then follows that

$$10.19 = 12.07 - 3.2638 + 2.4863 - 0.4713 - 0.7338 + 0.2288 - 0.2413 + 0.1138.$$

Similar results follow for the remaining set of observations.   ■

A Regression Model

Consider now the multiple regression model

$$y = \mu + \beta_1 x_1 + \beta_2 x_2 + \beta_3 x_3 + \beta_{12} x_1 x_2 + \beta_{13} x_1 x_3 + \beta_{23} x_2 x_3 + \beta_{123} x_1 x_2 x_3$$

where the x's are either $+1$ or -1 according to some assignment of the levels of the three factors. Our calculation of the β's geometrically produced a model that exactly predicts the observations at the corners of the cube.

We have a multiple regression model again, as we did in Section 6.2. Minitab and other computer statistical packages will produce the equation for y in terms of the x's exactly as we calculated it above.

Minitab will also produce an analysis of variance table for the regression; some details are shown in Appendix B. This is shown in Table 6.19 where the three-factor interaction serves as the error term; otherwise, the analysis of variance has an error term of zero since every one of the observations is exactly predicted by the model containing the three-factor interaction. Had there been more than one observation at each corner, we would need a model containing an error term such as

$$y_i = \mu + \beta_1 x_{1i} + \beta_2 x_{2i} + \beta_3 x_{3i} + \beta_{12} x_{1i} x_{2i} + \beta_{13} x_{1i} x_{3i}$$
$$+ \beta_{23} x_{2i} x_{3i} + \beta_{123} x_{1i} x_{2i} x_{3i} + \varepsilon_i.$$

We now show an example using this model.

		ANALYSIS OF VARIANCE			
Source	df	SS	MS	F	p
Regression	6	141.636	23.606	228.05	0.051
Error	1	0.104	0.104		
Total	7	141.740			

Table 6.19

Example 6.7.2   Buckner, Cammenga, and Weber [38] report on a designed experiment in Elimination of TiN Peeling During Exposure to CVD Tungsten Deposition Process Using Designed Experiments. Three factors, Temperature, Pressure, and Argon Flow, were studied at two levels each. The first observations of tungsten deposition rate are taken from the paper mentioned above. The next two sets of observations have been added in order to provide an example with replication; these data points do not substantially alter the conclusions given in the paper, but give us an example with replication. The data are shown in Table 6.20.

Argon Flow	Pressure	Temperature	Tungsten Deposition Rate
0	0.8	440	265, 301, 245
300	0.8	440	329, 300, 342
0	4.0	440	989, 1010, 994
300	4.0	440	1019, 1010, 1026
0	0.8	500	612, 589, 642
300	0.8	500	757, 784, 745
0	4.0	500	2236, 2310, 2289
300	4.0	500	2389, 2104, 2681

Table 6.20

The effects can be calculated geometrically in a manner exactly similar to that used in the previous example. These effects can also be found using a multiple regression model, which is the model used here.

Minitab gives the following multiple regression equation for the data in Table 6.20.

$$TungDepRate = 1082.00 + 41.83 ArgonFlow + 589.42 Pressure$$
$$+ 429.50 Temperature - 8.42 Argon \cdot Pressure$$
$$+ 23.33 Argon \cdot Temperature + 233.92 Pressure \cdot Temperature$$
$$- 0.25 Argon \cdot Pressure \cdot Temperature$$

The analysis of variance is given in Table 6.21.

ANALYSIS OF VARIANCE					
Source	df	SS	MS	F	p
Regression	7	14,135,151	2,019,307	185.15	0.000*
Error	16	174,497	10,906		
Total	23	14,309,648			

* Minitab gives this as 0.000; Mathematica calculates it as $4.22995 \cdot 10^{-14}$.

Table 6.21

The regression is of considerable significance here. Minitab also produces a table from which the significance, or lack of it, of each of the terms in the regression equation can be studied. This is given in Table 6.22. (For further research, the factors Pressure and Temperature would surely be retained while Argon might be dropped due to the lack of significance of this factor with the other two factors.)

Predictor	Coeff.	St.Dev.	t	p
Constant	1082.00	21.32	50.76	$4 \cdot 10^{-19}$
Argon	41.83	21.32	1.96	0.06765
Pressure	589.42	21.32	27.65	$6 \cdot 10^{-15}$
Temperature	429.50	21.32	20.15	$8.5 \cdot 10^{-13}$
Argon · Pressure	−8.42	21.32	−0.39	0.70168
Argon · Temperature	23.33	21.32	1.09	0.29186
Pressure · Temp	233.92	21.32	10.97	$7.5 \cdot 10^{-9}$
Argon · Press · Temp	−0.25	21.32	−0.01	0.99215

Table 6.22

Using the assumption that the error terms in the model are $NID(0, \sigma)$, it can be shown that the individual predictors are Student t variables, each with 1 degree of freedom.

Our estimate for σ^2 is the mean square for error in the analysis of variance. Since each predictor is a mean based on 24 observations, each has standard deviation

$$\sqrt{\frac{10,906}{24}} = 21.32.$$

It follows that a test for the hypothesis that the three-factor interaction, for example, is zero is

$$t_1 = \frac{-0.25 - 0}{\sqrt{\frac{10,906}{24}}} = -0.01.$$

The p-values in Table 6.22 are two-sided and were calculated using Mathematica. These results differ a bit from those found with Minitab. ■

We now show an analysis of variance approach to the factorial design.

Analysis of Variance

An analysis of variance partition of the total sum of squares,

$$SST = \sum_{i=1}^{2} \sum_{j=1}^{2} \sum_{k=1}^{2} \sum_{l=1}^{3} (y_{ijkl} - \bar{y}_{....})^2$$

can be derived in a manner similar to those done previously in this chapter. The data show a 2^3 factorial experiment with three observations for each combination of factor levels. The model we use is

$$y_{ijkl} = \mu + \alpha_i + \beta_j + \gamma_k + (\alpha\beta)_{ij} + (\alpha\gamma)_{ik} + (\beta\gamma)_{jk} + (\alpha\beta\gamma)_{ijk} + \varepsilon_{ijkl}$$

where μ is an overall mean, α_i, β_j, and γ_k represent the main effects of the first, second, and third factors, respectively, $(\alpha\beta)_{ij}$, $(\alpha\gamma)_{ik}$, and $(\beta\gamma)_{jk}$ symbolically represent the two-factor interactions (the effects are not multiplied together here), and in a similar way $(\alpha\beta\gamma)_{ijk}$ symbolically represents the three-factor interaction. The range for the indices i, j, and k is 1 to 2, while l ranges from 1 to 3. We will assume that the random errors, the ε_{ijkl}, are $NID(0, \sigma)$, meaning that they are independently distributed from a normal distribution with mean zero and variance σ^2.

The result is a sum of squares associated with each of the main effects and interactions. Minitab gives the result shown in Table 6.23.

The p-values here, which we calculated using Mathematica, should be exactly the same as those found using the regression model in Table 6.22 since $F[1, n] = t_n^2$. The slight differences are due to numerical round-off.

Each sum of squares in the analysis of variance derives from the effects that we calculated with the regression model. Each of these is a contrast. For example, the main effect for Argon is calculated as $1/2$ of the mean observations where Argon is at

Source	df	SS	MS	F	p
Argon	1	42,001	42,001	3.8512	0.067342
Pressure	1	8,337,888	8,337,888	764.52	$6 \cdot 10^{-15}$
Temperature	1	4,427,286	4,427,286	405.95	$8.5 \cdot 10^{-13}$
Argon · Pressure	1	1700	1700	0.1559	0.69817
Argon · Temp	1	13,067	13,067	1.1981	0.28991
Pressure · Temp	1	1,313,208	1,313,208	120.412	$7.4 \cdot 10^{-9}$
Argon · Pressure · Temp	1	1	1	$9.17 \cdot 10^{-5}$	0.99257
Error	16	174,497	10,906		
Total	23	14,309,648			

Table 6.23

the $+1$ level (or at 300) minus the mean of the observations where Argon is at the -1 level (or at 0):

$$\text{Argon} = \frac{1}{2}\left[\frac{329 + 300 + 342 + \cdots + 2681}{12} - \frac{265 + 301 + 245 + \cdots + 2289}{12} \right]$$

$$= 41.8\overline{3}.$$

The sum of squares associated with this contrast is then

$$SS\text{Argon} = \frac{(41.8\overline{3})^2}{24\left(\frac{1}{24}\right)^2} = 24(41.8\overline{3})^2 = 42{,}001.$$

The remaining sums of squares in the analysis of variance table can be calculated in a similar way.

In general, if a factor, say F, has f levels, then the sum of squares associated with it has $f - 1$ degrees of freedom. If another factor, say R, has r levels, then the sum of squares associated with the interaction of F and R has $(f - 1)(r - 1)$ degrees of freedom. The total sum of squares has $n - 1$ degrees of freedom if there are n observations in total. The degrees of freedom associated with the error is found by subtraction.

Some Visual Results

Minitab and other computer statistics packages produce some interesting and informative graphs for factorial experiments. It is possible to create graphs of the main effects and interactions and it is possible in many cases to detect significant effects visually. We show some graphs produced by Minitab; details for producing them can be found in Appendix B.

Figure 6.10 shows a plot of the main effects. The individual graphs show the change in the mean values as the level of the factor changes from -1 to $+1$. One can see that there is not much change for the factor Argon, but that there are substantial changes for the factors Pressure and Temperature. The actual measurement of the significance of

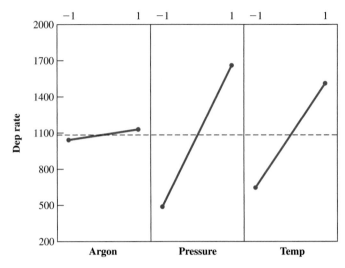

Figure 6.10 Main Effects Plot

these factors has been given in the analysis of variance table. Nonetheless the graph is informative.

Figure 6.11 shows a graph of the two-factor interactions. The dashed lines indicate the means at the $+1$ level of the factor shown on the vertical scale and the solid lines indicate the means at the -1 level of the factor shown on the vertical scale. The horizontal scale indicates the means at the -1 and $+1$ levels of the factor on the horizontal scale.

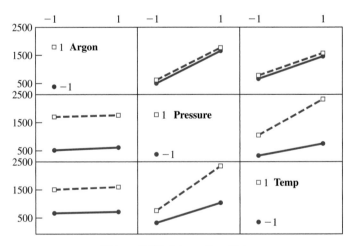

Figure 6.11 Interaction Plots

Parallel lines indicate no interaction; that is almost the case in the graph of the Argon · Pressure interaction as well as the Argon · Temperature interaction where the lines are close to being parallel. The graph of the Pressure · Temperature interaction, however, shows that there may be a significant interaction since the lines are far from parallel. As we have stated before, these conclusions might well be enough for an experimenter who can now safely eliminate Argon as a factor in future experiments.

The conclusions from both the plots of the main effects and those of the interactions can also be summarized in a **normal probability plot,** shown in Figure 6.12.

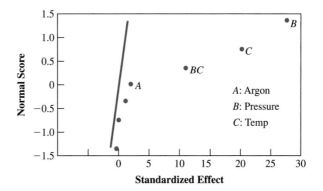

Figure 6.12 Normal Probability Plot of the Standardized Effects

The normal probability plot is constructed as follows. (It will be helpful to refer to Table 6.22.) First, the effects are standardized by dividing each by the standard deviation, 21.32 in this case. They are then arranged according to size in the StdEff column. We show these in Table 6.24.

Source	StdEff	i	Percentile	Normal Score
Argon · Pressure	−0.39	1	0.086207	−1.36449
Argon · Pressure · Temp	−0.01	2	0.224138	−0.75829
Argon · Temp	1.09	3	0.362069	−0.35293
Argon	1.96	4	0.500000	0
Pressure · Temp	10.97	5	0.637931	0.35293
Temperature	20.15	6	0.725862	0.758293
Pressure	27.65	7	0.913793	1.36449

Table 6.24

The column headed "i" gives the rank of each standardized effect. Percentiles are then assigned to each of the standardized effects. There are seven effects here and one might think that i/n would be the correct percentile to use for each of the effects. A correction is used for small samples, such as this one, and rather than use i/n, Minitab uses $(i - 3/8)/(n + 1/4)$. For the Argon · Temperature effect, this gives $(1 - 3/8)/(7 + 1/4) = 0.086207$.

The normal score is then the normal value that has that percentile of the data to its left on the standard $N(0, 1)$ curve. For the preceding example, this is -1.36449 since for a normal random variable Z, $P(Z < -1.36449) = 0.086207$. These scores are indicated in the column titled "Normal Score." The graph then plots these normal scores against the standardized effects.

If the data are selected from normal populations, the plot will approximate a straight line. Our graph, in Figure 6.12, shows that some of the points are more or less along a straight line while those apart from the straight line, namely, Pressure · Temperature, Temperature, and Pressure, are not along the line. This indicates that these may be significant effects. This is exactly the same conclusion we reached before, but this time we used graphs of the analysis.

Graphs are easily produced by computer statistical packages and should be an integral part of any statistical analysis of experimental data.

Comments

Factorial experiments can be conducted with factors at more than two levels, but we will not consider such experiments here. As the number of factors increases, the number of interactions also increases. A factorial experiment having s factors has $\binom{s}{2}$ two-factor interactions, $\binom{s}{3}$ three-factor interactions and, so on. There are then $2^s - 1$ main effects and interaction terms all together. While the main effects and perhaps second-order interactions are fairly simple to interpret, the higher order interactions are generally very difficult, if not impossible, to interpret. For this reason, such higher order interactions are not often computed and their sums of squares simply included in the error term in the model.

As the number of factors increases, the number of observations required increases as well. For example, a r^s experiment has s factors, each at r levels, so if we were to make n observations for each combination of factor levels, this would require $n \cdot r \cdot s$ observations, a number that may well be very large. It is a surprising fact that if some of the higher order interactions can be ignored, as we are suggesting, then, perhaps surprisingly, the number of observations can be greatly reduced by not observing all of the possible combinations of factor levels. The size, and so the cost, of the experiment is then greatly reduced while the loss of some information is then sustained. Such experiments are known as **fractional factorial** experiments. There are other experiments that are incomplete in the sense that not all the combinations of factor levels are observed. We will look at Latin Square designs and incomplete block designs first and then consider fractional factorial experiments in Section 6.10.

EXERCISES 6.7

1. For the data given in the first example in this section, produce main effect, and interaction plots and interpret them.

2. Sauter and Lenth [46] report on Experimental Design for Process Settings in Aircraft Manufacturing. In aircraft, the body of the plane, or the skin, as well as the wings, is

made of pieces of metal that are fastened together with rivets. An airframe then can contain many hundreds of holes drilled for the rivets. These holes must be drilled with precision. The object of this study was to investigate several factors influencing the quality of the holes drilled.

We will consider here the factors RPM (revolutions per minute of the drill at levels 6,000 RPM and 15,000 RPM), Air Temperature (at $30°$ and $50°$), and finally Lubricant Volume (in amounts 0.01 and 0.02 milliliters). The complete data set includes other factors and other levels of factors than those considered here, as well as other measurements on the drilled holes. The response we choose to analyze is the excess diameter in the drilled hole, which we abbreviate as XSDiameter. The data we will analyze are given below. Note that in this problem that there are multiple observations (three in this case) at each combination of factor levels.

RPM	Air Temperature	Lubricant Volume	XSDiameter
6,000	30	0.01	1.25
6,000	30	0.01	1.01
6,000	30	0.01	1.18
15,000	30	0.01	1.20
15,000	30	0.01	1.12
15,000	30	0.01	1.01
6,000	50	0.01	1.25
6,000	50	0.01	0.84
6,000	50	0.01	1.48
15,000	50	0.01	1.37
15,000	50	0.01	0.85
15,000	50	0.01	1.04
6,000	30	0.02	1.21
6,000	30	0.02	1.16
6,000	30	0.02	1.34
15,000	30	0.02	1.27
15,000	30	0.02	1.14
15,000	30	0.02	1.06
6,000	50	0.02	0.79
6,000	50	0.02	1.32
6,000	50	0.02	1.04
15,000	50	0.02	1.32
15,000	50	0.02	1.00
15,000	50	0.02	1.40

a. Analyze the data as a factorial experiment.

b. Show a regression model fit to the data.

3. A soft-drink manufacturer is concerned about the amount of soft-drink in a can. The factors influencing this amount are thought to be A: percent of carbonation in the soft drink, B: pressure in the filling device, and C: the speed of the production line. Each of these main factors occurs at two levels, which are indicated by $+1$ and -1. The following data are coded showing the deviation from the normal fill height of the can.

Deviation	A	B	C
−3	−1	−1	−1
−1	−1	−1	−1
0	1	−1	−1
1	1	−1	−1
−1	−1	−1	1
0	−1	−1	1
2	1	−1	1
1	1	−1	1
−1	−1	1	−1
0	−1	1	−1
2	1	1	−1
3	1	1	−1
1	−1	1	1
1	−1	1	1
6	1	1	1
5	1	1	1

a. Show the data at the corners of a cube and then find all the main effects and interactions geometrically.

b. Use a regression model to find the analysis of variance and interpret the results.

c. Test the main effects for significance.

4. An investigation in the automobile tire industry concerns the homogeneity of mixing a vinyl compound used in the manufacture of tires. The compound is placed in a vat and then mixed with heads driven by a motor. The response is the torque when the heads are running at constant speed. The main factors studied, each at two levels, are A: motor speed, B: temperature of the vat, C: whether or not the sample is preheated, and D: the weight of the sample. The data are as follows.

A	B	C	D	Response
-1	-1	-1	-1	1460
1	-1	-1	-1	1345
-1	1	-1	-1	990
1	1	-1	-1	1075
-1	-1	1	-1	1455
1	-1	1	-1	1330
-1	1	1	-1	1000
1	1	1	-1	1065
-1	-1	-1	1	2235
1	-1	-1	1	2070
-1	1	-1	1	1795
1	1	-1	1	1730
-1	-1	1	1	2540
1	-1	1	1	2100
-1	1	1	1	1700
1	1	1	1	1670

a. Find all the main effects and interactions geometrically. Since diagrams are impossible, use the plus and minus signs of the factors to determine these effects and interactions.

b. Show that factor C is not significant. Fit a regression model to the data ignoring this factor.

c. Show the analysis of variance for the factorial design, ignoring factor C.

5. The surface roughness of a part produced in a metal-cutting operation is being studied by a mechanical engineer. Three factors are studied, A: feed rate (in inches per minute), B: depth of cut, and C: tool angle in degrees). Each factor occurs at two levels and two replications are run. The data are as follows.

		Depth of Cut			
		0.025 in.		0.04 in.	
	Tool Angle	15°	25°	15°	25°
Feed Rate	20 in./min	9	11	9	10
		7	10	11	8
	30 in./min	10	10	12	16
		12	13	15	14

a. Show the data at the corners of a cube. Determine each of the main effects and interactions geometrically.

b. Fit a regression model to the data and state any conclusions that can be drawn from it.

6. A manufacturer of voltage regulators thinks that the quality of two regulators may differ. Factors in the study include A: setting station, B: test station, C: regulator, D: laboratory, and E: operator. A full 2^5 factorial experiment gave the following results.

A	B	C	D	E	Measure
−	−	−	−	−	14.2
+	−	−	−	−	13.6
−	+	−	−	−	18.1
+	+	−	−	−	17.0
−	−	+	−	−	18.6
+	−	+	−	−	14.9
−	+	+	−	−	20.2
+	+	+	−	−	18.6
−	−	−	+	−	13.9
+	−	−	+	−	14.6
−	+	−	+	−	21.4
+	+	−	+	−	18.6
−	−	+	+	−	19.0
+	−	+	+	−	14.0
−	+	+	+	−	16.6
+	+	+	+	−	17.1
−	−	−	−	+	16.2
+	−	−	−	+	21.9
−	+	−	−	+	16.6
+	+	−	−	+	16.7
−	−	+	−	+	18.9
+	−	+	−	+	20.2
−	+	+	−	+	26.1
+	+	+	−	+	24.0
−	−	−	+	+	23.2
+	−	−	+	+	18.9
−	+	−	+	+	13.7
+	+	−	+	+	18.0
−	−	+	+	+	21.3
+	−	+	+	+	22.4
−	+	+	+	+	13.9
+	+	+	+	+	18.6

a. Analyze the data using the main effects and second-order interactions in the linear model.

b. Test each of the main effects and second-order interactions for significance. Which of these effects are of greatest importance?

c. If only one factor could be chosen for the model, which one would that be? Why?

7. The percent conversion in a reactor is the subject of a study. The factors thought to be of significance are *A*: temperature, *B*: reactor, *C*: duration, and *D*: operator. The data follow.

A	B	C	D	Percent
−	−	−	−	26
+	−	−	−	31
−	+	−	−	42
+	+	−	−	36
−	−	+	−	18
+	−	+	−	14
−	+	+	−	12
+	+	+	−	22
−	−	−	+	19
+	−	−	+	31
−	+	−	+	42
+	+	−	+	36
−	−	+	+	18
+	−	+	+	41
−	+	+	+	40
+	+	+	+	39

a. Analyze the data using the main effects and second-order interactions in the linear model.

b. Test each of the main effects and second-order interactions for significance. Which of these effects are of greatest importance?

c. If only a few factors could be chosen for the model, which ones would be chosen? Why?

8. The quality of photographs taken are being judged as a function of the camera, film speed, and the presence or absence of a flash. The data are as follows.

A	B	C	Quality
−	−	−	81
+	−	−	84
−	+	−	75
+	+	−	68
−	−	+	92
+	−	+	83
−	+	+	72
+	+	+	91

a. Analyze the data using only the main effects in the linear model.

b. Test each of the main effects for significance. Which of these effects is of greatest importance?

6.8 Latin Square Designs

We now consider the first of designs that are incomplete in the sense that not every level of every factor is observed with every level of every other factor. When data are observed with every level of every factor with every level of every other factor, the data are said to be completely *crossed*. The factorial experiments considered previously in this chapter are examples of completely crossed data. It is possible, however, with properly designed experiments, to analyze data that are not completely crossed. We consider some of these designs in the next two sections before considering fractional factorial designs. We begin with a design that is called a **Latin square.**

An example of a Latin square is shown in Table 6.25.

The Latin square shown in Table 6.25 contains three factors (the numbers, the English letters, and the Greek letters) and has four rows and four columns. The rows represent a factor in the experiment as do the columns. The Greek letters in the square represent a third factor, but one whose four levels are not observed with each combination of levels of the other two factors. However, each Greek letter appears exactly once in each row and in each column. It is this fact that makes the square a Latin square. The combinations of factors indicate the observations that are to be made.

	A	*B*	*C*	*D*
1	α	β	γ	δ
2	δ	α	β	γ
3	γ	δ	α	β
4	β	γ	δ	α

Table 6.25

The square in Table 6.25 is a **cyclic square** because the rows are cyclic permutations of the first row, and, while cyclic Latin squares exist for any number of levels of the factors, Latin squares that are not cyclic can be constructed for different sizes of the square; one of these should be chosen at random before the experiment is performed. One way to do this is to construct the cyclic square of given order and then randomly permute both the rows and the columns. Note that this will not disturb the central property that the Greek letters appear exactly once in each row and column.

For example, a random number generator produced the following digits in the range from 1 to 4: 3, 2, 4, 1. We use these digits to rearrange the columns of the square. The random number generator then gave the digits 1, 3, 2, 4; these are then used to permute the rows. The result is the square shown in Table 6.26. The square then indicates the combinations of levels of the three factors to observe.

	A	*B*	*C*	*D*
1	γ	β	δ	α
2	α	δ	β	γ
3	β	α	γ	δ
4	δ	γ	α	β

Table 6.26

While the Latin square allows us to observe (in this case) three factors, the design does have considerable drawbacks, the major one being that interactions cannot be measured. Each factor must occur at exactly the same number of levels as well. Also, many statistical computer packages will not analyze the data directly. The total sum of squares can again be partitioned into independent sums of squares, one associated with each factor, as we have done for other designs. Each factor has $k - 1$ degrees of freedom if each factor has k levels.

We show an example.

Example 6.8.1   Snedecor and Cochran [30] show the data given in Table 6.27, which represents an experiment giving the yield in grams on plots of millet. The plants are arranged in a grid comprising 5 rows and 5 columns. It was thought that the field in which the plants were planted may have a gradient in fertility, which may occur along the rows of the field or along the columns, or, most usually, in an unknown direction. The Latin square allows the elimination of the effect of the gradient, regardless of its direction. The Greek letters indicate the primary factor of interest, namely, the spacings between the plants. These are: α (2 inches), β (4 inches), γ (6 inches), δ (8 inches), and ε (10 inches).

	A	B	C	D	E
1	β: 257	ε: 230	α: 279	γ: 287	δ: 202
2	δ: 245	α: 283	ε: 245	β: 280	γ: 260
3	ε: 182	β: 252	γ: 280	δ: 246	α: 250
4	α: 203	γ: 204	δ: 227	ε: 193	β: 259
5	γ: 231	δ: 271	β: 266	α: 334	ε: 338

Table 6.27

Now to analyze the data, we run one-way analyses on each factor. The results are combined in the analysis of variance table shown in Table 6.28.

			Analysis of Variance		
Source	df	SS	MS	F	p
Rows	4	13,601	3400	3.22	0.052
Columns	4	6,146	1537	1.46	0.275
Greek letters	4	4,157	1039	0.98	0.4544
Error	12	12,667	1056		
Total	24	36,571			

Table 6.28

Only the row factor here could possibly be considered to be significant, so we conclude specifically that the spacings between the plants is not a significant factor in predicting yields. Therefore our advice to millet growers might well be to plant the plants closely together in order to maximize yield. ∎

Orthogonal Latin Squares

It is sometimes possible to find Latin squares that are orthogonal to each other; that is, the intersections of rows and columns contain each pair of symbols exactly once. Such arrangements are called **Graeco-Latin squares.**

Table 6.29 shows one Graeco-Latin square for four treatments, each at four levels. We can then analyze each of the four treatments independently. The total sum of squares has 15 degrees of freedom; 12 of these are used by the treatments while the remaining 3 represent the degrees of freedom for the error. One must be careful not to run out of degrees of freedom for the error term if even more Latin squares are found orthogonal to each of the others. As the number of treatments increases, finding mutually orthogonal Latin squares becomes increasingly problematic and, in some cases, impossible.

	A	*B*	*C*	*D*
1	III γ	II β	IV δ	I α
2	II α	III δ	I β	IV γ
3	IV β	I γ	III α	II δ
4	I δ	IV α	II γ	III β

Table 6.29

6.9 Balanced Incomplete Block Designs

We now consider another design that is incomplete.

In Section 6.5, we examined blocking and the designs that are called block designs. In that section, the blocks were complete in the sense that every block contained every treatment. Often, however, the blocks are not of sufficient size to contain each of the treatments or the experimenter is concerned that the block is not homogeneous when the number of treatments is large and is concerned that this absence of homogeneity may introduce an unwanted feature into the experiment.

Following is an example of a balanced incomplete block experiment: We have six chemists who comprise the blocks in the experiment. We have four possibly different methods of measuring the amount of a certain chemical in a compound. The chemists, however, can use only two of the measuring techniques. The data are shown in Table 6.30.

Example 6.9.1 The chemists comprise the six blocks in the experiment. There are four treatments, or methods of measuring the component in the compound. We will summarize these statements by saying that $t = 4$ and $b = 6$. The balance in the experiment consists in the fact that every pair of treatments (there are $\binom{4}{2} = 6$ of them) appears in exactly one block. If the block size were larger, we could measure each of the pairs twice, or more, but this measure must be the same constant in each block. It is usually called λ. In this case, $\lambda = 1$. We call the number of treatments in a block k. Here $k = 2$. The quantity r is the number of times each treatment is replicated, so here $r = 3$.

The analysis of the data proceeds as follows. Some of the arithmetic involved can be done with Minitab, by first entering the data as they appear in Example 6.9.1, but using six columns (for the blocks) with two rows for each block, one row for each observation within the block. The one-way analysis will then produce the total sum of squares and the sum of squares for the blocks. These are $SSTotal = 570.92$ and $SSBlocks = 211.42$.

The sum of squares for the treatments cannot be found in the usual way since the data are incomplete. For this circumstance, the procedure is as follows. First, find the total for each of the treatments. These are given in Table 6.30. We abbreviate them as T_j, $j = 1, 2, 3, 4$. Second, for each treatment, find the sum of the treatments in all the blocks containing that treatment; these sums are called B_j, $j = 1, 2, 3, 4$. The data in Table 6.30 indicate, then, that

$$B_1 = 57 + 50 + 42 = 149$$

since treatment 1 occurs in blocks 1, 3, and 5.

		Treatment				Block (Chemist)
		1	**2**	**3**	**4**	**Totals**
Chemist	1	23	34			57
	2			14	29	43
	3	18		32		50
	4		16		14	30
	5	22			20	42
	6		18	31		49
Treatment Totals (T_j)		63	68	77	63	271

Table 6.30

In a similar way,

$$B_2 = 57 + 30 + 49 = 136,$$
$$B_3 = 43 + 50 + 49 = 142,$$

and, finally,

$$B_4 = 43 + 30 + 42 = 115.$$

Now we calculate for each treatment a quantity $Q_j = kT_j - B_j$. We find that

$$Q_1 = 2(63) - 149 = -23,$$
$$Q_2 = 2(68) - 136 = 0,$$
$$Q_3 = 2(77) - 142 = 12,$$

and

$$Q_4 = 2(63) - 115 = 11.$$

Finally, the sum of squares for the treatments, adjusted for the incompleteness of the data, is found by

$$SSTmts(Adjusted) = \frac{\sum_{j=1}^{t} Q_j^2}{\lambda kt}$$
$$= \frac{(-23)^2 + 0^2 + 12^2 + 11^2}{1 \cdot 2 \cdot 4}$$
$$= 99.25.$$

Now the analysis of variance table can be constructed; it is shown in Table 6.31. As it turns out, the treatments are not significant at all and can reasonably be judged to be equally effective in measuring the component in the compound. We have found the p-value for the blocks as well, but this is often not done (as we pointed out in Section 6.5).

ANALYSIS OF VARIANCE					
Source	df	SS	MS	F	p
Blocks	5	211.42	42.28	0.487	0.774837
Treatments (Adj)	3	99.25	33.08	0.381	0.775484
Error	3	260.25	86.75		
Total	11	570.92			

Table 6.31

Balanced incomplete blocks have become of great interest to experimenters, but, obviously, they are not always easy to construct given the number of treatments. Some restrictions that apply are:

1. $b \geq t$,

2. $tr = bk$,

3. $\lambda = \dfrac{r(k-1)}{t-1}$.

The reasons for these restrictions are:

1. The block size, b, must be large enough to observe all of the treatments, or more.

2. Both tr and bk represent the total number of observations.

3. Consider a specific treatment, say treatment i. It appears r times in blocks together with $k-1$ other treatments. So $r(k-1)$ represents the totality of other treatments that occur together with treatment i. However, λ represents the number of occurrences of each pair of treatments in the design. $\lambda(t-1)$ also represents the totality of other treatments that occur with a given treatment. Consequently, $\lambda(t-1) = r(k-1)$. Note also that λ must be a positive integer.

These conditions place considerable restriction upon the experimenter; nonetheless, incomplete block designs have become of great interest to scientific and engineering experimenters. Some work in constructing a design can be reduced by consulting a partial catalog of balanced incomplete block designs, which can be found in Cochran and Cox [39].

Latin square and incomplete block designs are two examples of designs that permit the analysis of incomplete data; that is, data that are not completely crossed. In the next section, we show another design that permits the analysis of incomplete data.   ■

6.10 Fractional Factorial Designs

We now show, perhaps surprisingly, the analysis of factorial experiments in which not each level of every factor is observed with each level of the remaining factors.

So far we have considered only factorial experiments where the factors are each at two levels only. These experiments are often used to screen out insignificant factors in preliminary experimentation by deliberately choosing both high and low levels of these factors. It is thought that if these prove to be insignificant, then the factor may be ignored in further experimentation. However, such 2^k experiments become large as k (the number of factors) increases. It takes 1024 observations for $k = 10$ and that allows for only one observation for every treatment combination. Cost, which is generally a direct function of the number of observations taken, is often very high in experiments of this size so there is motivation to reduce the size of the experiment. One way to do this is to reduce the number of factors studied, but, as we will see, the size of the experiment may be reduced without doing this. Interactions are another consideration. If $k = 10$, there are $2^{10} - 1 = 1023$ main effects and interactions, 968 of which involve three or more of the factors. Most, if not all, of these interactions may be very difficult, if not impossible, to interpret.

It is also generally true that, as the number of factors increases, the size of the interactions involving three or more factors is generally small in comparison to the size of the main effects and the two-factor interactions.

If it is not necessary to estimate some of these interactions, then the size of the experiment can be reduced dramatically. This is done by using only a portion, albeit a carefully selected portion, of the full factorial design. Such designs are called **fractional factorial designs.** We begin with an example.

Example 6.10.1   Consider the 2^3 factorial design using the factors A, B, and C, as shown in Figure 6.13. Each factor is considered to be at two levels, $+1$ and -1. The signs of the factors, in the order (A, B, C), are shown at the corners of the cube.

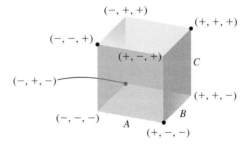

Figure 6.13 2^3 Factorial Design

Consider the three-factor interaction, ABC. The circled corners in Figure 6.13 are those for which the product ABC is positive; those not circled are those for which ABC is negative. The ABC effect is then found by taking the mean of the circled corners and subtracting the mean of the noncircled corners and dividing by 2.

We show the signs of the factors A, B, and C, as well as some purely symbolic observations using the letters s through z, in Table 6.32 for the full 2^3 factorial design.

Above the row of stars ($\star$) are the observations for which ABC is positive; below the row of stars are the observations for which ABC is negative. Suppose now that we have only the observations above the row of stars, that is, only the observations where ABC is positive. These are the observations at the circled corners in Figure 6.13.

A	B	C	Observations
+	−	−	s
−	+	−	t
−	−	+	u
+	+	+	v
⋆	⋆	⋆	⋆
+	+	−	w
−	−	−	x
+	−	+	y
−	+	+	z

Table 6.32

What effects can be estimated from this reduced set of observations? ABC cannot be estimated since we have only those observations for which ABC is positive. Consider the main effect A. It is estimated by one-half of the difference between the mean of the observations where A is positive and the mean of the observations where A is negative, or

$$\frac{1}{2}\left[\frac{(s+v)}{2} - \frac{(t+u)}{2}\right] = \frac{1}{4}[s+v-t-u].$$

Notice now that A and BC have exactly the same signs, so the contrast above used to estimate the A effect is exactly the same as that used to estimate the BC interaction. We say that the effects A and BC are *confounded* or that the effects A and BC are *aliases* of each other.

Further, consider the full set of observations and the sum of the effects $A + BC$. This total would be estimated by

$$\frac{1}{2}\left[\frac{s+w+y+v}{4} - \frac{u+x+t+z}{4}\right] + \frac{1}{2}\left[\frac{z+v+x+s}{4} - \frac{u+y+t+w}{4}\right]$$
$$= \frac{1}{4}[s+v-t-u],$$

which is exactly the calculation made to estimate either A or BC in the reduced set of observations.

The observations above the stars comprise a half fraction of the 2^3 factorial and are generally denoted by a 2^{3-1} factorial design. ∎

Generators and Confounding Patterns

The half fraction shown in Example 6.10.1 was generated by using the observations for which ABC is positive. The observations then have the product ABC as all $+1$'s. If we denote a column of $+1$'s as I, we could then abbreviate the observations in the one-half factorial as those observations for which

$$I = ABC.$$

This equation is called a **generator.**

Now it is also true, denoting the product of the elements of A with themselves as A^2, that $A^2 = I$. So we can multiply both sides of $I = ABC$ by A to get $A = A^2BC$, or $A = BC$. This gives the confounding pattern. It also indicates that there are other confounding patterns. If both sides of $I = ABC$ are multiplied by B, we obtain $B = AC$ and, multiplying through by C, $C = AB$.

It is easy to check that these effects are indeed confounded by the generator $I = ABC$. We conclude that the appropriate contrasts actually produce $A + BC$, $B + AC$, and $C + AB$.

A Geometric Fact

Consider again the observations in the 2^3 experiment for which ABC is positive. These are shown in the cube in Figure 6.13 with only the corners where ABC is positive circled.

Now observe, if the cube were to be collapsed along any of the three coordinate axes, that the result in each case would be a complete 2^2 factorial. Collapsing the cube along any coordinate axis eliminates that factor from the design. We conclude, in this case at least, that eliminating any single factor produces a full factorial experiment in the remaining factors. This is generally true of any fractional factorial design.

This can also be seen from Table 6.33, which shows the signs of the factors for which ABC is positive. Eliminating any one of the columns produces a full factorial experiment in the remaining factors.

A	B	C
+	−	−
−	+	−
−	−	+
+	+	+

Table 6.33

The fact that eliminating any one of the factors in a factorial 2^{k-1} design produces a full factorial in all the remaining factors is often used as a justification for using a 2^{k-1} design. If one of the factors proves to be unnecessary in the experiment, one has all the advantages of a full factorial experiment in the remaining factors.

A Caution

The 2^{k-1} design shown above was illustrated because both the geometry and the algebraic analysis are easy to see. Fractional factorial designs are frequently used if the higher order interactions can be ignored. However, it is usually important to be able to estimate both the main effects and at least some of the lower order interactions.

Our example confounds the main effects with the second-order interactions. This generally should not be done; the example was selected since it is a simple one in which to see the confounding patterns as well as the geometry. We proceed with examples that must necessarily have more than three factors in order to avoid this pitfall.

Example 6.10.2   Barnett, Czitrom, John, and Leon [34] report on Using Fewer Wafers to Resolve Confounding in Screening Experiments. Computer chips are created on wafers, which are round, shiny pieces of silicon. Chemicals are deposited on the wafers and removed with acid to reveal the chip's circuitry and then the wafers are flushed with water. The experiment involved a new etching process in producing wafers. Six factors—revolutions per minute (A), Pre-etch total flow (B), Pre-etch vapor flow (C), Etch total flow (D), Etch vapor flow (E), and Amount of oxide etched (F) were studied along with a number of responses. Each factor was observed at two levels. The data, showing only *LogEtU* here, are given in Table 6.34.

A	B	C	D	E	F	*LogEtU*
−	−	−	−	−	−	2.40
−	+	−	−	+	+	2.31
−	−	+	−	+	+	2.16
−	+	+	−	−	−	2.22
−	−	−	+	−	+	1.16
−	+	−	+	+	−	1.59
−	−	+	+	+	−	1.76
−	+	+	+	−	+	1.06
+	−	−	−	+	−	1.13
+	+	−	−	−	+	1.28
+	−	+	−	−	+	1.28
+	+	+	−	+	−	2.04
+	−	−	+	+	+	−0.22
+	+	−	+	−	−	3.71
+	−	+	+	−	−	4.26
+	+	+	+	+	+	0.41

Table 6.34

Since there are six factors, a full factorial experiment would require $2^6 = 64$ runs if every factor combination were to be observed only once. The data show a one-quarter fraction of the 2^6 design, or a 2^{6-2} design. This is produced by using the generators $E = ABC$ and $F = BCD$.

The alias structure, showing all the confounding patterns, is then

$I + ABCE + ADEF + BCDF;$

$A + BCE + DEF + ABCDF;$

$B + ACE + CDF + ABDEF;$

$C + ABE + BDF + ACDEF;$

$D + AEF + BCF + ABCDE;$

$E + ABC + ADF + BCDEF;$

$$F + ADE + BCD + ABCEF;$$
$$AB + CE + ACDF + BDEF;$$
$$AC + BE + ABDF + CDEF;$$
$$AD + EF + ABCF + BCDE;$$
$$AE + BC + DF + ABCDEF;$$
$$AF + DE + ABCD + BCEF;$$
$$BD + CF + ABEF + ACDE;$$
$$BF + CD + ABDE + ACEF;$$
$$ABD + ACF + BEF + CDE; \text{ and}$$
$$ABF + ACD + BDE + CEF.$$

Now we examine the significance, if any, of the individual effects. Minitab produces the effects as shown in Table 6.35. Minitab analyzes full as well as fractional factorial designs. The generators in this case are the default generators although Minitab allows other generators to be specified if the experimenter desires.

Term	Effect	SS
A	-0.0481	0.0370
B	0.0431	0.0297
C	0.1144	0.2094
D	-0.0681	0.0742
E	-0.3869	2.3951
F	-0.6044	5.8448
$A \cdot B$	0.0806	0.1039
$A \cdot C$	0.1469	0.3453
$A \cdot D$	0.3719	2.2130
$A \cdot E$	-0.5094	4.1518
$A \cdot F$	-0.4444	3.1599
$B \cdot D$	-0.0669	0.0716
$B \cdot F$	0.0419	0.0281
$A \cdot B \cdot D$	-0.0369	0.0218
$A \cdot B \cdot F$	-0.0081	0.0010

Table 6.35

Since the geometry would be four dimensional, the effects are now found from the plus and minus signs in Table 6.34. The A effect, for example, is $1/2$ of the mean of the eight observations where A is plus, minus the mean of the eight observations where A is minus. Each effect is based on 16 observations, so the sum of squares associated with each contrast is the contrast squared divided by $16(1/16)^2 = 1/16$. The sum of squares associated with each effect is then 16 times the square of the effect. Table 6.35 shows the effects and their associated sum of squares.

The analysis of variance on these data produces an error term of zero since the model exactly predicts each of the observations. In Table 6.36 we show an analysis of variance where the three-factor interactions (together with their confounding effects) have been used as an error term.

Term	Effect	df	SS	MS	F	p
A	−0.0481	1	0.0371	0.0371	3.2544	0.2130
B	0.0431	1	0.0297	0.0297	2.6049	0.2479
C	0.1134	1	0.2058	0.2058	18.0327	0.0512
D	−0.0681	1	0.0742	0.0742	6.5032	0.1255
E	−0.3869	1	2.3951	2.3951	209.909	0.0047
F	−0.6044	1	5.8448	5.8448	512.252	0.0019
A · B	0.0806	1	0.1039	0.1039	9.1097	0.0945
A · C	0.1469	1	0.3453	0.3453	30.2606	0.0315
A · D	0.3719	1	2.2130	2.2130	193.949	0.0051
A · E	−0.5094	1	4.1518	4.1518	363.875	0.0027
A · F	−0.4444	1	3.1599	3.1599	276.938	0.0036
B · D	−0.0669	1	0.0716	0.0716	6.2761	0.1292
B · F	0.0419	1	0.0281	0.0281	2.4619	0.2572
Error		2	0.0228	0.0114		
Total		15	18.6821			

Table 6.36

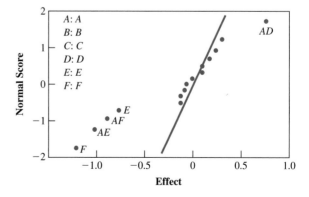

Figure 6.14 Normal Probability Plot of the Effects (Response is LogEtU, Alpha = 0.10)

Certainly, the factors E, F, AD, AE, and AF would be judged significant. These involve four of the main factors, so a subsequent investigation might involve these factors alone. This conclusion regarding the significance of the factors E, F, AD, AE, and AF is borne out by the normal probability plot, the main effects plot, and the interaction plot, shown in Figures 6.14, 6.15, and 6.16, respectively.

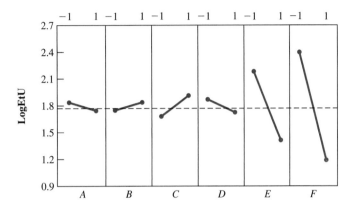

Figure 6.15 Main Effects Plot (data means) for LogEtU

One must be cautious when interpreting the interaction plots since other effects are confounded with them, leaving conclusions about individual interactions questionable. In any event, the pictures are of great value in supporting the conclusions made from the analysis of variance table. ∎

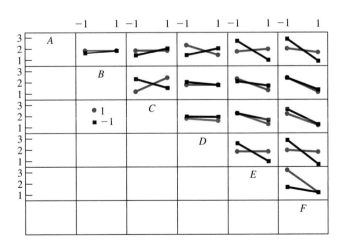

Figure 6.16 Interaction Plot (data means) for LogEtU

Resolution

We saw in Example 6.10.1 that the confounding pattern was $I = ABC$, while the confounding pattern in Example 6.10.2 was $I = ABCE = BCDF$. If the confounding pattern is of the form $I = A_1 A_2 \cdots A_k$ where k is the smallest order of an interaction in the defining relationship, then we say the design is of **resolution k**. Usually, k is expressed in Roman numerals. In Example 6.10.1, with the generator $I = ABC$, the design is then of resolution III, and in Example 6.10.2, where the generators are $I = ABCE = BCDF$, the design is of resolution IV. The higher the resolution, the higher the order of interactions that are confounded with each other.

A design of resolution III then confounds main effects with second-order interactions; a design of resolution IV confounds two-factor interactions with each other, but does not confound main effects with two-factor interactions. A design of resolution V confounds three-factor interactions with two-factor interactions, but does not confound main effects with either two- or three-factor interactions.

Factorial designs of resolution k are often denoted by the symbol 2_k^n where k is the resolution and is expressed in Roman numerals. Example 6.10.1 in the literature would be regarded as a 2_{III}^3 design and Example 6.10.2 would be described as a 2_{IV}^4 design.

Minitab allows the design of a fractional factorial experiment and reports the resolution of the design.

Variances

In Section 6.3 we pointed out that the weighing designs provided not only unbiased estimators, but also estimators that had smaller variances than one might have anticipated. The factorial designs enjoy similar properties. Note that all the main effects and interactions are contrasts and that each observation has variance σ^2.

Consider a 2^k factorial experiment. Each of the main effects and interactions is of the form

$$\frac{1}{2}\left[\frac{N_1}{2^{k-1}} - \frac{N_2}{2^{k-1}}\right]$$

where the numerators of each of the fractions contain 2^{k-1} terms. Each of these is a contrast. The variance of this contrast is then

$$\frac{1}{4}\left(\frac{1}{2^{k-1}}\right)^2 \cdot \left[2^{k-1} + 2^{k-1}\right] \cdot \sigma^2 = \frac{\sigma^2}{2^k},$$

a quantity that becomes small as k increases.

One then would have to have 2^{k-1} observations at both the plus and minus level in a one-factor-at-a-time experiment to achieve the same efficiency as the fractional factorial experiment and then one would not be able to estimate the interactions or the effects of any of the remaining factors.

Fractional factorial experiments clearly have tremendous advantages in reducing the size, and hence the cost, of factorial experiments.

EXERCISES 6.10

1. A study was made of five different formulations of an explosive mixture used in the manufacture of dynamite and the resultant explosive force. Five operators comprise another factor. Since interactions are not of interest, a cyclic Latin square design was used. English letters refer to operators, Latin numerals refer to batches, and Greek letters refer to formulations. The data are as shown in the accompanying table.

		Operators				
		A	*B*	*C*	*D*	*E*
	I	24(α)	20(β)	19(γ)	24(δ)	24(ε)
	II	17(β)	24(γ)	30(δ)	27(ε)	36(α)
Batches	III	18(γ)	38(δ)	26(ε)	27(α)	21(β)
	IV	26(δ)	31(ε)	26(α)	23(β)	22(γ)
	V	22(ε)	30(α)	20(β)	29(γ)	31(δ)

a. Analyze the data showing the analysis of variance. Write all the conclusions that can be made in a paragraph.
b. Show that only one of the factors is of significance. Show a one-way analysis of variance with this as the single factor.

2. A wear test on fabric measured the extent to which a fabric gave homogeneous results over its surface. The variables were: English letters: sample, Latin numerals: positions, and Greek letters: type of fabric. The data follow.

		Sample			
		A	*B*	*C*	*D*
	I	26.1(α)	20.0(δ)	32.4(γ)	20.7(β)
Position	II	22.3(β)	26.9(α)	21.5(δ)	35.0(γ)
	III	29.0(γ)	22.4(β)	26.3(α)	22.8(δ)
	IV	30.6(δ)	28.8(γ)	25.2(β)	26.4(α)

a. Analyze the data showing the analysis of variance. Write all the conclusions that can be made in a paragraph.
b. Show that only one of the factors is of significance. Show a one-way analysis of variance with this as the single factor.

3. A quality improvement team decides to use a designed experiment in order to study an injection molding process so that shrinkage might be reduced. There are six factors to be studied: *A*: mold temperature, *B*: screw speed, *C*: holding time, *D*: cycle time, *E*: gate size, and *F*: holding pressure. Each factor has two levels, but cost constraints dictate that only 16 runs can be performed. The team decides to use a fractional factorial experiment with confounding patterns $E = ABC$ and $F = BCD$. The data are as follows.

A	*B*	*C*	*D*	$E=ABC$	$F=BCD$	**Response**
−	−	−	−	−	−	6
+	−	−	−	+	−	10
−	+	−	−	+	+	32
+	+	−	−	−	+	60
−	−	+	−	+	+	4
+	−	+	−	−	+	15
−	+	+	−	−	−	26
+	+	+	−	+	−	60
−	−	−	+	−	+	8
+	−	−	+	+	+	12
−	+	−	+	+	−	34
+	+	−	+	−	−	60
−	−	+	+	+	−	16
+	−	+	+	−	−	5
−	+	+	+	−	+	37
+	+	+	+	+	+	52

a. Show that if any two columns are eliminated from the design, then the result is a full factorial design in the remaining four factors.
b. Show the entire alias structure.
c. Analyze the data using the main effects and second-order interactions only.

4. Box and Hunter [35] in a paper "The 2^{k-p} Fractional Factorial Designs," give an example of a new manufacturing unit that encountered considerable difficulty in a filtration stage. The group decided to investigate the following variables: *A*: water supply, *B*: temperature of filtration, *C*: hold-up time, *D*: recycle, *E*: rate of addition of caustic soda, and *F*: type of filter cloth. Each of these seven variables was studied at two levels, but a full factorial experiment was not possible, so the confounding pattern $D = AB$, $E = AC$, $F = BC$, and

$G = ABC$ was chosen in order to produce a fractional factorial design. The data are as follows.

A	B	C	D	E	F	G	Response
−	−	−	+	+	+	−	68.4
+	−	−	−	−	+	+	77.7
−	+	−	−	+	−	+	66.4
+	+	−	+	−	−	−	81.0
−	−	+	+	−	−	+	78.6
+	−	+	−	+	−	−	41.2
−	+	+	−	−	+	−	68.7
+	+	+	+	+	+	+	38.7

a. Determine the resolution of the design.
b. Analyze the data. Note that the factors A, C, and E appear to be those of primary significance.
c. Analyze the data using only the factors A, C, and E and state the conclusions that can be drawn.

5. The owner of a chain of dry-cleaning stores has five competing dry-cleaning solvents to test. Ten different fabrics are available, but since large samples of fabrics may differ with respect to their staining and dirt, each piece of fabric can only be cut into two pieces. It is then not possible to test each solvent with each type of fabric. The following data are gathered where the observations are measures of the relative cleanliness of the fabric after cleaning with the indicated solvent.

	Solvent				
Fabric	A	B	C	D	E
1	3	5			
2	2		9		
3	1			7	
4	3				8
5		4	4		
6		2		6	
7		9			8
8			10	2	
9			1		9
10				4	2

a. Explain why the design is a balanced incomplete block design. State the values of b, k, t, r, and λ.
b. Analyze the data and state the conclusions that can be drawn.

6. Seven types of lawn fertilizers are being compared with respect to their efficacy in nourishing lawn grass. Due to possible gradients in the plots of land available for the tests, only three fertilizers can be tested on a given plot of land. In the following data, the observations are relative measures of the quality of the fertilizer.

	Fertilizer						
Plot	A	B	C	D	E	F	G
1	3	5	2				
2	1			8	4		
3	2					7	1
4		6		9		8	
5		1			4		2
6			0	5			9
7			9		10	6	

a. Explain why the design is a balanced incomplete block design. State the values of b, k, t, r, and λ.
b. Analyze the data and state the conclusions that can be drawn.

7. A popular classroom experiment illustrating principles of experimental design involves the construction of paper airplanes. There are eight factors that can be altered for the airplanes: A: type of paper, B: body width, C: body length, D: wing length, E: presence or absence of a paper clip, F: fold, G: taped body, and H: taped wing. Each of these factors can be set at one of two levels. A full factorial experiment would involve $2^8 = 256$ observations if only one observation were made for each factor combination. The class decided to use a fractional factorial with $E = BCD$, $F = ACD, G = ABC$, and $H = ABD$ as the confounding pattern. The data follow. The observations are the mean of five distances the airplanes flew.

A	B	C	D	E	F	G	H	Response
−	−	−	−	−	−	−	−	3.114
+	−	−	−	−	+	+	+	4.217
−	+	−	−	+	−	+	+	3.523
+	+	−	−	+	+	−	−	4.002
−	−	+	−	+	+	+	−	4.185
+	−	+	−	+	−	−	+	4.004
−	+	+	−	−	+	−	+	3.175
+	+	+	−	−	−	+	−	3.024
−	−	−	+	+	+	−	+	3.319

A	B	C	D	E	F	G	H	Response
+	−	−	+	+	−	+	−	3.307
−	+	−	+	−	+	+	−	3.507
+	+	−	+	−	−	−	+	3.422
−	−	+	+	−	−	+	+	4.172
+	−	+	+	−	+	−	−	3.816
−	+	+	+	+	−	−	−	3.537
+	+	+	+	+	+	+	+	3.265

a. Determine the resolution of the design.
b. Show the full confounding pattern.
c. Explain why the main effects and two factor interactions cannot be used as factors in an analysis of variance.
d. Analyze the data using the main effects only and state the conclusions that can be drawn.

8. The quality of a product produced by a machine tool is known to be affected by four factors: A: cutting speed, B: time the machine has been used, C: tool design, and D: vibration level. In order to study the effects of each of these factors, a fractional factorial experiment was performed using the confounding pattern $D = ABC$. The data follow where the measurement is one of the quality of the product produced.

A	B	C	D	Quality
−	−	−	−	6.4
+	−	−	+	13.2
−	+	−	+	7.1
+	+	−	−	9.3
−	−	+	+	12.6
+	−	+	−	20.9
−	+	+	−	14.1
+	+	+	+	21.7

a. Determine the resolution of the design.
b. Show the full confounding pattern.
c. Explain why the main effects and two factor interactions cannot be used as factors in an analysis of variance.
d. Analyze the data using the main effects only and state the conclusions that can be drawn.

9. A market research firm would like to compare six different market survey instruments in ten different markets. The firm can only test three of the instruments in each market. The data following show a measure of the effect of the instrument in the indicated market.

		A	B	C	D	E	F
	1	10	4			6	
	2	8	3				2
	3	9		9	1		
	4	7		8			2
	5		6	4	6		
Area	6		0	3		9	
	7		5		4		8
	8			2		7	4
	9				6	5	5
	10	1			8	6	

a. Explain why the design is a balanced incomplete block design. State the values of b, k, t, r, and λ.
b. Analyze the data and state the conclusions that can be drawn.

Chapter Review

The design of experiments was the subject of this chapter. Carefully designed experiments can yield important information on the significance of factors in an experiment as well as their interactions with each other if the data are gathered properly in accordance with the principles discussed in this chapter.

Key Concepts

- Identifying the factors in an experiment
- Measuring the size of the main effects and the interactions geometrically
- Identifying the model as a multiple regression model
- Comparing weighing designs to determine unknown weights
- Identifying and analyzing a one-way classification
- Testing contrasts for significance
- Identifying a blocking factor in a two-way classification
- Identifying and analyzing a two-way classification
- Identifying and analyzing a 2^3 factorial design
- Identifying and analyzing a Latin square design
- Analyzing an incomplete block design
- Identifying the generators for a fractional factorial design
- Analyzing a fractional factorial design

Key Terms

Factor	A *factor* in an experiment is a cause for variation in experimental results.
Factor Levels	The *level* of a *factor* is an identification of the specific size of a factor.
Main Effect	A *main effect* is the size of the influence of a single factor.
Interaction	An *interaction* is the size of the influence of two or more main effects together.
One-Way Classification	A *one-way classification* is an arrangement of data in columns corresponding to the levels of a factor in an experiment.
Analysis of Variance Identity	The *analysis of variance identity* is the separation of sums of squares for an experimental design.
Contrast	A *contrast* is a linear expression involving the sums of the columns in a one-way classification.
Two-Way Classification	A *two-way classification* is an arrangement of data in which both rows and columns represent the levels of two main factors.
Block	A *block* is a factor in an experiment which is not analyzed for significance.

Factorial Experiment	In a *factorial experiment* every level of each factor is observed with every level of each of the remaining factors.
Fractional Factorial Experiment	In a *fractional factorial experiment,* not every level of each factor is observed with every level of each of the remaining factors.
Generator	A *generator* is an expression which determines how the main effects and interactions are confounded in a fractional factorial experiment.
Confounding	*Confounding* occurs when some main effects and interactions cannot be distinguished in a fractional factorial experiment.
Latin Square	A *Latin Square* is an incomplete design using Greek letters in a two-way table to represent a third factor.
Othogonal Latin Squares	Two or more *Latin squares* are *orthogonal* when every combination of letters occurs once and only once.
Incomplete Block Experiment	In an *incomplete block experiment* not every level of the factors is observed in every block.

Key Theorems/Facts

Finding Effects Geometrically	Data have been arranged on the corners of a square or at the corners of a cube. Main *effects* and *interactions* are found using the edges of the square or the planes passing through the cube.
Multiple Regression Model	The *multiple regression model* $$y = \mu + \beta_1 x_1 + \beta_2 x_2 + \beta_3 x_3 + \varepsilon$$ where $\varepsilon \sim N(0, \sigma)$ can be used in the design of experiments where the $x's$ are -1 or $+1$ and the $\beta's$ are the main effects or interactions. Many models involve more terms.
One-Way Classification	In a *one-way classification,* data are arranged in columns representing the levels of a single factor. The analysis of

variance identity

$$\sum_{i=1}^{k}\sum_{j=1}^{n_i}\left(x_{ij}-\overline{\overline{x}}\right)^2 = \sum_{i=1}^{k}\sum_{j=1}^{n_i}\left(x_{ij}-\overline{x}_i\right)^2 + \sum_{i=1}^{k}\sum_{j=1}^{n_i}\left(\overline{x}_i-\overline{\overline{x}}\right)^2$$

is the basis for the analysis of variance table which can be found on page 294. The identity is abbreviated as $SST = SSW + SSB$.

Testing Contrasts

A *contrast* or *comparison* among treatment totals is of the form

$$C_p = \sum_{i=1}^{k} c_{ip} T_i = c_{1p} T_1 + c_{2p} T_2 + \cdots + c_{kp} T_k$$

where $\sum_{i=1}^{k} n_i c_{ip} = 0$. The sum of squares associated with this contrast is $SSC_p = C_p^2 / \sum_{i=1}^{k} n_i c_{ip}^2$. It is a fact that the factor sum of squares in the analysis of variance is completely partitioned by a set of $(k-1)$ orthogonal contrasts where k is the number of treatments.

Two-Way Classification

In the *two-way classification*, both the rows and the columns represent factors of interest. The linear model for this situation is

$$x_{ijk} = \mu + \alpha_i + \beta_j + \delta_{ij} + \varepsilon_{ijk}$$

$$\text{for} \begin{cases} i = 1, 2, \ldots, c \\ j = 1, 2, \ldots, r \\ k = 1, 2, \ldots, n \end{cases}$$

The analysis of variance identity is

$$\sum_{i=1}^{c}\sum_{j=1}^{r}\sum_{k=1}^{n}(x_{ijk}-\overline{x}...)^2 = \sum_{i=1}^{c}\sum_{j=1}^{r}\sum_{k=1}^{n}[(\overline{x}_{i..}-\overline{x}...)^2 + (\overline{x}_{.j.}-\overline{x}...)^2$$
$$+ (x_{ijk}-\overline{x}_{ij.})^2 + (\overline{x}_{ij.}-\overline{x}_{i..}-\overline{x}_{.j.}+\overline{x}...)^2].$$

The analysis of variance table is shown on page 310.

Factorial Experiments

A *factorial experiment* with k factors each at two levels, can be modeled by

$$y_{ijkl} = \mu + \alpha_i + \beta_j + \gamma_k + (\alpha\beta)_{ij} + (\alpha\gamma)_{ik}$$
$$+ (\beta\gamma)_{jk} + (\alpha\beta\gamma)_{ijk} + \varepsilon_{ijkl}$$

where μ is an overall mean, α_i, β_j, and γ_k represent the main effects of the first, second, and third factors respectively, $(\alpha\beta)_{ij}$, $(\alpha\gamma)_{ik}$, and $(\beta\gamma)_{jk}$ symbolically represent the two-factor interactions (the effects are not multiplied together here) and $(\alpha\beta\gamma)_{ijk}$ similarly symbolically represents the three-factor interaction. An example can be found on page 316.

Latin Square Design

The defining characteristic of the *Latin square* is that each Greek letter appears exactly once in each row and each column in a two-way array. The main effects can be estimated, but it is not possible to estimate any of the interactions. Occasionally it is possible to analyze another factor if two orthogonal squares can be found. As the number of levels of each factor increases, this becomes increasingly difficult. An example can be found on page 330.

Fractional Factorial Design

A *fractional factorial design* consists of part of a full factorial experiment. The observations are determined by a defining relation or generator. A one-half fraction of the observations required for a full 2^3 experiment could be determined by $ABC = I$ where I is a column of $+1's$. The relation $ABC = I$ is called the *defining* relation for the fractional factorial. It is then possible to estimate the effects, but they are confounded with each other. For this example, the main effect A is confounded with the second order interaction, BC. These designs are generally used when the higher order interactions, which are difficult to interpret as a rule, are of little interest and can be ignored.

7 Statistical Process Control

7.1 Introduction

STATISTICS CAN IMPROVE THE QUALITY OF CONSUMER PRODUCTS produced and sold. This fact may be surprising, but it is true; it is the objective of this chapter to explain why this is so. We will explain how statistics can be used to monitor the production process and, after items are produced, increase the quality of items sold by sampling.

We have seen the theory of sampling used to test hypotheses, to determine confidence intervals, to find least squares lines, and to fit other functions, including multivariate ones, to data. Finally, we considered the design of industrial and scientific experiments. These applications are of great importance in science and engineering and explain the widespread use of statistics in these and many other fields.

As a final area of application of statistical methods and theory we turn to **statistical process control,** which refers to the application of statistical theory to manufacturing processes and product inspection. These methods, which were first applied in the late 1930s, increase the quality of produced material and have proved to be of great value to industries adopting them. Statistical process control continues to be an area of active research in statistics and many advances are likely in the near future. In this chapter, we offer only a brief sampling of the material known in this expanding field.

We begin with an example, which we will analyze in some detail.

7.2 An Example

Example 7.2.1 In order to monitor production processes and the characteristics of material produced by these processes, samples of the production are usually taken periodically. Joshi and Sprague in a paper "Obtaining and Using Statistical Process Control Limits in the

351

Semiconductor Industry" [51] report on a series of samples taken in wafer fabrication in the manufacture of semiconductors. Both the stability and the predictability of the production process are of interest to the manufacturer. Seventy-four samples of size 3 were taken from wafers, each of which contains several integrated circuits. Three different sites on the wafers were used, and the samples were taken at periodic time intervals. Table 7.1 shows some of the samples, their means, standard deviations, and ranges.

Lot No.	Site 1	Site 2	Site 3	Mean	St. Dev.	Range
1	3.26	3.09	3.20	3.18333	0.086217	0.17
2	3.27	3.24	3.12	3.21000	0.079373	0.15
3	3.19	3.17	3.18	3.18000	0.010000	0.02
⋮	⋮	⋮	⋮	⋮	⋮	⋮
39	3.06	3.14	3.17	3.12333	0.056862	0.11
40	3.81	3.89	3.30	3.66667	0.320052	0.59
41	3.21	3.25	3.30	3.25333	0.045092	0.09
⋮	⋮	⋮	⋮	⋮	⋮	⋮
72	2.97	2.89	2.95	2.93667	0.041633	0.08
73	3.11	3.11	3.11	3.11000	0.000000	0.00
74	3.02	3.00	2.99	3.00333	0.015275	0.03

Table 7.1

Unlike the situations encountered previously in this book, the order in which the samples were taken is of importance because variation in the process may be detected if the samples are considered in order as time goes on. Combining all the samples together would not show information on possible changes in the manufacturing process over time. We therefore show a graph of the means of the samples

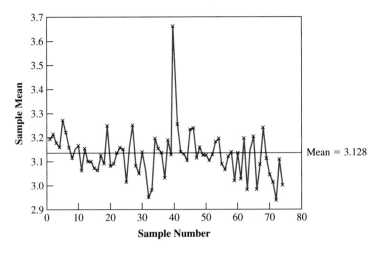

Figure 7.1 Sample Means for Example 7.2.1

in the order the samples were observed (Figure 7.1). The chart is used to examine the stability and predictability of the production process; we now explain how this is done.

The points in Figure 7.1 have been joined so that they are more easily visible, but there were no observations between the points. It appears that something decidedly different occurred at sample number 40, since it is at some distance from the remaining points. The question, of course, is whether it is significantly far removed from the remaining points to indicate that something has occurred in the process, that is, that the process may be out of control. This is a central question, which we now consider. ■

7.3 Control Chart for $\overline{X}$

Suppose the process is producing items with mean μ and standard deviation σ. One of our central problems is that neither of these numbers is known. Another problem, which we will discuss later, is that they may be changing as time goes on, but for the moment we suppose that they are constant but unknown.

Since we are considering the mean of, say, n observations per sample ($n = 3$ in Example 7.2.1), we will use the central limit theorem, which assures us that $\overline{X}$ is approximately $N(\mu, \sigma/\sqrt{n})$. If the probability distribution of X is normal, then the probability distribution of $\overline{X}$ is exactly normal; otherwise, the approximation of the probability distribution of $\overline{X}$ is somewhat dependent upon the probability distribution of X, which is usually unknown, and the sample size. The central limit theorem is known to work quite well with probability distributions commonly encountered in science and engineering, even if the sample size is moderate.

It is also clear from our discussion of the central limit theorem in Chapter 2 that the minimum variance unbiased estimator of μ is

$$\overline{\overline{X}} = \frac{\overline{x}_1 + \overline{x}_2 + \cdots + \overline{x}_k}{k}$$

where k is the number of samples considered and $\overline{x}_i$ is the mean of the ith sample. The value of k is 74 in Example 7.2.1.

So our estimate for μ in Example 7.2.1 is

$$\widehat{\mu} = \overline{\overline{X}} = \frac{3.18333 + 3.21000 + 3.18000 + \cdots + 2.93667 + 3.11000 + 3.00333}{74}$$

$$= 3.1276.$$

If we knew or could estimate σ, then we could determine whether sample number 40 was extreme or not. One way to do this would be to calculate the z score for the 40th observation.

We might also find a confidence interval for the true mean, μ, but since we don't know the standard deviation, we can't do that either. So it is necessary first to arrive at an estimate of the standard deviation, and then an estimate of the confidence interval is usually found as

$$\overline{\overline{X}} \pm 3 \frac{\widehat{\sigma}}{\sqrt{n}},$$

where $\widehat{\sigma}$ is an estimate of the unknown standard deviation. The multiplier 3 is commonly used in industry and produces about 0.0027 of the data outside the boundaries of the confidence interval. Since an observation outside the confidence interval occurs with a very small probability, it is then fairly safe to assume that an observation outside these limits did not arise by chance, but is due to some alteration in the production process. The boundaries are called **upper and lower control limits.** But we can't find the upper and lower control limits until we have some idea of what σ is.

If σ is unknown, and it generally is, there are three standard ways of estimating it. These use, respectively, the mean standard deviation of the samples, the mean range of the samples, and the pooled standard deviations of the samples. Each of these ways is considered now.

Table 7.1 exhibits some of the samples chosen, their means, standard deviations, and ranges. Minitab gives summary results for each of these sample statistics in Table 7.2. So the mean of the sample means is 3.1276, the mean of the sample standard deviations is 0.0618, and the mean of the sample ranges is 0.1155. As we shall now see, in addition to the pooled standard deviations, the mean of the sample standard deviations and the mean of the sample ranges can both be used to estimate the unknown σ.

Variable	N	Mean
Mean	74	3.1276
St. Dev.	74	0.0618
Range	74	0.1155

Table 7.2

Use of the Mean Standard Deviation to Estimate Sigma (σ)

If the sampling is done from a normal distribution with variance σ^2 and S^2 denotes the sample variance, we saw in our consideration of the chi-squared distribution that the variable $(n-1)S^2/\sigma^2$ follows a chi-squared distribution with $(n-1)$ degrees of freedom and so

$$E\left[\frac{(n-1)S^2}{\sigma^2}\right] = n-1.$$

From this it follows easily that $E(S^2) = \sigma^2$. Unfortunately, $E(S) \neq \sigma$. The proof of this is beyond the scope of this book, but for a demonstration, see Kinney [18]. It can be shown, however, that the expected value of the sample standard deviation, s, is a multiple of the unknown standard deviation, σ. The constant multiple, which depends on n, is usually denoted in statistical process control literature as c_4 so that

$$E(S) = c_4\sigma.$$

An estimate then of σ, namely $\widehat{\sigma}$, is then $\overline{s}/c_4$ where $\overline{s}$ is an estimate of the true mean of the standard deviations, $E(S)$.

n	c_4
2	0.797885
3	0.886227
4	0.921318
5	0.939986
6	0.951533
7	0.959369
8	0.965030
9	0.969311
10	0.972659
11	0.975350
12	0.977559
13	0.979406
14	0.980971
15	0.982316
16	0.983484
17	0.984506

Table 7.3

Table 7.3 gives some values for c_4 as a function of n, the sample size. Note that the values of c_4 approach 1 quite rapidly as the sample size increases. Production, however, usually employs small sample sizes, so it would be unwise to assume that the mean value for s and the unknown standard deviation, σ, were equal. In the case of Example 7.2.1, we have that $\overline{s} = 0.0618$, and since the value of c_4 for a sample of size 3 is 0.886227 we have

$$\widehat{\sigma} = \frac{\overline{s}}{c_4} = \frac{0.0618}{0.886227} = 0.0697.$$

The upper and lower control limits are then

$$\overline{\overline{X}} + 3\frac{\widehat{\sigma}}{\sqrt{n}} = 3.1276 + 3\frac{0.0697}{\sqrt{3}} = 3.24832$$

and

$$\overline{\overline{X}} - 3\frac{\widehat{\sigma}}{\sqrt{n}} = 3.1276 - 3\frac{0.0697}{\sqrt{3}} = 3.00688.$$

Since the mean for sample 40 is 3.66667, which is far outside the upper control limit, we conclude that this observation did not occur by chance and that there has been some alteration in the production process. Alternatively, the z score for the mean in sample 40 is

$$z = \frac{3.66667 - 3.1276}{0.0697/\sqrt{3}} = 13.3959,$$

which is far outside the 3 sigma limits.

Minitab can be used to produce the control chart showing all the means and the upper and lower control limits. This is shown in Figure 7.2, where σ has been estimated by $\bar{s}$. Joshi and Sprague [51] argue that in this case the control limits should be altered, producing fewer observations outside the control limits.

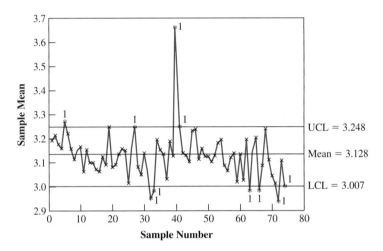

Figure 7.2 Control Chart for $\overline{X}$ Using $\bar{s}$ to Estimate σ

Figure 7.2 also shows any point that is outside the control limits by indicating such points with the number 1. We find, perhaps surprisingly, that the process is out of control in sample numbers 5, 27, 32, 33, 41, 63, 66, 72, and 74 in addition to sample number 40. Sample number 40 is so extreme that it is far from obvious that the process is out of control with some frequency. It may also be that the mean of the process is drifting downward after sample 55 or so; we will show how to investigate this possibility later in this chapter. For now we continue our discussion of methods for estimating σ.

Use of the Mean Range to Estimate Sigma (σ)

It may be intuitively clear that the range of a sample—that is, the difference between the maximum value and the minimum value in a sample—is an estimate of the standard deviation of the population from which the sample is selected. However, it is very difficult to find the exact probability distribution of the range of a sample except in fairly contrived situations. The probability distribution of the range is unknown even in samples selected from a normal distribution and can only be estimated numerically. It is known, however, that the true mean range of samples selected from a normal population is the product of a constant, d_2, depending upon the sample size, n, and the population standard deviation, σ, so

$$E(R) = d_2\sigma.$$

Estimating the true expected value of the range, $E(R)$, by $\overline{R}$, our estimate for σ, namely $\widehat{\sigma}$, becomes $\widehat{\sigma} = \overline{R}/d_2$.

Table 7.4 gives some values of d_2 for various sample sizes.

n	d_2
2	1.128
3	1.693
4	2.059
5	2.326
6	2.534
7	2.704
8	2.847
9	2.970
10	3.078

Table 7.4

In Example 7.2.1, our estimate for σ then becomes

$$\widehat{\sigma} = \frac{\overline{R}}{d_2} = \frac{0.1155}{1.693} = 0.0682.$$

This differs a bit from the estimate of 0.0697 for σ found using the mean standard deviation. One might expect that the range would not be as efficient an estimator for σ as the standard deviation since the range uses only two of the sample values in its calculation, regardless of the sample size, while the standard deviation uses all the sample values in its calculation. One argument for using the range is that the use of only two sample values makes it more easily computed on the production floor than the standard deviation.

Now the control limits are given by

$$\overline{\overline{X}} + 3\frac{\widehat{\sigma}}{\sqrt{n}} = 3.1276 + 3\frac{0.0682}{\sqrt{3}} = 3.2457$$

and

$$\overline{\overline{X}} - 3\frac{\widehat{\sigma}}{\sqrt{n}} = 3.1276 - 3\frac{0.0682}{\sqrt{3}} = 3.0095.$$

The resulting control chart showing these control limits and those samples outside them is shown in Figure 7.3. Note that in addition to the samples shown out of control in the control chart using the sample standard deviations to establish control limits, sample number 19 also indicates that the process is out of control.

We show one more method commonly used to estimate σ.

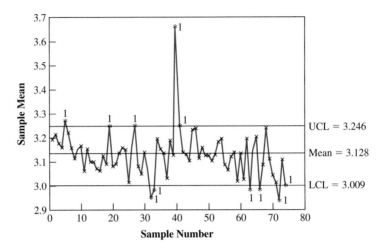

Figure 7.3 Control Chart for $\overline{X}$ Using $\overline{R}$ to Estimate σ

Use of the Pooled Standard Deviation to Estimate Sigma (σ)

Since we have the standard deviations of each of the samples, we may combine them to find a pooled estimate of the population variance. Recall that we used the pooled variance in Chapter 3, when we tested two samples for possible differences between their means. The pooled variance also arises in the analysis of variance test of the equality of several means. Denoting the sample sizes by n_i and the sample variances by s_i^2, the pooled variance, s_p^2, is found by

$$s_p^2 = \frac{\sum\limits_{i=1}^{k}(n_i-1)s_i^2}{\sum\limits_{i=1}^{k}(n_i-1)} = \frac{2(0.086217)^2 + 2(0.079373)^2 + \cdots + 2(0.015275)^2}{2(74)}$$

$$= 0.00624$$

and from this we find that $s_p = \sqrt{0.00624} = 0.0790$, which is our estimate of σ. This gives control limits as

$$\overline{\overline{X}} + 3\frac{\widehat{\sigma}}{\sqrt{n}} = 3.1276 + 3\frac{0.0790}{\sqrt{3}} = 3.2644$$

and

$$\overline{\overline{X}} - 3\frac{\widehat{\sigma}}{\sqrt{n}} = 3.1276 - 3\frac{0.0790}{\sqrt{3}} = 2.9908.$$

The resulting control chart is shown in Figure 7.4. The control chart in Figure 7.4 indicates fewer points where the process is out of control, but this is a consequence of the larger estimate of σ from this method.

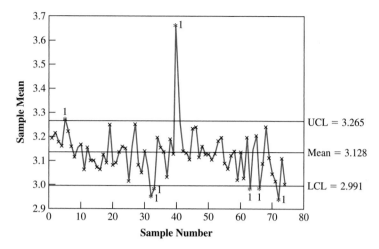

Figure 7.4 Control Chart for $\overline{X}$ Using the Pooled Standard Deviation to Estimate σ

The control chart for $\overline{X}$ is the chart usually constructed to obtain information on the central value for the process. Three methods for estimating σ have been given and the resulting control charts shown. Users employ different methods for estimating σ; any of the methods above can be used in Minitab.

The $\overline{X}$ chart aids in assessing whether the centering of the process is changing. We will show two control charts that are useful in assessing the variability of the process in Section 7.4.

EXERCISES 7.3

1. A production process takes periodic samples of size 5 and measures the mean of each sample. The following data were obtained, using 35 samples.

$$\sum_{i=1}^{35} \overline{x}_i = 700, \quad \sum_{i=1}^{35} R_i = 87, \quad \sum_{i=1}^{35} s_i = 43,$$

$$\sum_{i=1}^{35} s_i^2 = 90.6$$

 a. Estimate the unknown process standard deviation using the ranges of the samples and find 3 sigma control limits for a control chart on the process mean.
 b. Estimate the unknown process standard deviation using the mean standard deviation of the samples and find 3 sigma control limits for a control chart on the process mean.
 c. Estimate the unknown process standard deviation using the pooled standard deviation of the samples

and find 3 sigma control limits for a control chart on the process mean.

2. The following sample means, based on samples of size 4 each, were observed from a production process whose target mean is 50:54.2584, 53.6065, 53.6817, 48.1792, 49.2067, 33.5034, 49.5806, 66.1872, 51.8335, and 46.2587. Assuming that the process standard deviation is 10, plot a control chart for the process mean using 3 sigma control limits.

3. A plant producing steel washers is subject to periodic sampling. The following data arise from 22 samples, each of size 4.

$$\sum_{i=1}^{22} \overline{x}_i = 0.20732, \quad \sum_{i=1}^{22} R_i = 43.174,$$

$$\sum_{i=1}^{22} s_i = 19.1576$$

The actual data consist of only the last three digits of the measurement of the diameter of the washer around a target value of 48 so a diameter of 48.132 becomes 13.2.

a. Estimate the unknown process standard deviation using the ranges of the samples and find 3 sigma control limits for a control chart on the process mean.

b. Estimate the unknown process standard deviation using the mean standard deviation of the samples and find 3 sigma control limits for a control chart on the process mean.

4. The mean tensile strength of a metal part used in the production of recreational vehicles is measured by taking samples of size 5. The following data arise from 15 samples.

$$\sum_{i=1}^{15} \overline{x}_i = 3011.7, \qquad \sum_{i=1}^{15} R_i = 542.57,$$

$$\sum_{i=1}^{15} s_i = 216.65$$

a. Estimate the unknown process standard deviation using the ranges of the samples and find 3 sigma control limits for a control chart on the process mean.

b. Estimate the unknown process standard deviation using the mean standard deviation of the samples and find 3 sigma control limits for a control chart on the process mean.

5. The output voltage of a production process producing power supplies is measured periodically by taking samples of size 5 and measuring the mean output voltage. The following data arise from 15 samples.

$$\sum_{i=1}^{15} \overline{x}_i = 291.48, \qquad \sum_{i=1}^{15} R_i = 104.9,$$

$$\sum_{i=1}^{15} s_i = 42.57$$

a. Estimate the unknown process standard deviation using the ranges of the samples and find 3 sigma control limits for a control chart on the process mean.

b. Estimate the unknown process standard deviation using the mean standard deviation of the samples and find 3 sigma control limits for a control chart on the process mean.

6. A bottling machine producing "12-ounce" cans of soft drink actually either over- or under-fills most of the cans. To investigate this behavior, samples of size 4

are taken periodically. The following data arise from 12 samples.

$$\sum_{i=1}^{12} \overline{x}_i = 154.4, \qquad \sum_{i=1}^{12} R_i = 67.222,$$

$$\sum_{i=1}^{12} s_i = 30.184$$

a. Estimate the unknown process standard deviation using the ranges of the samples and find 3 sigma control limits for a control chart on the process mean.

b. Estimate the unknown process standard deviation using the mean standard deviation of the samples and find 3 sigma control limits for a control chart on the process mean.

7. The samples for the washer data considered in Exercise 3 are shown in the accompanying table. The X's represent individual observations within the samples. Show the $\overline{X}$ control chart, basing the estimate of σ on the ranges of the samples. In a paragraph write the conclusions that can be drawn.

X_1	X_2	X_3	X_4
−2.11024	−0.69463	0.34318	−1.01792
−0.52075	1.37703	−0.76598	0.38154
1.80180	−0.14795	−0.58011	1.17482
0.55523	−0.67479	1.84902	0.63193
1.56745	0.98114	−0.86182	0.75036
0.02337	1.52298	1.22669	0.12440
−0.31978	−0.52246	0.46028	0.57943
0.53701	0.90234	0.94606	−1.22357
−1.29825	0.75377	−0.35230	−0.55325
−1.95997	0.71155	1.36232	−0.39416
−1.08708	−0.44461	−1.16977	−0.68063
0.20731	−1.30404	0.74110	−0.18130
2.03153	−0.09185	−0.46208	−1.83936
−0.93059	−0.93595	0.13379	1.38784
−0.07153	−0.27240	−0.16571	−0.15060
−0.63105	−0.04508	−1.77947	−0.28878
−0.94942	0.59044	−0.73082	−0.48459
0.57437	−1.40332	1.76785	−0.23710
−0.02064	0.83584	0.64877	−0.79623
1.82781	−0.37235	−0.03772	0.10012
0.08966	−0.06080	0.90024	0.98140
−0.06907	−0.10470	−0.07521	−0.68040

8. The tensile strength data considered in Exercise 4 are shown in the accompanying table. The X's represent

individual observations within the samples. Show the $\overline{X}$ control chart, basing the estimate of σ on the ranges of the samples. In a paragraph write the conclusions that can be drawn.

X_1	X_2	X_3	X_4	X_5
220.096	187.514	172.980	205.145	191.689
199.003	191.027	193.664	187.956	202.231
190.872	179.868	179.698	190.021	173.077
205.409	166.825	210.916	183.505	203.800
222.440	182.194	208.125	208.275	194.132
196.243	199.970	215.465	203.887	217.297
206.035	215.909	201.503	221.127	225.110
203.601	211.452	216.604	228.101	228.958
188.646	192.222	192.291	223.076	180.735
219.907	195.659	217.615	192.283	197.598
198.841	212.588	219.276	185.226	270.181
244.470	196.643	186.389	169.941	199.178
186.179	173.554	189.570	206.685	183.111
191.080	188.587	179.311	211.072	206.888
210.953	196.268	206.814	197.001	207.048

9. The samples gathered for the output voltage study in Exercise 5 are shown in the accompanying table. The X's represent individual observations within the samples. Show the $\overline{X}$ control chart, basing the estimate of σ on the ranges of the samples. Write in a paragraph the conclusions that can be drawn.

X_1	X_2	X_3	X_4	X_5
25.2567	22.6834	19.4833	17.3201	23.5383
22.6905	21.6293	22.5641	17.2877	21.8713
23.6346	20.3334	19.0719	25.9775	24.3556
21.6303	16.0668	20.4386	22.2978	23.5658
21.6448	22.9044	21.3023	15.8440	18.0496
16.5162	20.4599	23.6605	21.1634	20.6381
20.0441	20.0656	23.7473	24.6286	21.3851
18.3265	14.0126	21.1410	20.8329	15.3966
20.2825	21.5897	14.6475	14.5659	16.1210
14.0698	16.5060	25.0523	18.8274	17.0733
16.2825	15.9153	21.6932	14.0221	21.1596
12.7565	15.5216	17.7131	15.9924	17.0627
18.1114	17.4626	20.8885	16.9436	21.2502
18.2458	21.9138	19.5323	16.8043	15.7171
18.5655	16.8316	18.5556	22.8881	13.3924

10. The samples gathered in the soft-drink study considered in Exercise 6 are shown in the accompanying table. The X's represent individual observations within the samples. Show the $\overline{X}$ control chart, basing the estimate of σ on the ranges of the samples. Write in a paragraph the conclusions that can be drawn.

X_1	X_2	X_3	X_4
7.7930	12.3721	10.0663	10.7014
13.3746	8.6707	8.8010	6.7189
11.3450	10.9383	14.4688	12.0347
19.0314	9.5583	12.5778	11.4288
10.6199	12.2877	15.6962	8.9779
12.1776	12.8380	16.7636	12.8449
13.2208	14.2227	17.5537	12.5686
14.6839	17.2049	11.4473	13.0109
12.9465	13.1961	14.1131	17.9824
11.5037	15.4819	12.1477	19.2913
17.5272	16.9569	13.0796	12.3379
12.7292	12.5073	9.8051	10.0142

11. The following data were chosen from, alternatively, the day and night shifts of a manufacturing plant. For each shift, two samples, each of size 5, were selected. The samples are shown vertically.

X_1	X_2	X_3	X_4	X_5
14.8920	15.9365	13.4949	14.5496	15.5497
14.9520	15.9306	14.9983	14.9890	13.7133
4.5829	3.1147	4.6282	5.1349	4.8918
4.0457	3.8453	3.4253	5.9240	4.3716
15.3099	16.0425	16.8373	14.5775	17.0275
14.7059	16.6440	15.9696	14.9017	13.6695
3.1230	4.0624	5.5327	5.6247	6.0296
3.8746	5.2849	2.9759	4.5091	5.0957
17.1224	13.8847	15.7040	13.4948	15.3460
16.1147	15.4782	14.0596	17.1788	16.4432
3.6861	4.6992	5.3667	5.4490	6.0853
4.9425	5.5813	5.5568	5.9032	4.1421
15.1610	14.8232	13.6976	16.2896	14.1040
17.0319	16.8216	15.9590	14.8614	15.1211
3.9634	5.2926	4.7284	4.2015	5.1077
5.3351	6.6772	5.3895	3.2959	5.9325
15.8979	12.8006	15.0052	15.4346	15.3764
15.3780	13.4501	17.2188	14.6185	15.0136
5.1162	5.2831	4.6647	3.9527	4.3069
4.6927	6.1578	5.8237	3.2184	5.8138

a. Show the $\overline{X}$ control chart using the sample ranges to estimate σ and comment on the result.

b. Show $\overline{X}$ control charts for the two shifts and comment on the results.

COMPUTER EXPERIMENT

Produce 1000 random observations from a standard normal probability distribution. Count the number of observations that are within one, two, or three standard deviations of the mean and compare these to the theoretical values.

7.4 Control Charts for R and s

We now turn to control charts concerning the variability of the production process. The range is investigated first.

Control Chart for Sample Ranges

In the previous section we saw that the sample ranges give an estimate for the process standard deviation, σ. We noted that $E(R) = d_2\sigma$ where d_2 is dependent only on the sample size, n. It can also be shown that the standard deviation of the range is also a multiple of the process standard deviation. The multiple, d_3, depends only upon n so we can write

$$\text{StDev}(R) = \sigma_R = d_3\sigma_x.$$

Table 7.5 gives some values for d_3.

n	d_3
2	0.853
3	0.888
4	0.880
5	0.864
6	0.848
7	0.833
8	0.820
9	0.808
10	0.797

Table 7.5

A control chart for the sample ranges can then be constructed using $\overline{R}$ as the central value and using the upper control limit (UCL) as

$$\text{UCL} = \overline{R} + 3\widehat{\sigma}_R = \overline{R} + 3d_3\widehat{\sigma}_x.$$

Estimating σ_x by $\overline{R}/d_2$, this becomes

$$\text{UCL} = \overline{R} + 3d_3 \frac{\overline{R}}{d_2}.$$

In a similar way, the lower control limit (LCL) is

$$\text{LCL} = \overline{R} - 3\widehat{\sigma}_R,$$

which is estimated as

$$\text{LCL} = \overline{R} - 3d_3 \frac{\overline{R}}{d_2}.$$

For the data in Example 7.2.1, $\overline{R} = 0.1155$ so

$$\text{UCL} = \overline{R} + 3d_3 \frac{\overline{R}}{d_2} = 0.1155 + 3 \cdot \frac{0.888}{1.693} \cdot 0.1155 = 0.2972$$

and

$$\text{LCL} = \overline{R} - 3d_3 \frac{\overline{R}}{d_2} = 0.1155 - 3 \cdot \frac{0.888}{1.693} \cdot 0.1155 = -0.0662.$$

Since the range of a sample must be positive, we take the lower control limit as 0. The control chart produced by Minitab is shown in Figure 7.5. Details for constructing this and other control charts can be found in Appendix B.

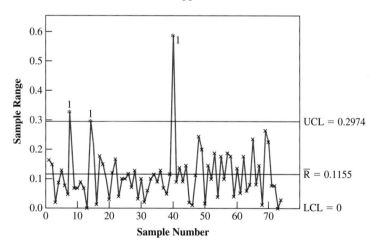

Figure 7.5 Control Chart for the Sample Range, Example 7.2.1

Control Chart for Standard Deviations

Now we consider the construction of a control chart for the sample standard deviations. We have used the fact that $E(S) = c_4 \sigma_x$ and we know that $E(S^2) = \sigma_x^2$. These facts can be used to find that

$$\text{Var}(S) = \sigma_S^2 = E(S^2) - [E(S)]^2 = \sigma_x^2 - (c_4 \sigma_x)^2$$

so that $\sigma_s = \sigma_x \sqrt{1 - c_4^2}$ where σ_s denotes the standard deviation of S.

A control chart may then be constructed by using a central line of $\bar{s}$ and upper and lower control limits as

$$\text{UCL} = \bar{s} + 3 \cdot \widehat{\sigma}_s$$

and

$$\text{LCL} = \bar{s} - 3 \cdot \widehat{\sigma}_s.$$

It is customary to estimate σ_s by

$$\widehat{\sigma}_s = \widehat{\sigma}_x \cdot \sqrt{1 - c_4^2} = \frac{\bar{s}}{c_4} \cdot \sqrt{1 - c_4^2}$$

so that

$$\text{UCL} = \bar{s} + 3 \cdot \frac{\bar{s}}{c_4} \cdot \sqrt{1 - c_4^2}$$

and

$$\text{LCL} = \bar{s} - 3 \cdot \frac{\bar{s}}{c_4} \cdot \sqrt{1 - c_4^2}.$$

For the data in Example 7.2.1, we find

$$\text{UCL} = 0.06180 + 3 \cdot \frac{0.06180}{0.886227} \cdot \sqrt{1 - 0.886227^2} = 0.1587$$

and

$$\text{LCL} = 0.06180 - 3 \cdot \frac{0.06180}{0.886227} \cdot \sqrt{1 - 0.886227^2} = -0.0351.$$

Again, since the standard deviation cannot be negative, zero is used for the lower control limit. The resulting control chart, produced by Minitab, is shown in Figure 7.6.

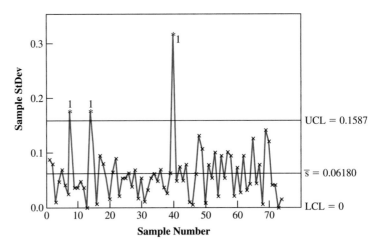

Figure 7.6 Control Chart for the Sample Standard Deviation, Example 7.2.1

EXERCISES 7.4

1. Refer to Exercise 1 of Section 7.3 on page 359.
   a. Find 3 sigma control limits for the control chart for the sample ranges.
   b. Find 3 sigma control limits for the control chart for the sample standard deviations.

2. Refer to Exercise 2 of Section 7.3 on page 359. For these data, $\overline{R} = 26.21$ and $\overline{s} = 8.41$.
   a. Find 3 sigma control limits for the control chart for the sample ranges.
   b. Find 3 sigma control limits for the control chart for the sample standard deviations.

3. For the data given in Exercises 3 and 7 of Section 7.3 on page 359:
   a. Find 3 sigma control limits for a control chart for the sample ranges.
   b. Find 3 sigma control limits for the control chart for the sample standard deviations.
   c. Plot the control chart for the sample ranges and show that all the points are within the control limits found in part (a).
   d. Plot the control chart for the sample standard deviations and show that all the points are within the control limits found in part (b).

4. For the data given in Exercises 4 and 8 of Section 7.3 on page 360:
   a. Find 3 sigma control limits for the control chart for the sample ranges.
   b. Find 3 sigma control limits for the control chart for the sample standard deviations.
   c. Plot the control chart for the sample ranges and show that all the points except one are within the control limits found in part (a).
   d. Plot the control chart for the sample standard devi-ations and show that all the points except one are within the control limits found in part (b).

5. For the data given in Exercises 5 and 9 of Section 7.3 on page 360:
   a. Find 3 sigma control limits for the control chart for the sample ranges.
   b. Find 3 sigma control limits for the control chart for the sample standard deviations.
   c. Plot the control chart for the sample ranges and show that all the points are within the control limits found in part (a).
   d. Plot the control chart for the sample standard deviations and show that all the points are within the control limits found in part (b).

6. For the data given in Exercises 6 and 10 of Section 7.3 on page 360:
   a. Find 3 sigma control limits for the control chart for the sample ranges.
   b. Find 3 sigma control limits for the control chart for the sample standard deviations.
   c. Plot the control chart for the sample ranges and show that all the points are within the control limits found in part (a).
   d. Plot the control chart for the sample standard deviations and show that all the points are within the control limits found in part (b).

7. For the data given in Exercise 11 of Section 7.3 on page 361:
   a. Plot the control chart for the sample ranges and show that all the points are within the control limits.
   b. Plot the control chart for the sample standard deviations and show that all the points are within the control limits.

7.5 Control Charts for Attributes

The control charts we have constructed for the variables $\overline{X}$, R, and S are called **control charts for variables** since the random variables involved, $\overline{X}$, R, and S, can take on values in an interval or intervals and are then continuous random variables. We now turn to control charts for random variables that are discrete. These are called **control charts for attributes.**

Example 7.5.1 Thirty samples of size 50 each are selected from a manufacturing process involving plating a metal. The data, in Table 7.6, show the number of parts that showed a plating defect.

Sample No.	Defects	Sample No.	Defects
1	1	16	5
2	6	17	4
3	5	18	1
4	5	19	6
5	4	20	15
6	3	21	12
7	2	22	6
8	2	23	3
9	4	24	4
10	6	25	3
11	2	26	3
12	1	27	2
13	3	28	5
14	1	29	7
15	4	30	4

Table 7.6

Interest might center on the number of defects in each sample and how this varies from sample to sample as the samples are taken in time. The control chart involved is usually called an **np chart.** We now show how this is constructed.

np Control Chart

Let the random variable X denote the number of parts showing plating defects. The random variable X is a binomial random variable since a part shows a defect or it does not show a defect and we assume, correctly or incorrectly, that the parts are produced independently and with constant probability of a defect. Since 50 observations were taken in each sample, X can take on integer values from 0 to 50. We know that the mean value of X is np and the standard deviation is $\sqrt{np(1-p)}$ where p is the true probability a part shows a plating defect. Reasonable control limits might then be from $\text{LCL} = \overline{X} - 3\sqrt{np(1-p)}$ to $\text{UCL} = \overline{X} + 3\sqrt{np(1-p)}$. We can find from the data that the total number of defective parts is 129, so the mean number of defective parts is $\overline{X} = 129/30 = 4.30$, but we don't know the value of p.

A reasonable estimate for p is the total number of defects divided by the total number of parts sampled, or $129/(30)(50) = 0.086$. This gives estimates for the control limits as

$$\text{LCL} = \overline{X} - 3\sqrt{n\widehat{p}(1-\widehat{p})} = 4.30 - 3\sqrt{50(0.086)(1-0.086)} = -1.6474$$

and

$$\text{UCL} = \overline{X} + 3\sqrt{n\widehat{p}(1-\widehat{p})} = 4.30 + \sqrt{50(0.086)(1-0.086)} = 10.25.$$

Since X, the number of parts showing defects, cannot be negative, the lower control limit is taken as 0. The resulting control chart, produced by Minitab, is shown in Figure 7.7.

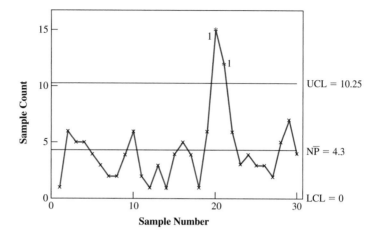

Figure 7.7 *np* Control Chart for Example 7.5.1 Using 3σ Limits

It appears that samples 20 and 21 show the process to be out of control. Except for these points, the process is in good control. Figure 7.8 shows the control chart using 2 sigma limits; none of the points, other than those for samples 20 and 21, are out of control. ∎

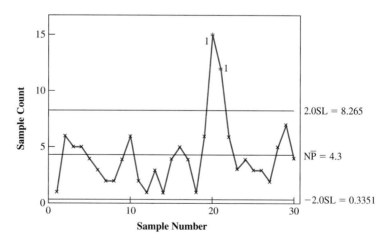

Figure 7.8 *np* Control Chart for Example 7.5.1 Using 2σ Limits

p Chart

The proportion of the production subject to defects is often of importance. We denote this random variable as p_s and we know that $p_s = X/n$. Since the random variable X is the number of parts showing defects in our example and since X is binomial and the sample size is n, it follows that

$$E(p_S) = E\left(\frac{X}{n}\right) = \frac{E(X)}{n} = \frac{np}{n} = p$$

and

$$\text{Var}(p_S) = \text{Var}\left(\frac{X}{n}\right) = \frac{\text{Var}(X)}{n^2} = \frac{np(1-p)}{n^2} = \frac{p(1-p)}{n}.$$

We see that control limits, using three standard deviations, are

$$\text{LCL} = p - 3\sqrt{\frac{p(1-p)}{n}}$$

and

$$\text{UCL} = p + 3\sqrt{\frac{p(1-p)}{n}}$$

but, of course, we don't know the value for p. A reasonable estimate for p would be the overall proportion defective considering all the samples. This is the estimate used in the previous section, namely 0.086. This gives control limits as

$$\text{LCL} = 0.086 - 3\sqrt{\frac{0.086(1 - 0.086)}{50}} = -0.032948$$

and

$$\text{UCL} = 0.086 + 3\sqrt{\frac{0.086(1 - 0.086)}{50}} = 0.2049.$$

Zero is used for the lower control limit. The resulting control chart is shown in Figure 7.9. This chart gives exactly the same information as the chart shown in Figure 7.8.

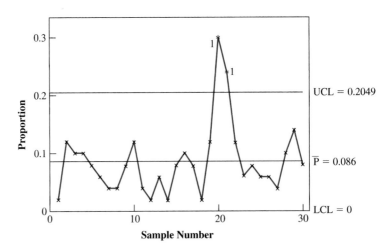

Figure 7.9 p Chart for Example 7.5.1

Sample Size to Detect a Change in the Process

Control charts are frequently used to detect changes in the production process. It may be quite important to detect such changes so we turn our attention to this question.

Suppose that a process is producing defective parts with probability p and if this probability changes to δp we want to be able to detect this with probability $1/2$ or greater. Assume, of course, that $0 \le \delta p \le 1$ so that $0 \le \delta \le 1/p$. Recall that the control limits for the p chart are

$$\text{LCL} = p - 3\sqrt{\frac{p(1-p)}{n}}$$

and

$$\text{UCL} = p + 3\sqrt{\frac{p(1-p)}{n}}.$$

Assume that the distribution can be reasonably well approximated by a normal distribution. Since we want to be able to detect the change with probability $1/2$, we then want the value δp to have a z score of 0, or more, for a normal distribution with mean $p + 3\sqrt{p(1-p)/n}$ and standard deviation $\sqrt{p(1-p)/n}$ so

$$\frac{\delta p - \left(p + 3\sqrt{\frac{p(1-p)}{n}} \right)}{\sqrt{\frac{p(1-p)}{n}}} \ge 0.$$

This simplifies to

$$n \ge \frac{9(1-p)}{p(\delta - 1)^2}.$$

For example, if $p = 0.01$ and we wish to detect a change to $p = 0.02$ (so $\delta = 2$), the formula indicates a sample of 891 or greater is needed. This enormous sample size only detects this change with probability $1/2$, so the detection process is not very sharp. Table 7.7 shows some values of n for various values of δ, assuming that $p = 0.01$.

Our process gives reasonable sample sizes only for large values of δ. This lack of responsiveness may be surprising, but it is, unfortunately, characteristic of the control charts we have considered up to this point. This lack can be corrected and we will soon discuss some possible corrections as well as other types of control charts. Before discussing these corrections, we show one more attribute control chart.

c Chart

Very often the probability that a production item is defective is quite small. In this case, the binomial distribution we have used before can be well approximated by the Poisson probability distribution. We pause here to discuss this distribution and its relation to the binomial distribution.

δ	n
1.25	14,256
1.50	3564
1.75	1584
2.00	891
2.25	571
2.50	396
2.75	291
3.00	223
3.25	176
3.50	143
3.75	118
4.00	99

Table 7.7

Poisson Distribution   It can be shown, if n is reasonably large and p is reasonably small, that the binomial probability distribution function,

$$f(x) = \binom{n}{x} p^x (1 - p)^{n-x} \quad x = 0, 1, \ldots, n,$$

can be well approximated by the Poisson probability distribution function,

$$p(x) = \frac{e^{-\lambda} \lambda^x}{x!} \quad x = 0, 1, 2, \ldots,$$

where in this case $\lambda = np$. It is known that if X is Poisson, then $E(X) = \lambda$ and $\mathrm{Var}(X) = \lambda$. For a demonstration of these facts, see Kinney [18].

We show some numerical results to suggest that the approximation is quite good in Table 7.8. Here we have chosen $n = 50$, a fairly large sample, and $p = 0.02$, not a particularly small value. The Poisson probabilities are then calculated with $\lambda = np = 50(0.02) = 1$.

X	Binomial Probability	Poisson Probability
0	0.36417	0.36788
1	0.37160	0.36788
2	0.18580	0.18394
3	0.06067	0.06131
4	0.01455	0.01533
5	0.00273	0.00307
6	0.00042	0.00051

Table 7.8

So we can approximate the binomial distribution, which we used to construct the np chart, by using a Poisson approximation with $E(X) = \lambda$ and $\mathrm{StDev}(X) = \sqrt{\lambda}$. The control limits then become $\lambda \pm 3\sqrt{\lambda}$. The resulting control chart for the data in Example 7.5.1 then has central value 4.30 and standard deviation $\sqrt{4.30}$ giving control limits

$$\mathrm{LCL} = 4.3 - 3\sqrt{4.30} = -1.92093$$

and

$$\mathrm{UCL} = 4.3 + 3\sqrt{4.30} = 10.52$$

so the control limits are set at 0 and 10.52. The control chart, called a c chart, is produced by Minitab and is shown in Figure 7.10.

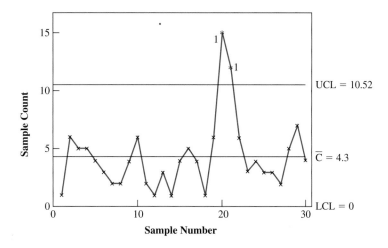

Figure 7.10 c Chart for Example 7.5.1

EXERCISES 7.5

1. A manufacturer of jewel cases for compact discs takes samples of 25 for each day's production and measures the number of jewel cases that have one or more defects. The data for a month are as follows: 0, 0, 1, 0, 0, 2, 1, 0, 0, 0, 0, 0, 0, 1, 0, 1, 2, 1, 0, 1, 1, 0, 1, 0, 0, 0, 0, 0, 3, 1.

Calculate the control limits for constructing a control chart for the number of defects per day for the data and show the control chart. What conclusions can be drawn from the control chart?

2. For the data in Exercise 1, calculate the control limits for constructing a control chart for the percentage of defects per day for the data and show the control chart. What conclusions can be drawn from the control chart?

3. For the data in Exercise 1, use the Poisson approximation to the binomial distribution to calculate control limits for a control chart for the number of defects per day for the data and show the control chart. What conclusions can be drawn from the control chart?

4. The numbers of defects in motherboards for personal computers based on 48 samples of size 30 each are as follows: 0, 0, 1, 0, 0, 2, 1, 0, 0, 0, 0, 0, 0, 1, 0, 1, 2, 1, 0, 1, 1, 0, 1, 0, 0, 0, 0, 0, 3, 1, 0, 0, 0, 0, 0, 0, 1, 0, 0, 2, 1, 0, 1, 0, 0, 0, 1, 0.

Calculate the control limits for constructing a control chart for the number of defects per sample for the data and show the control chart. What conclusions can be drawn from the control chart?

5. For the data in Exercise 4, calculate the control limits for constructing a control chart for the percentage of defects per sample for the data and show the control chart. What conclusions can be drawn from the control chart?

6. For the data in Exercise 4, use the Poisson approximation to the binomial distribution to calculate control limits for a control chart for the number of defects per sample for the data and show the control chart. What conclusions can be drawn from the control chart?

7. Weld defects occasionally occur in the manufacture of automobiles. A manufacturer has daily samples for one month, each sample being of size 20. The data are as follows: 1, 0, 2, 1, 1, 3, 2, 6, 1, 3, 2, 1, 2, 3, 3, 1, 1, 1, 2, 0, 2, 4, 1, 3, 3, 0, 4, 2, 2, 1.

Calculate the control limits for constructing a control chart for the number of defective welds per sample for the data and show the control chart. What conclusions can be drawn from the control chart?

8. For the data in Exercise 7, calculate the control limits for constructing a control chart for the percentage of defective welds per sample for the data and show the control chart. What conclusions can be drawn from the control chart?

9. For the data in Exercise 7, use the Poisson approximation to the binomial distribution to calculate control limits for a control chart for the number of defective welds per sample for the data and show the control chart. What conclusions can be drawn from the control chart?

7.6 Some Characteristics of Control Charts

The control charts we have considered offer great insight into production processes since they track the production process in a periodic manner. They have some disadvantages as well, as we saw when we investigated the sample size necessary to detect a change in the probability that the process produced an unsatisfactory item. We want to investigate some of these characteristics in this section.

First, consider the probability of a false reading, that is, an observation falling beyond the 3 sigma limits due entirely to chance rather than to a real change in the process. Assuming that the observations are normally distributed, an assumption justified for the sample mean if the sample size is moderate, then the probability an observation is outside the 3 sigma limits is 0.0027. Such observations are then very rare and when one occurs, we are unlikely to conclude that the observations occurred by chance alone. However, if

n	Probability
50	0.126444
100	0.236899
200	0.417677
300	0.555629
400	0.660900
500	0.741233
600	0.802534
700	0.849314
800	0.885011
900	0.912252
1000	0.933039

Table 7.9

50 observations are made, the probability that at least one of them is outside the 3 sigma limits is $1 - (1 - 0.0027)^{50} = 0.12644$. This probability increases rapidly, as the data in Table 7.9 indicate. (Here n represents the number of observations.) So an extreme observation becomes almost certain as the process continues although the process in reality has not changed.

Another disadvantage of the control charts we have considered is that they are not particularly sensitive to small changes in the process mean. Consider an example.

Example 7.6.1 Nails used in home construction are sold by weight. The manufacturing process has a target weight of 12 pounds but weights vary. Twenty-six samples, each of size 5, and some summary statistics for each sample, are shown in Table 7.10.

Sample	Obs. 1	Obs. 2	Obs. 3	Obs. 4	Obs. 5	Mean	Range
1	12.8017	11.8050	12.9699	11.4006	14.9355	12.7825	3.53489
2	10.0750	11.8384	10.5107	11.2108	12.0680	11.1406	1.99296
3	11.4807	12.5267	12.6365	13.2435	11.7668	12.3309	1.76275
4	11.9763	11.9154	12.0088	12.0427	11.6540	11.9195	0.38872
5	12.2800	13.8317	11.4395	11.6543	12.0066	12.2424	2.39214
6	12.6807	10.7560	11.2311	11.8808	11.1807	11.5458	1.92469
7	11.2999	12.3258	12.3257	10.6312	12.4620	11.8089	1.83079
8	11.4778	13.6298	12.8438	12.6503	13.1561	12.7516	2.15201
9	10.8776	12.1523	10.6718	12.4886	12.4829	11.7346	1.81672
10	12.7570	11.4802	12.2810	13.1466	11.8711	12.3072	1.66636
11	11.2367	11.1991	11.3911	12.9237	12.1216	11.7745	1.72464
12	12.8610	11.8035	11.8854	13.0624	11.5447	12.2314	1.51776
13	10.3506	11.2773	13.0499	12.4673	11.3356	11.6961	2.69938
14	14.3013	11.5217	13.1112	12.9900	11.1720	12.6193	3.12934
15	11.4053	11.1635	11.7243	11.5039	11.8008	11.5196	0.63728
16	13.7098	12.4313	13.0919	12.1028	13.6209	12.9913	1.60706
17	11.6445	11.5145	12.2125	11.8424	11.2738	11.6975	0.93870
18	12.2346	12.5764	12.8598	12.9737	14.1652	12.9619	1.93058
19	12.7700	12.3956	12.1397	14.2764	12.6625	12.8489	2.13667
20	12.0756	12.1342	12.6121	13.7196	13.5146	12.8112	1.64396
21	11.8571	11.1601	11.6446	12.1465	14.0117	12.1640	2.85156
22	11.7346	14.5614	12.8134	12.9276	12.2675	12.8609	2.82677
23	12.5122	13.2639	12.3683	12.6250	13.1471	12.7833	0.89559
24	12.2005	11.7287	10.3429	12.3568	13.3369	11.9932	2.99400
25	11.9767	14.0282	12.8509	13.0880	12.4684	12.8824	2.05151
26	10.0721	11.6773	12.5882	12.4494	11.9990	11.7572	2.51611

Table 7.10

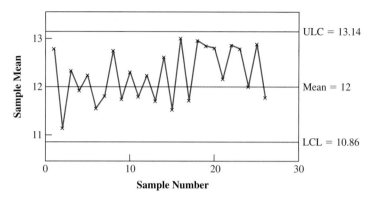

Figure 7.11 Control Chart for $\overline{X}$, Example 7.6.1

The control chart for $\overline{X}$ centered around the target value 12 is shown in Figure 7.11. The chart shows no particular unusual features. The process appears to be in control. However, the mean actually changed at sample 16. The first 15 samples were selected from a normal population with mean 12 and standard deviation 1, while samples numbered 16 through 26 were selected from a normal population with mean 12.5 and standard deviation 1. So the process mean has changed by $1/2$ of a standard deviation and our chart has not detected that in the following 11 samples. In fact, when more random samples are added to the set, no unusual points were found in an additional 60 points, but this will vary with the random sample selected. ∎

This raises the question of the length of time the chart takes to detect changes in the process mean. If the control limits are three standard deviations from the mean, then the probability an observation is outside the control limits is 0.0027. It can be shown then that the mean number of observations between such extreme limits is $1/0.0027 = 371$ observations. The mathematical details can be found by referring to the geometric probability distribution. See Kinney [18]. In this case, we call the **average run length,** or **ARL,** the expected number of observations between extreme observations. Here ARL $= 371$ observations. In general, the average run length is the reciprocal of the probability an observation exceeds the control limits.

Now suppose the process mean has increased by k standard deviations. The situation is depicted in Figure 7.12.

Since the lower limit can be safely ignored, here we see that we must find

$$P(\overline{x} > \mu + 3\sigma_{\overline{x}} \mid \mu_{\overline{x}} = \mu + k\sigma_{\overline{x}}) = P\left[z > \frac{(\mu + 3\sigma_{\overline{x}}) - (\mu + k\sigma_{\overline{x}})}{\sigma_{\overline{x}}}\right]$$

$$= P(z > 3 - k).$$

The average run lengths are then the reciprocals of these probabilities. For example, if the mean has changed by $1/2$ of a standard deviation, as it did in Example 7.6.1, then $k=0.5$ and $P(z>3-0.5=2.5)=0.00620967$ and the average run length is $1/0.00620967=162$ observations. Table 7.11 shows some average run lengths for various values of k.

So the $\overline{X}$ control chart is not rapidly responsive to small changes in the process mean. Only very large changes in the process mean produce small average run lengths. What

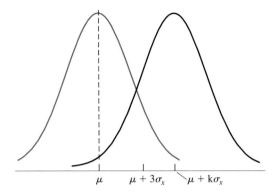

Figure 7.12 Process Mean Increased by k Sigma Units

k	Average Run Length
0.5	162
1	44
2	7
3	2
4	2

Table 7.11

other characteristics, other than noting points outside the control limits, could be made? If the process mean changes, it may be suspected that runs of points more than one or two standard deviations from the mean or runs of points on the same side of the center line may well occur. So it has been suggested that the $\overline{X}$ chart be supplemented by additional tests. Many of these tests, which became known as Western Electric Rules after they were first suggested in [57], are implemented by several computer statistics packages. We describe these tests now.

Some Additional Tests for Control Charts

Most of the tests presented here, except for test 1, examine series of points that do not exceed the usual 3 sigma control limits. We will show the calculation of each of the probabilities of the events in question, using first the default values used in Minitab and then expanding to other values. Since in general the probabilities of these events are quite small, the following situations are to be regarded as cautionary flags for the production process. The calculation of the probabilities involved in most of these tests relies upon the binomial or normal probability distributions. The default values of the constant k in each of these tests can easily be changed in Minitab.

Test 1: Points more than k sigma units from the center line.

We have used $k = 3$, but Table 7.12 shows probabilities with which a single point is more than k standard deviations from the target mean.

k	Probability
1.0	0.317311
1.5	0.133614
2.0	0.045500
2.5	0.012419
3.0	0.002700

Table 7.12

Test 2: k points in a row on the same side of the center line.

The probability that k points in a row are on the same side of the center line is $2\left(\frac{1}{2}\right)^k$. It is common to use $k = 9$. The probability that 9 points in a row are on the same side of the center line is

$$2\left(\frac{1}{2}\right)^9 = 0.00390625.$$

Table 7.13 shows this probability for $k = 7, 8, \ldots, 11$.

k	Probability
7	0.01563
8	0.00781
9	0.00391
10	0.00195
11	0.00098

Table 7.13

Test 3: k points in a row all increasing or decreasing.

Consider all the $k!$ arrangements of k different observations. Of these, only 2 have the points strictly increasing or decreasing, so the probability is $2/k!$

The default value here is $k = 6$ in Minitab and the probability that all six points are strictly increasing or decreasing is $2/6! = 0.00277778$. Probabilities for some other values of k are shown in Table 7.14.

k	Probability
5	0.01667
6	0.00278
7	0.00040
8	0.00005

Table 7.14

Test 4: k points in a row alternating up and down.

The default value for k is 14 in Minitab. For small values of k it is easy to enumerate the permutations of k integers for which the pattern is either up, down, up, ... or down, up, down, ... For example, if $k = 4$ the following ten permutations showing alternating up and down observations are easily found: (2, 1, 4, 3), (3, 1, 4, 2), (3, 2, 4, 1), (4, 1, 3, 2), (4, 2, 3, 1), (1, 3, 2, 4), (1, 4, 2, 3), (2, 3, 1, 4), (2, 4, 1, 3), and (3, 4, 1, 2) so the probability that four observations show an alternating up and down pattern is then $10/4! = 0.416667$, not a particularly rare event.

So we are interested in larger values for k. However, counting the permutations for increasing values of k becomes very difficult, if not practically impossible. Grimaldi [50] indicates that the probabilities we seek are twice the coefficients of powers of x in the power series expansion of $\sec(x) + \tan(x)$. Some of these probabilities are shown in Table 7.15.

k	Probability
5	0.266667
6	0.169444
7	0.107937
8	0.068700
9	0.043739
10	0.027845
11	0.017727
12	0.011285
13	0.007184
14	0.004574
15	0.002912
16	0.001854
17	0.001180
18	0.000751
19	0.000478
20	0.000304

Table 7.15

So the probability that 14 points show the alternating up and down pattern is roughly 0.005.

Test 5: At least k out of $k + 1$ points in a row more than 2 sigmas from the center line.

Since the probability that one point is more than 2 sigmas above the center line is 0.0227501 and since the number of observations outside these limits is a binomial random variable, the probability that at least k out of $k + 1$ observations are more than

2 sigmas from the center line and either above or below the center line is

$$2 \sum_{x=k}^{k+1} \binom{k+1}{x}(0.0227501)^x (0.9772499)^{k+1-x}$$

The quantity k is commonly chosen as 2. In that case,

$$2 \sum_{x=2}^{3} \binom{3}{x}(0.0227501)^x (0.9772499)^{3-x} = 0.0030583$$

Table 7.16 gives values of this probability for other values of k.

k	Probability
2	0.003058
3	0.000092
4	0.000003

Table 7.16

Test 6: At least k out of $k+1$ points in a row more than 1 sigma from the center line.

Since the probability that one point is more than one sigma above the center line is 0.158655 and since the number of observations outside these limits is a binomial random variable, the probability that at least k out of $k+1$ observations are more than 2 sigmas from the center line is

$$2 \sum_{x=k}^{k+1} \binom{k+1}{x}(0.158655)^x (1-0.158655)^{k+1-x}.$$

The value of k is commonly chosen as 4. In that case,

$$2 \sum_{x=4}^{5} \binom{5}{x}(0.158655)^x (1-0.158655)^{5-x} = 0.00553181.$$

Table 7.17 gives values of this probability for other values of k.

k	Probability
2	0.135054
3	0.028147
4	0.005532

Table 7.17

Test 7: k points in a row within one sigma of the center line.

Since the probability a single point is within one sigma of the center line is 0.682689, the probability that k points are within one sigma of the center line is $(0.682689)^k$.

The value of k is commonly chosen as 15. In that case, $(0.682689)^{15} = 0.00326095$. Table 7.18 gives this probability for other values of k.

k	Probability
12	0.010249
13	0.006997
14	0.004777
15	0.003261

Table 7.18

Test 8: k points in a row more than one sigma from the center line (on either side of the center line).

The probability a single point is within one sigma from the center line is 0.682689. Noting whether the point is above or below the center line, there are 2^k arrangements of k points. Of these, only 2 have all the points either above or below the center line, so the probability that k points in a row are each more than one sigma from the center line and on either side of the center line is

$$\left(1 - \frac{2}{2^k}\right)(1 - 0.682689)^k.$$

The value of k is commonly chosen as 8. In that case,

$$\left(1 - \frac{2}{2^8}\right)(1 - 0.682689)^8 = 0.00010197.$$

Table 7.19 gives this probability for other values of k.

k	Probability
6	0.000989
7	0.000319
8	0.000102
9	0.000032
10	0.000010

Table 7.19

Computer algebra systems allow us to calculate the probabilities for each of the eight tests with relative ease. One could even approximate the probability for the test in question and find a value for k. In test 4, for example, where we seek sequences of points that are alternately up and down, if we wanted a probability of approximately 0.01, Table 7.15 indicates that a run of 12 points is sufficient.

We embarked on this discussion of additional tests for the control chart because we discovered that the usual control chart is not rapidly responsive for small changes in the process mean. If we look at the control chart for $\overline{X}$ for the data in Example 7.6.1 and impose all the eight tests indicated here, Minitab gives the control chart shown in Figure 7.13 where control lines have been drawn at one standard deviation above or below the center line. Note that the tests are far from being independent.

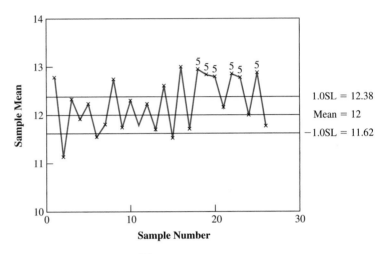

Figure 7.13 $\overline{X}$ Control Chart for Example 7.6.1

Figure 7.13 indicates that test 5 (testing if at least two out of three consecutive points are more than 2 sigmas from the center line) has failed at samples 18, 19, 20, 22, 23, and 25. While the original control chart, in Figure 7.11, indicated no changes in the data (although the process actually changed at sample number 15), the additional tests now indicate a change in the process at sample number 18.

The additional tests provide some increased sensitivity for the control chart, but they decrease its simplicity. Simplicity is a desirable feature when production workers monitor the process. Fortunately, there is another class of control charts that offer much increased sensitivity over the control charts considered previously and that are easy to interpret as well. They are called **cumulative sum charts,** and they are the topic of the next section. These charts are very attractive alternatives to the control charts on $\overline{X}$, R, and s.

EXERCISES 7.6

1. Use the data in Exercise 2 of Section 7.3 on page 359.
   a. Using $\overline{R} = 9.266$, show that test 1 fails for this data set for $k = 3$.
   b. Verify that 0.0027 of normal data falls outside 3 sigma control limits.

2. Use the data in Exercise 7 of Section 7.3 on page 360.
   a. Show that tests 2 and 6 both fail for this data set for the $\overline{X}$ chart.

b. Verify that the probability that seven consecutive points on the same side of the center line for normal data is 0.015625.

c. Verify that the probability that at least four out of five consecutive points occur more than one standard deviation from the center line of a control chart is 0.00553181.

3. Use the data in Exercise 9 of Section 7.3 on page 361.

a. Show that tests 1, 5, and 6 all fail for this data set for the $\overline{X}$ chart.

b. Verify that the probability that at least three out of four consecutive points more than two standard deviations from the center line is 0.0000925904.

4. Use the data in Exercise 10 of Section 7.3 on page 361.

a. Show that test 6 fails for this data set for the $\overline{X}$ chart.

b. Verify that the probability that at least two out of three consecutive points occur more than one standard deviation from the center line of a control chart is 0.135054.

5. Verify that the probability that five points in a row of a control chart are all either increasing or decreasing is 0.0166667.

6. Verify that the probability that 13 points in a row are all within one standard deviation of the center line of a control chart is 0.006996784.

7. Verify that the probability that six consecutive points in a control chart are all more than one standard deviation from the center line (on either side of the center line) is 0.000988823.

8. Show that test 1 fails repeatedly for the data given in Exercise 11 of Section 7.3 on page 361.

COMPUTER EXPERIMENT

Use the power series expansion of sec x + tan x to:
a. Verify the probabilities given in Table 7.15.
b. Determine the probability that 22 points show an alternating up, down, up . . . or down, up, down . . . pattern.

7.7 Cumulative Sum Control Charts

The control charts presented so far show individual points representing distinct separate samples. One might think that *all* the samples taken to date should be considered, which, perhaps might improve the sensitivity of the control chart. It is possible to use all the previously taken samples in what we call a **cumulative sum chart.** One consequence of this is that the cumulative sum chart is in fact more rapidly responsive to small changes in the process mean.

We return to the data in Example 7.6.1 on page 373. The cumulative sum chart is constructed, as its name indicates, by accumulating data from all the previous samples in a fairly simple manner. When the sums reach a certain point, the chart indicates that a change in the production process has occurred. Cumulative sum charts exist in various configurations and for various random variables. We show only one fairly simple implementation using sample means here; other cumulative sum charts can be found in some of the references in the Bibliography.

To construct our chart, we need to calculate both **upper** and **lower sums,** denoted by S_H and S_L, respectively. The following process is to be followed:

Assuming the process has target mean μ, standard deviation σ, and sample size n, calculate the normal z score for each of the observed means by

$$z = \frac{\overline{x} - \mu}{\sigma / \sqrt{n}}.$$

In Example 7.6.1, $\mu = 12$, $\sigma = 1$, and $n = 5$. In practice, μ is the target value, but σ is most often unknown. The standard deviation can be estimated in several ways; we will estimate σ by using the mean of the ranges of the samples. Here the mean range is $\overline{R} = 1.983$ and the corresponding value for d_2 is 2.326 so $\widehat{\sigma} = 1.983/2.326 = 0.85254$ and this gives $\widehat{\sigma}_{\overline{x}} = 0.85254/\sqrt{5} = 0.3813$.

Table 7.20 indicates values of $\overline{x}$ and the corresponding values of z for the data, as well as other quantities to be explained subsequently. The z values should always be studied by themselves; this provides an $\overline{X}$ chart in tabular form. Values of $|z| \geq 3$ would be of immediate interest since these are the usual control values on a simple $\overline{X}$ chart. So part of our process here constructs an $\overline{X}$ chart in tabular form. Suppose we desire to detect a change of $\delta\sigma$ in the production process. It is customary to choose a value of a constant k as $1/2$ of the multiple of σ we would like to detect, or $\delta/2$. Here we choose $k = 1/2$ so we seek to detect changes of one standard deviation in the process.

Sample	$\overline{x}$	z	S_H	S_L
1	12.7825	2.05224	1.55224	0
2	11.1406	−2.25393	0	1.7539
3	12.3309	0.86770	0.36770	0.3862
4	11.9195	−0.21125	0	0.0975
5	12.2424	0.63576	0.13576	0
6	11.5458	−1.19106	0	0.6911
7	11.8089	−0.50114	0	0.6922
8	12.7516	1.97111	1.47111	0
9	11.7346	−0.69595	0.2752	0.1960
10	12.3072	0.80562	0.5808	0
11	11.7745	−0.59152	0	0.0915
12	12.2314	0.60692	0.10692	0
13	11.6961	−0.79696	0	0.2970
14	12.6193	1.62406	1.12406	0
15	11.5196	−1.26000	0	0.7600
16	12.9913	2.59988	2.09988	0
17	11.6975	−0.79325	0.8066	0.2933
18	12.9619	2.52277	2.8294	0
19	12.8489	2.22620	4.5556	0
20	12.8112	2.12750	6.1831	0
21	12.1640	0.43008	6.1132	0
22	12.8609	2.25780	7.8710	0
23	12.7833	2.05427	9.4253	0
24	11.9932	−0.01791	8.9073	0
25	12.8824	2.31427	10.7216	0
26	11.7572	−0.63671	9.5849	0.1367

Table 7.20

The values of S_H are found using the following recursive relationship:
We begin with $S_0 = 0$ and then set

$$S_{H_i} = \max[0, \ z_i - k + S_{H_{i-1}}]$$

where $\max[a, b]$ is the larger of the quantities a and b. Note that if the z score exceeds k, then that excess is added to the previous value of S_H. This process clearly detects an increasing mean since, in that case, the deviations are positive. Here then is how the first few values of S_H are calculated:

$$S_{H_1} = \max[0, \ 2.05224 - 0.5 + 0] = \max[0, 1.55224] = 1.55224.$$
$$S_{H_2} = \max[0, \ -2.25393 - 0.5 + 1.55224] = \max[0, -1.2017] = 0.$$
$$S_{H_3} = \max[0, 0.86770 - 0.5 + 0] = \max[0, 0.36770] = 0.36770.$$

How are these values to be interpreted? If the values of $\bar{x}$ are consistently above the target value μ, then the values of z_i will become large. If 0.5 is subtracted from these and the previous value of S_H is added to this, the values of S_H will also tend to become large. If the process mean increases, as it does in this example, then the values of $\bar{x}$ will tend to be above the target value and eventually will exceed a prescribed limit. That is what happens in this example.

On the other hand, if the process mean has not increased, then the sample means will tend to be both above and below the target value. Negative values of z will tend to produce values of 0 for S_H; this also occurs in this example, explaining the large numbers of zeros for S_H in the first 15 samples, where we know the target mean has not changed. In the last 11 samples, however, the values of S_H tend to be large.

Now how large should we allow S_H to become before concluding that the process mean has changed? The choice of this limit, called the **decision interval, h,** then becomes important. A general rule is that we should put a limit on S_H at h units; h is almost always chosen as 4 or 5. In this case, we choose $h = 4$. This decision is important here and arises again when we consider V-masks.

If we refer to Table 7.20, we see that this limit is exceeded at sample number 19. This is a far more rapid response than was given by the $\overline{X}$ chart. The entries in Table 7.20 can be produced easily with Mathematica. The procedure is shown in Appendix A.

Another way to express the limit is to use $\sigma_{\bar{x}}$ units. This is what Minitab does. We saw previously that $\widehat{\sigma}_{\bar{x}} = 0.3813$. Then $h \cdot \widehat{\sigma}_{\bar{x}} = h \cdot \widehat{\sigma}_x / \sqrt{n} = 4 \cdot 0.3813 = 1.5252$. This is the value then for S_H, when S_H is expressed in units of $\sigma_{\bar{x}}$, that indicates a probable change in the process mean. After the process has been allowed to run for a time, we see that this value is first exceeded at sample number 19.

The cumulative sum technique can be done easily on a chart of the values of S_H, as we showed in Table 7.16. This, in fact, is a very acceptable way in which to construct a cumulative sum chart. A graph is also useful and can be produced by many statistical computer packages. Minitab produces cumulative sum charts; we show the cumulative sum chart for the data in Example 7.6.1 in Figure 7.14. Note that the vertical scale is in units of $\widehat{\sigma}_{\bar{x}}$, so corresponding to $h = 4$, is $4 \cdot \widehat{\sigma}_{\bar{x}} = 4 \cdot 0.3813 = 1.5252$. Figure 7.14 shows a change in the process at sample 19, consistent with the data in Table 7.20.

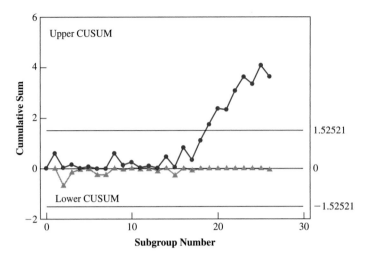

Figure 7.14 CUSUM Chart for Example 7.6.1

Figure 7.14 also shows the values of the lower cumulative sums, the values of S_L. These, which are used to detect **decreases** in the mean of the production process, are calculated in the following way:

$$S_{L_i} = \max[0, -z_i - k + S_{L_{i-1}}]$$

where we begin with $S_{L_0} = 0$. We show how the first few values of S_L are calculated.

$$S_{L_1} = \max[0, -2.05224 - 0.5 + 0] = 0$$
$$S_{L_2} = \max[0, 2.25393 - 0.5 + 0] = 1.7539$$
$$S_{L_3} = \max[0, -0.86770 - 0.5 + 1.7539] = 0.3862$$
$$S_{L_4} = \max[0, 0.21125 - 0.5 + 0.3862] = 0.0975$$

The values of S_{L_i} indicate when the process mean has decreased. When the process mean decreases, the samples means can be expected to be less than the target mean, producing negative values for z, and hence positive values for S_{L_i}. Minitab changes the sign of each S_{L_i} and indicates a negative control limit (in this case -1.5252) so that the values of both S_{H_i} and S_{L_i} can be seen in the same graph. In this example, the mean has surely not decreased.

V-Mask Cumulative Sum Charts

The cumulative sum process can very easily be carried out in the tabular form in which we have presented it. This in fact is the most desirable way in which to construct the control chart; the graph simply makes the table values visual and is technically unnecessary. One other type of graph is popular; this graph is known as a **V-mask.**

Note, in the cumulative sum chart, that whenever S_H or S_L becomes 0, then all the information in the previous samples is lost and the process essentially begins again so it is only in a general sense that all the previous data are used. We now use all the previous data in another way.

The V-mask control chart differs from the cumulative sum control chart we have presented here in some fundamental ways, so this chart is not always equivalent to our cumulative sum control chart. The basic idea is to chart the cumulative sums of the departures of the observed means from the target mean, μ, without resetting the sum to zero when the sum becomes negative or is less than the previous cumulative sum. Let

$$S_n = \sum_{j=1}^{n} (\overline{X}_j - \mu)$$

so S_n is the sum of deviations from the target mean, μ.

If the sample means are normally distributed, say, $N(\mu, \sigma)$, then $S_n \sim N(0, \sqrt{n}\,\sigma)$. Now suppose, at some point, say at the mth sample, that the mean of the process changes from μ to $\mu + \delta$. Now

$$S_n = \sum_{j=1}^{n} (\overline{X}_j - \mu) = \sum_{j=1}^{m} (\overline{X}_j - \mu) + \sum_{j=m+1}^{n} (\overline{X}_j - \mu)$$

and

$$\sum_{j=m+1}^{n} (\overline{X}_j - \mu) \sim N[(n-m)\delta, \sqrt{(n-m}\,\sigma)].$$

This means that a plot of S_n will change from having the slope 0 to having slope δ. Figure 7.15 shows a typical situation where we have plotted some cumulative sums and the sample number.

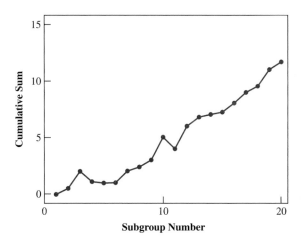

Figure 7.15 A Typical Cumulative Sum Chart

Suppose that we begin at sample number m and wish to detect whether or not the shift in slope begins there. The quantity δ is usually expressed in terms of standard deviations so we want to be able to detect a shift of $\delta = k\sigma_{\overline{x}}$ units in the process mean, for some constant k. This constant is usually referred to as the **reference value**. So we want to be able to distinguish between a slope of 0 and one of k (assuming now that all the units are in terms of standard deviations). It would then be reasonable to start at some point, say the point (m, S_m), and draw a line with slope $k/2$. However, one ought to give the process some slack and so it might be reasonable to begin at the point $(m, S_m + h)$ where again h is expressed in units of the standard deviation. The quantity h is usually called the **decision interval.** For the moment we will take $h = 4$, but we will return to a discussion of the value of h later in this section. We show the situation in Figure 7.16. If any of the plotted points become above the line we have drawn, this indicates that the slope, and hence the mean, has increased. We see in the picture that this has occurred.

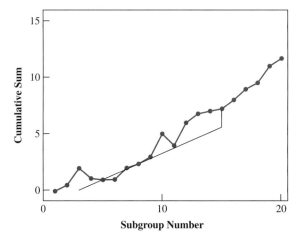

Figure 7.16 Part of a V-Mask Control Chart

It would be just as reasonable to go to some point beyond sample m, construct a line with slope $k/2$ from the point $(m, S_m - h)$, and see if any of the points previous to this one lie *outside* the line. This situation is shown in Figure 7.17, which returns to the nail data we have been considering.

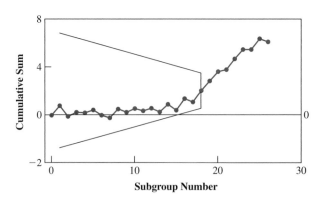

Figure 7.17 A V-Mask Control Chart for Example 7.6.1

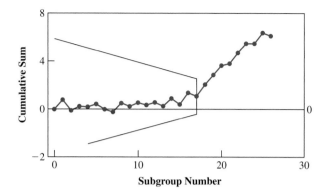

Figure 7.18 V-Mask for Example 7.6.1 Centered on Sample 17

In Figure 7.17, a line of slope $-k/2$ has been drawn through the point $(m, S_m + h)$. If the process mean has decreased, we would expect some points to be *above* this line. If the two lines shown were to be extended to the right, they would meet and form a letter V resting on its side. This is the origin of the name V-mask.

The angle at which the top and bottom lines create the V is determined by the values of h and k. The point of the V is seldom shown; Figure 7.18 shows a V-mask chart produced by Minitab for the data given in Example 7.6.1. It is possible to move the V; Figure 7.18 shows the V-mask positioned at the 17th observation. Up to that point, no observation lies outside the V. However, if we move the V to be positioned at the 19th observation, which we show in Figure 7.19, then we see that the observation at the 15th sample lies outside the V-mask. Why is this? Since the process mean increased, and since the shape of the V-mask remains constant, the V-mask must move upward, producing some points *below* the V-mask. This is an indication that the process mean has **increased.**

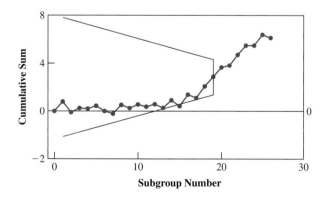

Figure 7.19 V-Mask for Example 7.6.1 Centered on Sample 19

In a similar way, points *above* the V-mask indicate that the process mean has **decreased.**

In this example, the V-mask shows the process has changed by the 19th sample, which is exactly the conclusion reached in both the tabular and graphic displays of the cumulative sum chart.

Cumulative sum charts deserve consideration due to their responsiveness to small changes in the mean of the production process. For this reason, they should be used in place of the usual $\overline{X}$ chart.

Now we turn to the values of the average run length for this type of chart.

Average Run Length

It is possible to determine the **average run length (ARL)** for both the cumulative sum charts and the V-mask designs we have discussed, but it is beyond the scope of this book to show the details. Table 7.21 shows some average run lengths for V-mask charts as functions of the parameter h and D, the actual displacement of the mean. The average run length then is the mean number of observations taken before the control chart will detect a change of $D\sigma$ units in the mean of the process. The parameter k has been set at $1/2$.

			h		
D	2	3	4	5	6
0	19.3	58.8	168	465	1280
0.25	15.2	36.2	74.2	139	249
0.50	9.63	17.2	26.6	38.0	513
0.75	6.27	9.67	13.3	17.0	20.9
1.00	4.49	6.40	8.38	10.4	12.4
1.50	2.74	3.75	4.75	5.75	6.74
2.00	1.99	2.68	3.34	4.01	4.67
2.50	1.58	2.12	2.62	3.11	3.62
3.00	1.32	1.77	2.19	2.57	2.98
4.00	1.07	1.31	1.71	2.01	2.24
5.00	1.006	1.07	1.31	1.69	1.95

Source: Lucas [52]

Table 7.21

We see here that a change of 2 sigma units with $h = 4$ gives an ARL of 3.34, while the $\overline{X}$ chart has an ARL of 7. The V-mask ARL can of course be decreased by decreasing h.

Usually, control charts are constructed with the ARL as a primary consideration. This, in fact, is the basis for choosing the parameters h and k. Although h has been chosen as an integer in Table 7.21, it need not be an integer.

EXERCISES 7.7

1. Regard consecutive pairs of data from Exercise 2 of Section 7.3 as samples of size 2 on page 359.
   a. Compute the four values for the means of these samples and then calculate the values of S_H and S_L for a cumulative sum control chart using $k = 1/2$. Recall that the target mean is 50 and that $\sigma = 10$.
   b. Compute both the upper and lower CUSUM limits and plot the CUSUM charts.

2. Use the data concerning the tensile strength of a material given in Exercise 8 of Section 7.3 on page 360.
   a. Show that the $\overline{X}$ chart indicates that the process is in control.
   b. Calculate both upper and lower CUSUM limits for the data.
   c. Assuming a target mean of 200 and estimating σ using the ranges of the samples, plot the CUSUM control chart and find when the process goes out of control.
   d. Use $h = 4$ and $k = 1/2$ to plot a V-mask control chart and find when the process goes out of control.

3. Use the data concerning output voltage given in Exercise 9 of Section 7.3 on page 361.
   a. Show that the $\overline{X}$ chart indicates that the process is in control.
   b. Calculate both upper and lower CUSUM limits for the data.
   c. Assuming a target mean of 19 and estimating σ using the ranges of the samples, plot the CUSUM control chart and find when the process goes out of control.
   d. Use $h = 4$ and $k = 1/2$ to plot a V-mask control chart and find when the process goes out of control.

4. Use the data concerning the filling of soft-drink cans given in Exercise 10 of Section 7.3 on page 361.
   a. Show that the $\overline{X}$ chart indicates that the process is in control.
   b. Calculate both upper and lower CUSUM limits for the data.
   c. Plot the CUSUM control chart and find when the process goes out of control.
   d. Use $h = 4$ and $k = 1/2$ to plot a V-mask control chart and find if the process ever goes out of control.

5. Use the data given in Exercise 11 of Section 7.3 on page 361.
   a. Show that the $\overline{X}$ chart indicates that the process is in control.
   b. Calculate both upper and lower CUSUM limits for the data.
   c. Plot the CUSUM control chart and find when the process goes out of control.
   d. Use $h = 4$ and $k = 1/2$ to plot a V-mask control chart and find if the process ever goes out of control.

6. Use the data for the second shift only given in Exercise 11 of Section 7.3 on page 361.
   a. Calculate both upper and lower CUSUM limits for the data.
   b. Assuming a target mean of 4.5 and estimating σ using the ranges of the samples, plot the CUSUM control chart and show that the process is in control.
   c. Use $h = 4$ and $k = 1/2$ to plot a V-mask control chart and show that the process is in control.
   d. Comment on the apparent trend shown in the control chart.

7.8 Acceptance Sampling

Control charts have been used widely and have improved the quality of products sold in many industries. However, if ordinary $\overline{X}$ charts are used, the process can drift out of control and remain out of control for some time without the manufacturer being aware of this, as we have seen. Even if the process has really not changed, or if it changes very little, it can still produce unacceptable numbers of product that do not meet the buyer's specifications. For this reason, produced product is often inspected after it has been produced and before it is sent to the buyer. Usually a sample is selected and the items in the sample are tested or measured. (Buyers also often inspect the product, so the principles shown here apply then as well.)

We show the principles involved through an example.

Example 7.8.1   A small lot of 20 items contains 5 that do not meet the manufacturer's specifications. A sample of size 4 is chosen. Since there is little value in selecting an item from the lot more than once, the sampling here is of course done without replacement.

There are $\binom{20}{4} = 4845$ different samples. Since there are 15 acceptable items, there are $\binom{15}{4} = 1365$ samples that contain none of the defective items. Regarding each of the $\binom{20}{4}$ samples as equally likely, we find that the probability the sample contains none of the defective items is

$$\frac{\binom{15}{4}}{\binom{20}{4}} = \frac{91}{323} = 0.281734.$$

There is then a considerable chance that the sample reveals none of the defective items in the lot despite the fact that $5/20 = 25\%$ of the lot is defective. However, this probability steadily decreases as the sample size increases.

Now if the sample reveals any defectives, they should be replaced with acceptable items before the lot is shipped to the buyer. What effect does this process, which is called **rectifying the sample,** have on the quality of the product sold? It obviously increases the quality of the product sold, but it is not clear by how much, so we seek a specific measure of this increase in quality.

We first make the process a bit more formal by introducing some calculations. Let X denote the random variable, which is the number of defective items in the sample. The possible values X can assume are then 0, 1, 2, 3, and 4. Then a sample containing x defectives and $4 - x$ good items has probability

$$P(X = x) = \frac{\binom{5}{x} \cdot \binom{15}{4-x}}{\binom{20}{4}} \quad \text{where } x = 0, 1, 2, 3, 4.$$

This probability distribution may be recognized as the hypergeometric probability distribution, which we considered in Chapter 2. Assume that any defective items found in the sample are replaced by acceptable items. Now we can construct Table 7.22 where we show the values of X, the probabilities, and the resulting proportion of defective items sold to the buyer.

X	% Defective Sold	Probability
0	$5/20 = 0.25$	$91/323 = 0.2817$
1	$4/20 = 0.20$	$455/969 = 0.4696$
2	$3/20 = 0.15$	$70/323 = 0.2167$
3	$2/20 = 0.10$	$10/323 = 0.0310$
4	$1/20 = 0.05$	$1/969 = 0.0010$

Table 7.22

What then is the mean value of the "% Defective Sold"? This is called the **average outgoing quality** (although it might more appropriately be called the average outgoing lack of quality) and is denoted by AOQ. In this case, we find that

$$\text{AOQ} = \frac{5}{20} \cdot \frac{91}{323} + \frac{4}{20} \cdot \frac{455}{969} + \frac{3}{20} \cdot \frac{70}{323} + \frac{2}{20} \cdot \frac{10}{323} + \frac{1}{20} \cdot \frac{1}{969} = \frac{1}{5}.$$

So the sampling scheme has reduced the percentage of bad product sold to the buyer from $5/20 = 25\%$ to $1/5 = 20\%$ on average. That is not a great gain, but it may be important for both the manufacturer and the buyer. As the sample size increases, the gain is of course much greater.

This raises the possibility—if the sample contains any defective items at all—of inspecting the entire lot and replacing any defective items found by good items. This process is not possible if the sampling is destructive, but if it is possible, the process is called **rectifying the lot.** It is probably clear, when this is possible, that greater gains are made in the quality of the product produced than when only the sample can be rectified.

Suppose the lot is accepted only if the sample contains no defective items. To calculate the average outgoing quality in this case, note that either the buyer receives 100% good product (if the sample shows one or more defective items) or the percentage defective in the lot (when the sample fails to reveal any defective product). So, in our example, recalling that the probability the sample contains no defective items is $91/323$, we find that

$$\text{AOQ} = \frac{0}{20} \cdot \left(1 - \frac{91}{323} \right) + \frac{5}{20} \cdot \frac{91}{323}$$

$$= \frac{91}{1292} = 0.0704334.$$

Now we have made a substantial gain over the original quality of the lot. The size of the gain is of course highly dependent upon the sample size. The proportion of defective items in the lot is also of great significance. We will investigate this later. ∎

Average Outgoing Quality Limit

Suppose the lot of N items contains D items that do not meet the manufacturer's specifications and we call these defectives for short. Suppose also that a random sample of size n is selected without replacement and that the lot is accepted only if the sample contains no defective items. Then the buyer will receive a lot with 0 defectives or a lot with D defectives. So the average outgoing quality is

$$\text{AOQ} = \frac{0}{N} \cdot \left[1 - \frac{\binom{N-D}{n}}{\binom{N}{n}} \right] + \frac{D}{N} \cdot \frac{\binom{N-D}{n}}{\binom{N}{n}}$$

$$= \frac{D}{N} \cdot \frac{\binom{N-D}{n}}{\binom{N}{n}}.$$

The behavior of this quantity may be surprising. Consider a specific example where the size of the lot, N, is 2000 and the sample size, n, is 100. We have then that

$$\text{AOQ} = \frac{D}{2000} \cdot \frac{\binom{2000 - D}{100}}{\binom{2000}{100}}.$$

A graph of AOQ as a function of D is shown in Figure 7.20. The figure shows that AOQ increases for a time, reaches a maximum, and then decreases. The reason for this is that the quantity D/N is steadily increasing and the quantity

$$\frac{\binom{N - D}{n}}{\binom{N}{n}}$$

is steadily decreasing while both are in the range from 0 to 1. Under these circumstances, the product will reach a maximum. This is shown in Figure 7.20 where the maximum value of AOQ can be calculated as 0.00356866 when $D = 19$. This maximum value of the average outgoing quality is called the **average outgoing quality limit** and is denoted by AOQL.

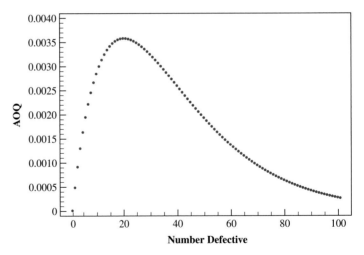

Figure 7.20 Average Outgoing Quality as a Function of the Number Defective in the Population

While the average outgoing quality in this case cannot exceed 0.00356866, this does not mean that poor lots cannot be produced and escape the filter the sampling process provides.

Operating Characteristic Curves

What is the effect of the sample size on the probability the lot is accepted? Suppose that a lot of 500 items contains D defectives and that a sample of n items is selected without replacement. Suppose again that the lot is accepted only if the sample contains no defective items. Then the probability the lot is accepted is

$$\frac{\binom{500-D}{n}}{\binom{500}{n}}.$$

Figure 7.21 shows a number of curves, called **operating characteristic curves,** that show the probability the lot is accepted as a function of the sample size, n. The curves become sharper with increasing sample size.

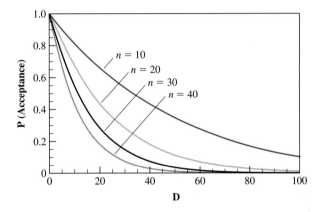

Figure 7.21 Operating Characteristic Curves for Various Sample Sizes

Producer's and Consumer's Risks

There are two risks involved in the acceptance sampling plans we have considered. The producer would like to guard against the rejection of a lot that is good while the consumer would like to guard against the acceptance of a lot that is bad. Neither of these risks can be avoided with certainty and the words "good" and "bad" must be defined, of course. It is possible in some cases to find operating characteristic curves that have properties satisfying both the consumer and the producer.

For example, suppose the producer would like to have a lot with 2 defectives accepted with probability 0.90 while the consumer would like to have a lot with 55 defectives rejected with probability 0.95. If the lot contains 500 items, then we would like to find a sample size n so that

$$\frac{\binom{498}{n}}{\binom{500}{n}} = 0.90 \quad \text{and} \quad \frac{\binom{445}{n}}{\binom{500}{n}} = 0.05.$$

We find that $n = 25$ satisfies both of these conditions since

$$\frac{\binom{498}{25}}{\binom{500}{25}} = 0.902405 \quad \text{and} \quad \frac{\binom{445}{25}}{\binom{500}{25}} = 0.050278.$$

Setting both of these conditions in general produces a difficult numerical problem, which may have no solution. A computer algebra system is of immense aid here.

Double Sampling

It is not uncommon for a lot to be accepted if a sample contains no more than, say, c defective items and to be rejected if the sample contains, say, d defective items where $d > c$. If the number of defectives falls between c and d, then occasionally a second sample is selected and the lot rejected if the *total* number of defective items found in both samples exceeds some quantity, say e, and the lot is accepted otherwise. The calculations involved will be illustrated with a specific example.

Example 7.8.2   A lot of 2000 batteries contains 100 unacceptable batteries. A sample of 50 is taken and the lot is accepted if the sample contains no more than 1 unacceptable battery. If the sample contains 2 or 3 unacceptable batteries, an additional sample of 50 is taken. The lot is accepted if the total number of defective batteries in both samples is at most 5; otherwise, the lot is rejected. Assume that none of the batteries found in the first sample are replaced before the second sample is selected and so none of these are available when the second sample is selected. We want the probability the lot is accepted.

Let X denote the number of unacceptable batteries in the first sample. Then the probability the lot is accepted on the basis of the first sample is

$$P(X \leq 1) = \sum_{x=0}^{1} \frac{\binom{100}{x} \cdot \binom{1900}{50-x}}{\binom{2000}{50}} = 0.275604.$$

Now if the first sample has 2 or 3 unacceptable batteries, then the second sample is taken. The lot now contains 1950 batteries of which $100 - X$ are unacceptable while $1900 - (50 - X) = 1850 + X$ are good batteries. If the lot is to be accepted, then the second sample of size 50 must be such that the total number of defective batteries selected in the two samples is at most 5. Let Y denote the number of defective batteries found in the second sample. We find that the probability the lot is accepted on the basis of the second sample is

$$\frac{\binom{100}{2} \cdot \binom{1900}{48}}{\binom{2000}{50}} \cdot \sum_{y=0}^{3} \frac{\binom{98}{y} \cdot \binom{1852}{50-y}}{\binom{1950}{50}} + \frac{\binom{100}{3} \cdot \binom{1900}{47}}{\binom{2000}{50}} \cdot \sum_{y=0}^{2} \frac{\binom{97}{y} \cdot \binom{1853}{50-y}}{\binom{1950}{50}}$$

$$= 0.320896.$$

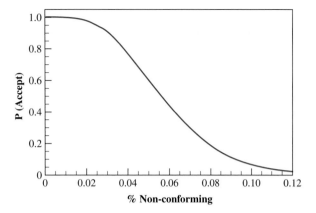

Figure 7.22 Probability of Acceptance in a Double Sampling Plan

The probability the lot is accepted is then

$$0.275604 + 0.320896 = 0.5965$$

and, assuming that none of the unacceptable batteries found in the samples are sold, the average outgoing quality is

$$\sum_{x=0}^{1} \frac{100-x}{2000} \frac{\binom{100}{x} \cdot \binom{1900}{50-x}}{\binom{2000}{50}} + \frac{\binom{100}{2} \cdot \binom{1900}{48}}{\binom{2000}{50}} \cdot \sum_{y=0}^{3} \frac{98-y}{2000} \frac{\binom{98}{y} \cdot \binom{1852}{50-y}}{\binom{1950}{50}}$$

$$+ \frac{\binom{100}{3} \cdot \binom{1900}{47}}{\binom{2000}{50}} \cdot \sum_{y=0}^{2} \frac{97-y}{2000} \frac{\binom{97}{y} \cdot \binom{1853}{50-y}}{\binom{1950}{50}} = 0.029078. \qquad \blacksquare$$

Example 7.8.2 can be generalized to the case where $p\%$ of the batteries in the lot of 2000 batteries are defective. Figure 7.22 shows the probability the lot is accepted by this double sampling plan as a function of p.

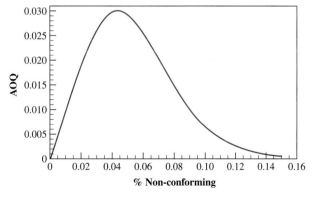

Figure 7.23 Average Outgoing Quality in a Double Sampling Plan

Figure 7.23 shows the average outgoing quality as a function of p. Again the graph shows a maximum, or what we have called the average outgoing quality limit. This occurs when $p = 0.0431$ and the average outgoing quality limit is 0.0300113. Since in our example the lot contained 5% defective batteries, this shows that the double sampling plan is effective in reducing the percent defective product sold to the consumer.

Average Sampling Number

The total number of items inspected is an important consideration when a single or a double sampling plan is constructed. We calculate the expected number of items sampled through an example.

Example 7.8.3 Suppose a lot of 2000 items contains 100 defective items. If a sample of 50 shows at most 1 defective item, the lot is accepted and shipped to the buyer; otherwise, the entire lot is inspected. What is the expected number of items sampled? This is called the **average sampling number.** We sample 50 items with probability

$$\sum_{x=0}^{1} \frac{\binom{100}{x} \cdot \binom{1900}{50-x}}{\binom{2000}{50}} = 0.275604$$

and the entire lot of 2000 items is then inspected with probability $1 - 0.275604 = 0.724396$, so the mean number of items inspected is

$$50(0.275604) + 2000(0.724396) = 1462.6,$$

which means on average this single sampling plan will sample a fairly large proportion of the lot.

Now extend this situation to a double sampling plan. Suppose again that an initial sample of 50 is taken and the lot is accepted if the sample contains no more than 1 unacceptable item and is rejected (and the entire lot is inspected) if the number of unacceptable items exceeds 3. If the sample contains 2 or 3 unacceptable items, an additional sample of 50 is taken. The lot is accepted if the total number of defective items in both samples is at most 5; otherwise, the lot is rejected and the entire lot is inspected. We want to find the expected number of items sampled.

There are three possibilities: The total number of items sampled is 50, 100, or 2000.

1. The probability that 50 items are sampled is the probability the lot is accepted by the first sample. This is

$$\sum_{x=0}^{1} \frac{\binom{100}{x} \cdot \binom{1900}{50-x}}{\binom{2000}{50}} = 0.275604.$$

2. The probability that 100 items are sampled is the probability the first sample has 2 or 3 defective items and the total number of defective items in the first and second

samples is at most 5. The probability of this is

$$
\frac{\binom{100}{2}\cdot\binom{1900}{48}}{\binom{2000}{50}}\cdot\sum_{y=0}^{3}\frac{\binom{98}{y}\cdot\binom{1852}{50-y}}{\binom{1950}{50}}+\frac{\binom{100}{3}\cdot\binom{1900}{47}}{\binom{2000}{50}}\cdot\sum_{y=0}^{2}\frac{\binom{97}{y}\cdot\binom{1853}{50-y}}{\binom{1950}{50}}
$$

$$
= 0.320896.
$$

3. There are two possibilities for which the entire lot of 2000 items is inspected. First, if the lot is rejected by the first sample, then all 2000 items in the lot are inspected. This has probability

$$
1-\sum_{x=0}^{3}\frac{\binom{100}{x}\cdot\binom{1900}{50-x}}{\binom{2000}{50}}=0.238056.
$$

The second possibility under which all 2000 items in the lot are inspected is that the first sample shows 2 or 3 defective items and the second sample produces then a total of 6 or more defective items. This has probability

$$
\frac{\binom{100}{2}\cdot\binom{1900}{48}}{\binom{2000}{50}}\cdot\left[1-\sum_{x=0}^{3}\frac{\binom{98}{x}\cdot\binom{1852}{50-x}}{\binom{1950}{50}}\right]
$$

$$
+\frac{\binom{100}{3}\cdot\binom{1900}{47}}{\binom{2000}{50}}\cdot\left[1-\sum_{x=0}^{2}\frac{\binom{97}{x}\cdot\binom{1853}{50-x}}{\binom{1950}{50}}\right]=0.165444.
$$

So the probability that all 2000 items are inspected is $0.238056 + 0.165444 = 0.4035$. Then we sample 50 items with probability 0.275604, we sample 100 items with probability 0.320896, and we sample 2000 items with probability 0.4035. Note that the sum of these probabilities is

$$
0.275604 + 0.320896 + 0.4035 = 1,
$$

as it should be. So the mean number of items sampled, or the average sampling number, is

$$
50 \cdot 0.275604 + 100 \cdot 0.320896 + 2000 \cdot 0.4035 = 852.87.
$$

This is much less than the average sampling number for the single sampling plan. ∎

The sample sizes as well as the criteria used to reject a sample and inspect the entire lot clearly influence the average sampling number. As a result, this becomes a consideration in the design of a sampling plan since costs are usually involved with each item sampled.

EXERCISES 7.8

A computer should be used to solve these problems.

1. A small manufacturing plant produces 32 items in a day. These are subject to the following inspection plan: A sample of 2 items is chosen and the lot is accepted if the sample contains no defective items. Find the probability the lot is accepted if in fact it contains 3 defective items.

2. A lot of 100 television sets contains 10 that do not meet the customer's specifications. A random sample of 10 television sets is selected and the lot is accepted if at most 2 of the sampled sets do not meet the customer's specifications.
   a. Find the probability the lot is accepted.
   b. Use the binomial distribution to approximate the probability in part (a).

3. A bag containing 2000 washers contains 300 that are not of the proper size. A random sample of 50 washers is chosen; the lot is accepted if the sample contains at most 1 washer not of the proper size.
   a. Find the probability the lot is accepted.
   b. Approximate the probability in part (a) using the binomial distribution.

4. A lot of 50 items contains 4 items that are defective. A random sample of 2 items is selected and the lot is accepted if the sample contains no defectives.
   a. Find the probability the lot is accepted.
   b. Assuming that the sample is rectified when any defective items are found in it, find the average outgoing quality.
   c. Plot the operating characteristic curve assuming that the lot contains D defective items.

5. In Exercise 1, find the average outgoing quality and the average outgoing quality limit if the lot contains D defective items.

6. In Exercise 3, find the average outgoing quality and the average outgoing quality limit if the lot contains D defective items.

7. A lot of 500 items is subject to sampling inspection. The lot is accepted only if it contains no defective items. The manufacturer would like to accept a lot containing 1 defective item with probability 0.95; the buyer would like to accept the lot with probability 0.05 if the lot contains 60 defective items. Find the sample size that will accomplish this.

8. A manufacturer subjects a day's production of 300 items to a sampling inspection plan that accepts a lot only if the sample contains no defective items. The manufacturer would like a lot with 4 defective items accepted with probability 0.93 while a buyer would like a lot containing 20 defective items accepted with probability 0.71. Find the sample size that will accomplish this.

9. Suppose a lot of 400 items contains 15 that do not meet the customer's specifications. A sample of 15 items is selected and the lot is accepted if the sample contains no defective items. The lot is rejected if the sample contains 3 or more defective items. If the sample contains 1 or 2 defective items, a second sample of size 20 is chosen and the lot is accepted only if the total number of defective items in the two samples selected is 3 or less.
   Assume that items sampled in the first sample are not replaced in the lot if a second sample is selected.
   a. Find the probability the lot is accepted.
   b. Use the binomial distribution to approximate the probability in part (a).

10. Find the average sampling number for the situation in Exercise 1, assuming a rejected lot is subject to 100% inspection with any defective items replaced by acceptable items.

11. Find the average sampling number for the situation in Exercise 3 assuming a rejected lot is subject to 100% inspection with any defective items replaced by acceptable items.

12. Find the average outgoing quality and the average outgoing quality limit for Exercise 9 assuming that any defective items are replaced with acceptable items when they are found and that a rejected lot is subject to 100% inspection.

13. A lot of 750 automobile tires contains 75 tires that do not meet the customer's specifications. A sample of 100 tires is chosen and the lot is accepted if at most 1 of these tires is defective. If 2 or 3 of these tires are defective, then a second sample of 100 tires is chosen and the lot accepted if the total number of defective tires in the two samples combined is at most 6. Otherwise, the lot is rejected.
   a. Find the probability the lot is accepted.
   b. Use the binomial distribution to approximate the probability in part (a).

c. Find the average outgoing quality.

d. Find the average outgoing quality limit assuming that the lot contains D defective tires.

e. Find the expected number of items sampled.

14. Find the average sampling number for the situation in Exercise 9 assuming a rejected lot is subject to 100% inspection with any defective items replaced by acceptable items.

15. A lot of 500 items contains 50 that do not meet the customer's specifications. A sample of 40 is chosen and the lot is accepted if at most 1 of these items is defective.

If 2 or 3 of these items are defective, then a second sample of 40 items is chosen and the lot is accepted if the total number of defective items in the two samples combined is at most 5. Otherwise, the lot is rejected.

a. Find the probability the lot is accepted.

b. Use the binomial distribution to approximate the probability in part (a).

c. Find the average outgoing quality.

d. Find the average outgoing quality limit assuming that the lot contains D defective items.

e. Find the expected number of items sampled.

Chapter Review

This chapter was concerned with statistical process control—the application of statistical methods to production and manufacturing. Two major topics were discussed: control charts and acceptance sampling.

Key Concepts

- Creating a control chart for sample means
- Estimating the unknown process standard deviation using 1) sample ranges, 2) sample standard deviations, or 3) pooling the sample standard deviations
- Creating a control chart for sample ranges
- Creating a control chart for sample standard deviations
- Using the Western Electric Rules to improve the sensitivity of a control chart
- Creating and analyzing an np control chart
- Creating and analyzing a p control chart
- Creating and analyzing a c control chart
- Creating a cumulative sum control chart
- Creating a V-mask control chart
- Analyzing a single sample acceptance sampling plan
- Analyzing a double sample acceptance sampling plan
- Finding the average outgoing quality limit in both single and double sampling plans
- Finding the average sampling number in both single and double sampling plans

Key Terms

Control Chart for Variables A *control chart for variables* is a time series graph of a statistic, usually the mean, range, or standard deviation, calculated from periodic samples of a production process.

Control Limits	*Control limits* are numerical limits created to detect extreme values of a measured statistic. Values in excess of the control limits often indicate a change in the production process.
Control Chart for Attributes	A *control chart for attributes* is a time series graph of the number or proportion of unacceptable items in periodic samples from a production process.
Cumulative Sum Control Chart	A *cumulative sum control chart* uses some or all samples previously observed from a production process.
V-Mask Control Chart	A *V-Mask control chart* is a control chart exhibiting a V-mask showing when the process mean (or some other statistic) exceeded control limits.
Acceptance Sampling Plan	An *acceptance sampling plan* is a plan for drawing one or more samples from the production of a process and then using the sample to either accept or reject the total production.
Average Outgoing Quality	The *average outgoing quality (AOQ)* is the mean percentage of unacceptable product delivered to a customer.
Average Outgoing Quality Limit	The *average outgoing quality limit (AOQL)* is the largest value the average outgoing quality attains regardless of the quality of the production process.
Average Sampling Number	The *average sampling number* is the expected number of items sampled using either a single or a double sampling plan.

Key Theorems/Facts

Estimating (σ)	Three methods for *estimating* σ, the unknown process standard deviation, by $\hat{\sigma}$ are: 1) $\hat{\sigma} = \overline{R}/d_2$ where $\overline{R}$ is the mean of the sample ranges. A table of values of d_2 can be found on page 357. 2) $\hat{\sigma} = \overline{s}/c_4$ where $\overline{s}$ is the mean of the sample standard deviations. A table of values of c_4 can be found on page 355.

3) Using the pooled standard deviation, s_p as $\hat{\sigma}$.

Control Limits for $\overline{X}$

Control limits for $\overline{X}$ are $\overline{\overline{X}} \pm 3(\hat{\sigma}/\sqrt{n})$

Control Limits for R

Control limits for R are $\overline{R} \pm 3d_3(\overline{R}/\overline{d}_2)$. A table of values of d_3 can be found on page 362.

Control Limits for s

Control limits for the sample standard deviation are usually set at

$$\overline{s} \pm 3 \cdot (\overline{s}/c_4) \cdot \sqrt{1 - c_4^2}.$$

***np* Chart Control Limits**

An *np chart* is a control chart on the number of unacceptable parts produced by a process. The *control limits* are usually set at $\overline{X} \pm 3\sqrt{np(1-p)}$ where $\overline{X}$ is the mean number of defective parts produced using all the samples and p is estimated by the total proportion of defective parts produced using all the samples.

***p* Chart Control Limits**

The *p chart* is a control chart showing how the proportion of defective parts produced by the process is changing over time. *Control limits* are usually set at $p \pm 3\sqrt{p(1-p)/n}$ where p is estimated by the proportion of defective parts produced using all the samples.

Poisson Probability Distribution Function

The *Poisson* approximation is based on the fact that the binomial probability distribution function,

$$f(x) = \binom{n}{x} p^x (1-p)^{n-x} \quad x = 0, 1, \ldots, n,$$

can be well approximated by the *Poisson probability distribution function,*

$$f(x) = \frac{e^{-\lambda}\lambda^x}{x!} \quad x = 0, 1, 2, \ldots.$$

***c* Chart Control Limits**

A control chart on the number of unacceptable items is called a *c chart*. The chart *uses* the Poisson approximation to the binomial distribution. *Control limits* are $\overline{X} \pm 3\sqrt{\overline{X}}$.

Sample Size to Detect a Change in the Process Proportion Defective

The *sample size, n,* necessary *to detect a change in the proportion of defective* parts produced by the process with probability 1/2 when this proportion changes from p to δp where

$$0 \leq \delta \leq 1/p \text{ is } n \geq \frac{9(1 - p)}{p(\delta - 1)^2}.$$

Western Electric Rules

The *Western Electric Rules* are designed to increase the sensitivity of the control chart. They are:

Test 1. Points more than k sigma units from the center line.

Test 2. k points in a row on the same side of the center line.

Test 3. k points in a row all increasing or decreasing.

Test 4. k points in a row alternating up and down.

Test 5. At least k out of $k + 1$ points in a row more than 2 sigma units from the center line.

Test 6. At least k out of $k + 1$ points in a row more than 1 sigma unit from the center line.

Test 7. k points in a row within 1 sigma unit of the center line.

Test 8. k points in a row more than 1 sigma unit from the center line (on either side of the center line).

Probabilities associated with each of these tests are given in Section 7.6 for various values of k for each test.

Cumulative Sum Control Chart

For each sample, let $z = (\overline{x} - \mu)/(\hat{\sigma}/\sqrt{n})$. If we wish to detect a change in the process mean of $\delta\sigma_{\overline{x}}$, we generally choose the constant k as $\delta/2$. Then let $S_0 = 0$ and $S_{H_i} = \max[0, z_i - k + S_{H_{i-1}}]$. Once S_H

reaches a threshold value of h, say, this indicates an increase in the process mean. The quantity h is generally chosen as 4 or 5. Decreases in the process mean can also be detected by letting $S_{L_0} = 0$ and $S_{L_i} = \max[0, -z_i - k + S_{L_{i-1}}]$.

V-Mask Control Charts

Let $S_n = \sum_{j=1}^{n}(\overline{X}_j - \mu)$. The V-mask is constructed by drawing a line of slope $k/2$ from the point $(m, S_m - h)$ and a line of slope $-k/2$ from the point $(m, S_m + h)$. Points falling outside these lines indicate, respectively, an increase or a decrease in the process mean.

Average Outgoing Quality

Suppose the lot of N items contains D items that do not meet the manufacturer's specifications. Suppose also that a random sample of size n is selected without replacement and that the lot is accepted only if the sample contains no defective items. The *average outgoing quality* is

$$AOQ = \frac{D}{N} \cdot \frac{\binom{N-D}{n}}{\binom{N}{n}}.$$

Average Outgoing Quality Limit

The quantity

$$AOQ = \frac{D}{N} \cdot \frac{\binom{N-D}{n}}{\binom{N}{n}}$$

can be shown to have an upper limit. This is called the *average outgoing quality limit*.

Double Sampling Plans

Suppose a lot is accepted if a sample contains no more than say c defective items and is rejected if the sample contains say d defective items where $d > c$. In a *double sampling plan* a second sample is selected if the number of defectives falls between c and d and the lot is rejected if the *total* number of defective items found in both samples exceeds some quantity, say e, and the lot is accepted otherwise. We can calculate both the average outgoing quality and the average outgoing quality limit for a double sampling plan.

Appendix A
Using *Mathematica*

This is a brief explanation of how the computer algebra system *Mathematica* was used to produce some of the graphs and to do some of the calculations in the text. It is not our purpose here to give the complete syntax for each command used; rather, we illustrate how some of the resuts in the text were produced. As we will see, *Mathematica's* capabilities far exceed those of any pocket calculator and produce results far beyond the capability of any printed table. Version 4.1 of *Mathematica* was used to produce the results in this Appendix and those in the text.

The following set of commands comprise what *Mathematica* calls a **notebook** and will run exactly as they are written here, with the exception of the random samples which will differ each time the notebook is evaluated. The commands here are frequently followed by a semicolon which prevents the output from appearing. These should not be included if it is desired to view the output.

Mathematica, because of its power, has been used in the text primarily for calculation with probability distributions, including noncentral and bivariate distributions, and it has been used for graphs because of the ease with which these can be produced and because of the wide array of graphic options *Mathematica* presents to the user.

Mathematica contains two packages of programs on statistics and graphics that are not available unless called. We begin with two commands that allow commands in these packages to be used. Then we show some calculations and graphs from various parts of the text. Bold type indicates input to *Mathematica* (also identified by the symbols such as *In[22]:=*) while output appears in ordinary type (also preceded by symbols such as *Out[22]:=*).

In[1]:= **<< Statistics 'Master'**

In[2]:= **<< Graphics 'Master'**

405

Calculations

Mathematica does calculations with exact arithmetic, so calculations can be done to any degree of accuracy. Several calculations in Chapter 2 were done using factorials and binomial coefficients, so we begin with these.

In[3]:= **10!**

Out[3]= **3628800**

A group of factorials can be calculated by producing a table of values. The symbol n here represents the variable and the domain of values of n is from 2 to 20 with a default increment of 1.

In[4]:= **Table[n!, {n, 2, 20}]**

Out[4]= **{2, 6, 24, 120, 720, 5040, 40320, 362880, 3628800, 39916800, 479001600, 6227020800, 87178291200, 1307674368000, 20922789888000, 355687428096000, 6402373705728000, 121645100408832000, 2432902008176640000}**

Binomial coefficients can be found easily. Binomial[n, r] represents the usual symbol $\binom{n}{r}$, which can also be used.

In[5]:= **Binomial[10, 3]**

Out[5]= **120**

This shows that there are 120 samples of size 3 that can be selected from a population of 10 distinct items. Ratios using binomial coefficients can also be found directly. Here is the probability a sample of size 34 from a population of 100 items arises from a subset of 80 items:

In[6]:= **Binomial[80, 34]/Binomial[100, 34]**

Out[6]= $\frac{19837688}{261494459199}$

Here we receive an exact answer. This can be evaluated numerically in one of two ways (% represents the immediately preceding result):

In[7]:= **% // N**

Out[7]= **0.0000758627**

A second way to evaluate the answer is:

In[8]:= **Binomial[80, 34]/Binomial[100, 34] // N**

Out[8]= **0.0000758627**

A hypergeometric probability was calculated in Example 2.6.1. This can be done as follows:

In[9]:= **Binomial[10, 3]*Binomial[90, 12]/Binomial[100, 15]**

Out[9]= $\frac{616980915}{4755579521}$

In[10]:= **% // N**

Out[10]= **0.129738**

A complete table of values of the hypergeometric distribution can also be found:

In[11]:= **t1 = Table[Binomial[10, x]*Binomial[90, 15 - x]/**
     Binomial[100, 15], {x, 0, 10}] // N

Out[11]= **{0.180769, 0.35678, 0.291911, 0.129738, 0.0344874,**
     0.00569042, 0.000585434, 0.0000367171,
     1.32713 $\times 10^{-6}$, 2.45764 $\times 10^{-8}$, 1.7348 $\times 10^{-10}$}

The output is more easily read if the values of the random variable X are also printed:

In[12]:= **Table[{x, Binomial[10, x]*Binomial[90, 15 - x]/**
     Binomial[100, 15]}, {x, 0, 10}] // N

Out[12]= **{{0., 0.180769}, {1., 0.35678}, {2., 0.291911},**
     {3., 0.129738}, {4., 0.0344874}, {5., 0.00569042},
     {6., 0.000585434}, {7., 0.0000367171},
     {8., 1.32713$\times 10^{-6}$}, {9., 2.45764$\times 10^{-8}$},
     {10., 1.7348$\times 10^{10}$}}

TableForm provides an even easier way to show the result:

In[13]:= **Table[{x, Binomial[10, x]*Binomial[90, 15 - x]/**
     Binomial[100, 15]}, {x, 0, 10}] // N // TableForm

Out[13]/Table Form=

0.	0.180769
1.	0.35678
2.	0.291911
3.	0.129738
4.	0.0344874
5.	0.00569042
6.	0.000585434
7.	0.0000367171
8.	1.32713×10^{-6}
9.	2.45764×10^{-8}
10.	1.7348×10^{-10}

These values can also be shown graphically by using the command **ListPlot:**

In[14]:= **ListPlot[t1, AxesOrigin $\rightarrow$ {0.69, -0.005},**
     DefaultFont $\rightarrow$ {"Helvetica-Bold", 12},
     AxesLabel $\rightarrow$ {x, Freq},
     Prolog $\rightarrow$ AbsolutePointSize[4]]

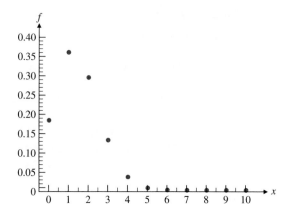

Out[14]= **-Graphics-**

The command **ListPlot,** as well as all other graphics commands in *Mathematica*, has a number of options, some of which are shown in the command above.

Probability Distributions

All the usual probability distributions, both discrete and continuous, are built into *Mathematica*. We show how calculations can be made with the hypergeometric, binomial, normal, chi-squared, Student *t* and *F* probability distributions as we considered them in Sections 2.7, 2.8, 2.11, 3.4, and 3.7. We show how some bivariate normal probability distributions are used in Section 2.12. Some calculations using the Poisson distribution, used in Chapter 7, are also given.

Hypergeometric Distribution

In the previous section, we calculated the probability that a sample of size 34 from a population of 100 items, 80 of which had a special characteristic, all shared the special characteristic. We used the ratio of binomial coefficients giving the probability, but we could as easily have used the following:

In[15]:= **PDF[HypergeometricDistribution[34, 80, 100], 34] // N**

Out[15]= **0.0000758627**

Here PDF stands for Probability Distribution Function (the cumulative distribution function, CDF, is also available); the general syntax is

$$\text{PDF[HypergeometricDistribution } [n, n_{\text{succ}}, n_{\text{tot}}], x]$$

where n is the sample size, n_{succ} is the number of special items in the population, n_{tot} is the population size, and x is the number of special items in the sample.

Binomial Probability Distribution

The binomial distribution can be used by specifying the values of n and p. Here we use the parameters given in Examples 2.7.2 and 2.7.3, where $n = 1000$ and $p = 0.8$. We first find the probability that $X = 822$.

In[16]:= **PDF[BinomialDistribution[1000, 0.8], 822]**
Out[16]= **0.00694972**

The probability that X is at least 780 is

In[17]:= **1 - CDF[BinomialDistribution[1000, 0.8], 779]**
Out[17]= **0.946143**

These calculations could also be made directly:

In[18]:= **Binomial[1000, 822]*((0.8)^822)*(0.2)^178**
Out[18]= **0.00694972**

In[19]:= **Sum[Binomial[1000, x]*((0.80)^x)*(0.2)^(1000 - x),**
{x, 780, 1000}]
Out[19]= **0.946143**

Means and variances of probability distributions can be found directly, or by using functions that are defined in *Mathematica*. Here are three ways to calculate the mean of a binomial distribution.

In[20]:= **Mean[BinomialDistribution[1000, 0.80]]**
Out[20]= **800**

In[21]:= **Sum[x*PDF[BinomialDistribution[1000, 0.80], x],**
{x, 0, 1000}]
Out[21]= **800**

Note that the random variable (X here) must be specified for the PDF. A final calculation to produce the mean is

In[22]:= **Sum[x*Binomial[1000, x]*((0.80)^x)*(0.2)^(1000-x),**
{x, 0, 1000}]
Out[22]= **800**

Variances can be done similarly. We only show

In[23]:= **Variance[BinomialDistribution[1000, 0.80]]**
Out[23]= **160**

The graph in Figure 2.10 was produced as follows. We calculate first the tick marks to be used on the horizontal axis since only some of the values of the probability distribution can be shown easily. A table of pairs of values is produced. The first component is the position on the x axis while the second component is the value to be printed at that position.

In[24]:= `tix1 = Table[{i, i + 7}, {i, 1, 8}]`

Out[24]= `{{1, 8}, {2, 9}, {3, 10}, {4, 11}, {5, 12},`
`{6, 13}, {7, 14}, {8, 15}}`

In[25]:= `ListPlot[Table[PDF[BinomialDistribution[15, 0.80],`
`x], {x, 8, 15}], DefaultFont → {"Helvetica-Bold",`
`12}, Ticks → {tix1, Automatic},`
`AxesLabel → {x, Freq},`
`Prolog → AbsolutePointSize[4],`
`AxesOrigin → {0.68, 0}]`

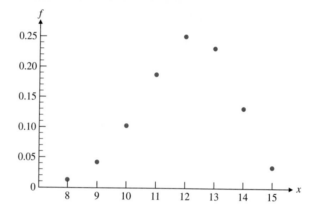

Out[25]= `-Graphics-`

Normal Probability Distribution

The normal probability distribution is considered in Section 2.8. It is specified by
NormalDistribution$[\mu, \sigma]$.

In Example 2.8.1, we calculated the probability that the normal random variable X
exceeded 550 for a N[500,100] distribution. This can be done by *Mathematica* in several
ways, some of which we now show.

In[26]:= `1 - CDF[NormalDistribution[500, 100], 550] // N`

Out[26]= `0.308538`

More accuracy can also be used by specifying the precision desired. We calculate the
above probability to 20 decimal places:

In[27]:= `N[1 - CDF[NormalDistribution[500, 100], 550], 20]`

Out[27]= `0.30853753872598689636`

Another calculation is:

In[28]:= `Integrate[PDF[NormalDistribution[500, 100], x],`
`{x, 550, Infinity}] // N`

Out[28]= `0.308538`

The mathematical expression for the normal probability distribution function could also be used with the same result.

There is no need for the z transformation when *Mathematica* is available, but calculations can be made using it.

Here are the calculations given in Example 2.8.4, each to 20 decimal places. In that example, we established the "2/3, 95, 99" Rule.

In[29]:= **N[CDF[NormalDistribution[0, 1], 1] -**
CDF[NormalDistribution[0, 1], -1], 20]
Out[29]= **0.68268949213708589717**

In[30]:= **N[CDF[NormalDistribution[0, 1], 2] -**
CDF[NormalDistribution[0, 1], -2], 20]
Out[30]= **0.95449973610364158560**

In[31]:= **N[CDF[NormalDistribution[0, 1], 3] -**
CDF[NormalDistribution[0, 1], -3], 20]
Out[31]= **0.99730020393673981095**

Inverse calculations can also be made by specifying the probability distribution and the desired cumulative probability.

In[32]:= **Quantile[NormalDistribution[0, 1], 0.975]**
Out[32]= **1.95996**

Or a nonstandard normal distribution can be used:

In[33]:= **Quantile[NormalDistribution[500, 100], 0.975]**
Out[33]= **695.996**

The quantiles for arbitrary values can also be found. Using the usual tables would be quite difficult.

In[34]:= **Quantile[NormalDistribution[0, 1], 0.3865432]**
Out[34]= **-0.28834**

In Section 3.8, the *p*-value for a test was calculated. In that example, the test value of a sample mean is 23.72, based on a sample of size 25, possibly from a normal distribution with mean 22 and known standard deviation 5. The *p*-value for this test is then

In[35]:= **2 (1 - CDF[NormalDistribution[22, 5/Sqrt[25]],**
23.72])
Out[35]= **0.0854324**

Chi-Squared Probability Distribution

The chi-squared probability distribution is specified by its number of degrees of freedom. In Example 2.11.2, we sought the probability that a chi-squared random variable with 8 degrees of freedom exceeded 12. This was done as follows.

In[36]:= **1 - CDF[ChiSquareDistribution[8], 12] // N**

Out[36]= **0.151204**

In Example 2.11.3, we calculated a confidence interval. The critical values of the chi-squared distribution can be calculated as follows.

In[37]:= **Quantile[ChiSquareDistribution[8], 0.975]**

Out[37]= **17.5345**

In[38]:= **Quantile[ChiSquareDistribution[8], 0.025]**

Out[38]= **2.17973**

Figure 2.21 was produced using these instructions:

In[39]:= **Plot[PDF[ChiSquareDistribution[4], x], {x, 0, 12},**
        DefaultFont → {"Helvetica-Bold", 12},
        AxesLabel → {x, Freq}]

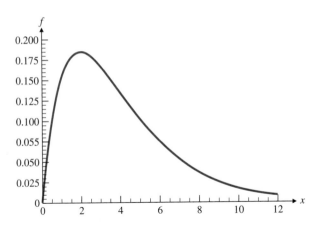

Out[39]= **-Graphics-**

The command **Plot** has a number of other options, in addition to those used here.

Student *t* Distribution

The Student *t* distribution is completely specified by its number of degrees of freedom. In Example 3.4.1, we calculated a 95% confidence interval. The upper 97.5% point on the Student *t* distribution with 9 degrees of freedom is found as follows.

In[40]:= **Quantile[StudentTDistribution[9], 0.975]**

Out[40]= **2.26216**

Cumulative probabilities are calculated using the CDF function.

In[41]:= **CDF[StudentTDistribution[24], 1.62352]**

Out[41]= **0.941229**

It is interesting to compare the critical points on the Student t distribution with its normal approximation. Usually the normal approximation is used for 30 or more degrees of freedom, but the actual Student t values differ significantly from the normal values as the following calculations show.

In[42]:= **Quantile[StudentTDistribution[30], 0.975]**

Out[42]= **2.04227**

This is comparable to the normal value of 1.95996 usually used. It is also instructive to show a table of these values for numbers of degrees of freedom not found in printed tables.

In[43]:= **values = Table[{n, Quantile[StudentTDistribution
        [n], 0.975]}, {n, 30, 50}];**

In[44]:= **headings = TableHeadings → {None, {StyleForm["n",
        FontWeight → "Bold"], StyleForm["Quantile",
        FontWeight → "Bold"]}}**

Out[44]= **TableHeadings → {None, {n, Quantile}}**

In[45]:= **TableForm[values, headings]**

Out[45]//Table Form=

n	Quantile
30	2.04227
31	2.03951
32	2.03693
33	2.03452
34	2.03224
35	2.03011
36	2.02809
37	2.02619
38	2.02439
39	2.02269
40	2.02108
41	2.01954
42	2.01808
43	2.01669
44	2.01537
45	2.0141
46	2.0129
47	2.01174
48	2.01063
49	2.00958
50	2.00856

This shows that the normal approximation may not be a suitable one when very accurate work is required. Computer algebra systems are of great value when these calculations are required. The following calculation in this regard is interesting since it shows that the normal value of 1.96996 is not reached until 239 degrees of freedom for the Student *t* distribution.

In[46]:= **Table[{n, Quantile[StudentTDistribution[n],**
      0.975]}, {n, 238, 239]}

Out[46]= **{{238, 1.96998}, {239, 1.96994}}**

In Section 3.14, the noncentral *t* distribution was used. Both the degrees of freedom and the noncentrality parameter must be specified. The calculation in Example 3.14.4 was done this way.

In[47]:= **CDF[NoncentralStudentTDistribution[19, 2.2361],**
      1.72913]

Out[47]= **0.304839**

This calculation is very difficult to do without a computer.

F Distribution

F distributions are specified by two degrees of freedom. We show some calculations and a graph from Section 3.7. The critical values for F[14, 24] in Example 3.7.1 were found using the following commands.

In[48]:= **Quantile[FRatioDistribution[14, 24], 0.975]**

Out[48]= **2.46766**

In[49]:= **Quantile[FRatioDistribution[14, 24], 0.025]**

Out[49]= **0.358576**

In[50]:= **Plot[PDF[FRatioDistribution[14, 24], x],**
      {x, 0, 4}, DefaultFont →
      {"Helvetica-Bold", 12}, AxesLabel → {x, F}]

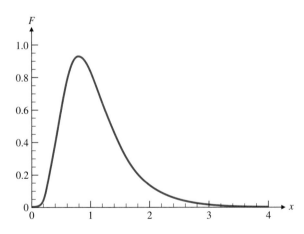

Out[50]= **-Graphics-**

One very useful feature of *Mathematica* is its ability to go far beyond the capability of any printed table. For example, suppose a test yields a value for F[33, 26] = 2.435. This combination of degrees of freedom is unlikely to be found in any printed table. Using *Mathematica* we find the *p*-value for the test is

In[51]:= **1 - CDF[FRatioDistribution[33, 26], 2.435]**

Out[51]= **0.0110342**

Bivariate Normal Distribution

Calculations and plots can be made for bivariate normal distributions. A variance-covariance matrix must first be defined, which we denote by r below. In this example, the mean vector is {0, 0} and the vector of standard deviations is {1, 1}, while the correlation coefficient is 0.6. The covariance is then $\rho \sigma_x \sigma_y = 0.6 \cdot 1 \cdot 1 = 0.6$.

```
In[52]:= r = {{1, 0.6}, {0.6, 1}}
         ndist = MultinormalDistribution[{0, 0}, r]
         pdfcn = PDF[ndist, {x, y}]
      Plot3D[pdfcn, {x, -2, 2}, {y, -2, 2},
         PlotPoints → 40,
         DefaultFont → {"Helvetica-Bold", 12},
         AxesLabel → {X, Y, " "}]
```

Out[52]= **{{1, 0.6}, {0.6, 1}}**

Out[53]= **MultinormalDistribution [{0, 0}, {{1, 0.6}, {0.6, 1}}]**

Out[54]= $\mathbf{0.198944e^{\frac{1}{2}(-x(1.5625\ x - 0.9375\,y) - y(-0.9375\,x + 1.5625\,y))}}$

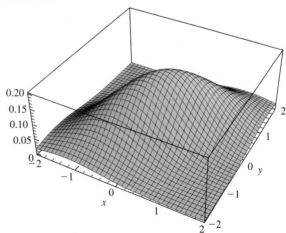

Out[55]= **-SurfaceGraphics-**

The output from these simultaneous commands is the variance-covariance matrix, the multinormal distribution showing the mean vector and the variance-covariance matrix, the probability distribution function in algebraic form, and finally the three-dimensional graph.

Now we show the calculation of a probability, using a numerical integration.

In[56]:= **NIntegrate[pdfcn, {x, 0, 1}, {y, 1, 1.6}]**
Out[56]= **0.0461609**

This calculation would be very difficult without a computer algebra system.

Poisson Distribution

We used the Poisson distribution in constructing c charts in Section 7.5. The Poisson distribution is completely specified by its parameter, which is its mean value. The Poisson probabilities in Table 7.8 were calculated using a Poisson distribution with parameter 1.

In[57]:= **PDF[PoissonDistribution[1], 4] // N**
Out[57]= **0.0153283**

The complete set of values given in Table 7.8 is calculated as follows:

In[58]:= **Table[PDF[PoissonDistribution[1], x],
          {x, 0, 6}] // N**
Out[58]= **{0.367879, 0.367879, 0.18394, 0.0613132,
          0.0153283, 0.00306566, 0.000510944}**

Simulation

Mathematica can be used to simulate situations and select random samples. Three examples are shown here.

Example A.1   We show how the random samples in Section 2.3 were chosen to illustrate the distribution of sample means; this in fact was an early indication in the text of the presence of normality in sampling and the central limit theorem.

The samples are produced first. These are all the possible samples of size 3, chosen without replacement, from the set of integers $\{1,2,\ldots,10\}$. The next command produces all the samples, each in numerical order from the smallest element of the sample to the largest. This ordering has no effect upon the distribution of the means of the samples; use of the ordering is simply an easy way to produce the samples.

In[59]:= **samples = Flatten[Table[{i, j, k}, {i, 1, 8},
          {j, i + 1, 10}, {k, j + 1, 10}], 2];**

Now we check that we have in fact all the $\binom{10}{3} = 120$ samples.

In[60]:= **Length[samples]**
Out[60]= **120**

Next we calculate the mean of each of the samples and then count the number of these means in the set $\{2, 7/3, 8/3, \ldots, 9\}$.

In[61]:= **avgs1 = Table[Mean[samples[[i]]], {i, 1, 120}];**

In[62]:= **freqs1 = BinCounts[avgs1, {2 - 1/3, 9, 1/3}]**

Out[62]= **{1, 1, 2, 3, 4, 5, 7, 8, 9, 10, 10, 10, 10, 9, 8, 7, 5, 4, 3, 2, 1, 1}**

Finally, we need a list of all the distinct means in the table of means.

In[63]:= **distinctmeans = Union[avgs1]**

Out[63]= $\{2, \frac{7}{3}, \frac{8}{3}, 3, \frac{10}{3}, \frac{11}{3}, 4, \frac{13}{3}, \frac{14}{3}, 5, \frac{16}{3}, \frac{17}{3}, 6, \frac{19}{3}, \frac{20}{3}, 7, \frac{22}{3}, \frac{23}{3}, 8, \frac{25}{3}, \frac{26}{3}, 9\}$

Here is the histogram.

In[64]:= **BarChart[Transpose[{freqs1, distinctmeans}]]**

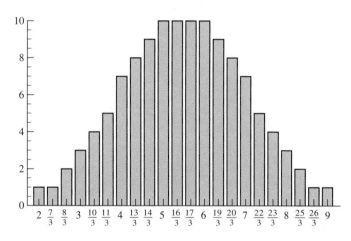

Out[64]= **-Graphics-**

This illustrates that the distribution of sample means is far from uniform. Later in the text we recognized the distribution as being approximately normal as the central limit theorem showed. We continue now with another example of simulation. ∎

Example A.2 We show how the samples in Section 2.10 were produced and used to illustrate the central limit theorem. Example A.1 used samples from a discrete distribution; this example uses samples from a continuous distribution.

We select 200 samples, each of size 5, from an exponential distribution with mean value 4; the parameter of the exponential distribution in *Mathematica* is 1/4. Most of the output here is suppressed.

The samples are generated first.

In[65]:= **data = RandomArray[ExponentialDistribution[1/4], {200, 5}];**

In[66]:= **avgs2 = Table[Mean[data[[i]]], {i, 1, 200}];**

In[67]:= **Max[avgs2]**

Out[67]= **9.00649**

In[68]:= **Min[avgs2]**

Out[68]= **0.742806**

In[69]:= **diffmeans = Table[i, {i, 0.8, 9.8, 0.8}]**

Out[69]= **{0.8, 1.6, 2.4, 3.2, 4., 4.8, 5.6, 6.4, 7.2, 8.,**
8.8, 9.6}

In[70]:= **freqs2 = BinCounts[avgs2, {0.8, 9.8, 0.8}]**

Out[70]= **{9, 15, 48, 36, 34, 28, 10, 11, 2, 4, 2, 1}**

In[71]:= **bchart = BarChart[Transpose[{freqs2, diffmeans}],**
DefaultFont → {"Helvetica-Bold", 12},
AxesLabel → {X, f}, BarSpacing → -0.2]

Mean

Out[71]= **-Graphics-**

The histogram differs markedly from the probability distribution from which the samples were selected, which is exponential. The normal approximation appears to be an acceptable one; the quality of the normal approximation is dependent upon both the sample size and the number of samples generated. ∎

As a final example, we examine the probability distribution of the sample variance.

Example A.3   We select 500 samples, each of size 5, from a standard normal distribution. We calculate the variance of each sample and then show a histogram of these variances. The commands, similar to those used previously, are shown here.

In[72]:= **samples = RandomArray[NormalDistribution[0, 1],**
{500, 5}];

In[73]:= **vars = Table[Variance[samples[[i]]], {i, 1, 500}];**
freqs3 = BinCounts[vars, {0, Max[vars], 0.25}]

Out[74]= **{54, 89, 109, 78, 51, 31, 33, 21, 11, 9, 4, 2, 5,**
2, 1}

In[75]:= **Apply[Plus, freqs3]**

Out[75]= **500**

Now we must produce a table of horizontal values for the histogram.

In[76]:= **values = Table[i, {i, 0, Max[vars], 0.25}]**

Out[76]= **{0, 0.25, 0.5, 0.75, 1., 1.25, 1.5, 1.75, 2.,**
2.25, 2.5, 2.75, 3., 3.25, 3.5}

In[77]:= **hist = BarChart[Transpose[{freqs3, values}],**
PlotRange → {0, Max[freqs3]}]

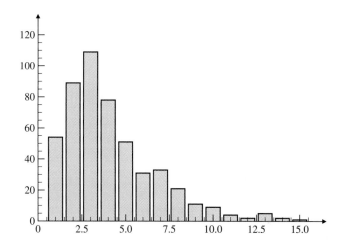

Out[77]= **-Graphics-**

In[78]:= **curve = Plot[600 PDF[ChiSquareDistribution[4], x],**
{x, 0, 14}];

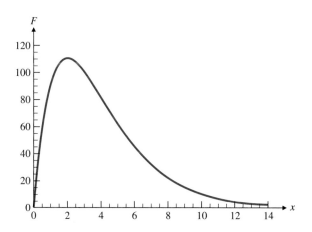

We anticipate that the histogram is well-approximated by the chi-squared curve. We now show the histogram and the chi-squared curve superimposed. It would appear that the chi-squared curve approximates the histogram quite well.

In[79]:= **Show[curve, hist]**

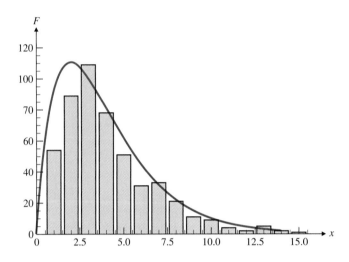

Out[79]= **-Graphics-**

Acceptance Sampling

We used *Mathematica* quite frequently in our consideration of acceptance sampling in Chapter 7. We show here how the graphs in Figures 7.20 and 7.23 were produced.

In Figure 7.20, we have a single sampling plan in which a lot of 2000 items contains D items that are defective. A sample of size 50 is chosen and the lot accepted only if none of the items in the sample are defective. The function f[D] below gives the size of the average outgoing quality for this sampling plan.

In[80]:= **f[D_] := (D/2000)*Binomial[2000 - D, 100]/
       Binomial[2000, 100]**

In[81]:= **ListPlot[Table[f[D], {D, 0, 100}], Frame → True,
       DefaultFont → {"Helvetica-Bold", 12},
       FrameLabel → {No.Defective, AOQ}]**

Out[81]= **-Graphics-**

The graph, found on the next page, can be traced to find the average outgoing quality limit is 0.0035866, which occurs when D = 19.

Mathematica can also be used to produce a table of values near the maximum of the curve; this provides an alternative to tracing the graph and approximating the maximum from the graph.

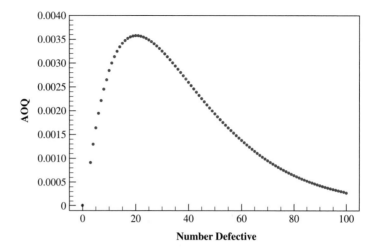

```
In[82]:= Table[{D, f[D]}, {D, 15, 25}] // N

Out[82]= {{15., 0.00346505}, {16., 0.00350985},
         {17., 0.00354125}, {18., 0.00356048},
         {19., 0.00356866}, {20., 0.00356686},
         {21., 0.00355605}, {22., 0.00353714},
         {23., 0.00351097}, {24., 0.00347831},
         {25., 0.00343987}}
```

Figure 7.23 shows the average outgoing quality in a double sampling plan. Here a lot of 2000 items has p% defective items. If at most 1 defective item is found in an initial sample of size 50, then the lot is accepted. If the initial sample contains 2 or 3 defective items, then a second sample of size 50 is selected and the lot is accepted if the total number of defective items in both samples is no more than 5. Otherwise, the lot is rejected.

The function avout q[p] gives the average outgoing quality as a function of p.

```
In[83]:= avoutq[p_] := Sum[((2000 p - x)/2000)*Binomial
         [2000 p, x]*Binomial[2000 - 2000 p, 50 - x]/
         Binomial[2000, 50], {x, 0, 1}] + (Binomial
         [2000 p, 2]*Binomial[2000 - 2000 p, 48]/
         Binomial[2000, 50])*
         Sum[((2000 p - 2 - x)/2000)*Binomial
         [2000 p - 2, x]*Binomial[1952 - 2000 p, 50 - x]/
         Binomial[1950, 50], {x, 0, 3}] + (Binomial
         [2000 p, 3]*Binomial[2000 - 2000 p, 47]/
         Binomial[2000, 50])*
         Sum[((2000 p - 3 - x)/2000)*Binomial
         [2000 p - 3, x]*Binomial[1953 - 2000 p, 50 - x]/
         Binomial[1950, 50], {x, 0, 2}]

In[84]:= Plot[avoutq[p], {p, 0, 0.15}, Frame → True,
         FrameLabel → {"%Non-Conforming", "AOQ"},
         DefaultFont → {"Helvetica-Bold", 12}]
```

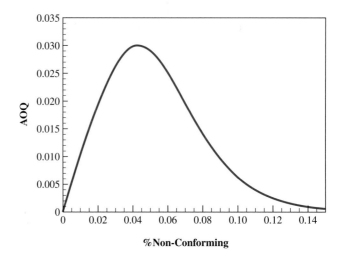

Out[84]= **-Graphics-**

The average outgoing quality can be found by tracing the graph or by a table of values near the maximum. Here the average outgoing quality limit is 0.0300113 when p = 0.0431.

Cumulative Sum Control Charts

We used the data in Example 7.6.1 to show the construction of a cumulative sum chart in tabular form. We use the data now to show how Table 7.20 was constructed.

We first enter the means of the samples.

In[85]:= **means = {12.7825, 11.1406, 12.3309, 11.9195,**
 12.2424, 11.5458, 11.8089, 12.7516, 11.7346,
 12.3072, 11.7745, 12.2314, 11.6961, 12.6193,
 11.5196, 12.9913, 11.6975, 12.9619, 12.8489,
 12.8112, 12.1640, 12.8609, 12.7833, 11.9932,
 12.8824, 11.7572};

Then the z score is computed for each of the means.

In[86]:= **zs = (means - 12)/0.3813**

Out[86]= **{2.05219, -2.25387, 0.867821, -0.21112, 0.63572,**
 -1.19119, -0.50118, 1.97115, -0.69604, 0.805665,
 -0.591398, 0.606871, -0.79701, 1.62418, -1.2599,
 2.59979, -0.793339, 2.52269, 2.22633, 2.12746,
 0.430108, 2.2578, 2.05429, -0.0178337, 2.31419,
 -0.636769}

Now we define the upper and lower sums S_H and S_L, as recursive functions as follows.

In[87]:= **S$_H$[0] = 0**

Out[87]= **0**

In[88]:= `S`$_H$`[n_] := Max[0, zs[[n]] - 1/2 + S`$_H$`[n - 1]]`

In[89]:= `S`$_L$`[0] = 0`
Out[89]= `0`

In[90]:= `SL[n_] := Max[0, -zs[[n]] - 1/2 + S`$_L$`[n - 1]]`

In[91]:= `exhibit = Table[{i, zs[[i]], S`$_H$`[i], S`$_L$`[i]},`
`{i, 1, Length[means]}];`

In[92]:= `captions = TableHeadings → {None,`
`{StyleForm["Sample", FontWeight →`
`"Bold"], StyleForm["z", FontWeight → "Bold"],`
`StyleForm["S`$_H$`", FontWeight → "Bold"],`
`StyleForm["S`$_L$`", FontWeight → "Bold"]}};`

In[93]:= `TableForm[exhibit, captions]`
Out[93]//TableForm=

Sample	z	S_H	S_L
1	2.05219	1.55219	0
2	-2.25387	0	1.75387
3	0.867821	0.367821	0.386048
4	-0.21112	0	0.0971676
5	0.63572	0.13572	0
6	-1.19119	0	0.691188
7	-0.50118	0	0.692368
8	1.97115	1.47115	0
9	-0.69604	0.275111	0.19604
10	0.805665	0.580776	0
11	-0.591398	0	0.0913978
12	0.606871	0.106871	0
13	-0.79701	0	0.29701
14	1.62418	1.12418	0
15	-1.2599	0	0.7599
16	2.59979	2.09979	0
17	-0.793339	0.806452	0.293339
18	2.52269	2.82914	0
19	2.22633	4.55547	0
20	2.12746	6.18293	0
21	0.430108	6.11303	0
22	2.2578	7.87084	0
23	2.05429	9.42512	0
24	-0.0178337	8.90729	0
25	2.31419	10.7215	0
26	-0.636769	9.58471	0.136769

Here is a plot of the cumulative sums, S_H:

In[94]:= `ListPlot[Table[S`$_H$`[i], {i, 1, 26}],`
 `      DefaultFont → {"Helvetica-Bold", 12},`
 `      Prolog → AbsolutePointSize[4],`
 `      AxesLabel → {Sample, S`$_H$`}]`

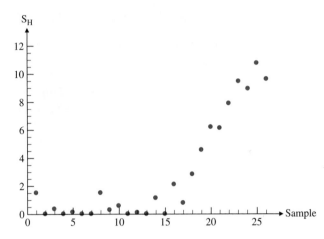

Out[94]= `-Graphics-`

Finally

This Appendix shows a very small sample of the capabilities of *Mathematica* in doing probability calculations and creating graphs. The general availability of computer algebra systems such as *Mathematica* may soon render tables of values obsolete. The approximation of one probability distribution by another, such as approximating the binomial distribution by the normal distribution, may well become an interesting, if useless, fact since the binomial probabilities are now easily computed.

Much more information on the general capabilities of *Mathematica* can be found in Wolfram [4] as well as in the supplementary publication, *Mathematica* 4 Standard Add-On Packages, which contains detailed information on all the probability distributions available in *Mathematica* and a description of the statistical capabilities of *Mathematica* as well.

Appendix B
Using Minitab

This is a brief explanation of how some of the calculations and graphs in the text were produced using Minitab. Minitab is a widely used computer statistical program. It is easy to use and has fairly wide capabilities for the statistical analysis of data. This appendix was prepared with Release 13.1 of Minitab.

The opening screen appears as shown in Figure B.1 when Minitab is accessed. The top portion of the screen is called the **Session Window** and the bottom portion of the screen is called the **Worksheet.** Calculated results appear in the Session Window while data are entered in the Worksheet. Data are entered as columns in the Worksheet, much in the manner of a spreadsheet.

Chapter 1   Graphs and Statistics

The Worksheet shown in Figure B.1 has the data from Example 1.1.1 on page 2 entered into columns 2 and 3. These columns have been labeled "Site 3" and "Site 8." It is good practice to title the columns in general; otherwise Minitab refers to them as C1, C2, and so on, giving no information at all on the content of the column.

We first produced some descriptive statistics in the text for the data from Sites 3 and 8. To do this, click first on Stat, which brings up a pull-down menu, and then click on Display Descriptive Statistics. (We will abbreviate this sequence of commands by *Stat→Display DescriptiveStatistics*. Other sequences of commands will be similarly abbreviated.) The dialog box titled Display Descriptive Statistics shown in Figure B.2 will appear in a separate window. Here we have indicated that we wish descriptive statistics from Sites 3 and 8. These will then appear in the Session window. This is shown in the Session window in Figure B.2.

Minitab can produce a number of different graphs. For example, to show boxplots of the data, click on Graph and then Boxplot. Indicate that two boxplots are to be produced, showing in the top menu in Figure B.3, both Site 3 and Site 8 for the *Y* variable.

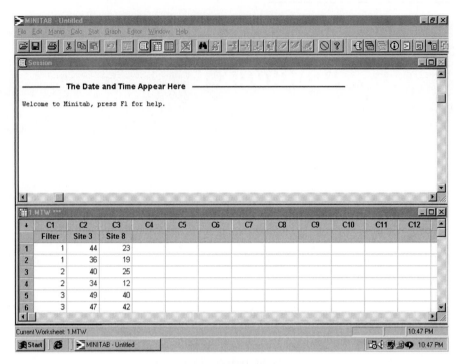

Figure B.1

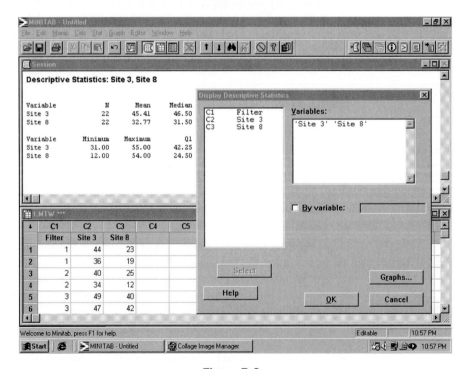

Figure B.2

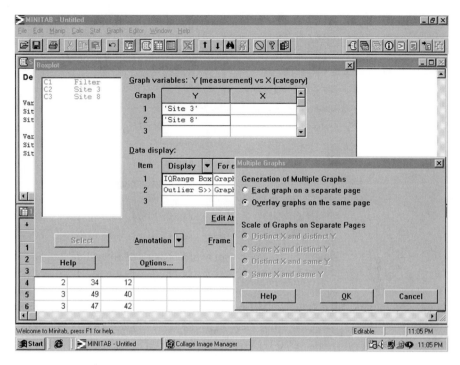

Figure B.3

Then click on Frame and then Multiple Graphs to produce the menu in the lower right corner of Figure B.3. Indicate that both graphs are to be overlaid on the same page. The resulting graphs are shown in Figure B.4. Dotplots and other graphs are produced in a similar way.

Chapter 2   Random Variables and Probability Distributions

Minitab has a number of standard probability distributions available; these can be found by *Calc→Probability Distributions*. Two calculations are shown here.

We calculated values of the hypergeometric distribution in Example 2.6.1 on page 43. Values of the independent variable, X, are entered into the column labeled "X." So here we have entered the values $0, 1, \ldots , 10$. Then the dialog box for the hypergeometric distribution appears as shown in the lower right corner of Figure B.5. We have asked for the probabilities for the hypergeometric distribution with samples of size 15 selected from a population of 100 items, 10 of which are special. We have indicated the input column as "X" and have selected the output to be placed in the column labeled "Hyper." This gives us the values of the hypergeometric distribution. These could now be plotted by *Graph→Plot* and filling in the resulting dialog box.

As a second example, we calculate $P(9.9 < X < 10.5)$ for a $N(10.1, 0.20)$ variable. This probability was calculated in Example 2.8.2 on page 64. Here the path is *Calc→Probability Distributions→Normal*. Figure B.6 shows the dialog box and the

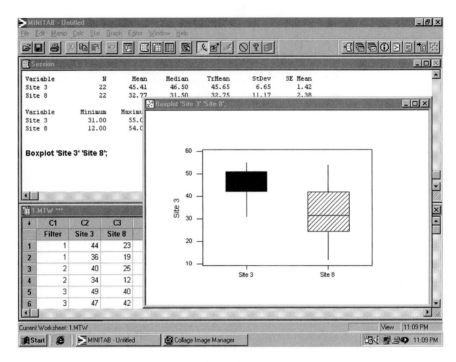

Figure B.4

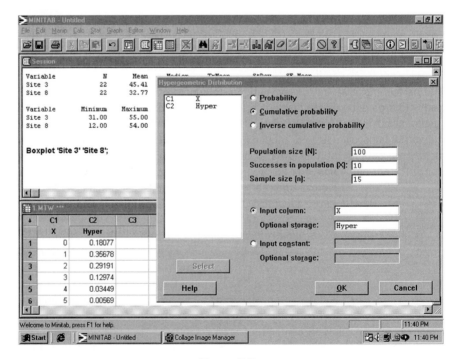

Figure B.5

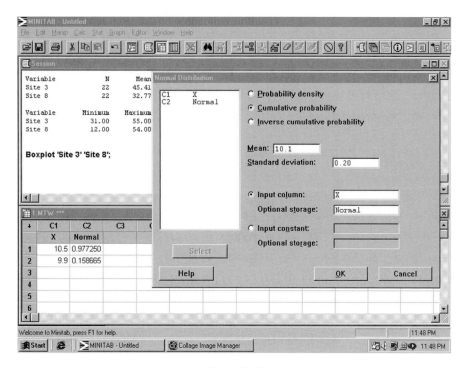

Figure B.6

appropriate entries. Notice that we have chosen the values of 9.9 and 10.5 for *X* and have placed the output in a column labeled "Normal." The resulting probability is then $0.977250 - 0.158655 = 0.818595$.

Inverse probability calculations can also be made easily in this menu.

Chapter 3   Estimation and Hypothesis Testing

In Example 3.4.1 on page 115, we need the upper 97.5% point on the t_9 distribution. The path for this menu is *Calc→Probability Distributions→t*. Figure B.7 shows the menu and the values filled in for the resulting critical *t* value. When an Optional storage column is not indicated, the critical value is returned in the Session window.

Now we show two examples of confidence intervals and hypothesis testing. First, consider Example 3.5.1 on page 118 where we wished to find a confidence interval for the difference between two proportions. The path here is *Stat→BasicStatistics→2 Proportions.* In this case, either the actual data, as they are entered into columns of the worksheet, or summarized data can be used, although for many analyses the present configuration of Minitab cannot be used for data that have been summarized.

The results, shown in the Session window, now include the confidence interval as well as the hypothesis test.

Figure B.9 shows the results of using the data in Example 3.13.1 on page 144. The data were entered into two columns of the Worksheet, labeled "*X*" and "*Y*." The path *Stat→BasicStatistics→2 Sample t* produces the dialog box in the lower left corner of

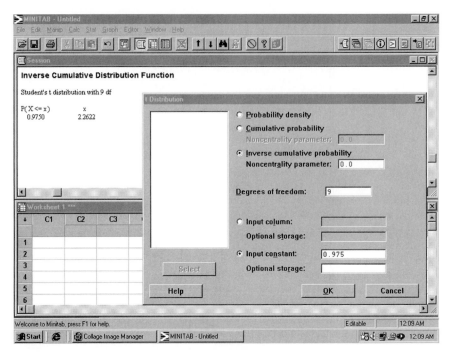

Figure B.7

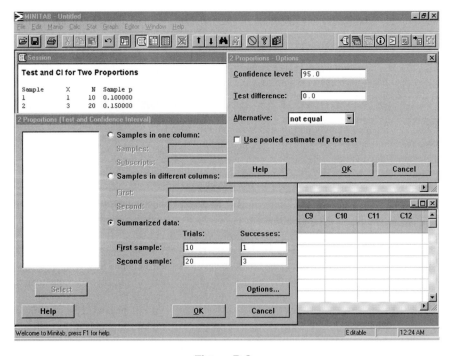

Figure B.8

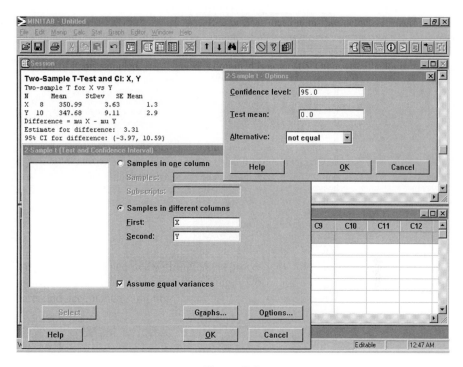

Figure B.9

Figure B.9. Note that the choice of equal variances has been made; Minitab's default here is unequal variances so the Welch approximation is provided. Then if the Options box is selected, the dialog box in the upper right corner of Figure B.9 appears. Choices can be made for the confidence level, the test mean, and the alternative hypothesis. The results are shown in the Session window.

Chapter 4   Simple Linear Regression

Regression routines are found by the path *Stat→Regression→Regression*. Generally the single independent variable is entered as X and the dependent variable as Y in the Worksheet, although other names are often used. The basic dialog box is partially shown at the bottom left in Figure B.10. We have entered X in the Response field and Y in the Predictor field for the data in Example 4.1.1 on page 173. We have also used the Results button, shown in the upper right corner of Figure B.10. Here we have asked for the full table of fits and residuals in addition to the regression equation, table of coefficients, R-squared, the analysis of variance, the sequential sums of squares and the unusual observations. Under Storage in this dialog box, we have asked for the values of Cook's distance, which will appear in the Worksheet, labeled Cook1. (The results of these requests are given in Figure B.11.) These results are those given in the text in Tables 2 and 3. Notice that the value of R^2 is also produced along with its adjusted value.

Figure B.12 shows the regression line as well as confidence bands around each predicted value. This is produced using the path *Stat→Regression→FittedLinePlot*.

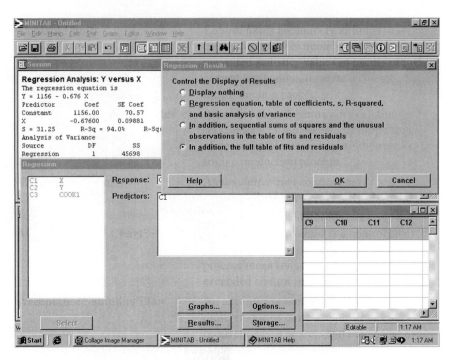

Figure B.10

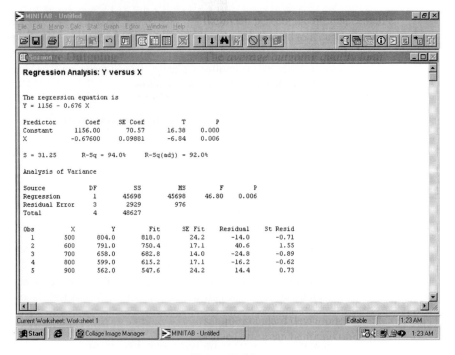

Figure B.11

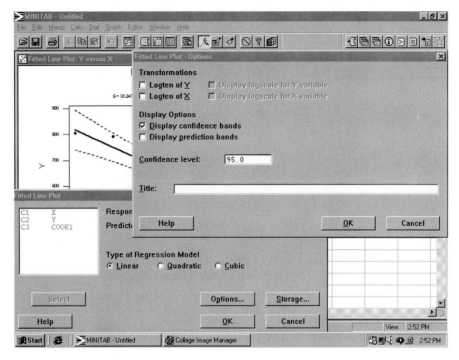

Figure B.12

The Linear model is chosen from this dialog box, which appears in the lower left corner of Figure B.12. Then if Options is selected, one can ask for confidence bands to be displayed.

Finally, Figure B.13 shows the calculation of a regression line passing through the origin. The data here are from Example 4.8.1 on page 209. In the regression dialog box, one clicks on Fit Intercept, which will remove the default check mark there. Some of the results can be seen in the upper left corner of Figure B.13.

Chapter 5   Multiple Linear Regression

Multiple regression differs from simple linear regression in Minitab only in the fact that two or more predictor variables are selected. The path then remains *Stat→Regression→Regression*. Figure B.14 shows both the Regression dialog box and the Results dialog box and Figure B.15 shows the results as they appear in the Session window for the SAT data in Example 5.7.1 on page 247.

It is important to know the correlations between the independent variables selected. The correlation matrix can be found using the path *Stat→BasicStatistics→Correlation*. Figure B.16 shows the Correlation dialog box and the result in the Session window.

The stepwise regression routines are found by *Stat→Regression→Stepwise*. The Stepwise dialog box is shown in the upper left corner of Figure B.17. The Methods button should be clicked in order to select the particular stepwise routine desired and to enter the value of Alpha to Enter or Alpha to Remove. Figure B.17 shows these dialog boxes

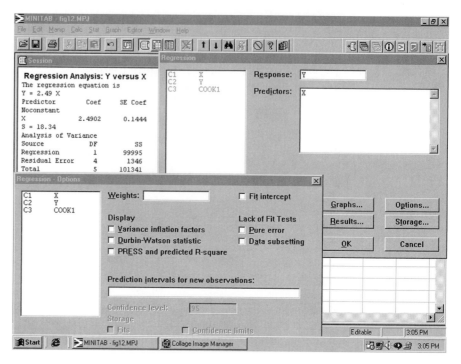

Figure B.13

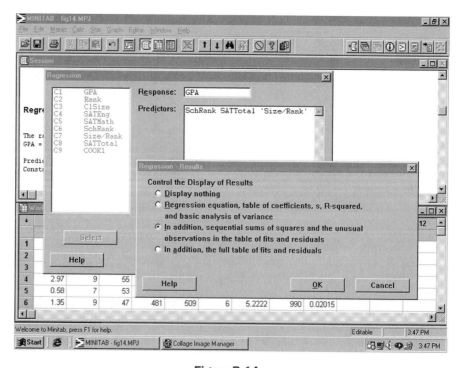

Figure B.14

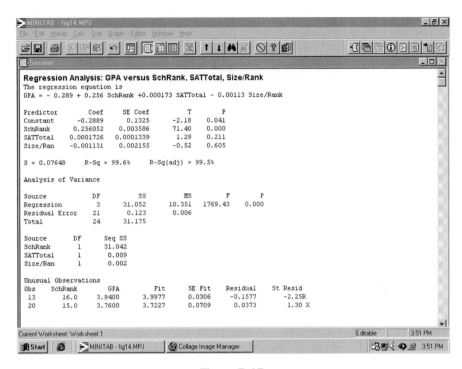

Figure B.15

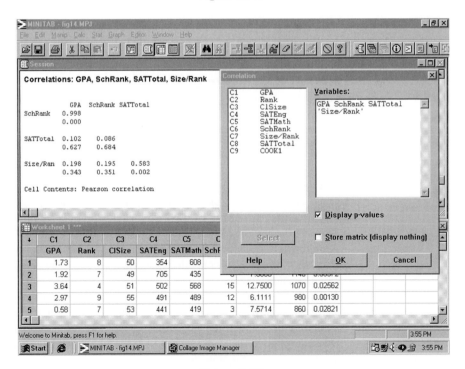

Figure B.16

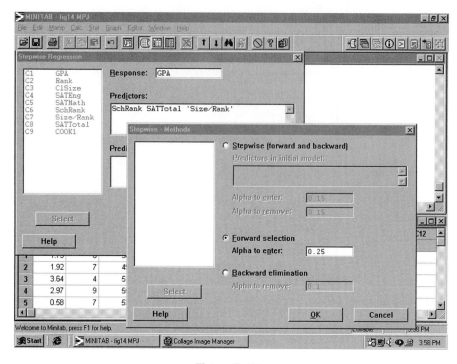

Figure B.17

as they were used to do the Forward stepwise method for the SAT data of Example 5.7.1 on page 247.

Chapter 6   Design of Science and Engineering Experiments

Example 6.4.1 on page 290 shows data from four levels of a catalyst. We wish to analyze the data to determine if the levels of the catalyst differ with respect to the concentration of a component in a liquid solution. The data are first entered into the Worksheet. We then select the path $Stat \rightarrow ANOVA \rightarrow One\text{-}way(Unstacked)$ since the data appear in separate columns. It is possible to place all the data in a single column. In that case select the path $Stat \rightarrow ANOVA \rightarrow One\text{-}way$. The dialog box, shown in Figure B.18, asks for the columns containing the responses. The result is shown in the Session window in Figure B.18.

Entering the data for a two-way classification is somewhat more complex than entering the data for a one-way classification. We refer to the data in Table 6.14 on page 311. Minitab requires a map of the data so that the origin of the data (in terms of the row and column in which they occur) is known to the program. In the Worksheet in Figure B.19, the column labeled "Row" gives the row index and the column labeled "Column" gives the column index. Then the appropriate responses are entered into the column labeled "Response." Finally, the path $Stat \rightarrow ANOVA \rightarrow Two\text{-}way$ will produce the Two-way Analysis of Variance dialog box shown in the right corner of Figure B.19. The analysis of variance table appears in the Session window.

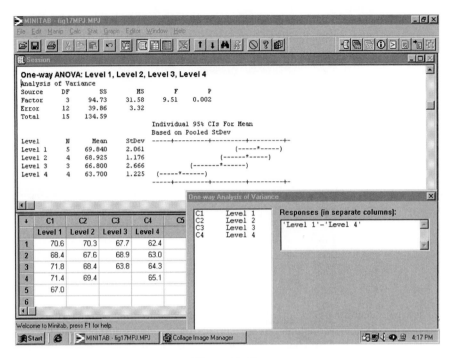

Figure B.18

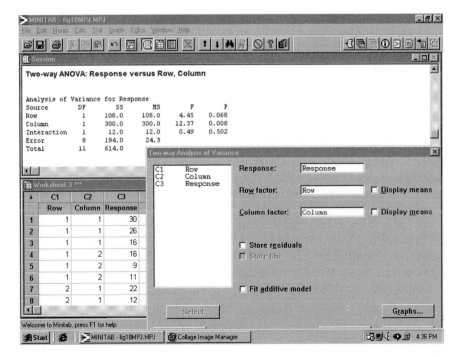

Figure B.19

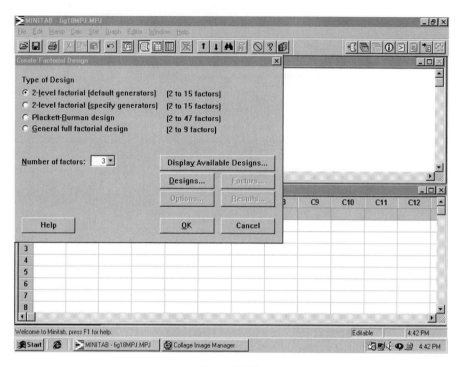

Figure B.20

Factorial and fractional factorial designs can be created and analyzed by Minitab. We show here the analysis of a full factorial experiment, using the data in Example 6.6.1 on page 310. Choices will be indicated for the analysis of a fractional factorial experiment as well.

First follow the path *Stat→DOE→Factorial→Create Factorial Design*. The resulting dialog box is shown in Figure B.20. We select a 2-level factorial using default generators and three factors. We then choose Designs. The dialog box is shown in Figure B.21. Only two choices—a full factorial or a half fraction—are available for this 2^3 design, but normally many other choices are available.

Now consider the data in Table 6.18 on page 316. The Worksheet will now appear as it is shown in Figure B.22. In this example, we use A for the factor Bulk Gas Flow, B for the factor CF_4 Flow, and C for the factor Power. The appropriate data are then entered into a column we have labeled "Selectivity." The data are also shown in Figure B.22.

Now choose the path *Stat→DOE→Factorial→Analyze Factorial Design*. The dialog box, shown in Figure B.23, asks for the response and then produces the analysis of variance shown in Figure B.22.

In Table 6.19, on page 320 in the text, we have combined the sums of squares for the Main Effects and the 2-Way Interactions so as to provide a nonzero Residual Error.

Figures 6.10 and 6.11 are produced using the path *Stat→DOE→Factorial→ FactorialPlots* while the normal probability plot shown in Figure 6.12 on page 325 is produced from the Graphs menu in the Analyze Factorial Design menu.

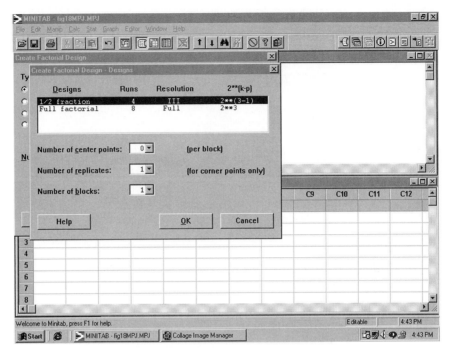

Figure B.21

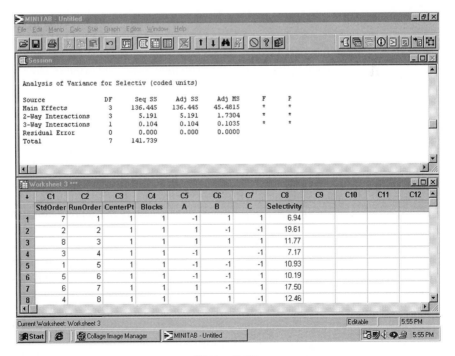

Figure B.22

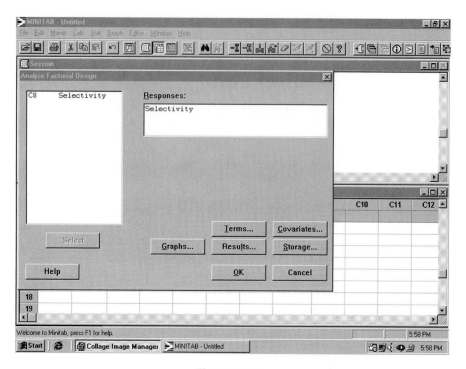

Figure B.23

Chapter 7   Statistical Process Control

A large variety of the most commonly occurring control charts is available in Minitab. We show the construction of some of these here.

Example 7.2.1 on page 351 was used to construct control charts for $\overline{X}$, R, and s. The data are entered as Sites 1, 2, and 3 into the Worksheet. Some of these data points are shown in Figure B.24. Minitab will calculate control limits based on the sample ranges, the sample variances, or the pooled standard deviation. If one wanted to check the limits, the values of these sample statistics could be calculated. In order to do this, note that the samples comprise the rows in Figure B.24. Under the Calc menu, choose Row Statistics. This brings up the Row Statistics dialog box shown in Figure B.24. One then selects the statistic desired and places the values in the appropriate column. (The variance was created using the Calc menu.) These columns can then be described as we have done previously, producing the values in Table 7.2 on page 354.

The row statistics need not be calculated in order to produce control charts using these statistics. In Figure B.25, we have used the path *Stat→ControlCharts→Xbar* to bring up the Xbar dialog box shown in the upper left corner of Figure B.25. In this case, since our samples are in rows, we select the "Subgroup across rows of:" option and then use the $\overline{X}$-Estimation of Mu and Sigma dialog box. This, shown in the bottom right corner, allows us to choose the method for estimating the unknown σ. We have selected the use of Rbar and this produces Figure 7.3 on page 358. Control charts using the other methods for estimating σ are also shown in Chapter 7.

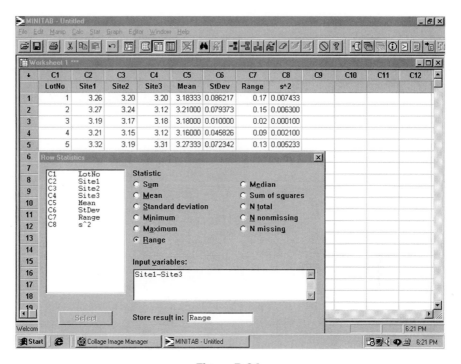

Figure B.24

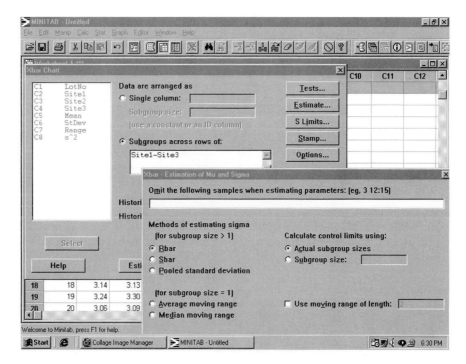

Figure B.25

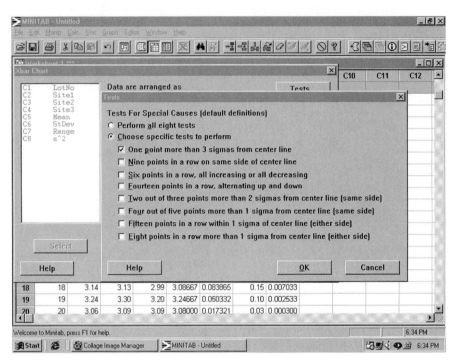

Figure B.26

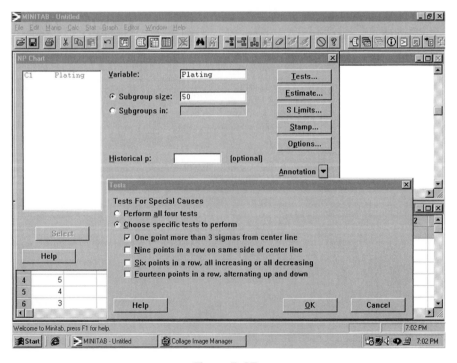

Figure B.27

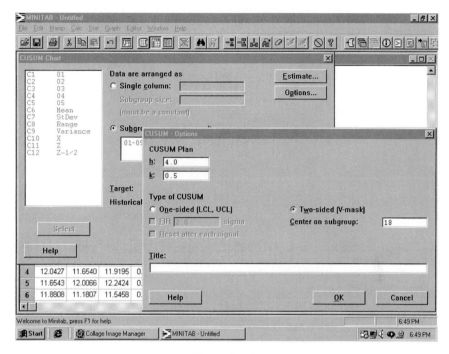

Figure B.28

It is also possible to ask Minitab to perform the Western Electric tests which we considered later in the chapter. If it is desired to change the default values of k, follow the path *Stat→ControlCharts→DefineTests*. The tests desired are specified by selecting the Tests button under the Xbar Chart menu. Figures B.26 and B.27 show two of the Test dialog boxes.

The construction of control charts for R and s follows very similar procedures to that for the $\overline{X}$ chart.

Several attribute control charts can also be constructed. We show the np chart for the plating defects data considered in Example 7.5.1. on page 365. The data are first entered into the Worksheet. The path *Stat→ControlCharts→np* produces the NP Chart menu shown in Figure 7.8 on page 367.

Here we have indicated where the data can be found and the sample size. The Tests submenu allows the performance of several of the Western Electric Rules.

The remaining material concerning control charts in Chapter 7 considers cumulative sum control charts. The path here is *Stat→ControlCharts→CUSUM*. This brings up the dialog box in the upper left corner of Figure B.28. One then chooses Options in order to select the type of control chart desired, the values of h and k (if needed), or the subgroup on which to center the V-mask.

Where to Learn More

This appendix is necessarily a very brief introduction to Minitab. The program has many more capabilities than we have been able to describe here. For much more information, consult the Minitab User's Guide [2] supplied with the program.

Appendix C
Tables

Table 1 Areas under the standard normal curve from $-\infty$ to z

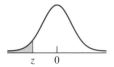

	Second decimal place in z								
z	0.09	0.08	0.07	0.06	0.05	0.04	0.03	0.02	0.01	0.00
-3.9										0.0000†
-3.8	0.0001	0.0001	0.0001	0.0001	0.0001	0.0001	0.0001	0.0001	0.0001	0.0001
-3.7	0.0001	0.0001	0.0001	0.0001	0.0001	0.0001	0.0001	0.0001	0.0001	0.0001
-3.6	0.0001	0.0001	0.0001	0.0001	0.0001	0.0001	0.0001	0.0001	0.0002	0.0002
-3.5	0.0002	0.0002	0.0002	0.0002	0.0002	0.0002	0.0002	0.0002	0.0002	0.0002
-3.4	0.0002	0.0003	0.0003	0.0003	0.0003	0.0003	0.0003	0.0003	0.0003	0.0003
-3.3	0.0003	0.0004	0.0004	0.0004	0.0004	0.0004	0.0004	0.0005	0.0005	0.0005
-3.2	0.0005	0.0005	0.0005	0.0006	0.0006	0.0006	0.0006	0.0006	0.0007	0.0007
-3.1	0.0007	0.0007	0.0008	0.0008	0.0008	0.0008	0.0009	0.0009	0.0009	0.0010
-3.0	0.0010	0.0010	0.0011	0.0011	0.0011	0.0012	0.0012	0.0013	0.0013	0.0013
-2.9	0.0014	0.0014	0.0015	0.0015	0.0016	0.0016	0.0017	0.0018	0.0018	0.0019
-2.8	0.0019	0.0020	0.0021	0.0021	0.0022	0.0023	0.0023	0.0024	0.0025	0.0026
-2.7	0.0026	0.0027	0.0028	0.0029	0.0030	0.0031	0.0032	0.0033	0.0034	0.0035
-2.6	0.0036	0.0037	0.0038	0.0039	0.0040	0.0041	0.0043	0.0044	0.0045	0.0047
-2.5	0.0048	0.0049	0.0051	0.0052	0.0054	0.0055	0.0057	0.0059	0.0060	0.0062
-2.4	0.0064	0.0066	0.0068	0.0069	0.0071	0.0073	0.0075	0.0078	0.0080	0.0082
-2.3	0.0084	0.0087	0.0089	0.0091	0.0094	0.0096	0.0099	0.0102	0.0104	0.0107
-2.2	0.0110	0.0113	0.0116	0.0119	0.0122	0.0125	0.0129	0.0132	0.0136	0.0139
-2.1	0.0143	0.0146	0.0150	0.0154	0.0158	0.0162	0.0166	0.0170	0.0174	0.0179
-2.0	0.0183	0.0188	0.0192	0.0197	0.0202	0.0207	0.0212	0.0217	0.0222	0.0228
-1.9	0.0233	0.0239	0.0244	0.0250	0.0256	0.0262	0.0268	0.0274	0.0281	0.0287
-1.8	0.0294	0.0301	0.0307	0.0314	0.0322	0.0329	0.0336	0.0344	0.0351	0.0359
-1.7	0.0367	0.0375	0.0384	0.0392	0.0401	0.0409	0.0418	0.0427	0.0436	0.0446
-1.6	0.0455	0.0465	0.0475	0.0485	0.0495	0.0505	0.0516	0.0526	0.0537	0.0548
-1.5	0.0559	0.0571	0.0582	0.0594	0.0606	0.0618	0.0630	0.0643	0.0655	0.0668
-1.4	0.0681	0.0694	0.0708	0.0721	0.0735	0.0749	0.0764	0.0778	0.0793	0.0808
-1.3	0.0823	0.0838	0.0853	0.0869	0.0885	0.0901	0.0918	0.0934	0.0951	0.0968
-1.2	0.0985	0.1003	0.1020	0.1038	0.1056	0.1075	0.1093	0.1112	0.1131	0.1151
-1.1	0.1170	0.1190	0.1210	0.1230	0.1251	0.1271	0.1292	0.1314	0.1335	0.1357
-1.0	0.1379	0.1401	0.1423	0.1446	0.1469	0.1492	0.1515	0.1539	0.1562	0.1587
-0.9	0.1611	0.1635	0.1660	0.1685	0.1711	0.1736	0.1762	0.1788	0.1814	0.1841
-0.8	0.1867	0.1894	0.1922	0.1949	0.1977	0.2005	0.2033	0.2061	0.2090	0.2119
-0.7	0.2148	0.2177	0.2206	0.2236	0.2266	0.2296	0.2327	0.2358	0.2389	0.2420
-0.6	0.2451	0.2483	0.2514	0.2546	0.2578	0.2611	0.2643	0.2676	0.2709	0.2743
-0.5	0.2776	0.2810	0.2843	0.2877	0.2912	0.2946	0.2981	0.3015	0.3050	0.3085
-0.4	0.3121	0.3156	0.3192	0.3228	0.3264	0.3300	0.3336	0.3372	0.3409	0.3446
-0.3	0.3483	0.3520	0.3557	0.3594	0.3632	0.3669	0.3707	0.3745	0.3783	0.3821
-0.2	0.3859	0.3897	0.3936	0.3974	0.4013	0.4052	0.4090	0.4129	0.4168	0.4207
-0.1	0.4247	0.4286	0.4325	0.4364	0.4404	0.4443	0.4483	0.4522	0.4562	0.4602
-0.0	0.4641	0.4681	0.4721	0.4761	0.4801	0.4840	0.4880	0.4920	0.4960	0.5000

† For $z \leq -3.90$, the areas are 0.0000 to four decimal places.

Table 1 (*continued*)

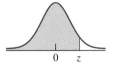

					Second decimal place in *z*					
z	0.00	0.01	0.02	0.03	0.04	0.05	0.06	0.07	0.08	0.09
0.0	0.5000	0.5040	0.5080	0.5120	0.5160	0.5199	0.5239	0.5279	0.5319	0.5359
0.1	0.5398	0.5438	0.5478	0.5517	0.5557	0.5596	0.5636	0.5675	0.5714	0.5753
0.2	0.5793	0.5832	0.5871	0.5910	0.5948	0.5987	0.6026	0.6064	0.6103	0.6141
0.3	0.6179	0.6217	0.6255	0.6293	0.6331	0.6368	0.6406	0.6443	0.6480	0.6517
0.4	0.6554	0.6591	0.6628	0.6664	0.6700	0.6736	0.6772	0.6808	0.6844	0.6879
0.5	0.6915	0.6950	0.6985	0.7019	0.7054	0.7088	0.7123	0.7157	0.7190	0.7224
0.6	0.7257	0.7291	0.7324	0.7357	0.7389	0.7422	0.7454	0.7486	0.7517	0.7549
0.7	0.7580	0.7611	0.7642	0.7673	0.7704	0.7734	0.7764	0.7794	0.7823	0.7852
0.8	0.7881	0.7910	0.7939	0.7967	0.7995	0.8023	0.8051	0.8078	0.8106	0.8133
0.9	0.8159	0.8186	0.8212	0.8238	0.8264	0.8289	0.8315	0.8340	0.8365	0.8389
1.0	0.8413	0.8438	0.8461	0.8485	0.8508	0.8531	0.8554	0.8577	0.8599	0.8621
1.1	0.8643	0.8665	0.8686	0.8708	0.8729	0.8749	0.8770	0.8790	0.8810	0.8830
1.2	0.8849	0.8869	0.8888	0.8907	0.8925	0.8944	0.8962	0.8980	0.8997	0.9015
1.3	0.9032	0.9049	0.9066	0.9082	0.9099	0.9115	0.9131	0.9147	0.9162	0.9177
1.4	0.9192	0.9207	0.9222	0.9236	0.9251	0.9265	0.9279	0.9292	0.9306	0.9319
1.5	0.9332	0.9345	0.9357	0.9370	0.9382	0.9394	0.9406	0.9418	0.9429	0.9441
1.6	0.9452	0.9463	0.9474	0.9484	0.9495	0.9505	0.9515	0.9525	0.9535	0.9545
1.7	0.9554	0.9564	0.9573	0.9582	0.9591	0.9599	0.9608	0.9616	0.9625	0.9633
1.8	0.9641	0.9649	0.9656	0.9664	0.9671	0.9678	0.9686	0.9693	0.9699	0.9706
1.9	0.9713	0.9719	0.9726	0.9732	0.9738	0.9744	0.9750	0.9756	0.9761	0.9767
2.0	0.9772	0.9778	0.9783	0.9788	0.9793	0.9798	0.9803	0.9808	0.9812	0.9817
2.1	0.9821	0.9826	0.9830	0.9834	0.9838	0.9842	0.9846	0.9850	0.9854	0.9857
2.2	0.9861	0.9864	0.9868	0.9871	0.9875	0.9878	0.9881	0.9884	0.9887	0.9890
2.3	0.9893	0.9896	0.9898	0.9901	0.9904	0.9906	0.9909	0.9911	0.9913	0.9916
2.4	0.9918	0.9920	0.9922	0.9925	0.9927	0.9929	0.9931	0.9932	0.9934	0.9936
2.5	0.9938	0.9940	0.9941	0.9943	0.9945	0.9946	0.9948	0.9949	0.9951	0.9952
2.6	0.9953	0.9955	0.9956	0.9957	0.9959	0.9960	0.9961	0.9962	0.9963	0.9964
2.7	0.9965	0.9966	0.9967	0.9968	0.9969	0.9970	0.9971	0.9972	0.9973	0.9974
2.8	0.9974	0.9975	0.9976	0.9977	0.9977	0.9978	0.9979	0.9979	0.9980	0.9981
2.9	0.9981	0.9982	0.9982	0.9983	0.9984	0.9984	0.9985	0.9985	0.9986	0.9986
3.0	0.9987	0.9987	0.9987	0.9988	0.9988	0.9989	0.9989	0.9989	0.9990	0.9990
3.1	0.9990	0.9991	0.9991	0.9991	0.9992	0.9992	0.9992	0.9992	0.9993	0.9993
3.2	0.9993	0.9993	0.9994	0.9994	0.9994	0.9994	0.9994	0.9995	0.9995	0.9995
3.3	0.9995	0.9995	0.9995	0.9996	0.9996	0.9996	0.9996	0.9996	0.9996	0.9997
3.4	0.9997	0.9997	0.9997	0.9997	0.9997	0.9997	0.9997	0.9997	0.9997	0.9998
3.5	0.9998	0.9998	0.9998	0.9998	0.9998	0.9998	0.9998	0.9998	0.9998	0.9998
3.6	0.9998	0.9998	0.9999	0.9999	0.9999	0.9999	0.9999	0.9999	0.9999	0.9999
3.7	0.9999	0.9999	0.9999	0.9999	0.9999	0.9999	0.9999	0.9999	0.9999	0.9999
3.8	0.9999	0.9999	0.9999	0.9999	0.9999	0.9999	0.9999	0.9999	0.9999	0.9999
3.9	1.0000†									

† For $z \geq 3.90$, the areas are 1.0000 to four decimal places.

Table 2   Student t Distribution
$P(t > t_\alpha) = \alpha$

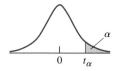

df	$t_{0.10}$	$t_{0.05}$	$t_{0.025}$	$t_{0.01}$	$t_{0.005}$	df
1	3.078	6.314	12.706	31.821	63.657	1
2	1.886	2.920	4.303	6.965	9.925	2
3	1.638	2.353	3.182	4.541	5.841	3
4	1.533	2.132	2.776	3.747	4.604	4
5	1.476	2.015	2.571	3.365	4.032	5
6	1.440	1.943	2.447	3.143	3.707	6
7	1.415	1.895	2.365	2.998	3.499	7
8	1.397	1.860	2.306	2.896	3.355	8
9	1.383	1.833	2.262	2.821	3.250	9
10	1.372	1.812	2.228	2.764	3.169	10
11	1.363	1.796	2.201	2.718	3.106	11
12	1.356	1.782	2.179	2.681	3.055	12
13	1.350	1.771	2.160	2.650	3.012	13
14	1.345	1.761	2.145	2.624	2.977	14
15	1.341	1.753	2.131	2.602	2.947	15
16	1.337	1.746	2.120	2.583	2.921	16
17	1.333	1.740	2.110	2.567	2.898	17
18	1.330	1.734	2.101	2.552	2.878	18
19	1.328	1.729	2.093	2.539	2.861	19
20	1.325	1.725	2.086	2.528	2.845	20
21	1.323	1.721	2.080	2.518	2.831	21
22	1.321	1.717	2.074	2.508	2.819	22
23	1.319	1.714	2.069	2.500	2.807	23
24	1.318	1.711	2.064	2.492	2.797	24
25	1.316	1.708	2.060	2.485	2.787	25
26	1.315	1.706	2.056	2.479	2.779	26
27	1.314	1.703	2.052	2.473	2.771	27
28	1.313	1.701	2.048	2.467	2.763	28
29	1.311	1.699	2.045	2.462	2.756	29
30	1.310	1.697	2.042	2.457	2.750	30
35	1.306	1.690	2.030	2.438	2.724	35
40	1.303	1.684	2.021	2.423	2.704	40
50	1.299	1.676	2.009	2.403	2.678	50
60	1.296	1.671	2.000	2.390	2.660	60
70	1.294	1.667	1.994	2.381	2.648	70
80	1.292	1.664	1.990	2.374	2.639	80
90	1.291	1.662	1.987	2.369	2.632	90
100	1.290	1.660	1.984	2.364	2.626	100
1000	1.282	1.646	1.962	2.330	2.581	1000

1.282	1.645	1.960	2.326	2.576
$z_{0.10}$	$z_{0.05}$	$z_{0.025}$	$z_{0.01}$	$z_{0.005}$

Table 3 Chi-Squared Distribution
$P(\chi^2 > \chi_\alpha^2) = \alpha$

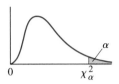

df	$\chi_{0.995}^2$	$\chi_{0.99}^2$	$\chi_{0.975}^2$	$\chi_{0.95}^2$	$\chi_{0.90}^2$
1	0.000	0.000	0.001	0.004	0.016
2	0.010	0.020	0.051	0.103	0.211
3	0.072	0.115	0.216	0.352	0.584
4	0.207	0.297	0.484	0.711	1.064
5	0.412	0.554	0.831	1.145	1.610
6	0.676	0.872	1.237	1.635	2.204
7	0.989	1.239	1.690	2.167	2.833
8	1.344	1.646	2.180	2.733	3.490
9	1.735	2.088	2.700	3.325	4.168
10	2.156	2.558	3.247	3.940	4.865
11	2.603	3.053	3.816	4.575	5.578
12	3.074	3.571	4.404	5.226	6.304
13	3.565	4.107	5.009	5.892	7.042
14	4.075	4.660	5.629	6.571	7.790
15	4.601	5.229	6.262	7.261	8.547
16	5.142	5.812	6.908	7.962	9.312
17	5.697	6.408	7.564	8.672	10.085
18	6.265	7.015	8.231	9.390	10.865
19	6.844	7.633	8.907	10.117	11.651
20	7.434	8.260	9.591	10.851	12.443
21	8.034	8.897	10.283	11.591	13.240
22	8.643	9.542	10.982	12.338	14.041
23	9.260	10.196	11.689	13.091	14.848
24	9.886	10.856	12.401	13.848	15.659
25	10.520	11.524	13.120	14.611	16.473
26	11.160	12.198	13.844	15.379	17.292
27	11.808	12.879	14.573	16.151	18.114
28	12.461	13.565	15.308	16.928	18.939
29	13.121	14.256	16.047	17.708	19.768
30	13.787	14.953	16.791	18.493	20.599
40	20.707	22.164	24.433	26.509	29.051
50	27.991	29.707	32.357	34.764	37.689
60	35.534	37.485	40.482	43.188	46.459
70	43.275	45.442	48.758	51.739	55.329
80	51.172	53.540	57.153	60.391	64.278
90	59.196	61.754	65.647	69.126	73.291
100	67.328	70.065	74.222	77.930	82.358

Table 3 *(continued)*

$\chi^2_{0.10}$	$\chi^2_{0.05}$	$\chi^2_{0.025}$	$\chi^2_{0.01}$	$\chi^2_{0.005}$	df
2.706	3.841	5.024	6.635	7.879	*1*
4.605	5.991	7.378	9.210	10.597	*2*
6.251	7.815	9.348	11.345	12.838	*3*
7.779	9.488	11.143	13.277	14.860	*4*
9.236	11.070	12.833	15.086	16.750	*5*
10.645	12.592	14.449	16.812	18.548	*6*
12.017	14.067	16.013	18.475	20.278	*7*
13.362	15.507	17.535	20.090	21.955	*8*
14.684	16.919	19.023	21.666	23.589	*9*
15.987	18.307	20.483	23.209	25.188	*10*
17.275	19.675	21.920	24.725	26.757	*11*
18.549	21.026	23.337	26.217	28.300	*12*
19.812	22.362	24.736	27.688	29.819	*13*
21.064	23.685	26.119	29.141	31.319	*14*
22.307	24.996	27.488	30.578	32.801	*15*
23.542	26.296	28.845	32.000	34.267	*16*
24.769	27.587	30.191	33.409	35.718	*17*
25.989	28.869	31.526	34.805	37.156	*18*
27.204	30.143	32.852	36.191	38.582	*19*
28.412	31.410	34.170	37.566	39.997	*20*
29.615	32.671	35.479	38.932	41.401	*21*
30.813	33.924	36.781	40.290	42.796	*22*
32.007	35.172	38.076	41.638	44.181	*23*
33.196	36.415	39.364	42.980	45.559	*24*
34.382	37.653	40.647	44.314	46.928	*25*
35.563	38.885	41.923	45.642	48.290	*26*
36.741	40.113	43.195	46.963	49.645	*27*
37.916	41.337	44.461	48.278	50.994	*28*
39.087	42.557	45.722	49.588	52.336	*29*
40.256	43.773	46.979	50.892	53.672	*30*
51.805	55.759	59.342	63.691	66.767	*40*
63.167	67.505	71.420	76.154	79.490	*50*
74.397	79.082	83.298	88.381	91.955	*60*
85.527	90.531	95.023	100.424	104.213	*70*
96.578	101.879	106.628	112.328	116.320	*80*
107.565	113.145	118.135	124.115	128.296	*90*
118.499	124.343	129.563	135.811	140.177	*100*

Table 4   *F* Ratio Distribution
$$P(F > F_\alpha) = \alpha$$

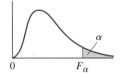

		dfn								
dfd	α	1	2	3	4	5	6	7	8	9
	0.10	39.86	49.50	53.59	55.83	57.24	58.20	58.91	59.44	59.86
	0.05	161.45	199.50	215.71	224.58	230.16	233.99	236.77	238.88	240.54
1	*0.025*	647.79	799.50	864.16	899.58	921.85	937.11	948.22	956.66	963.28
	0.01	4052.2	4999.5	5403.4	5624.6	5763.6	5859.0	5928.4	5981.1	6022.5
	0.005	16211	20000	21615	22500	23056	23437	23715	23925	24091
	0.10	8.53	9.00	9.16	9.24	9.29	9.33	9.35	9.37	9.38
	0.05	18.51	19.00	19.16	19.25	19.30	19.33	19.35	19.37	19.38
2	*0.025*	38.51	39.00	39.17	39.25	39.30	39.33	39.36	39.37	39.39
	0.01	98.50	99.00	99.17	99.25	99.30	99.33	99.36	99.37	99.39
	0.005	198.50	199.00	199.17	199.25	199.30	199.33	199.36	199.37	199.39
	0.10	5.54	5.46	5.39	5.34	5.31	5.28	5.27	5.25	5.24
	0.05	10.13	9.55	9.28	9.12	9.01	8.94	8.89	8.85	8.81
3	*0.025*	17.44	16.04	15.44	15.10	14.88	14.73	14.62	14.54	14.47
	0.01	34.12	30.82	29.46	28.71	28.24	27.91	27.67	27.49	27.35
	0.005	55.55	49.80	47.47	46.19	45.39	44.84	44.43	44.13	43.88
	0.10	4.54	4.32	4.19	4.11	4.05	4.01	3.98	3.95	3.94
	0.05	7.71	6.94	6.59	6.39	6.26	6.16	6.09	6.04	6.00
4	*0.025*	12.22	10.65	9.98	9.60	9.36	9.20	9.07	8.98	8.90
	0.01	21.20	18.00	16.69	15.98	15.52	15.21	14.98	14.80	14.66
	0.005	31.33	26.28	24.26	23.15	22.46	21.97	21.62	21.35	21.14
	0.10	4.06	3.78	3.62	3.52	3.45	3.40	3.37	3.34	3.32
	0.05	6.61	5.79	5.41	5.19	5.05	4.95	4.88	4.82	4.77
5	*0.025*	10.01	8.43	7.76	7.39	7.15	6.98	6.85	6.76	6.68
	0.01	16.26	13.27	12.06	11.39	10.97	10.67	10.46	10.29	10.16
	0.005	22.78	18.31	16.53	15.56	14.94	14.51	14.20	13.96	13.77
	0.10	3.78	3.46	3.29	3.18	3.11	3.05	3.01	2.98	2.96
	0.05	5.99	5.14	4.76	4.53	4.39	4.28	4.21	4.15	4.10
6	*0.025*	8.81	7.26	6.60	6.23	5.99	5.82	5.70	5.60	5.52
	0.01	13.75	10.92	9.78	9.15	8.75	8.47	8.26	8.10	7.98
	0.005	18.63	14.54	12.92	12.03	11.46	11.07	10.79	10.57	10.39
	0.10	3.59	3.26	3.07	2.96	2.88	2.83	2.78	2.75	2.72
	0.05	5.59	4.74	4.35	4.12	3.97	3.87	3.79	3.73	3.68
7	*0.025*	8.07	6.54	5.89	5.52	5.29	5.12	4.99	4.90	4.82
	0.01	12.25	9.55	8.45	7.85	7.46	7.19	6.99	6.84	6.72
	0.005	16.24	12.40	10.88	10.05	9.52	9.16	8.89	8.68	8.51
	0.10	3.46	3.11	2.92	2.81	2.73	2.67	2.62	2.59	2.56
	0.05	5.32	4.46	4.07	3.84	3.69	3.58	3.50	3.44	3.39
8	*0.025*	7.57	6.06	5.42	5.05	4.82	4.65	4.53	4.43	4.36
	0.01	11.26	8.65	7.59	7.01	6.63	6.37	6.18	6.03	5.91
	0.005	14.69	11.04	9.60	8.81	8.30	7.95	7.69	7.50	7.34

Table 4 (*continued*)

					dfn					
10	12	15	20	24	30	40	60	120	α	dfd
60.19	60.71	61.22	61.74	62.00	62.26	62.53	62.79	63.06	*0.10*	
241.88	243.91	245.95	248.01	249.05	250.10	251.14	252.20	253.25	*0.05*	
968.63	976.71	984.87	993.10	997.25	1001.41	1005.60	1009.80	1014.02	*0.025*	*1*
6055.8	6106.3	6157.3	6208.7	6234.6	6260.6	6286.7	631.9	6339.4	*0.01*	
24224	24426	24630	24836	24940	25044	25148	25253	25359	*0.005*	
9.39	9.41	9.42	9.44	9.45	9.46	9.47	9.47	9.48	*0.10*	
19.40	19.41	19.43	19.45	19.45	19.46	19.47	19.48	19.49	*0.05*	
39.40	39.41	39.43	39.45	39.46	39.46	39.47	39.48	39.49	*0.025*	*2*
99.40	99.42	99.43	99.45	99.46	99.47	99.47	99.48	99.49	*0.01*	
199.40	199.42	199.43	199.45	199.46	199.47	199.47	199.48	199.49	*0.005*	
5.23	5.22	5.20	5.18	5.18	5.17	5.16	5.15	5.14	*0.10*	
8.79	8.74	8.70	8.66	8.64	8.62	8.59	8.57	8.55	*0.05*	
14.42	14.34	14.25	14.17	14.12	14.08	14.04	13.99	13.95	*0.025*	*3*
27.23	27.05	26.87	26.69	26.60	26.50	26.41	26.32	26.22	*0.01*	
43.69	43.39	43.08	42.78	42.62	42.47	42.31	42.15	41.99	*0.005*	
3.92	3.90	3.87	3.84	3.83	3.82	3.80	3.79	3.78	*0.10*	
5.96	5.91	5.86	5.80	5.77	5.75	5.72	5.69	5.66	*0.05*	
8.84	8.75	8.66	8.56	8.51	8.46	8.41	8.36	8.31	*0.025*	*4*
14.55	14.37	14.20	14.02	13.93	13.84	13.75	13.65	13.56	*0.01*	
20.97	20.70	20.44	20.17	20.03	19.89	19.75	19.61	19.47	*0.005*	
3.30	3.27	3.24	3.21	3.19	3.17	3.16	3.14	3.12	*0.10*	
4.74	4.68	4.62	4.56	4.53	4.50	4.46	4.43	4.40	*0.05*	
6.62	6.52	6.43	6.33	6.28	6.23	6.18	6.12	6.07	*0.025*	*5*
10.05	9.89	9.72	9.55	9.47	9.38	9.29	9.20	9.11	*0.01*	
13.62	13.38	13.15	12.90	12.78	12.66	12.53	12.40	12.27	*0.005*	
2.94	2.90	2.87	2.84	2.82	2.80	2.78	2.76	2.74	*0.10*	
4.06	4.00	3.94	3.87	3.84	3.81	3.77	3.74	3.70	*0.05*	
5.46	5.37	5.27	5.17	5.12	5.07	5.01	4.96	4.90	*0.025*	*6*
7.87	7.72	7.56	7.40	7.31	7.23	7.14	7.06	6.97	*0.01*	
10.25	10.03	9.81	9.59	9.47	9.36	9.24	9.12	9.00	*0.005*	
2.70	2.67	2.63	2.59	2.58	2.56	2.54	2.51	2.49	*0.10*	
3.64	3.57	3.51	3.44	3.41	3.38	3.34	3.30	3.27	*0.05*	
4.76	4.67	4.57	4.47	4.41	4.36	4.31	4.25	4.20	*0.025*	*7*
6.62	6.47	6.31	6.16	6.07	5.99	5.91	5.82	5.74	*0.01*	
8.38	8.18	7.97	7.75	7.64	7.53	7.42	7.31	7.19	*0.005*	
2.54	2.50	2.46	2.42	2.40	2.38	2.36	2.34	2.32	*0.10*	
3.35	3.28	3.22	3.15	3.12	3.08	3.04	3.01	2.97	*0.05*	
4.30	4.20	4.10	4.00	3.95	3.89	3.84	3.78	3.73	*0.025*	*8*
5.81	5.67	5.52	5.36	5.28	5.20	5.12	5.03	4.95	*0.01*	
7.21	7.01	6.81	6.61	6.50	6.40	6.29	6.18	6.06	*0.005*	

Table 4 (*continued*)

dfd	α	dfn 1	2	3	4	5	6	7	8	9
9	*0.10*	3.36	3.01	2.81	2.69	2.61	2.55	2.51	2.47	2.44
	0.05	5.12	4.26	3.86	3.63	3.48	3.37	3.29	3.23	3.18
	0.025	7.21	5.71	5.08	4.72	4.48	4.32	4.20	4.10	4.03
	0.01	10.56	8.02	6.99	6.42	6.06	5.80	5.61	5.47	5.35
	0.005	13.61	10.11	8.72	7.96	7.47	7.13	6.88	6.69	6.54
10	*0.10*	3.29	2.92	2.73	2.61	2.52	2.46	2.41	2.38	2.35
	0.05	4.96	4.10	3.71	3.48	3.33	3.22	3.14	3.07	3.02
	0.025	6.94	5.46	4.83	4.47	4.24	4.07	3.95	3.85	3.78
	0.01	10.04	7.56	6.55	5.99	5.64	5.39	5.20	5.06	4.94
	0.005	12.83	9.43	8.08	7.34	6.87	6.54	6.30	6.12	5.97
11	*0.10*	3.23	2.86	2.66	2.54	2.45	2.39	2.34	2.30	2.27
	0.05	4.84	3.98	3.59	3.36	3.20	3.09	3.01	2.95	2.90
	0.025	6.72	5.26	4.63	4.28	4.04	3.88	3.76	3.66	3.59
	0.01	9.65	7.21	6.22	5.67	5.32	5.07	4.89	4.74	4.63
	0.005	12.23	8.91	7.60	6.88	6.42	6.10	5.86	5.68	5.54
12	*0.10*	3.18	2.81	2.61	2.48	2.39	2.33	2.28	2.24	2.21
	0.05	4.75	3.89	3.49	3.26	3.11	3.00	2.91	2.85	2.80
	0.025	6.55	5.10	4.47	4.12	3.89	3.73	3.61	3.51	3.44
	0.01	9.33	6.93	5.95	5.41	5.06	4.82	4.64	4.50	4.39
	0.005	11.75	8.51	7.23	6.52	6.07	5.76	5.52	5.35	5.20
13	*0.10*	3.14	2.76	2.56	2.43	2.35	2.28	2.23	2.20	2.16
	0.05	4.67	3.81	3.41	3.18	3.03	2.92	2.83	2.77	2.71
	0.025	6.41	4.97	4.35	4.00	3.77	3.60	3.48	3.39	3.31
	0.01	9.07	6.70	5.74	5.21	4.86	4.62	4.44	4.30	4.19
	0.005	11.37	8.19	6.93	6.23	5.79	5.48	5.25	5.08	4.94
14	*0.10*	3.10	2.73	2.52	2.39	2.31	2.24	2.19	2.15	2.12
	0.05	4.60	3.74	3.34	3.11	2.96	2.85	2.76	2.70	2.65
	0.025	6.30	4.86	4.24	3.89	3.66	3.50	3.38	3.29	3.21
	0.01	8.86	6.51	5.56	5.04	4.69	4.46	4.28	4.14	4.03
	0.005	11.06	7.92	6.68	6.00	5.56	5.26	5.03	4.86	4.72
15	*0.10*	3.07	2.70	2.49	2.36	2.27	2.21	2.16	2.12	2.09
	0.05	4.54	3.68	3.29	3.06	2.90	2.79	2.71	2.64	2.59
	0.025	6.20	4.77	4.15	3.80	3.58	3.41	3.29	3.20	3.12
	0.01	8.68	6.36	5.42	4.89	4.56	4.32	4.14	4.00	3.89
	0.005	10.80	7.70	6.48	5.80	5.37	5.07	4.85	4.67	4.54
16	*0.10*	3.05	2.67	2.46	2.33	2.24	2.18	2.13	2.09	2.06
	0.05	4.49	3.63	3.24	3.01	2.85	2.74	2.66	2.59	2.54
	0.025	6.12	4.69	4.08	3.73	3.50	3.34	3.22	3.12	3.05
	0.01	8.53	6.23	5.29	4.77	4.44	4.20	4.03	3.89	3.78
	0.005	10.58	7.51	6.30	5.64	5.21	4.91	4.69	4.52	4.38

Table 4 *(continued)*

				dfn						
10	12	15	20	24	30	40	60	120	α	dfd
2.42	2.38	2.34	2.30	2.28	2.25	2.23	2.21	2.18	*0.10*	
3.14	3.07	3.01	2.94	2.90	2.86	2.83	2.79	2.75	*0.05*	
3.96	3.87	3.77	3.67	3.61	3.56	3.51	3.45	3.39	*0.025*	*9*
5.26	5.11	4.96	4.81	4.73	4.65	4.57	4.48	4.40	*0.01*	
6.42	6.23	6.03	5.83	5.73	5.62	5.52	5.41	5.30	*0.005*	
2.32	2.28	2.24	2.20	2.18	2.16	2.13	2.11	2.08	*0.10*	
2.98	2.91	2.85	2.77	2.74	2.70	2.66	2.62	2.58	*0.05*	
3.72	3.62	3.52	3.42	3.37	3.31	3.26	3.20	3.14	*0.025*	*10*
4.85	4.71	4.56	4.41	4.33	4.25	4.17	4.08	4.00	*0.01*	
5.85	5.66	5.47	5.27	5.17	5.07	4.97	4.86	4.75	*0.005*	
2.25	2.21	2.17	2.12	2.10	2.08	2.05	2.03	2.00	*0.10*	
2.85	2.79	2.72	2.65	2.61	2.57	2.53	2.49	2.45	*0.05*	
3.53	3.43	3.33	3.23	3.17	3.12	3.06	3.00	2.94	*0.025*	*11*
4.54	4.40	4.25	4.10	4.02	3.94	3.86	3.78	3.69	*0.01*	
5.42	5.24	5.05	4.86	4.76	4.65	4.55	4.45	4.34	*0.005*	
2.19	2.15	2.10	2.06	2.04	2.01	1.99	1.96	1.93	*0.10*	
2.75	2.69	2.62	2.54	2.51	2.47	2.43	2.38	2.34	*0.05*	
3.37	3.28	3.18	3.07	3.02	2.96	2.91	2.85	2.79	*0.025*	*12*
4.30	4.16	4.01	3.86	3.78	3.70	3.62	3.54	3.45	*0.01*	
5.09	4.91	4.72	4.53	4.43	4.33	4.23	4.12	4.01	*0.005*	
2.14	2.10	2.05	2.01	1.98	1.96	1.93	1.90	1.88	*0.10*	
2.67	2.60	2.53	2.46	2.42	2.38	2.34	2.30	2.25	*0.05*	
3.25	3.15	3.05	2.95	2.89	2.84	2.78	2.72	2.66	*0.025*	*13*
4.10	3.96	3.82	3.66	3.59	3.51	3.43	3.34	3.25	*0.01*	
4.82	4.64	4.46	4.27	4.17	4.07	3.97	3.87	3.76	*0.005*	
2.10	2.05	2.01	1.96	1.94	1.91	1.89	1.86	1.83	*0.10*	
2.60	2.53	2.46	2.39	2.35	2.31	2.27	2.22	2.18	*0.05*	
3.15	3.05	2.95	2.84	2.79	2.73	2.67	2.61	2.55	*0.025*	*14*
3.94	3.80	3.66	3.51	3.43	3.35	3.27	3.18	3.09	*0.01*	
4.60	4.43	4.25	4.06	3.96	3.86	3.76	3.66	3.55	*0.005*	
2.06	2.02	1.97	1.92	1.90	1.87	1.85	1.82	1.79	*0.10*	
2.54	2.48	2.40	2.33	2.29	2.25	2.20	2.16	2.11	*0.05*	
3.06	2.96	2.86	2.76	2.70	2.64	2.59	2.52	2.46	*0.025*	*15*
3.80	3.67	3.52	3.37	3.29	3.21	3.13	3.05	2.96	*0.01*	
4.42	4.25	4.07	3.88	3.79	3.69	3.58	3.48	3.37	*0.005*	
2.03	1.99	1.94	1.89	1.87	1.84	1.81	1.78	1.75	*0.10*	
2.49	2.42	2.35	2.28	2.24	2.19	2.15	2.11	2.06	*0.05*	
2.99	2.89	2.79	2.68	2.63	2.57	2.51	2.45	2.38	*0.025*	*16*
3.69	3.55	3.41	3.26	3.18	3.10	3.02	2.93	2.84	*0.01*	
4.27	4.10	3.92	3.73	3.64	3.54	3.44	3.33	3.22	*0.005*	

Table 4 (*continued*)

dfd	α	1	2	3	4	5	6	7	8	9
						dfn				
	0.10	3.03	2.64	2.44	2.31	2.22	2.15	2.10	2.06	2.03
	0.05	4.45	3.59	3.20	2.96	2.81	2.70	2.61	2.55	2.49
17	*0.025*	6.04	4.62	4.01	3.66	3.44	3.28	3.16	3.06	2.98
	0.01	8.40	6.11	5.18	4.67	4.34	4.10	3.93	3.79	3.68
	0.005	10.38	7.35	6.16	5.50	5.07	4.78	4.56	4.39	4.25
	0.10	3.01	2.62	2.42	2.29	2.20	2.13	2.08	2.04	2.00
	0.05	4.41	3.55	3.16	2.93	2.77	2.66	2.58	2.51	2.46
18	*0.025*	5.98	4.56	3.95	3.61	3.38	3.22	3.10	3.01	2.93
	0.01	8.29	6.01	5.09	4.58	4.25	4.01	3.84	3.71	3.60
	0.005	10.22	7.21	6.03	5.37	4.96	4.66	4.44	4.28	4.14
	0.10	2.99	2.61	2.40	2.27	2.18	2.11	2.06	2.02	1.98
	0.05	4.38	3.52	3.13	2.90	2.74	2.63	2.54	2.48	2.42
19	*0.025*	5.92	4.51	3.90	3.56	3.33	3.17	3.05	2.96	2.88
	0.01	8.18	5.93	5.01	4.50	4.17	3.94	3.77	3.63	3.52
	0.005	10.07	7.09	5.92	5.27	4.85	4.56	4.34	4.18	4.04
	0.10	2.97	2.59	2.38	2.25	2.16	2.09	2.04	2.00	1.96
	0.05	4.35	3.49	3.10	2.87	2.71	2.60	2.51	2.45	2.39
20	*0.025*	5.87	4.46	3.86	3.51	3.29	3.13	3.01	2.91	2.84
	0.01	8.10	5.85	4.94	4.43	4.10	3.87	3.70	3.56	3.46
	0.005	9.94	6.99	5.82	5.17	4.76	4.47	4.26	4.09	3.96
	0.10	2.96	2.57	2.36	2.23	2.14	2.08	2.02	1.98	1.95
	0.05	4.32	3.47	3.07	2.84	2.68	2.57	2.49	2.42	2.37
21	*0.025*	5.83	4.42	3.82	3.48	3.25	3.09	2.97	2.87	2.80
	0.01	8.02	5.78	4.87	4.37	4.04	3.81	3.64	3.51	3.40
	0.005	9.83	6.89	5.73	5.09	4.68	4.39	4.18	4.01	3.88
	0.10	2.95	2.56	2.35	2.22	2.13	2.06	2.01	1.97	1.93
	0.05	4.30	3.44	3.05	2.82	2.66	2.55	2.46	2.40	2.34
22	*0.025*	5.79	4.38	3.78	3.44	3.22	3.05	2.93	2.84	2.76
	0.01	7.95	5.72	4.82	4.31	3.99	3.76	3.59	3.45	3.35
	0.005	9.73	6.81	5.65	5.02	4.61	4.32	4.11	3.94	3.81
	0.10	2.94	2.55	2.34	2.21	2.11	2.05	1.99	1.95	1.92
	0.05	4.28	3.42	3.03	2.80	2.64	2.53	2.44	2.37	2.32
23	*0.025*	5.75	4.35	3.75	3.41	3.18	3.02	2.90	2.81	2.73
	0.01	7.88	5.66	4.76	4.26	3.94	3.71	3.54	3.41	3.30
	0.005	9.63	6.73	5.58	4.95	4.54	4.26	4.05	3.88	3.75
	0.10	2.93	2.54	2.33	2.19	2.10	2.04	1.98	1.94	1.91
	0.05	4.26	3.40	3.01	2.78	2.62	2.51	2.42	2.36	2.30
24	*0.025*	5.72	4.32	3.72	3.38	3.15	2.99	2.87	2.78	2.70
	0.01	7.82	5.61	4.72	4.22	3.90	3.67	3.50	3.36	3.26
	0.005	9.55	6.66	5.52	4.89	4.49	4.20	3.99	3.83	3.69

Table 4 *(continued)*

				dfn						
10	12	15	20	24	30	40	60	120	α	dfd
2.00	1.96	1.91	1.86	1.84	1.81	1.78	1.75	1.72	*0.10*	
2.45	2.38	2.31	2.23	2.19	2.15	2.10	2.06	2.01	*0.05*	
2.92	2.82	2.72	2.62	2.56	2.50	2.44	2.38	2.32	*0.025*	*17*
3.59	3.46	3.31	3.16	3.08	3.00	2.92	2.83	2.75	*0.01*	
4.14	3.97	3.79	3.61	3.51	3.41	3.31	3.21	3.10	*0.005*	
1.98	1.93	1.89	1.84	1.81	1.78	1.75	1.72	1.69	*0.10*	
2.41	2.34	2.27	2.19	2.15	2.11	2.06	2.02	1.97	*0.05*	
2.87	2.77	2.67	2.56	2.50	2.44	2.38	2.32	2.26	*0.025*	*18*
3.51	3.37	3.23	3.08	3.00	2.92	2.84	2.75	2.66	*0.01*	
4.03	3.86	3.68	3.50	3.40	3.30	3.20	3.10	2.99	*0.005*	
1.96	1.91	1.86	1.81	1.79	1.76	1.73	1.70	1.67	*0.10*	
2.38	2.31	2.23	2.16	2.11	2.07	2.03	1.98	1.93	*0.05*	
2.82	2.72	2.62	2.51	2.45	2.39	2.33	2.27	2.20	*0.025*	*19*
3.43	3.30	3.15	3.00	2.92	2.84	2.76	2.67	2.58	*0.01*	
3.93	3.76	3.59	3.40	3.31	3.21	3.11	3.00	2.89	*0.005*	
1.94	1.89	1.84	1.79	1.77	1.74	1.71	1.68	1.64	*0.10*	
2.35	2.28	2.20	2.12	2.08	2.04	1.99	1.95	1.90	*0.05*	
2.77	2.68	2.57	2.46	2.41	2.35	2.29	2.22	2.16	*0.025*	*20*
3.37	3.23	3.09	2.94	2.86	2.78	2.69	2.61	2.52	*0.01*	
3.85	3.68	3.50	3.32	3.22	3.12	3.02	2.92	2.81	*0.005*	
1.92	1.87	1.83	1.78	1.75	1.72	1.69	1.66	1.62	*0.10*	
2.32	2.25	2.18	2.10	2.05	2.01	1.96	1.92	1.87	*0.05*	
2.73	2.64	2.53	2.42	2.37	2.31	2.25	2.18	2.11	*0.025*	*21*
3.31	3.17	3.03	2.88	2.80	2.72	2.64	2.55	2.46	*0.01*	
3.77	3.60	3.43	3.24	3.15	3.05	2.95	2.84	2.73	*0.005*	
1.90	1.86	1.81	1.76	1.73	1.70	1.67	1.64	1.60	*0.10*	
2.30	2.23	2.15	2.07	2.03	1.98	1.94	1.89	1.84	*0.05*	
2.70	2.60	2.50	2.39	2.33	2.27	2.21	2.14	2.08	*0.025*	*22*
3.26	3.12	2.98	2.83	2.75	2.67	2.58	2.50	2.40	*0.01*	
3.70	3.54	3.36	3.18	3.08	2.98	2.88	2.77	2.66	*0.005*	
1.89	1.84	1.80	1.74	1.72	1.69	1.66	1.62	1.59	*0.10*	
2.27	2.20	2.13	2.05	2.01	1.96	1.91	1.86	1.81	*0.05*	
2.67	2.57	2.47	2.36	2.30	2.24	2.18	2.11	2.04	*0.025*	*23*
3.21	3.07	2.93	2.78	2.70	2.62	2.54	2.45	2.35	*0.01*	
3.64	3.47	3.30	3.12	3.02	2.92	2.82	2.71	2.60	*0.005*	
1.88	1.83	1.78	1.73	1.70	1.67	1.64	1.61	1.57	*0.10*	
2.25	2.18	2.11	2.03	1.98	1.94	1.89	1.84	1.79	*0.05*	
2.64	2.54	2.44	2.33	2.27	2.21	2.15	2.08	2.01	*0.025*	*24*
3.17	3.03	2.89	2.74	2.66	2.58	2.49	2.40	2.31	*0.01*	
3.59	3.42	3.25	3.06	2.97	2.87	2.77	2.66	2.55	*0.005*	

Table 4 *(continued)*

		dfn								
dfd	α	1	2	3	4	5	6	7	8	9
	0.10	2.92	2.53	2.32	2.18	2.09	2.02	1.97	1.93	1.89
	0.05	4.24	3.39	2.99	2.76	2.60	2.49	2.40	2.34	2.28
25	*0.025*	5.69	4.29	3.69	3.35	3.13	2.97	2.85	2.75	2.68
	0.01	7.77	5.57	4.68	4.18	3.85	3.63	3.46	3.32	3.22
	0.005	9.48	6.60	5.46	4.84	4.43	4.15	3.94	3.78	3.64
	0.10	2.91	2.52	2.31	2.17	2.08	2.01	1.96	1.92	1.88
	0.05	4.23	3.37	2.98	2.74	2.59	2.47	2.39	2.32	2.27
26	*0.025*	5.66	4.27	3.67	3.33	3.10	2.94	2.82	2.73	2.65
	0.01	7.72	5.53	4.64	4.14	3.82	3.59	3.42	3.29	3.18
	0.005	9.41	6.54	5.41	4.79	4.38	4.10	3.89	3.73	3.60
	0.10	2.90	2.51	2.30	2.17	2.07	2.00	1.95	1.91	1.87
	0.05	4.21	3.35	2.96	2.73	2.57	2.46	2.37	2.31	2.25
27	*0.025*	5.63	4.24	3.65	3.31	3.08	2.92	2.80	2.71	2.63
	0.01	7.68	5.49	4.60	4.11	3.78	3.56	3.39	3.26	3.15
	0.005	9.34	6.49	5.36	4.74	4.34	4.06	3.85	3.69	3.56
	0.10	2.89	2.50	2.29	2.16	2.06	2.00	1.94	1.90	1.87
	0.05	4.20	3.34	2.95	2.71	2.56	2.45	2.36	2.29	2.24
28	*0.025*	5.61	4.22	3.63	3.29	3.06	2.90	2.78	2.69	2.61
	0.01	7.64	5.45	4.57	4.07	3.75	3.53	3.36	3.23	3.12
	0.005	9.28	6.44	5.32	4.70	4.30	4.02	3.81	3.65	3.52
	0.10	2.89	2.50	2.28	2.15	2.06	1.99	1.93	1.89	1.86
	0.05	4.18	3.33	2.93	2.70	2.55	2.43	2.35	2.28	2.22
29	*0.025*	5.59	4.20	3.61	3.27	3.04	2.88	2.76	2.67	2.59
	0.01	7.60	5.42	4.54	4.04	3.73	3.50	3.33	3.20	3.09
	0.005	9.23	6.40	5.28	4.66	4.26	3.98	3.77	3.61	3.48
	0.10	2.88	2.49	2.28	2.14	2.05	1.98	1.93	1.88	1.85
	0.05	4.17	3.32	2.92	2.69	2.53	2.42	2.33	2.27	2.21
30	*0.025*	5.57	4.18	3.59	3.25	3.03	2.87	2.75	2.65	2.57
	0.01	7.56	5.39	4.51	4.02	3.70	3.47	3.30	3.17	3.07
	0.005	9.18	6.35	5.24	4.62	4.23	3.95	3.74	3.58	3.45
	0.10	2.79	2.39	2.18	2.04	1.95	1.87	1.82	1.77	1.74
	0.05	4.00	3.15	2.76	2.53	2.37	2.25	2.17	2.10	2.04
60	*0.025*	5.29	3.93	3.34	3.01	2.79	2.63	2.51	2.41	2.33
	0.01	7.08	4.98	4.13	3.65	3.34	3.12	2.95	2.82	2.72
	0.005	8.49	5.79	4.73	4.14	3.76	3.49	3.29	3.13	3.01
	0.10	2.75	2.35	2.13	1.99	1.90	1.82	1.77	1.72	1.68
	0.05	3.92	3.07	2.68	2.45	2.29	2.18	2.09	2.02	1.96
120	*0.025*	5.15	3.80	3.23	2.89	2.67	2.52	2.39	2.30	2.22
	0.01	6.85	4.79	3.95	3.48	3.17	2.96	2.79	2.66	2.56
	0.005	8.18	5.54	4.50	3.92	3.55	3.28	3.09	2.93	2.81

Table 4 (*continued*)

				dfn						
10	12	15	20	24	30	40	60	120	α	dfd
1.87	1.82	1.77	1.72	1.69	1.66	1.63	1.59	1.56	*0.10*	
2.24	2.16	2.09	2.01	1.96	1.92	1.87	1.82	1.77	*0.05*	
2.61	2.51	2.41	2.30	2.24	2.18	2.12	2.05	1.98	*0.025*	25
3.13	2.99	2.85	2.70	2.62	2.54	2.45	2.36	2.27	*0.01*	
3.54	3.37	3.20	3.01	2.92	2.82	2.72	2.61	2.50	*0.005*	
1.86	1.81	1.76	1.71	1.68	1.65	1.61	1.58	1.54	*0.10*	
2.22	2.15	2.07	1.99	1.95	1.90	1.85	1.80	1.75	*0.05*	
2.59	2.49	2.39	2.28	2.22	2.16	2.09	2.03	1.95	*0.025*	26
3.09	2.96	2.81	2.66	2.58	2.50	2.42	2.33	2.23	*0.01*	
3.49	3.33	3.15	2.97	2.87	2.77	2.67	2.56	2.45	*0.005*	
1.85	1.80	1.75	1.70	1.67	1.64	1.60	1.57	1.53	*0.10*	
2.20	2.13	2.06	1.97	1.93	1.88	1.84	1.79	1.73	*0.05*	
2.57	2.47	2.36	2.25	2.19	2.13	2.07	2.00	1.93	*0.025*	27
3.06	2.93	2.78	2.63	2.55	2.47	2.38	2.29	2.20	*0.01*	
3.45	3.28	3.11	2.93	2.83	2.73	2.63	2.52	2.41	*0.005*	
1.84	1.79	1.74	1.69	1.66	1.63	1.59	1.56	1.52	*0.10*	
2.19	2.12	2.04	1.96	1.91	1.87	1.82	1.77	1.71	*0.05*	
2.55	2.45	2.34	2.23	2.17	2.11	2.05	1.98	1.91	*0.025*	28
3.03	2.90	2.75	2.60	2.52	2.44	2.35	2.26	2.17	*0.01*	
3.41	3.25	3.07	2.89	2.79	2.69	2.59	2.48	2.37	*0.005*	
1.83	1.78	1.73	1.68	1.65	1.62	1.58	1.55	1.51	*0.10*	
2.18	2.10	2.03	1.94	1.90	1.85	1.81	1.75	1.70	*0.05*	
2.53	2.43	2.32	2.21	2.15	2.09	2.03	1.96	1.89	*0.025*	29
3.00	2.87	2.73	2.57	2.49	2.41	2.33	2.23	2.14	*0.01*	
3.38	3.21	3.04	2.86	2.76	2.66	2.56	2.45	2.33	*0.005*	
1.82	1.77	1.72	1.67	1.64	1.61	1.57	1.54	1.50	*0.10*	
2.16	2.09	2.01	1.93	1.89	1.84	1.79	1.74	1.68	*0.05*	
2.51	2.41	2.31	2.20	2.14	2.07	2.01	1.94	1.87	*0.025*	30
2.98	2.84	2.70	2.55	2.47	2.39	2.30	2.21	2.11	*0.01*	
3.34	3.18	3.01	2.82	2.73	2.63	2.52	2.42	2.30	*0.005*	
1.71	1.66	1.60	1.54	1.51	1.48	1.44	1.40	1.35	*0.10*	
1.99	1.92	1.84	1.75	1.70	1.65	1.59	1.53	1.47	*0.05*	
2.27	2.17	2.06	1.94	1.88	1.82	1.74	1.67	1.58	*0.025*	60
2.63	2.50	2.35	2.20	2.12	2.03	1.94	1.84	1.73	*0.01*	
2.90	2.74	2.57	2.39	2.29	2.19	2.08	1.96	1.83	*0.005*	
1.65	1.60	1.55	1.48	1.45	1.41	1.37	1.32	1.26	*0.10*	
1.91	1.83	1.75	1.66	1.61	1.55	1.50	1.43	1.35	*0.05*	
2.16	2.05	1.94	1.82	1.76	1.69	1.61	1.53	1.43	*0.025*	120
2.47	2.34	2.19	2.03	1.95	1.86	1.76	1.66	1.53	*0.01*	
2.71	2.54	2.37	2.19	2.09	1.98	1.87	1.75	1.61	*0.005*	

Bibliography

WHERE TO LEARN MORE

References here are grouped by chapters in the book; groupings are approximate because references for one chapter may shed some light on material in other chapters. There is now a large literature on probability and statistics; the list here is necessarily a small sampling of the writing that is available.

Computing

1. Velleman, Paul, *ActivStats,* Addison-Wesley, Boston, 1998.
2. Minitab, *Minitab User's Guide,* Release 13, Minitab Inc., 1999.
3. Wickham-Jones, Tom, *Mathematica Graphics,* Springer-Verlag, New York, 1994.
4. Wolfram, Stephen, *The Mathematica Book,* Wolfram Media, Champaign, Illinois, 1999.

CHAPTER 1 GRAPHS AND STATISTICS

5. Cleveland, William S., *The Elements of Graphing Data,* Wadsworth, Monterey, California, 1985.
6. Czitrom, Veronica, and P. D. Spagon, *Statistical Case Studies for Industrial Improvement,* American Statistical Association and Society for Industrial and Applied Mathematics, Alexandria, Virginia, 1997.
7. Hall, Kathryn, and Steven Carpenter, Sampling to Meet a Variance Specification: Clean Room Qualification, *Statistical Case Studies for Industrial Improvement,* American Statistical Association and Society for Industrial and Applied Mathematics, Alexandria, Virginia, 1997.
8. Hoaglin, David C., Frederick Mosteller, and John W. Tukey, *Exploring Data Tables, Trends, and Shapes,* John Wiley & Sons, New York, 1985.

9. Mosteller, Frederick, and John W. Tukey, *Data Analysis and Regression,* Addison-Wesley, Boston, 1977.

10. Peck, Roxy, L. D. Haugh, and A. Goodman, *Statistical Case Studies, A Collaboration Between Academe and Industry,* American Statistical Association and Society for Industrial and Applied Mathematics, Alexandria, Virginia, 1998.

11. Saville, David J., and Graham D. Wood, *Statistical Methods: The Geometric Approach,* Springer, New York, 1991.

12. Tukey, John W., *Exploratory Data Analysis,* Addison-Wesley, Boston, 1977.

CHAPTER 2   RANDOM VARIABLES AND PROBABILITY DISTRIBUTIONS

13. Feller, William, *An Introduction to Probability and Its Applications,* Volume 1, Third Edition, John Wiley & Sons, New York, 1968.

14. Hoel, Paul G., *Introduction to Mathematical Statistics,* Fourth Edition, John Wiley & Sons, New York, 1971.

15. Hogg, Robert V., and Allen T. Craig, *Introduction to Mathematical Statistics,* Fifth Edition, Prentice Hall, Upper Saddle River, New Jersey, 1995.

16. Johnson, Norman L., Samuel Kotz, and N. Balakrishnan, *Continuous Univariate Distributions,* Volume 1, Second Edition, John Wiley & Sons, New York, 1994.

17. Johnson, Norman L., Samuel Kotz, and N. Balakrishnan, *Continuous Univariate Distributions,* Volume 2, Second Edition, John Wiley & Sons, New York, 1995.

18. Kinney, John J., *Probability: An Introduction With Statistical Applications,* John Wiley & Sons, New York, 1997.

CHAPTER 3   ESTIMATION AND HYPOTHESIS TESTING

19. Bartoszynski, Robert, and Magdalena Niewiadomska-Bugaj, *Probability and Statistical Inference,* John Wiley & Sons, New York, 1996.

20. Mitchell, Teresa, Victor Hegemann, and K. C. Liu, GRR Methodology for Destructive Testing and Quantitative Assessment of Gauge Capability for One-Side Specifications, *Statistical Case Studies for Industrial Improvement,* American Statistical Association and Society for Industrial and Applied Mathematics, Alexandria, Virginia, 1997.

21. Rao, C. Radhakrishna, *Linear Statistical Inference and Its Applications,* Second Edition, John Wiley & Sons, New York, 1973.

22. Welch, B. L., The Significance of the Difference Between Means When the Population Variances are Unequal, *Biometrika* 29 (1938), 350–362.

CHAPTERS 4 AND 5   SIMPLE AND MULTIPLE LINEAR REGRESSION

23. Belseley, D. A., E. Kuh and R. E. Welsch, *Regression Diagnostics: Identifying Influential Data and Sources of Collinearity,* John Wiley & Sons, New York, 1980.

24. Cook, R. D., Detection of Influential Observations in Linear Regression, *Technometrics* 19 (1972), 15–18.

25. Cook, R. Dennis, and Sanford Weisberg, *An Introduction to Regression Graphics,* John Wiley & Sons, New York, 1994.

26. Cook, R. Dennis, and Sanford Weisberg, *Applied Regression Including Computing and Graphics,* John Wiley & Sons, New York, 1999.

27. Daniel, Cuthbert, and Fred S. Wood, *Fitting Equations to Data,* John Wiley & Sons, New York, 1971.

28. Draper, Norman, and Harry Smith, *Applied Regression Analysis,* Second Edition, John Wiley & Sons, New York, 1981.

29. Hocking, Ronald R., *Methods and Applications of Linear Models,* John Wiley & Sons, New York, 1996.

30. Snedecor, George, and William G. Cochran, *Statistical Methods,* Sixth Edition, Iowa State University Press, Ames, Iowa, 1967.

31. *Statistical Abstract of the United States,* U.S. Bureau of the Census, Washington, 1994.

32. *World Almanac and Book of Facts, 2000,* Perimedia Reference Inc., Mahwah, New Jersey, 1999.

CHAPTER 6   DESIGN OF EXPERIMENTS

33. Banerjee, K. S., *Weighing Designs for Chemistry, Medicine, Economics, Operations Research, Statistics,* Marcel Dekker, New York, 1975.

34. Barnett, Joel, Veronica Czitrom, Peter W. M. John, and Ramon V. Leon, Using Fewer Wafers to Resolve Confounding in Screening Experiments, *Statistical Case Studies for Industrial Improvement,* American Statistical Association and Society for Industrial and Applied Mathematics, Alexandria, Virginia, 1997.

35. Box, George E. P., and J. S. Hunter, The 2^{k-p} Fractional Factorial Designs, Parts I and II, *Technometrics* 3 (1966), 311–357, 449–458.

36. Box, George E. P., and Norman R. Draper, *Empirical Model-Building and Response Surfaces,* John Wiley & Sons, New York, 1987.

37. Box, George E. P., William G. Hunter, and J. Stuart Hunter, *Statistics for Experimenters,* John Wiley & Sons, New York, 1978.

38. Buckner, James, David J. Cammenga, and Ann Weber, Elimination of TiN Peeling During Exposure to CUD Tungsten Deposition Process Using Designed Experiments, *Statistical Case Studies for Industrial Improvement,* American Statistical Association and Society for Industrial and Applied Mathematics, Alexandria, Virginia, 1997.

39. Cochran, William G., and Gertrude M. Cox, *Experimental Designs,* Second Edition, John Wiley & Sons, New York, 1957.

40. Czitrom, Veronica, John Sniegowski, and Larry D. Haugh, Improving Integrated Circuit Manufacture Using a Designed Experiment, *Statistical Case Studies, A Collaboration Between Academia and Industry,* American Statistical Association and Society for Industrial and Applied Mathematics, Alexandria, Virginia, 1997.

41. Heyl, Paul R., A Redetermination of the Constant of Gravitation, *Journal of Research of the Bureau of Standards* 5 (1930), 1243–1250.

42. Hinkelmann, Klaus, and Oscar Kempthorne, *Design and Analysis of Experiments,* Volume 1, John Wiley & Sons, New York, 1994.

43. Lin, Lawrence I-Kuei, and Robert W. Stephenson, Validating an Assay of Viral Contamination, *Statistical Case Studies, A Collaboration Between Academia and Industry,*

American Statistical Association and Society for Industrial and Applied Mathematics, Alexandria, Virginia, 1997.

44. Mason, Robert L., Richard Gunst, and James L. Hess, *Statistical Design and Analysis of Experiments,* John Wiley & Sons, New York, 1989.

45. Reeve, Russell, and Francis Giesbrecht, Chemical Assay Validation, *Statistical Case Studies, A Collaboration Between Academia and Industry,* American Statistical Association and Society for Industrial and Applied Mathematics, Alexandria, Virginia, 1997.

46. Sauter, Roger M., and Russell V. Lenth, Experimental Design for Process Settings in Aircraft Manufacturing, *Statistical Case Studies, A Collaboration Between Academia and Industry,* American Statistical Association and Society for Industrial and Applied Mathematics, Alexandria, Virginia, 1997.

47. Sloane, Neil J., Multiplexing Methods in Spectroscopy, *Mathematics Magazine* 5 (1979), No.2, 71–80.

CHAPTER 7 STATISTICAL PROCESS CONTROL

48. Duncan, Acheson J., *Quality Control and Industrial Statistics,* Fifth Edition, Richard D. Irwin, Inc., Homewood, Illinois, 1986.

49. Grant, Eugene L., and Richard S. Leavenworth, *Statistical Qualiy Control,* Sixth Edition, McGraw-Hill, New York, 1988.

50. Grimaldi, Ralph P., *Discrete and Combinatorial Mathematics,* Fourth Edition, Addison-Wesley, Boston, 1999.

51. Joshi, Madhukar, and Kimberley Sprague, Obtaining and Using Statistical Process Control Limits in the Semiconductor Industry, *Statistical Case Studies for Industrial Improvement,* American Statistical Association and Society for Industrial and Applied Mathematics, Alexandria, Virginia, 1997.

52. Lucas, James M., The Design and Use of V-Mask Control Schemes, *Journal of Quality Technology* 8 (1976), 1–12.

53. Lucas, James M., A Modified "V" Mask Control Scheme, *Technometrics* 15 (1973), 833–847.

54. Lucas, James M., Combined Shewhart-CUSUM Quality Control Schemes, *Journal of Quality Technology* 14 (1982), 51–59.

55. Lucas, James M., and Ronald B. Crosier, Fast Initial Response for CUSUM Quality-Control Schemes: Give Your CUSUM a Head Start, *Technometrics* 24 (1982), 199–205.

56. Patnaik, P. B., The Use of Mean Range as an Estimator of Variance in Statistical Tests, *Biometrika* 37 (1950), 78–87.

57. Western Electric, *Statistical Quality Control Handbook,* Western Electric Corporation, Indianapolis, Indiana, 1956.

Answers for the Odd-Numbered Exercises

CHAPTER 1

Exercises 1.1

1. b) Sample A appears to be more variable than sample B while sample A appears to be centered about a larger value than sample B.

   c) While sample A appears to last longer, sample B is less variable and would be the more reliable gear to choose.

Exercises 1.3

1. a) There are 28 samples: $\{(1, 2), (1, 3), \ldots, (7, 8)\}$.

   b) The histogram is uniform with height 1 for each of the values $1, 2, \ldots, 8$.

   c) The means range from $3/2$ to $15/2$ in steps of $1/2$. The histogram is symmetric around $9/2$ and its appearance is somewhat "bell-shaped."

   d) The ranges assume the values $\{1, 2, 3, 4, 5, 6, 7\}$ with frequencies $\{7, 6, 5, 4, 3, 2, 1\}$ so the histogram shows steadily declining frequencies, unlike the histogram for the mean values.

Exercises 1.4

1. a) The boxplot for Year 1 shows $Q_1 = 35$, $Q_2 = 57.5$, and $Q_3 = 72.75$ while the boxplot for Year 2 shows $Q_1 = 42.75$, $Q_2 = 53.5$, and $Q_3 = 59.75$ so the boxplot for Year 2 is considerably less variable than the boxplot for Year 1.

   b) The stem-and-leaf displays convey much the same information as the boxplots.

3. a) The histogram has frequencies $\{6, 5, 5, 3, 1, 3, 6, 1\}$ for values $\{0, 1, 2, 3, 4, 5, 6, 7\}$.

   b) The boxplot shows a range of about 4 units and has no outliers.

5. The histogram shows frequencies $\{1, 3, 2, 4, 2, 2, 1, 2, 1, 1, 4, 2, 2, 1, 2, 1, 1\}$ for times $\{21.0, 21.5, 22.0, \ldots, 25.5\}$.

7. b) The observation 23 is an outlier.

9. a) $Q_1 = 90$, $Q_3 = 111$
b) Outliers would occur at values greater than $111 + 1.5(111 - 90) = 142.5$ or at values less than $90 - 1.5(111 - 90) = 58.5$.
11. a) $Q_1 \doteq 4$, $Q_3 \doteq 13$
b) It appears that about 40% of the data lies above the median and 60% lies below the median.
c) The graph would have a long right tail.

Exercises 1.5

1. a) $\bar{x} = 72.315$, $s^2 = 14.8302$
b) Eleven observations are inside the interval; Chebyshev's Inequality indicates that some of the data are inside this interval.
3. a) $\bar{x} = 70.85$, $s^2 = 262.1161$
b) All the observations are inside the interval. Chebyshev's Inequality indicates that at least $3/4$ of the data are inside the interval.
5. a) $\bar{x} = 16, 67$, $s^2 = 15.8404$
b) $k = 1$

CHAPTER 2

Exercises 2.2

1. a) There are 10 samples: $\{(1, 2), (1, 3), \ldots, (4, 5)\}$.
b) There are 15 samples; those given in part (a) plus samples $(1, 1), (2, 2), \ldots, (5, 5)$.
3. $\binom{24}{3} = 2024$
5. $52! = 8.0658 \cdot 10^{67}$
7. a) $\binom{10}{6} = 210$
b) $\dfrac{\binom{9}{5}}{\binom{10}{6}} = 0.6$
c) 0.4
9. 2^n
11. $\binom{15}{6} \cdot \binom{4}{3} = 20{,}020$

Exercises 2.3

1. a) There are $\binom{6}{3} = 20$ samples.
b) The mean values range from 2 to 5 in steps of $1/3$.
c) $\mu = 3.5$
d) $\sigma_x^2 = 35/12$, $N = 6, n = 3$ so $\sigma_{\bar{x}}^2 = 7/12$.

Exercises 2.4

1. There are 10 samples: $\{(a, b), (a, c), \ldots, (d, e)\}$.
3. The sample (d, e) is the only sample containing all underfilled pills. So the probability is $1/10$.

5. There are now nine points in the sample space. Six of these have one acceptable item and one unacceptable item, so the probability is 2/3.

7. 1/2

9. 1/6

11. 98%

Exercises 2.5

1. a) 17/36
 b) 1/3
 c) 4/9
 d) $P(A \cap B) = 1/9 \neq P(A) \cdot P(B)$ so A and B are not independent.

3. a) 0.038
 b) 10/19

5. a) 3/8
 b) 2/3

7. $20000/29999 \doteq 2/3$

9. a) $P(A) + P(B)$
 b) $P(A) + P(\overline{A}) \cdot P(B)$

Exercises 2.6

1. a) 0.163244
 b) 0.20297

x	$P(X = x)$
c) 0	0.122760
1	0.333587
2	0.340684
3	0.163244
4	0.0366467
5	0.0030783

3. a) $\displaystyle\sum_{x=1}^{6} f(x) = 6k + 5k + 4k + 3k + 2k + k = 21k = 1$ so $k = 1/21$.

 b) $\mu = 8/3; \sigma^2 = 980/441$

5. a) $\displaystyle\sum_{x=0}^{\infty} \frac{x \cdot e^{-\lambda} \cdot \lambda^x}{x!} = \sum_{x=1}^{\infty} \frac{e^{-\lambda} \cdot \lambda^x}{(x-1)!} = \lambda \cdot e^{-\lambda} \cdot e^{\lambda} = \lambda$

 b) $E[X \cdot (X-1)] = \lambda^2$ so $\sigma^2 = E[X \cdot (X-1)] + E[X] - (E[X])^2 = \lambda$.

7. a) $\displaystyle\int_0^{\infty} \frac{1}{4} x \cdot e^{-x/4} dx = 4$

 b) $E(X^2) = \displaystyle\int_0^{\infty} \frac{1}{4} x^2 \cdot e^{-x/4} dx = 32$ so $\sigma^2 = 32 - 4^2 = 16$ giving $\sigma = 4$.

 c) $P(X \geq 4) = \displaystyle\int_4^{\infty} \frac{1}{4} e^{-x/4} dx = e^{-1} = 0.3679$

 d) $P(X \geq 4 | X \geq 1) = \dfrac{e^{-1}}{e^{-1/4}} = e^{-3/4} = 0.4724$

9. a) $\displaystyle\sum_{x=1}^{\infty}\left(\frac{1}{2}\right)^x = 1$ and $\left(\frac{1}{2}\right)^x > 0$ for $x \geq 1$.

b) $P(X > 3 | X > 2) = \dfrac{P(X > 3)}{P(X > 2)} = \dfrac{\left(\frac{1}{2}\right)^3}{\left(\frac{1}{2}\right)^2} = \dfrac{1}{2}$

11. a) $f(x) \geq 0$ and $\displaystyle\sum_{x=1}^{\infty}\frac{1}{3}\left(\frac{2}{3}\right)^{x-1} = 1$.

b) $P(X > 4) = 16/81$

Exercises 2.7

1. a) 0.103182
  b) 0.938949
  c) 0.250139
3. a) 0.248423
  b) 0.869236
  c) 0.375073
5. a) 0.0279816
  b) 0.0493686
  c) $1 - \left(\dfrac{1}{2}\right)^{30} = 1.0$
7. a) $P(X = x) = \dbinom{20}{x} \cdot (0.6)^x \cdot (0.4)^{20-x}$, $x = 0, 1, \ldots, 20$
9. $n = 27$, $p = 0.6$
11. 0.950701

Exercises 2.8

1. a) 0.47725
  b) 0.47725
  c) 0.1359051
  d) 0.818595
  e) 0.9545
  f) 0.9545
  g) 0.9545
3. a) 0.841345
  b) 0.158655
  c) 0.624655
  d) $(335.5, 664.5)$
5. 95.221%
7. a) 0.0967102
  b) 3.82087 lb.
9. a) 1.3420
  b) 2.93054
11. a) 31/32
  b) 0.000112451
13. 6.05

Exercises 2.9

1. a) 0.0421247
   b) 0.983133
3. 0.00317846
5. a) 0.0938135
   b) 0.546907
7. 0.0575326

Exercises 2.10

1. a) 0.234681
   b) 0.818595
   c) 0.0401
   d) 0
3. a) 0.139509
   b) 0.0648179
5. 385
7. 25

Exercises 2.11

1. 0.191208
3. 228.456
5. 0.0351735

Exercises 2.12

1. a) $f(x, y) \geq 0$ and $\int_2^4 \int_1^2 \frac{1}{2} dx\, dy = 1$.
   b) $f(x) = 1$, $1 < x < 2$ and $g(y) = 1/2$, $2 < y < 4$.
   c) 1/4

3. a) $f(x) = \dfrac{1}{\sqrt{2\pi}} e^{-\frac{x^2}{2}}$, $-\infty < x < \infty$. $g(y)$ is similar.

   b) 0.651627
5. a) $k = 1$
   b) $f(x) = x/2$, $0 < x < 2$; $g(y) = 2 - 2y$, $0 < y < 1$
   c) 2/3
7. a) $f(x, y) > 0$ and the sum of the six probabilities is 1.

   b) $f(x) = \begin{cases} 1/2, & x = 1 \\ 1/3, & x = 2 \\ 1/6, & x = 3 \end{cases}$ and $g(y)$ has the same distribution.

   c) $P(X = 3, Y = 3) = 0 \neq P(X = 3) \cdot P(Y = 3) = \dfrac{1}{6} \cdot \dfrac{1}{6}$, so X and Y are not independent.

CHAPTER 3

Exercises 3.2

1. a) Group I
   b) Choose Group I if the sample contains 2 or more items with the factor.

3. a) $L(\mu) = \dfrac{e^{-n\mu}\mu^{\sum\limits_{i=1}^{n} x_i}}{\prod\limits_{i=1}^{n} x_i!}$

5. a) $L(\sigma) = \left(\dfrac{1}{\sigma\sqrt{2\pi}}\right)^{n} e^{-\frac{1}{2\sigma^2}\sum\limits_{i=1}^{n}(x_i-\mu)^2}$

 b) $\hat{\mu} = \bar{x}, \hat{\sigma}^2 = \frac{1}{n}\sum\limits_{i=1}^{\mu}(x_i - \bar{x})^2$

7. x/n

9. 213.75

Exercises 3.3

1. (58.2, 77.8)
3. (55.46, 61.34)
5. (0.6342, 0.7783)
7. (2.0421, 2.3329)
9. a) (0.5264, 0.6736)
 b) 188

Exercises 3.4

1. a) (16.3001, 18.4999)
 b) (16.3596, 18, 4404)
3. (1810.63, 1944.37)
5. (17.3768, 17.4232)
7. 0.0251307
9. a) (99.5396, 99.7324) and (99.5308, 99.7412).
 b) The true percent of oxygen delivered by the two suppliers may be equal.

Exercises 3.5

The results given here are the two-sided 90% confidence interval using the sample proportions followed by the two-sided 90% confidence interval using the pooled sample proportion.

1. (−0.330058, 0.0633913), (−0.317198, 0.05053)
3. (−0.0519767, 0.0881836), (−0.0521418, 0.0883487). Yes, the proportions are likely to be equal.
5. (−0.212651, −0.007349), (−0.20873, −0.011267)
7. a) (−0.16267, 0.111287), (−0.164681, 0.113298)
 b) No
9. 301

Exercises 3.7

1. a) 0.394772
 b) 0.45387
 c) 0.349073
 d) 0.427225

3. 0.172908
5. (359.124, 8261.08)
7. 0.0017001
9. a) (0.43454, 10.421)
　b) Yes

Exercises 3.8

1. a) 0.198179
　b) $X \geq 16$
　c) 0.455369
3. a) 0.027114
　b) 0.601977
5. a) $\overline{X} > 104.6979$
　b) 0.830373
7. a) 0.0228
　b) 0.1587
9. a) Accept H_0
　b) 0.89904
　c) 0.14236

Exercises 3.10

1. a) Yes
　b) 0.382342
3. a) Yes
　b) 0.193319
5. a) No
　b) 0.0426209

Exercises 3.11

1. Accept H_0
3. Accept H_0
5. Accept H_0
7. Accept H_0
9. Accept H_0

Exercises 3.12

1. a) Reject H_0 if $\overline{x} < 340 - (76.8931/\sqrt{n})$ where n is the sample size.
　b) Reject H_0
3. $\ln(x_1 x_2 \ldots x_n) \leq k$
5. $\sum_{i=1}^{n} x_i \leq k$
7. a) $x \geq k$
　b) $\sum_{i=1}^{n} x_i \geq k$

9. a) $x \geq 0.95$
   b) 0.9025
11. $x_1 + x_2 \leq k$

Exercises 3.13

1. a) Accept H_0
   b) Reject H_0
3. Accept H_0
5. Reject H_0
7. Reject H_0
9. Reject H_0
11. Reject H_0: mean breaking strengths are equal. The first manufacturer makes the stronger product if $\alpha = 0.05$.

Exercises 3.14

1. 43
3. 1083
5. a) 115
   b) $n_1 = 112$, $n_2 = 117$
7. 0.00902748

Exercises 3.15

1. a) Accept the hypothesis of independence.
   b) Accept H_0
3. Accept H_0
5. Accept H_0
7. Yes, the data are exponential.

CHAPTER 4

Exercises 4.2

1. e) The least squares line is $y = 2.00 + 2.86x$ and $\sum_{i=1}^{6}(y_i - \widehat{y}_i)^2 = 11.1426$.
3. e) The least squares line is Moisture $= 0.49 + 0.272$ Humidity and
$$\sum_{i=1}^{12}(y_i - \widehat{y}_i)^2 = 11.341.$$

Exercises 4.5

1. a) $\sum_{i=1}^{6}(\widehat{y}_i - \overline{y})^2 = 142.86$, $\sum_{i=1}^{6}(y_i - \widehat{y}_i)^2 = 11.14$, and $\sum_{i=1}^{6}(y_i - \overline{y})^2 = 154.000$.
   b) $F(1,4) = 51.28$ giving a p-value of 0.002, so the hypothesis is rejected.
3. a) $\sum_{i=1}^{5}(\widehat{y}_i - \overline{y})^2 = 24.649$, $\sum_{i=1}^{5}(y_i - \widehat{y}_i)^2 = 0.343$, and $\sum_{i=1}^{5}(y_i - \overline{y})^2 = 24.992$.
   b) $F(1,3) = 215.59$, giving a p-value of $6.85209 \cdot 10^{-4}$ so the hypothesis is rejected.
5. a) The least squares regression line is

$$\text{Viscosity} = 87.5 - 0.361\,\text{Temperature}.$$

b) $\sum_{i=1}^{8}(\widehat{y}_i - \overline{y})^2 = 136.98$, $\sum_{i=1}^{8}(y_i - \widehat{y}_i)^2 = 1.24$, and $\sum_{i=1}^{8}(y_i - \overline{y})^2 = 138.22$.

$F(1, 6) = 664.28$, giving a p-value of $2.249 \cdot 10^{-7}$ so the hypothesis is rejected.

7. a) The least squares regression line is $y = 153 - 6.32x$.

b) $\sum_{i=1}^{9}(\widehat{y}_i - \overline{y})^2 = 770.26$, $\sum_{i=1}^{9}(y_i - \widehat{y}_i)^2 = 33.96$, and $\sum_{i=1}^{9}(y_i - \overline{y})^2 = 804.22$.

$F(1, 7) = 158.77$, giving a p-value of $4.57895 \cdot 10^{-6}$ so the hypothesis is rejected.

9. The degrees of freedom are 1, 10, and 11, respectively. The sum of squares due to regression is 741.125, giving $F(1, 10) = 30.65$ with a p-value of $2.488 \cdot 10^{-4}$.

11. Add the fractions and expand $\sum_{i=1}^{n}(x_i - \overline{x})^2$.

Exercises 4.6

1. $z = 1.3899$ so the p-value is 0.164559, so the hypothesis cannot be rejected.

3. $-0.5725 \le \rho \le -0.0793$

5. $z = -3.5271$ so the p-value is $4.202227 \cdot 10^{-4}$, so the hypothesis is rejected.

7. $z = 0.3777$ so the p-value is 0.705654, so the hypothesis cannot be rejected.

9. Use the fact that $\widehat{y}_i - \overline{y} = \widehat{\beta}(x_i - \overline{x})$.

Exercises 4.7

1. 672.66, (592, 753.3)

3. a) (391.93, 532.67)

b) (440.62, 483.98)

5. a) (109.463, 120.6969)

b) (112.9770, 117.1830)

7. a) The analysis of variance test gives $F(1, 8) = 17.00$ while the t-test for the hypothesis $H_0: \beta = 0$ gives $t_8 = 4.1225 = \sqrt{17.00}$, so the two tests are equivalent.

b) (45.7298, 58.8942)

c) (31.6032, 73.0208)

d) (−1.3535, −0.3825)

e) $t = 0.6269$ giving a p-value of 0.553796, so the hypothesis cannot be rejected.

Exercises 4.8

1. a) $y = 3.05x$

b) $F(1, 4) = 1383.05$ giving a p-value of $3.12166 \cdot 10^{-6}$, so the hypothesis is rejected.

3. For the regression line through the origin, $F(1, 19) = 4664.24$ giving a p-value of 0 while the regression line with an intercept gives $F(1, 18) = 2241.88$ also giving a p-value of 0.

b) Either regression line is an excellent fit for the data.

Exercises 4.11

1. The residual plot shows random variation.

3. b) The value of Cook's d for the first data point is 0.044261, which is not significant.

5. The residual plot shows random variation.

Exercises 4.12

1. $\widehat{a} = 3.0957$, $\widehat{b} = -1.96$

3. $\widehat{a} = 0.0510$. The analysis of variance gives $F(1, 14) = 46.67$ giving a p-value of $8.17222 \cdot 10^{-6}$ so the fit is an excellent one.

CHAPTER 5

Exercises 5.2

1. c) $\widehat{\beta}_0 = 6.00$, $\widehat{\beta}_1 = 15.00$, $\widehat{\beta}_2 = -3.00$. The regression plane is completely determined by the three data points given, so the points all lie exactly on the regression plane.

3. c) $\widehat{\beta}_0 = 2.6$, $\widehat{\beta}_1 = 1$, $\widehat{\beta}_2 = 1$. The sum of squares of the residuals is 5.20.

5. c) $\widehat{\beta}_0 = 0$, $\widehat{\beta}_1 = 0$, $\widehat{\beta}_2 = 0$

   d) Each of the predicted values is 0.

   e) The $X'Y$ matrix must be the 0 vector.

7. c) $\widehat{\beta}_0 = -13.2$, $\widehat{\beta}_1 = 1$, $\widehat{\beta}_2 = 10.8$

Exercises 5.3

1. a) $\sum_{i=1}^{4}(\widehat{y}_i - \bar{y})^2 = 40$, $\sum_{i=1}^{4}(y_i - \widehat{y}_i)^2 = 16$, and $\sum_{i=1}^{4}(y_i - \bar{y})^2 = 56$.

   b) $F(2, 1) = 1.25$, giving a p-value of 0.535, so the regression is not significant.

   c) $\overrightarrow{y_i - \widehat{y}_i} = (2, -2, 2, -2)$ and $\overrightarrow{\widehat{y}_i - \bar{y}_i} = (-4, 2, 4, -2)$. The dot product of these vectors is 0.

3. a) $\sum_{i=1}^{5}(\widehat{y}_i - \bar{y})^2 = 4.167$, $\sum_{i=1}^{5}(y_i - \widehat{y}_i)^2 = 15.833$, and $\sum_{i=1}^{5}(y_i - \bar{y})^2 = 20$.

   b) $F(2, 2) = 0.26$, giving a p-value of 0.792, which is not significant.

   c) $\overrightarrow{y_i - \widehat{y}_i} = (-1, -1, 19/6, 2/3, -33/18)$ and $\overrightarrow{\widehat{y}_i - \bar{y}_i} = (0, 0, 15/18, -5/3, 15/18)$. The dot product of these vectors is 0.

5. a) $\sum_{i=1}^{10}(\widehat{y}_i - \bar{y})^2 = 2315.1$, $\sum_{i=1}^{10}(y_i - \widehat{y}_i)^2 = 61.8$, and $\sum_{i=1}^{10}(y_i - \bar{y})^2 = 2376.9$.

   c) $F(2, 7) = 131.20$, giving a p-value of $2.8278 \cdot 10^{-6}$ so the regression is very significant.

Exercises 5.4

1. a) $\begin{bmatrix} 4 & 0 & 0 \\ 0 & 8 & 8 \\ 0 & 8 & 16 \end{bmatrix}$

   b) The t values for these hypotheses are 5.00, 1.41, and 1.50, respectively, with p-values 0.126, 0.393, and 0.374 so none of the hypotheses can be rejected.

3. a) $\begin{bmatrix} 1.9133 & 0 & -0.3299 \\ 0 & 0.7917 & 0 \\ -0.3299 & 0 & 0.3299 \end{bmatrix}$

   b) The t values are 1.75, 0, and -0.73, respectively, giving p-values of 0.223, 1, and 0.544 so none of the hypotheses can be rejected.

5. a)
$$\begin{bmatrix} 15.335 & -0.22717 & -0.201755 \\ -0.22717 & 5.87018 \cdot 10^{-3} & 7.51506 \cdot 10^{-4} \\ -0.201755 & 7.51506 \cdot 10^{-4} & 5.36063 \cdot 10^{-3} \end{bmatrix}$$

b) The t values are 58.61, 15.30, and 7.31, respectively, giving p-values of 0 so none of the hypotheses can be accepted.

7. a) $\mathrm{Var}(\widehat{\beta_2}) = 0.26$ giving the confidence interval $(-1.19349, 3.19349)$.

b) $\widehat{y} = 8.60$ and $\mathrm{Var}(\widehat{y}) = 2.5133$. The confidence interval is $(1.7802, 15.41981)$.

9. a) $(-3.2390, 3.2390)$

b) $(-7.3204, 7.3204)$

Exercises 5.6

1. b) Students/Computer $= 97.7 - 17.8\,\mathrm{Year} + 0.858\,\mathrm{Year}^2$

d) Students/Computer $= 71.7 - 5.78\,\mathrm{Year}$

e) In part (b), $F(2, 12) = 42.85$ giving a p-value of $3.433 \cdot 10^{-6}$. In (d), $F(1, 13) = 25.49$ giving a p-value of $2.223 \cdot 10^{-4}$, so either model provides an excellent fit.

3. In Exercise 1, part (b), $R^2 = 0.877$ and $R^2_{\mathrm{Adj}} = 0.857$. In Exercise 1, part (d), $R^2 = 0.662$ and $R^2_{\mathrm{Adj}} = 0.636$, which, based on the correlation coefficient alone, would indicate that the quadratic model is the better one.

5. b) $y = 0.0254 + 2.9469 \log x$ with $F(1, 19) = 414.99$ giving a p-value of 0. However, the model with no intercept is $\log_{10} y = 2.97313$ with $F(1, 20) = 1004.20$, an even more extreme value.

Exercises 5.8

1. a) Torque $= -4.59 + 4.60\,\mathrm{Diameter} + 3.12\,\mathrm{Penetration} - 0.0474\,\mathrm{Temp}$

b) $F(3, 20) = 41.95$, giving a p-value of 0, so the regression is a very good one.

d) None of the values of Cook's d exceed the critical value of $F_{0.50}(3, 20) = 0.8162$.

e) The t values are $-1.21, 10.18, 3.51$, and -2.62, respectively. The corresponding p-values are $0.239, 0, 0.001$, and 0.016 indicating that a model without an intercept might be appropriate.

f) $\widehat{y} = 30.186$ with $\mathrm{Var}(\widehat{y}) = 2.38843$ giving the confidence interval $(26.9622, 33.40987)$.

3. a) Pages $= -16.8 + 27.0\,\mathrm{Magazine} + 5.49\,\mathrm{YrCode} + 1.17\,\mathrm{FullAds}$

b) $F(3, 20) = 8.42$ giving a p-value of 0.001, so the regression is a very good fit.

d) The maximum value of Cook's d is 1.52378 for the 21st data point. It is the only data point to exceed the critical value $F_{0.50}(3, 20) = 0.8162$.

e) The t values are $-0.71, 3.29, 1.08$, and 3.75, respectively. The corresponding p-values are $0.488, 0.004, 0.292$, and 0.001.

f) $\widehat{y} = 123.06$ with $\mathrm{Var}(\widehat{y}) = 48.2218$ giving the confidence interval as $(108.575, 137.545)$.

5. a) Price $= 3818 + 7.57\,\mathrm{HD} - 715\,\mathrm{Floppy} + 702\,\mathrm{CPUCode} + 504\,\mathrm{SpeedCode}$

b) $F(4, 30) = 12.40$, giving a p-value 0 so the regression is a very good fit.

d) None of the values of Cook's d exceed the critical value $F_{0.50}(4, 30) = 0.858437$.

e) The t values are $5.07, 2.22, -1.93, 3.76$, and 2.74. The corresponding p-values are $0, 0.021, 0.063, 0.001$, and 0.010.

f) $\widehat{y} = 5108$ with $\mathrm{Var}(\widehat{y}) = 73811.3$ giving the confidence interval as $(4553.15, 5662.85)$.

7. a) *ScoreGap* $= -5.55 + 3.14$ *SeedGap*
The F value is $F(1, 30) = 43.05$ giving a p-value of $2.9382 \cdot 10^{-7}$ so the regression line is an excellent fit for the data.

b) Following are the regression lines and analysis of variance tables for each region.
ScoreGapEast $= -24.1 + 5.28$ *SeedGap*. $F(1, 6) = 43.65$ with a p-value of $5.785 \cdot 10^{-4}$ so the regression line is an excellent fit.
ScoreGapMidWest $= -0.08 + 2.74$ *SeedGap*. $F(1, 6) = 14.01$ with a p-value of $9.5886 \cdot 10^{-3}$ so the regression line is an excellent fit.
ScoreGapMid East $= -1.35 + 2.68$ *SeedGap*. $F(1, 6) = 8.81$ with a p-value of 0.025 so the regression line is an excellent fit.
ScoreGapWest $= -14.8 + 3.32$ *SeedGap*. $F(1, 6) = 6.98$ with a p-value of 0.038 so the regression line is an excellent fit.

c) *ScoreGap2* $= -3.62 + 3.04$ *SeedGap*. $F(1, 14) = 6.23$ with a p-value of 0.026 so the regression line is an excellent fit.

d) For the regression line in part a), the critical value of Cook's d is 0.466155. None of the data points exhibit unusual influence. For the regression lines in part b), the critical value of Cook's d is 0.51489. The points are not unduly influential except for two points in the data from the first round in the East, those with seed gaps of 15 and 13 which produced score gaps of 72 and 29 points respectively.

e) The regression lines are quite good predictors of the score gaps indicating that the higher seeded team is very likely to beat the lower seeded team.
Since the first round produced so few upsets, perhaps the tournament should begin with only the top 16 teams.

9. a) *Salary* $= 9852 + 3550$ *Gender* $+ 7936$ *RankCode* $- 393$ *DeptCode* $+ 0.192$ *Begin\$*

b) $F(4, 6) = 9.26$ giving a p-value of 0.010, so the regression equation is a very good fit for the data.

d) None of the values of Cook's d exceed the critical value of 0.941913.

e) The t values are $1.27, 1.29, 4.96, -0.80$, and 0.76, respectively, with corresponding p-values of $0.252, 0.245, 0.003, 0.454$, and 0.477 so only β_2 appears to be significant.

11. b) $y = 0.603x - 0.0709x^3 + 0.00153x^5$

c) $F(3, 23) = 66.21$ giving a p-value of 0 so the sinusoidal fit is a very good one.

Exercises 5.9

1. a) The regressions in order are:
1) Torque $= -3.560 + 4.60$ Diameter
2) Torque $= -9.805 + 4.60$ Diameter $+ 3.12$ Penetration
3) Torque $= -4.587 + 4.60$ Diameter $+ 3.12$ Penetration $- 0.041$ Temperature

b) The backward regression gives the same results as in part (a).

3. a) The forward procedure gives two regressions:
1) Pages $= 36.952 + 1.13$ FullAds and
2) Pages $= -3.548 + 1.13$ FullAds $+ 27.0$ Magazine

b) The backward procedure gives Pages $= -16.765 + 5.5$ YrCode $+ 1.17$ FullAds $+ 27.0$ Magazine; the next two results are those given in part (a), with the second regression given next.

c) The line Pages $= 36.952 + 1.13$ FullAds has the smallest p-value, 0.00034415, of the three regressions calculated.

5. a) The forward procedure gives these regressions:

    1) Price $= 1802 + 13.5$ HD

    2) Price $= 2149 + 10.7$ HD $+ 498$ CPUCode

    3) Price $= 2695 + 6.6$ HD $+ 723$ CPUCode $+ 475$ SpeedCode

    4) Price $= 3818 + 7.6$ HD $+ 702$ CPUCode $+ 504$ SpeedCode $- 715$ Floppy

b) The backward procedure gives regressions (4) and (3) above.

d) The smallest p-value is $4.54911 \cdot 10^{-6}$ when all four independent variables are included in the model.

Exercises 5.10

1. a) The maximum appears to be near $x = 30$.

b) The maximum appears to be near $y = 30$.

c) The quadratic surface is $Q = -876 + 59.1237x + 59.1151y - 0.986429x^2 - 0.985286y^2$. The analysis of variance table gives $F(4, 20) = 3589.64$ with a corresponding p-value of 0.

d) The maximum point is $(29.9686, 29.999)$.

e) The least squares plane is $Q = -504.444 + 30.7333x + 30.8667y$.

f) The lines perpendicular to the contours have slope approximately 1, so the values of Q for the points $(10, 10)$, $(20, 20)$, ... should be investigated, showing the maximum at $(30, 30)$.

CHAPTER 6

Exercises 6.2

1. The linear model is Time $= 34 - 2.5$ Sp $- 5$ RAM $- 0.5$ Sp $\cdot$ RAM. The coeffcients are the effects.

3. a) The linear model is Sales $= 92.1 + 7.12$ S $- 5.13$ T $+ 1.88$ S $\cdot$ T. The coeffcients are the effects.

d) Store 1 should be more heavily stocked early in the day.

Exercises 6.3

1. c) The variance-covariance matrix is $\sigma^2 \begin{bmatrix} 1 & -1 \\ -1 & 2 \end{bmatrix}$ so the estimates are correlated.

d) $\widehat{\beta} = \begin{bmatrix} w_2 \\ w_1 - w_2 \end{bmatrix}$

3. The matrix $(X'X)^{-1}$ in either case is $\begin{bmatrix} \dfrac{1}{2} & \dfrac{1}{4} & \dfrac{1}{4} \\[2mm] \dfrac{1}{4} & \dfrac{1}{2} & \dfrac{1}{4} \\[2mm] \dfrac{1}{4} & \dfrac{1}{4} & \dfrac{1}{2} \end{bmatrix}$.

5. Use the design matrix
$$\begin{bmatrix} 1 & 1 & 1 & 1 & 1 & 1 & 1 & 1 \\ 1 & -1 & 1 & -1 & 1 & -1 & 1 & -1 \\ 1 & 1 & -1 & -1 & 1 & 1 & -1 & -1 \\ 1 & -1 & -1 & 1 & 1 & -1 & -1 & 1 \\ 1 & 1 & 1 & 1 & -1 & -1 & -1 & -1 \\ 1 & -1 & 1 & -1 & -1 & 1 & -1 & 1 \\ 1 & 1 & -1 & -1 & -1 & -1 & 1 & 1 \\ 1 & -1 & -1 & 1 & -1 & 1 & 1 & -1 \end{bmatrix}.$$

Exercises 6.4

1. a) SSTmts $= 62$, SSError $= 32$, SSTotal $= 94$.
  b) $F(3, 11) = 7.10$ giving a p-value of 0.006.
  c) The contrasts are $8T_A + 8T_B - 7T_C - 7T_D = 120$, $5T_A - 2T_B = -30$, and $T_C - T_D = 16$ with sums of squares 17.1429, 12.8571, and 32 respectively.

3. b) SSTmts $= 199.3$, SSError $= 942.3$, SSTotal $= 1141.6$. $F(3, 14) = 0.99$ giving a p-value of 0.427, which is not significant. The storage periods do not influence the number of bulbs blooming.
  c) Possible contrasts are: $11T_1 + 11T_3 - 7T_2 - 7T_4 = 415$, $3T_1 - 4T_3 = -44$, and $6T_2 - 5T_4 = -131$ with associated sums of squares of 124.2605, 23.0476, and 52.003. Note that these contrasts are orthogonal, so the treatment sum of squares is completely partitioned.

5. b) Using the reflective indices as the response and the values 1, 2, 3, and 4 for the independent variable, the regression line is Ref $= 179 - 17.1$ Ind. The analysis of variance gives $F(1, 18) = 1.71$ with a p-value of 0.208, so the regression is not significant, showing that the reflective index is not a function of the type of paint used.
  c) For the one-way classification, SSTmts $= 38424$, SSError $= 45440$, SSTotal $= 83864$. $F(3, 16) = 4.51$ giving a p-value of 0.018. Note that the tests in parts (b) and (c) are quite different.
  d) Possible contrasts are: $9T_1 + 9T_2 - 11T_3 - 11T_4 = 215$, $6T_1 - 5T_2 = 1555$, and $5T_3 - 4T_4 = 2365$ with associated sums of squares of 23.346, 7327.3485, and 31073.4772. Note that these contrasts are orthogonal, so the treatment sum of squares is completely partitioned.

7. a) SSTmts $= 2534.6$, SSError $= 13.00$, SSTotal $= 2547.6$. $F(4, 5) = 243.71$ giving a p-value of 0.
  b) Voltage influences the length of life of the bulbs.

9. a) SSTmts $= 2.987$, SSError $= 8.233$, SSTotal $= 11.219$. $F(4, 50) = 4.53$ giving a p-value of 0.003.
  c) The boxplots indicate that the following orthogonal contrasts might be of interest: $T_1 - T_3 = 0.4$, $T_2 - T_5 = -0.5$, $T_4 - \frac{1}{2}(T_2 + T_5) = -6.85$, and $T_1 + T_3 - \frac{2}{3}(T_2 + T_4 + T_5) = -2.133$ with associated sums of squares of 0.0073, 0.0114, 2.8438, and 0.1241, which completely partitions the treatment sum of squares.

11. The t test in Chapter 3 uses
$$t_k = \frac{\bar{x}_1 - \bar{x}_2}{s_p \sqrt{\dfrac{1}{n_1} + \dfrac{1}{n_2}}}$$

where

$$s_p^2 = \frac{(n_1 - 1)s_1^2 + (n_2 - 1)s_2^2}{n_1 + n_2 - 2}$$

and

$$k = n_1 + n_2 - 2$$

while the analysis of variance test uses

$$F(1, n_1 + n_2 - 2) = \frac{\sum\limits_{i=1}^{2} n_i(\overline{x}_i - \overline{\overline{x}})^2}{\left[\sum\limits_{j=1}^{n_1}(x_{1j} - \overline{x}_1)^2 + \sum\limits_{j=1}^{n_2}(x_{2j} - \overline{x}_2)^2\right]/(n_1 + n_2 - 2)}.$$

Show that the denominator in this fraction is s_p^2 while the numerator expanded becomes

$$\frac{(\overline{x}_1 - \overline{x}_2)^2}{\frac{1}{n_1} + \frac{1}{n_2}}$$

showing that $t_k^2 = F(1, k)$. Use the fact that these distributions can be shown to be equal. Note that the F test is one-sided while the t test is two-sided.

Exercises 6.6

1. **a)** SSBlocks $= 32.95$, SSSubjects $= 36.30$, SSError $= 15.30$, SSTotal $= 84.55$.
    $F(4, 12) = 12.37$ giving a p-value of 0.008.
   b) SSSubjects $= 36.30$, SSError $= 48.25$, SSTotal $= 84.55$. $F(4, 15) = 2.82$ giving a p-value of 0.063. When the influence of the block is removed, the subjects become very significant, while if blocking is not used the subjects would probably not be judged significant.
3. **a)** SSRows $= 608$, SSColumns $= 185$, SSInteraction $= 43$, SSError $= 12261$,
    SSTotal $= 13098$. Testing the rows, columns, and interaction, respectively, give the following F values: 0.79, 0.24, and 0.06 with p-values 0.386, 0.630, and 0.815, respectively, so none of the factors is significant.
5. **a)** SSPlants $= 0.3333$, SSShifts $= 0.3800$, SSInteraction $= 0.3033$, SSError $= 0.5100$.
    F values are 3.92, 2.24, and 3.57 respectively giving p-values 0.095, 0.188, and 0.095 so none of these factors is significant.
   b) SSShifts $= 0.3800$, SSError $= 1.4500$ giving $F(7, 9) = 1.1794$ with p-value 0.3508, SSPlants $= 0.3333$, SSError $= 1.4967$ giving $F(1, 10) = 2.227$ with p-value 0.1665. Neither factor alone is of significance.
7. **a)** SSTemps $= 1033$, SSAlloys $= 3179$, SSError $= 1477$, SSTotal $= 5689$
   b) Testing the temperatures with the influence of the alloys removed gives $F(5, 10) = 1.40$ with a p-value 0.304, so the temperatures do not in fact produce different tensile strengths.
9. **a)** SSTerrain $= 23.82$, SSCars $= 30.22$, SSError $= 57.80$, SSTotal $= 111.84$. Testing the cars gives $F(3, 6) = 1.05$ giving a p-value of 0.438 so the cars do not differ significantly with respect to gasoline mileage.
   b) Testing the terrains gives $F(2, 6) = 1.24$ giving a p-value of 0.355 so the terrains do not differ significantly with respect to gasoline mileage.

Exercises 6.7

3. b) The effects are the coefficients in the regression equation

$$\text{Deviation} = 1.00 + 1.50A + 1.12B + 0.875C + 0.375AB$$
$$+ 0.125AC + 0.250BC + 0.250ABC.$$

c) The sums of squares for the main effects, A, B, and C, respectively, are: 36.000, 20.250, and 12.250 giving p-values of 0, 0, and 0.002 so all the main effects are significant. None of the interactions are significant.

5. b) The effects are the coefficients in the regression equation

$$\text{Response} = 11.2 + 1.69A + 0.812B + 0.437C + 0.687AB$$
$$+ 0.062AC - 0.313BC + 0.563ABC.$$

A is the only significant factor with p-value 0.003.

7. b) Only the factor B shows some significance, giving a p-value of 0.045.

c) Probably only the factor B would be chosen for further studies.

Exercises 6.10

1. a) The sums of squares for the factors Batches, Operators, and Formulations are 150, 68, and 330, respectively.

b) These factors have p-values of 0.141942, 0.434316, and 0.0236834 so only the formulations would be regarded as significant.

3. b) The generators are $E = ABC$ and $F = BCD$. Some of the alias structure is

$$I = ABCE = ADEF = BCDF, \ A = BCE = DEF, \ AB = CE = ACDF = BDEF.$$

There are many more such expressions in the alias structure.

c) SSMainEffects $= 5858.37$, SS2$-$WayInteractions $= 705.94$ giving p-values of 0.0472 and 0.357526 so the main effects are significant.

5. a) $t = 5, \lambda = 1, b = 10, k = 2, r = 4$. Note that $\lambda = r(k-1)/(t-1)$.

b) SSFabrics $= 43.4$ and SSSolvents $= 56.6$ giving p-values of 0.903628 and 0.423939 so neither factor has significance.

7. a) The design has resolution IV.

b) The design generators are: $E = BCD$, $F = ACD$, $G = ABC$, and $H = ABD$. Two of the alias equations are:

$$I = ABCG = ABDH = ABEF = ACDF = ACEH = ADEG = AFGH = BCDE$$
$$= BCFH = BDFG = BEGH = CDGH = CEFG = DEFH$$

and $A = BCG = BDH = BEF = CDF = CEH = DEG = FGH.$

c) The two-factor interactions are confounded with each other.

d) SSMainEffects $= 2.2097$ and SSError $= 0.2878$ giving $F(8, 7) = 6.7188$ with a p-value 0.0105217 so the main effects are significant.

9. a) $t = 6, \lambda = 2, b = 10, k = 3, r = 5$. Note that $\lambda = r(k-1)/(t-1)$.

b) SSMarketArea $= 20.53$, SSInstruments $= 49.44$ giving p-values of 0.983275 and 0.442176, so neither factor is of significance.

CHAPTER 7

Exercises 7.3

1. a) (18.5662, 21.4338)
   b) (18.2465, 21.7535)
   c) (17.8414, 22.1586)
3. a) (−1.420, 1.439)
   b) (−1.4084, 1.4272)
5. a) (15.422, 23.443)
   b) (15.3813, 23.4827)
7. The sample means are never outside the 3σ control limits.
9. The sample means show a definite downward trend but none are outside the 3σ control limit.
11. a) The control chart has $UCL = 11.53$ and $LCL = 8.570$. Each of the data points from Shift 1 exceed the UCL while each of the data points from Shift 2 are exceeded by the LCL.
   b) For Shift 1: $UCL = 16.93$, $LCL = 13.65$.
   For Shift 2: $UCL = 6.128$, $LCL = 3.489$.
   All control limits have been based on the sample ranges.

Exercises 7.4

1. a) (0, 5.2557)
   b) (0, 2.5665)
3. a) (0, 4.4788)
   b) (0, 1.9733)
5. a) (0, 14.7005)
   b) (0, 5.9288)
7. a) $UCL = 5.4528$, $LCL = 0$
   b) $UCL = 2.126$, $LCL = 0$

Exercises 7.5

1. $UCL = 2.6997$, $LCL = 0$. Only sample number 29 with 3 defects is outside the control limits.
3. $UCL = 2.724$, $LCL = 0$. Only sample number 29 with 3 defects is outside the control limits.
5. $UCL = 0.02456$, $LCL = 0$. Only sample number 29 with 3 defects is outside the control limits.
7. $UCL = 5.898$, $LCL = 0$. Only the eighth sample with 6 defects is outside the control limits.
9. $UCL = 6.105$, $LCL = 0$.

Exercises 7.6

1. a) The control limits are 43.8792 and 57.38, so Test 1 fails for the eighth sample.
   b) If z is a $N(0, 1)$ variable, $P(-3 \leq z \leq 3) = 0.9973$.
3. a) Test 1 fails on the 12th sample; test 5 fails on samples 9, 10, and 11; test 6 fails on samples 13, 14, and 15.
   b) $2 \sum_{x=2}^{3} \binom{3}{x} (0.0227501)^x (0.9771499)^{3-x} = 0.003058$

5. $2/5! = 2/120 = 0.016667$

7. $\left(1 - \dfrac{2}{2^6}\right)(1 - 0.682689)^6 = 9.88823 \times 10^{-4}$

Exercises 7.7

1. a) The values for S_H are: $0.0561395, 0, 0, 0.614952, 0$. The values for S_L are: $0, 0,$ $0.722573, 0, 0$.
  b) The process goes out of control at the third sample (using a limit of 4 on both S_H and S_L).
3. b) The limits are ± 5.3474.
  c) Some values of S_H for the first 8 samples are: $1.47805, 2.63914, 4.88783, 5.73419,$ $5.94407, 6.55683, 8.28153,$ and 6.99004. The corresponding values for S_L are: $0, 0, 0, 0, 0, 0, 0,$ and 0.291484.
5. b) The CUSUM chart shows that the process is always in control (using a limit of 4 on both S_H and S_L).
  d) The V-mask control chart does not show the process out of control.

Exercises 7.8

1. 0.818548
3. a) 0.002672
  b) 0.00290545
5. $AOQ(D) = \left(\dfrac{D}{32}\right)\dfrac{\binom{32-D}{2}}{\binom{32}{2}}.$
  The average outgoing quality limit is 0.145539 when $D = 10$ or 11.
7. A sample of 23 approximately satisifes the conditions.
9. a) 0.96228
  b) 0.958477
11. 1995
13. a) 0.0001786707
  b) 0.000356106
  c) $1.9385 \cdot 10^{-5}$
  d) The average outgoing quality limit is 0.0398371 when $D = 9$.
  e) 750
15. a) 0.1650
  b) 0.1778
  d) The average outgoing quality limit is 0.03543 when $D = 27$.
  e) 442

Index